FORGOTTEN LEGIONS

This book is dedicated to the countless millions of civilians and soldiers who lost their lives during World War II. A special prayer is sent to the children of that war. May it never happen again.

FORGOTTEN LEGIONS

Obscure Combat Formations of the Waffen-SS

ANTONIO J. MUNOZ

PALADIN PRESS
BOULDER, COLORADO

Forgotten Legions:
Obscure Combat Formations of the Waffen-SS
by Antonio J. Munoz

ISBN 0-87364-646-0
Printed in the United States of America

Published by Paladin Press, a division of
Paladin Enterprises, Inc., P.O. Box 1307,
Boulder, Colorado 80306, USA.
(303) 443-7250

Direct inquires and/or orders to the above address.

CONTENTS

LIST OF MAPS

TABLES & CHARTS

PHOTOGRAPHS

FOREWORD

I have been a fan of Tony Munoz' since I read his first article in *Siegrunen* in the mid-1980s. When I was given the opportunity to write the introduction for his first book on the Waffen-SS, I accepted with alacrity, because I knew I would be amply rewarded—I would get to read the book! When I did so, I was certainly not disappointed. *Forgotten Legions* is the best English-language book on the armed SS that has been released in several years.

The author has crammed an incredible amount of information and detail into a single volume and has presented it in a lucid and exciting manner. Here the reader will be introduced to the unique SS parachute battalion and learn how it jumped into the heart of Partisan-controlled Yugoslavia, only narrowly missing Tito. He or she will be carried along with Otto Skorzeny—Mussolini's rescuer—and his special-purpose panzer brigade in the Battle of the Bulge and with the elite SS commandos in their successful coup against the Hungarian government—at the very moment it was attempting to defect from the Axis and join the Allies. The reader will find out what it was like to be surrounded with the SS fortress regiment in the doomed city of Breslau, fighting in the snow-covered ruins against odds of a dozen to one or more, or to be with a veteran SS infantry unit, fiercely resisting massive Soviet tank attacks with its last grenades as the Eastern Front crumbled.

This book has all of the elements of an exciting war novel: ambushes, firefights, glider assaults behind enemy lines, forced marches into the night, courage, cowardice, self-sacrifice, betrayal, desertion, and tattered bands of superbly trained soldiers isolated miles behind Russian lines, desperately trying to reach friendly territory as the Third Reich died. One has to occasionally remind oneself that this is not a novel; it is a true story with an impressive amount of painstaking research behind it. Much of what is found here has never been published anywhere else before—in any language. *Forgotten Legions* is, in fact, one of those rare books that will appeal to World War II buffs and serious scholars alike. It is a solid and important contribution to the literature of Nazi Germany and the history of the Waffen-SS.

Antonio J. Munoz is indeed to be commended for a job well done.

Samuel W. Mitcham, Jr., PhD
Statesboro, Georgia
September 1990

ACKNOWLEDGMENTS

The author would like to express his heartfelt thanks to everyone who contributed to making this book possible, in particular, to the following individuals whose assistance was so critical that, without their help, this book could not have been written.

GENERAL

Mr. Kurt Kahofer, Mr. Larry Caffo, and Mrs. Brigitte Steiner, for their critical assistance in translating the multitude of documents and dozens of reports and testimonials. Mr. Harry Reilly, of the U.S. National Archives, whose great help and long patience eased much of my research.

Mr. R. Landwehr, Mr. Robert Tyler, Mr. Michael J. Belles, Esq., Dr. Einar Himma, Mr. Kenneth Nieman, Mr. Gerald Smoak, Mr. Justin Horgan, and countless others who selflessly donated or lent part or whole collections of photographs, documents, and other memorabilia for this work.

Mr. Eric Weisman, who labored for dozens of hours to illustrate this work.

Finally, I wish to thank my wife, Margarita, and my two little daughters, Martha and Victoria, from whom I took away a lot of time to work on the book. I hope that one day they will forgive and understand.

PHOTOGRAPHIC

I wish to thank the following individuals and organizations who contributed photographic documentation for this work:

Mr. Robert Tyler, Mr. Michael J. Belles, Dr. Einar Himma, Mr. Kenneth Nieman, Mr. Gerald Smoak, Mr. Justin Horgan, Mr. Andreas Meyer, Mr. Walter Redeker, Mr. Werner Schrödl, Mr. Erhart Knöfel, Munin Verlag, Bundesarchiv, Interfoto MTI.

TEXTUAL

Extracts came from the following works:

Neither Fear Nor Hope: The Wartime Memoirs of the German Defender of Cassino by General Frido von Senger und Etterlin. Presidio Press, 1989.

Fallschirmjäger der Waffen-SS Im Bild by Siegfried Milius and Adolf Kunzmann. Munin Verlag, 1986.

Die deutschen Lastensegler-Verbande 1937-1945 by Georg Schlaug. Stuttgart: Motorbuch Verlag, 1945.

Skorzeny's Special Missions, London: Robert Hale, 1987.

Encyclopaedia Britannica, Inc., Britannica Centre, 310 South Michnigan Ave., Chigaco, Il.

AUTHOR'S NOTE

In 1989, the free world witnessed the miracle of revolution take hold in Eastern Europe. Some of the most detestable regimes came tumbling down. While Leipzigers demonstrated by the thousands in East Germany, West Berliners helped their East Berlin brothers tear down the hated Berlin Wall.

In Romania, a dejected and abused populace rose up in defiance of the repressive regime of Nicolae Ceausescu. Thousands of them were murdered in cold blood by Ceausescu's dreaded Securitate. Still, the demonstrations continued, and thousands more died before the Romanian army finally sided with its countrymen and began to fight the regime.

Yugoslavia, ruled by Joseph Broz Tito from World War II until his death in 1980, is currently in the the news. The Croatian, Slovenian, Macedonian, and ethnic Albanian populations are trying to free themselves from the Serbian-run Communist Central Government. Many Serbians oppose the old guard that wishes to continue to run the country as Tito had done. Much blood has been shed, and many feel that Yugoslavia will be the country to break apart. The old problems once again resurface.

Poles, Hungarians, Czechs, Lithuanians, and Latvians also joined in the freedom movement and have now declared their independence from the Communist regimes that had ruled over them. All over the world, people followed the transformation of dictatorships into struggling democracies. What most people do not know is that the roots of the these independence movements go back almost fifty years, when the world was embroiled in World War II.

When Hitler and Stalin signed their nonaggression pact in August 1939, the Soviet Union swallowed up the small Baltic countries of Estonia, Latvia, and Lithuania. Stalin's regime had already destroyed the liberty enjoyed by Ukrainians from 1919 to 1921, as well as Turkestan's fleeting taste of freedom. Stalin banned religious beliefs and local languages and substituted ineffective collectivization methods for private ownership. Individual freedoms disappeared, and the terror of the knock in the middle of the night began. Against this backdrop of repression, Germany invaded the Soviet Union in June 1941. Is it any wonder, then, that hundreds of thousands of Eastern Europeans, Russians, and other Slavic peoples rushed to the aid of the Germans?

What makes this resistance even more remarkable is the fact that these citizens took up arms against Stalinism, even after experiencing the brutal, racist Nazi policies instituted by bigoted, nearsighted men in Berlin. While men such as Paul Hausser and Felix Steiner espoused a Pan-European, anti-Communist crusade that welcomed any who wished to join the ranks of the Waffen-SS, Hitler and his henchmen threw away the only chance Eastern Europe and Russia had of eliminating Stalin and Communism. The excesses of the Nazi regime forever tarnished the escutcheon of the Waffen-SS. At a time when many were striving to rid Europe of Communism, the Waffen-SS represented the best chance for such an anti-Communist movement to succeed. It might have succeeded, had it not been for the fact that the Waffen-SS, like the rest of the Wehrmacht, was defending a regime as detestable and oppressive as Stalin's.

Still, the hatred felt by the Russians and Eastern Europeans for Stalin and Communism was so great that hundreds of thousands of them flocked to join the German army. Perhaps the war might have taken a different turn in the East if Germany's policies had been more moderate. Perhaps the Western Allies might have linked up with the Red Army on Russian soil, rather than in central Europe. Perhaps millions of Eastern Europeans would have been spared half a century of Communist repression.

Antonio J. Munoz, 1991

INTRODUCTION

The Waffen-SS was the child of three parents: the Leibstandarte-SS Adolf Hitler (LSSAH), the SS-Totenkopfverbände (SSTKV, or Death's Head regiments), and the SS-Verfügungstruppen (VT). The LSSAH evolved from a battalion that had been raised in 1933 to protect the life of Germany's newly appointed chancellor, Adolf Hitler.

By the Anschluss with Austria in 1938, the LSSAH had evolved to a reinforced motorized infantry regiment. In 1940 the unit was enlarged to brigade level. In the fall of 1941, barely five months into the Russian campaign, the LSSAH was elevated to divisional level, though losses in the field prevented the motorized unit from attaining real divisional strength until the summer of 1942, when it was returned to Russia after being reorganized during the spring as a panzergrenadier division in France.

It was not until October 1943 that Hitler authorized the formation of SS panzer divisions. At that time, the LSSAH was one of only seven SS divisions that eventually became armored.

The SS-Totenkopfverbände were created to run the concentration camps established in 1934 under SS auspices. There were a total of eighteen Totenkopfstandarten formed. The first three were detached from concentration camp duty in 1939 and used to form a frontline combat division. This motorized formation became known as the SS-Totenkopf-Division. Its development paralleled that of the LSSAH from 1942 onward. The other Totenkopfstandarten were disbanded and/or used whole to form other Waffen-SS units.

The SS-Verfügungstruppen (SSVT) contained the LSSAH for a time. In 1935 two SS-Standarten (SS regiments) named Gemania and Deutschland were raised and combined with LSSAH and called the SS-Verfüsgungstruppen, the name initially given to the standing armed formations of the SS. Eventually, that name would be changed to Waffen-SS (Armed SS) in 1940. The Germania and Deutschland regiments were formed as motorized infantry regiments, like the LSSAH.

After the Anschluss with Austria in 1938, a fourth regiment, Der Führer, was raised in Austria. Before the outbreak of World War II, on 1 September 1939, the Leibstandarte-SS Adolf Hitler Regiment was separated from the SSVT and made an independent unit, while the Der Führer Regiment had taken its place in the SSVT. Therefore, before the outbreak of hostilities, the SS units under arms were the the LSSAH, the SSVT, and SSTKV (or SS-Totenkopfverbände).

At the same time the first three Totenkopf regiments were organized into the Totenkopf Division in the fall of 1939, another SS division made up of militarized members of the German National Police Force was also formed. This formation was not motorized until well after the French campaign and became known as the SS Polizei (Police) Division. The Polizei Division was always treated as a second child, having to make do with hand-me-downs from the older siblings. When the LSSAH, Das Reich, Totenkopf, and other panzergrenadier divisions were being converted to panzer divisions, the Polizei was just becoming a panzergrenadier division.

The majority of the SS-Totenkopfstandarten were manned by over- and underage part-time volunteers of the Allgemeine-SS (General-SS). The General-SS was basically an auxiliary SS organization that had grown from the small SS cadres existing in every German city in the 1930s. These cadres had at first been organized to protect Nazi party officials when the SA (Sturmabteilung), or Storm Troops, became too large and unruly to control. It was not until after the top SA leadership was liquidated by a distrustful Hitler in 1934 that the embryonic SS organization began to grow by leaps and bounds. Frightened members of the SA began to join the General-SS for fear of a repetition of the events of June 1934. So fast did recruitment grow that Himmler had to put a cap on it for a while.

Two SS Death's Head cavalry regiments were raised in 1940, later formed into a brigade, and then expanded to become the Florian Geyer SS Cavalry Division. Two additional Totenkopfstandarten were used to form SS-Kampfgruppe Nord (North) in the first half of 1941. The battle group went to northern Finland just before the beginning of the Russo-German war and later became the 6th SS Mountain Division Nord in 1942.

When the French campaign was over, the SS-Verfügungs-Division (which had been formed from the Deutschland, Der Führer, and Germania SS regiments) was redesignated SS Division Das Reich. Additionally, the Germania SS Regiment was withdrawn, and in its place, the 11th SS-Totenkopf-Standarte was added. This Death's Head regiment was soon redesignated the Langemarck. At this time, the Waffen-SS went through a dramatic change. Up until then, only ethnic Germans had been allowed to join the Waffen-SS (in small numbers). But for the first time, foreigners (albeit, "Germanic" volunteers) were allowed to wear the double thunderbolt insignia.

THE FOREIGN VOLUNTEER MOVEMENT BEGINS

Some scholars would have us believe that the Waffen-SS was always a homogeneous political organization, espousing the thinking of the Reichsführer-SS Heinrich Himmler's flights of Nordic fancy. Nothing could be further from the truth. There were many men of influence within the Waffen-SS who opposed Himmler and his theories, men such as SS Generals Paul Hausser and Felix Steiner, who believed that the Waffen-SS should have been the instrument to unite all of Europe against the Communist threat from the east.

Himmler's expanding SS divisions restricted the amount of time he could devote to each part of the empire. Although he strove to control the front-line Waffen-SS divisions, the thousands of kilometers' distance between him and the units made his influence on the Waffen-SS superficial—especially as the war progressed and the differentiation between SS trooper and Heer landser (army soldier) disappeared under the weight of enemy action. From the beginning, Waffen-SS formations had been under army control. In fact, it wasn't until 1944 that the first SS army was formed, and that was a mixed army/ SS affair.

Granted, some Waffen-SS units were briefly attached to the Command Headquarters of the National SS Leader (Kommando Stab Reichsführer-SS), but these were temporary affairs. The only exception was the bodyguard battalion that provided permanent security to his command post. It, too, had been formed from Death's Head units.

As time passed, men with opposing views about military roles began to assume more control within the combat units of the Waffen-SS. It was natural that the men under their command would eventually begin to see things in the same light as their commanding officers—especially when they compared their fighting generals with the "armchair" one back in Berlin. The combat troops of the Waffen-SS began to view Himmler as the chief policeman and not as their ultimate superior officer, which, technically, he was.

Another opponent of Himmler's was SS General Gottlob Berger, whose allegiance lay with Martin Bormann, Hitler's personal secretary. Hitler stayed in power by dividing his strength among his subordinates, ensuring that they would fight among themselves for his favor and rendering them too weak individually to oppose him.

Thus, Berger, as the head of the Waffen-SS recruiting office, could push successfully for the changes he felt were necessary to keep the ever-increasing Waffen-SS up to strength. Berger first suggested allowing foreign ethnic Germans (volksdeutsche) into the Waffen-SS at a time when Himmler wanted to enlarge his SS "army," but opponents in the German army pressed Hitler to allow them the vast majority of the German recruits. Berger's successful move opened the door to foreign recruitment wide enough that it would forever remain open.

For Berger, finding enough men to fill the constantly thinning ranks of the Waffen-SS units was a full-time job. He did not have time to ponder the validity or stupidity of Himmler's racial theories. By the end of the summer of 1940, Berger was able to convince Himmler of the need to expand the Waffen-SS even further by the influx of "Germanic" volunteers. No longer would volunteers need to be German or even ethnic German. The Waffen-SS could recruit from the so-called Germanic races.

Norwegian and Danish volunteers were organized into a new SS regiment, called Nordland (North Land), while the Flemish volunteers from Belgium were grouped into SS Regiment Westland. The Dutch were at first sprinkled between both regiments and in the first foreign-manned SS division, titled Wiking. Wiking was formed by grouping together the Nordland, Westland, and Germania regiments.

Eventually, the Dutch contribution to the Waffen-SS warranted the formation of two SS divisions (23rd and 34th); while the Flemish made up the 27th SS; the Walloons the 28th; and the Danes and Norwegians the 11th SS divisions. Later, the French were also allowed to join the Waffen-SS and were grouped into the 33rd SS Division.

Meanwhile, the recruitment of men into the Waffen-SS grew ever greater. By the end of 1940, the Waffen-SS could count on 100,000 men. By June 1941 that number had surpassed 150,000, and by the close of 1941, it had reached 222,000. On 1 May 1941, just about one month before Hitler invaded Russia, two SS motorized brigades were formed from four SS Death's Head regiments. The 8th and 10th Death's Head regiments made up the 1st SS Motorized Infantry Brigade, while the 4th and 14th SS Death's Head regiments merged to become the 2nd SS Motorized Infantry Brigade.

On 22 June 1941 the Waffen-SS had the following divisional strengths:

- Leibstandarte-SS Adolf Hitler, 10,796
- Das Reich, 19,021
- Wiking, 19,377
- Totenkopf, 18,754
- Kampfgruppe Nord, 10,573.

Other SS regiments included SS-Standarte Nordwest, 2,500 men, and SS-Freiwilligen-Legion Flandern, 875 men. SS Freikorps Danmark was 480-men strong in 1941 but, by the first half of 1942, had increased to 1,164 men.

Although 1943 was the turning point in the development and expansion of the Waffen-SS into what would eventually be the fourth branch of the Wehrmacht (armed forces), 1942 had a few significant developments for the Waffen-SS. During that year, Kampfgruppe (Battle Group) Nord became a division, the SS Cavalry Brigade was expanded to division-level, and a new SS division, the 7. SS-Freiwilligen-Gebirgsjäger Division Prinz Eugen was formed. This new formation had a German cadre but was manned almost exclusively by ethnic Germans from the Kingdom of Croatia, a separate state created when Yugoslavia was partitioned after its defeat by Germany, Italy, Hungary, and Bulgaria. This volkdeutsche division operated exclusively in the Balkans, where it was raised, and for most of its history fought Tito's Partisans.

More and more, the Waffen-SS began to slip away from Himmler's clutches. Fighting a death struggle thousands of kilometers from home, the Waffen-SS began to see itself more and more like the regular German army units. The Waffen-SS formations, with the exception of the units under Himmler's direct command headquarters, were always controlled by the army. Even after the Waffen-SS formed SS corps headquarters to control the increasing number of SS divisions, the SS corps themselves were almost exclusively under army control.

As quickly as the German army conquered a country, the Waffen-SS would be there, recruiting volunteers. First it was the Germanic peoples of Northwestern Europe. But before the war had reached the halfway mark, even the so-called "untermensche" (or "subhumans") were wearing the SS uniform. Throughout the war, two distinctly different camps developed within Germany: those opposed to recruiting foreigners, especially the "subhumans," and those who saw the Waffen-SS as the perfect instrument in which to merge the diverse European community into one movement to stop Communism. Aside from half-promises of independence once the war was over, not much else was offered to the volunteers. In spite of this, thousands of Europeans and Russians continued to die for Germany up until the end of the war. In spite of the narrow and racist policy espoused by Hitler, Himmler, and the rest of the Nazi fanatics, hundreds of thousands of non-Germans rallied to join the ranks of the Waffen-SS. Whether they were misguided idealists, base criminals, or unsung anti-Communist heroes, the men of the Waffen-SS ensured, by their military actions and esoteric nature, their place in the annals of military history, right alongside that little-known and often forgotten elite unit, the Praetorian Guard.

PART ONE

Illustration: Eric Weisman.

PARACHUTE BATTALION: THE HISTORY OF SS-FALLSCHIRMJÄGER BATAILLON 500/600

CHAPTER ONE

Caesar's spirit, ranging for revenge,
With Ate by his side, come hot from hell,
Shall these confines, with a monarch's voice
Cry, "Havoc!" and let slip the dogs of war.

William Shakespeare
Julius Caesar

The year was 1943. On all fronts, Hitler's armies were being put on the defensive by the resurgent Allied war machine. Once thought to be invincible, Hitler's armies now waged a war of attrition that clearly was going against them. A nation of eighty million was fighting the world.

Although the German Wehrmacht retained much of its punch and was still a formidable foe, it was not the same force that had overrun Poland in a few weeks, Denmark in a day, and France and the low countries in ten days' time (the rest of the campaign was just a mopping-up operation). The panzer divisions in particular still had much of their earlier élan. Although valuable, experienced men had been lost by the hundreds of thousands, the army still had enough seasoned soldiers to break in the vast number of new recruits being called to the colors.

Most German generals in the armed forces high command believed that the Wehrmacht should have remained in a defensive posture on all fronts during the summer of 1943. Colonel General Heinz Guderian was one of those. He believed that Germany could frustrate the Soviet Union's field army by letting its armored formations batter their heads against a mobile German defense line. Guderian realized that Germany still possessed the Soviet territory that would allow the German army room to maneuver and, when necessary, withdraw, to properly employ mobile defensive tactics. Additionally, the Soviet war machinery was still inflexible and rudimentary when making combined arms operations. This added advantage helped the German defensive tactics while causing grievous losses to the Red Army.

Hitler would hear nothing about staying on the defensive. Although it was virtually impossible for the Wehrmacht to attack along the entire Eastern Front, Hitler was bent on launching another summer offensive. The salient that had been formed by the Red Army during its winter and spring offensives in central Russia encompassed a wide, bulgelike area, whose center was near the city of Kursk. This "bulge" jutted into German-held territory like a chancre. Hitler was afraid that this "chancre" would soon burst open, flooding the surrounding territory with Soviet forces. It was for this reason (and others that only Hitler himself could have answered) that the German army was forced to begin preparations for an offensive directed at snipping the Soviet intrusion off.

In March 1943, the Germans were too weak to accomplish such a mission. In April, their forces were a little better off, and most historians now agree that would have been the best time to launch an attack. The Soviet army had yet to build the rings and rings of defense lines the German attack would encounter in July 1943. It would be those heavy defense lines and the vast number of Soviet reserves that would defeat Hitler's summer offensive. On 5 July 1943, he launched Operation Zitadelle, which turned out to be a bloody draw between the world's largest tank armies. While the Soviet juggernaut quickly replaced those bloody losses, the Germans were not able to replace theirs fully. So great had been the loss, that even one entire Panzer division (the 18th) had to be demoted to an artillery formation, while every other German armored division would never again attain full table-of-organization (TOE) strength.

From this point on the German army was on the defensive. Only at local focal points would the Germans gain temporary numerical superiority or win a tactical victory. No longer would entire German army groups sweep forward unopposed over the vastness of the Russian steppes. Hitler now faced the consequences of one of his greatest blunders. He opted to blame his generals for the military disasters that began to befall the Wehrmacht with more frequency.

Hitler's rage increased with every new defeat. His SS formations appeared to be the only ones immune to what he called the "defeatist germ" he believed his generals had passed on to the troops in the field. He believed that only his politically indoctrinated combat troops had the will to win.

In this Hitler may have been partly correct. Fanaticism often does blind an individual to reality. It makes him the perfect soldier, sacrificing himself for a cause that might be already lost. Although the Waffen-SS possessed many men of such thought, it did not have a monopoly on fanaticism, for whatever reason. One has to remember that the Nazis had been in power for a total of ten years by 1943, certainly plenty of time to attempt an indoctrination of an entire nation, especially the easily influenced youth.

Illustration: Eric Weisman.

Still, the notion that it was only his SS units who were carrying the fight to the enemy became firmly imbedded in Hitler's brain. This idea was reinforced as he began to use his prized SS armored divisions as mobile "fire brigades" all along the crumbling Eastern Front. Those SS divisions were always replenished first and foremost before the Heer (army) panzer divisions. Additionally, they always got the best men and equipment. This fact alone made the SS formations appear to operate with much greater success than their army brothers. Hitler, of course, never mentioned this fact.

On 8 September 1943 Italy was invaded on the mainland. The Italian government sued for peace. Benito Mussolini, the Fascist leader of Italy, had been removed by his own Fascist Grand Council and placed under house arrest. Hitler had been anxiously watching the developments south of the Brenner Pass. In spite of the cries of continued loyalty by the new Badoglio government in Rome, Hitler was too cynical to fall for such empty statements.

Ever since the spring of 1943, Hitler had been moving more German divisions into Italy on the pretext that they were to support the German and Italian units in Sicily and southern Italy. Some of those units remained not only in central but northern Italy, menacingly close to every Italian army formation. When Italy surrendered, Hitler's divisions moved not only to block the Allied advance from the south, but to disarm the Italian army everywhere.

Some Italian units resisted briefly, while others simply laid down their arms. Others took to the hills to join the partisans, while some remained loyal to *Il Duce*. A large segment of the Italian population was still opposed to surrender and actually still backed Mussolini. The Fascist Party, however, was in disarray. Without Mussolini, it was incapable of forging the supporters of Fascism who still remained in every level of Italian society. Hitler knew this only too well. He planned to make the Duce the rallying point for every Italian who still believed in a Fascist solution.

Hitler saw this as the way for the Italians to help him bolster his defenses against the Allies in Italy. With the fortunes of war drastically altered against Germany, Hitler needed to gather as many allies as he could to his side. The problem he faced was that Mussolini was under house arrest, beyond Hitler's grasp. At any moment, the Italian government could consider the Duce as too dangerous to keep alive. For this reason, it was imperative that German intelligence locate the Fascist leader and make plans to rescue him.

Soon German intelligence did locate him, at a skiing lodge miles south of Rome. The lodge was located high atop a mountain and was accessible only by cable car. Not surprisingly of the eight capable German officers who were presented to Hitler to lead the rescue mission, he chose an SS officer.

At six feet, five inches, SS-Hauptsturmführer Otto Skorzeny stood out like a sore thumb. This giant was to prove his worth many times over. With a handpicked company of SS commandos and a company of Luftwaffe paratroopers, Skorzeny made a glider assault on the mountain resort-turned-fortress on 12 September 1943. The operation was so unorthodox and the Italians so taken by surprise that the Gran Sasso hotel defenders gave up the Fascist leader without a fight. A pro-Fascist Italian general accompanying the Germans had dissuaded his countrymen from resisting Skorzeny's troops.

The daring raid was publicized throughout the world, giving the Germans a propaganda victory that Dr. Goebbels, as head of the German Propaganda Ministry, trumpeted for weeks on end. The manner in which the operation had been planned and carried out was so daring and fantastic that it could have been scripted in Hollywood.

After this SS success (the Luftwaffe troops were conveniently forgotten, much to Luftwaffe General Kurt Student's chagrin), Hitler's mind was firmly made up. On the ground, only his "fire brigades" (his Waffen-SS troops) still had the will to win. It had been an SS commando force, led by a dedicated and fanatical officer, that had carried the day in Germany's favor. Only his SS troops, Hitler reasoned, were capable of staving off defeat and once again clutching the seeds of victory for Germany.

Thus, the Waffen-SS began to expand at a prodigious rate: from a strength of 350,000 men in 1943 (280,000 field and 70,000 reserve) to a peak of 950,000 men in the fall of 1944. One of the new Waffen-SS units formed was the SS Parachute Battalion. No longer able to trust his army or air force, Hitler firmly believed that the Waffen-SS was the instrument that would save the day.

Again Hitler was partly correct, for after the expansion of the Waffen-SS, Germany did stave off defeat for another two years, but that feat was the result of determined army units just as much as Waffen-SS formations. Turning the tide of war in Germany's favor was far beyond the means of any military formation that Hitler could have formed or expanded. Hitler could have picked a ready-trained Luftwaffe parachute battalion and given it a refresher course in commando training. He could have even used a commando battalion from the Brandenburg Division (a formation specializing in espionage and infiltration tactics). Instead, he chose to form a parachute battalion made up exclusively of Waffen-SS men.

Creating a completely new unit from scratch meant that it would be some time before it was combat ready. This delayed the employment of the formation until the summer of 1944. By that time, no power on Earth could hold up the defeat of the Third Reich. This SS Parachute Battalion, the experiment that was to help Hitler regain the initiative by changing the tide of the war, arrived only in time to delay Germany's inevitable collapse. Like the ME-262 jet or the V-1 and V-2 rockets, the battalion was just too little, too late.

The men who made up the SS Parachute Battalion had to pay a high

SS-FALLSCHIRMJÄGER-BATAILLON 500/600
1943 – 1945

This general reference map shows the movements of the SS-Fallschirmjäger-Bataillon 500/600 between 1943 and 1945.

price. A total of 3,000 to 3,500 men served in the battalion from one time to another. When the war ended, 180 had survived unscathed. What better indicator of how heavily engaged the battalion was during its brief existence than these heavy losses?

Fighting for a cause that was already lost, the battalion members often went beyond the call of duty. They fought with bravery and fatalistic ferocity. They committed no war crimes and behaved in keeping with the normal standards of war. Most of the men who served in the battalion were volunteers. Certainly they were not the stereotypical Nazi fighting for the enslavement of the world.

If anything, they were fighting against the dramatic Allied announcement at Casablanca in January 1943, to accept nothing less than the unconditional surrender of Germany. The agreement had far-reaching repercussions. To the German nation, it was tantamount to suicide. Germans believed there was nothing left for them to do but continue the war under Hitler and the Nazis and hope for a stalemate.

From 1943, the overwhelming majority of the German troops no longer believed that they could win the war. They continued to fight on for their homes, families, and their country's continued existence—not for Nazism. The bravery and accomplishments of the SS paratroopers are what set them apart from the regular German forces. For this reason, at least, their full story deserves to be told.

CHAPTER TWO

Form, form, riflemen form!

Lord Alfred Tennyson

On Hitler's direct instructions, the order to form SS-Fallschirmjäger Bataillon was passed down to the Führungshauptamt der Waffen SS (Command Headquarters of the Waffen-SS). The unit was to be made up of volunteers of existing Waffen-SS divisions, with a percentage of the battalion's men being minor disciplinary cases.

These disciplinary problems resulted from Heinrich Himmler's strict Draconian rules for the Waffen-SS. What would have been a minor infraction on the part of an Allied soldier was cause for court-martial proceedings for a member of the Waffen-SS. For example, reporting for duty one half hour late (as trivial as this sounds) could land an individual in front of a court-martial. Sleeping on post and drunk and disorderly behavior were other types of relatively minor infractions punishable by court-martial.

Accusations of serious offenses such as rape, murder, looting, and poaching were also decided by court-martial, and, if found guilty, the perpetrators were sent to a special penal unit. This penal battalion (later, a brigade) was under the command of a Dr. Oskar Dirlewanger, a convicted sex offender whose character would set the standard for his repulsive Sonderverbande Dirlewanger. This unit would make itself infamous throughout the world for its bloodthirsty activities against the Polish Home Army during the Warsaw uprising of August 1944.

The newly created SS-Fallschirmjäger-Bataillon 500 received only court-martialed officers and men who hoped to distinguish themselves enough to have their sentences commuted and their former ranks restored. One such case was that of SS-Rottenführer (Corporal) Paul Firley, who had been a member of 1st Company/SS Engineer-Replacement-Battalion Dresden before being sent sent in the summer of 1941 to the SS Division Nord in Finland. It was there that he got into trouble by failing to show up for guard duty.

Findley was broken in rank and sent to a penal unit stationed at Matzkau. In the fall of 1943, such minor disciplinary cases were transferred to the newly formed SS Parachute Battalion. Firley would later distinguish himself in combat and be decorated for wounds received. Like most disciplinary cases in the battalion, he was restored in rank and given the choice of remaining with his new unit or returning to his old command. Firley chose to stay with the Parachute Battalion, as did most men. As time wore on, the number of disciplinary cases in the SS Parachute Battalion decreased. By the fall of 1944, the battalion was made up almost exclusively of volunteers.

The disciplinary cases were overseen by Section-III, an integral part of the SS Parachute Battalion, which contained an SS lawyer with a clerical staff for keeping records on the probationary cases.

Initial formation of the battalion was carried out at Chlum in Czechoslovakia. Command of the battalion was at first given to SS-Sturmbannführer Herbert Gilhofer. The thirty-three-year-old Gilhofer had been the commander of SS-Panzergrenadier-Regiment 21 (10th SS Panzer Division Frundsberg). After failing to carry out an order, he was removed from his command and sent to the SS Parachute Battalion. His tour with the battalion lasted until mid-April 1944. In 1945, he returned to the Frundsberg Division—this time as commander of III/SS-Panzergrenadier-Regiment 22.

Training began in earnest during fall 1943. That November parachute training shifted to Madarushka-Banja near Sarajevo, Yugoslavia, at the Luftwaffe's Paratroop School Number 3. The school later relocated to Papa, Hungary, with the SS paratroopers finishing their training there. Troopers jumped from German Ju-52 transports (nicknamed "Auntie-

Ju's") and Italian Savoias transports manned by Italian pilots still loyal to Mussolini.

After training was completed, the battalion moved back to Yugoslavia. It now had a strength of just over 1,000 men (nearly a regiment) and was organized in the following manner:

- Battalion Staff Company
  - One supply company
  - One reporting squad
  - One signals platoon
  - One motorcycle dispatch squad
  - One maintenance platoon
  - One parachute maintenance platoon
- 1st Parachute Rifle Company
  - Three parachute rifle platoons plus a communications squad
- 2nd Parachute Rifle Company
  - Same as above
- 3rd Parachute Rifle Company
  - Same as above. NOTE: Each parachute rifle platoon contained three rifle squads, three light submachine-gun squads, and a mortar squad.
- 4th Heavy Weapons Parachute Company
  - One machine-gun platoon
  - One mortar platoon
  - One flamethrower platoon
  - One antitank platoon

It is interesting to note that the SS Parachute Battalion was organized quite differently from a regular Luftwaffe parachute battalion. The main difference was the relatively self-contained structure of this SS unit. The SS Parachute Battalion had its own training company, which the Luftwaffe battalion did not, and while the table-of-organization (T/O) strength of the Luftwaffe battalion headquarters listed 138 men on paper, the SS equivalent was around 267 men—almost twice as many.

Again, because the SS unit was self-contained, it needed more personnel for battalion support services. The Heavy Weapons Company of the SS Battalion was a hybrid. The Luftwaffe battalion had either a parachute machine-gun company of 205 men or a 120mm mortar company of 163 men. The Luftwaffe battalion had a table strength of two LG-40 75mm antitank guns, while the SS unit had four. In addition, the SS Battalion had two or three flamethrowers, something which was only found in the upper echelons of the Luftwaffe parachute division. Motorization for the SS Battalion included about one hundred trucks and about thirty motorcycles, while the standard motor pool for the Luftwaffe battalion was eighty-one trucks and nineteen motorcycles. Finally, a strength of more than 1,000 SS paratroopers was quite high when you consider that a regular Luftwaffe parachute battalion had a T/O strength of only 853 men. This discrepancy is even more apparent if you consider that not every battalion was up to T/O strength.

The battalion was committed to anti-Partisan drives in the areas of Uzice in Serbia, Tuzla in Bosnia-Herzegovina, Montenegro, and Macedonia. Around the middle of April 1944, under the strictest of security measures, the battalion returned to Madarushka-Banja, where it was "interned." No one was allowed to leave the base or to write home. Rumors, ranging from the very bland to the most colorful, spread like wild fire.

Sturmbannführer Gilhofer was relieved of duty and transferred from the battalion, and SS-Haupsturmführer Kurt Rybka took over. This twenty-seven-year-old captain would lead the battalion in its first and only air drop on the mountain stronghold of Marshal Joseph Broz Tito, the Communist Party leader who headed the Yugoslav Partisans.

Partisan activity against the Axis occupation forces in Yugoslavia had increased steadily. The number and boldness of these attacks had risen from a negligible amount in late 1941 to almost continuous firefights by the beginning of 1944. Even after numerous and concerted anti-Partisan drives, the Axis occupation forces had been unable

SS-FALLSCHIRMJÄGER-BATAILLON 500 IN YUGOSLAVIA, 1943 – 1944

SS-Hauptstrumführer Kurt Rybka, the commander of SS-Fallschirmjäger-Bataillon 500 at Drvar, as he appeared in 1944. Photo: Munin Verlag.

The air-transportable German Luftwaffe LG-40 antitank gun. This 75mm gun is shown here being used for gun practice by men of the SS Parachute Battalion 500, somewhere in Yugoslavia. Photo: Munin Verlag.

to reduce, let alone destroy, the Partisan movement. Each time an anti-Partisan offensive was launched, it seemed as if Tito's Partisans would be crushed, but, unfortunately for the Axis, Tito was a shrewd commander who used the extensive mountain terrain of Yugoslavia to his advantage.

Not only did he avoid destruction of his forces, but he managed to increase their size and number so that by 1944, the German army High Command stopped referring to them as "bandit groups" and started calling them Partisan brigades or divisions. Many of the tactics employed by Tito were later used by other Communist guerrilla movements. Tito would always hit the enemy at its weakest points, using an overwhelming force to guarantee success. If the enemy moved in a superior or equal force against him, he would withdraw immediately.

In this way, he slowly drained the enemy of its strength while increasing his own

Officers of the SS-Fallschirmjäger-Bataillon 500 at parachute training school (top). The two officers in the center of the photograph wearing the khaki summer uniform are Luftwaffe parachute instructors. Photo: Munin Verlag.

SS-Untersturmführer Eric Brörup, shown here on the far right (above). This picture was taken at parachute training school somewhere in Hungary, 1944. Photo: Munin Verlag.

from captured stocks and new recruits who were impressed by the Partisans' success. This tactic also kept the occupation troops off-balance and on the defensive, trying to shift their meager forces from one place to another, and it is well-known that the Germans and their Axis allies never had sufficient forces garrisoning the Balkans. Only during the latter part of 1944, when the war there turned conventional, did a sufficient number of Axix troops begin to operate.

The German High Command, long aware of the danger it faced in Yugoslavia, had compiled an extensive dossier on Tito. A Croatian by birth, Tito's interest in the Communist Party had begun when he was a young man working in a factory. He longed for the day when Yugoslavia would follow the Bolshevik example. He had been personally involved in the murder of Yugoslav Communist Leader Gorkic in 1936. By the time the Spanish Civil War broke out, Tito had already placed himself at the head of the Yugoslav Communist Party, and he began to recruit his fellow countrymen for that war. He later moved to Paris, where he organized a similar "secret railway" of recruits. Though he denied ever having been in Spain, he had in fact visited the International Brigades on several occasions.

Tito's credentials were well suited for leading a Communist uprising. Those who see Tito as a patriot should note that his wartime rival, Draza Mihailović, the Chetnik leader (whose allegiance was to the exiled King of Yugoslavia), had begun operating against the Germans long before Tito. In fact, the Yugoslav Communists had been instructed by Moscow that under no conditions were they to take any actions against the Germans. Of course, this changed after Hitler invaded the Soviet Union.

The Germans rationalized that if Tito could be done away with, then perhaps his Partisan movement could be destroyed. Tito's field strength at the end of 1943 has been estimated at between 90,000 and 150,000 men (depending on which source you believe). By the summer of 1944, this had increased to between 150,000 and 250,000 men (again, depending on who is telling the story). The exact numbers are irrelevant, for they were large enough to pose a more serious threat to the German forces in 1944 than they had in 1943.

In April 1944, Tito's headquarters began receiving disturbing reports from agents in the German and Croatian areas. (Croatia had declared its independence after the fall of Yugoslavia; with Italian backing, a Fascist state had been set up with Ante Pavelić as head of the government.) All of these reports pointed to an impending seventh anti-Partisan campaign being planned by the Germans and their allies, this time with an emphasis on Partisan headquarters (which meant Tito).

The reports also indicated the increasing combat readiness of the German 5. SS-Freiwilligen-Gebirgs-Armee-Korps (5th SS Volunteer Mountain Army Corps) and the Wehrmacht 15. Gebirgs-Armee-Korps (15th Mountain Corps). Early on, ULTRA messages intercepted by British MI6 had revealed that the Germans were discussing a large-scale operation near Drvar (in western Bosnia).

(By way of digression, ULTRA was the British wartime operation for decoding intercepted German operational radio traffic. As early as 1938, the Polish secret service had got hold of a German army ciphering machine, quite by accident. The secrets of its workings were recorded and studied before the machine was sent to its original destination, the German Embassy in Warsaw, without the Germans ever realizing that their ciphering system had been broken. After Germany invaded Poland, the British got a working model and immediately began intercepting messages. ULTRA was the most closely guarded secret of the war. One commander, Marshal the Viscount Montgomery, gained his reputation as a skillful commander based on the daily messages that MI6 fed him. ULTRA was the single most important factor, other than Hitler's constant meddling in military matters, that contributed to the Allied victory in Europe. It is ironic that Poland, the first country to suffer Hitler's military aggression, was responsible for providing the Allies with the key to winning the war.)

Other ULTRA messages passed on to Tito indicated (as early as 22 and 23 May) that for an important operation in the following days, 300 men were to be accommodated at the airfields at Biha´c and Banja-Luka.

Although British intelligence was uncertain who these 300 men were or to what unit they belonged, there is no doubt that the message referred to the SS Parachute Battalion. A further decoded message aroused the alarm of the British intelligence officers. It stated that Fliegerführer Kroatien (Flight Command Croatia) was to receive temporary aircraft reinforcements for the operation. On 24 May, another decrypted message disclosed that an operation was scheduled to take place the following day. On that day, one more message revealed the most alarming news of all: the German air force in Zagreb had notified an SS parachute battalion that "all is ready for boarding." Fortunately for the Germans, this last message was decoded too late to make any difference in the tactical surprise the parachute drop achieved.

Some in the Yugoslav government still insist, to this day, that this last and most important message was withheld by orders of British Prime Minister Winston Churchill. The Yugoslavs argue that Churchill wanted Yugoslavia to be ruled by any other of government other than Communism. Getting

Picket duty for the men of SS Parachute Battalion 500 during an anti-Partisan drive in the wild mountainous countryside of Yugoslavia, sometime in 1944 (top and above). Photos: Munin Verlag.

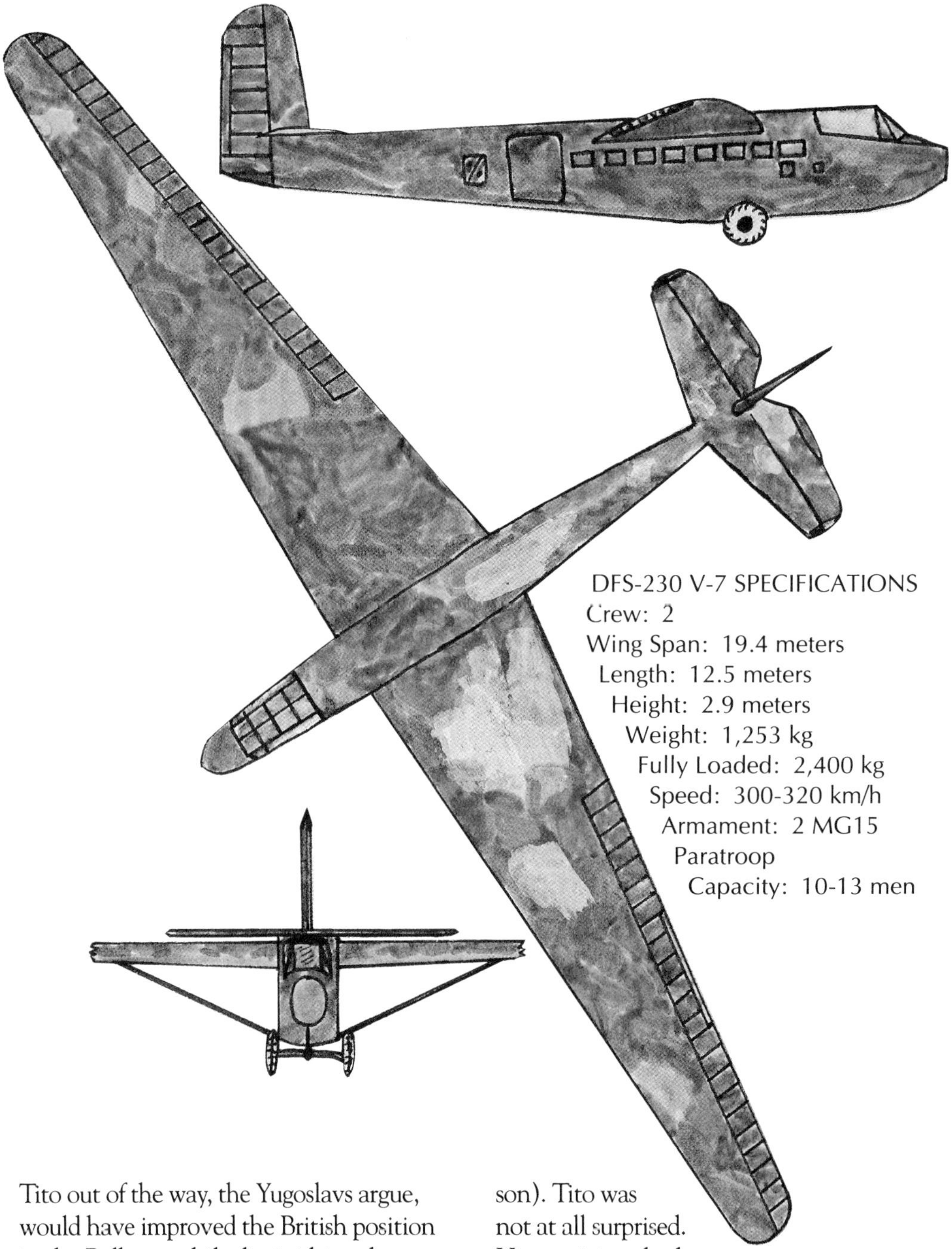

German glider used at Drvar: the DFS-230 A-1 Glider. Illustration: Antonio Munoz.

Tito out of the way, the Yugoslavs argue, would have improved the British position in the Balkans, while diminishing the Soviet Union's. Whether this is all nonsense or has a basis in fact has never been clearly disclosed. Of course, the British continue to deny the accusation.

Enough information was available to deduce that Tito's headquarters at Drvar was to be subjected to an air raid at the very least. This warning was passed on to Marshal Tito himself by the head of the British military mission at Drvar, Major Randolph Churchill (Winston Churchill's son). Tito was not at all surprised. His suspicions had been aroused not only by his own reports but also by low-flying German reconnaissance aircraft that had flown over Drvar during the previous few days.

It was obvious that the reconnaissance flights had been made in anticipation of a bombing mission by the German and Croatian air forces, perhaps heralding the start of another German anti-Partisan drive. What better way to start an offensive than by knocking out the enemy headquarters with its important officers and communications equipment? Tito firmly believed that the impending bombing mission would herald another concerted German drive against his forces. He would be proved correct in his assumption, but with one major element missing—the use of the SS Parachute Battalion. Surprisingly, this little bit of information failed to reach the ears of Tito's extensive spy network in the German camp. This nonobservance almost cost Tito his life.

In his autobiography, Otto Skorzeny wrote that he had been given the mission of "neutralizing" Tito by Hitler himself. It is typical of the German system of duplicity of command that Skorzeny was given the same mission as the SS Parachute Battalion. As it turned out, this entanglement of command was to work against the German effort to eliminate Tito.

It seems that the commanding officer of the Wehrmacht in southeastern Europe, General Loehr, had plans of his own on how to deal with Marshal Tito. He had planned a seventh anti-Partisan offensive, beginning with a bombing raid on Tito's headquarters at Drvar, followed by a lightning parachute drop to kill or capture the Partisan leader. Skorzeny had another idea.

"Get Tito—alive or dead!" was the challenge tossed to Skorzeny. Leaving the Croatian capital of Zagreb, he drove all the way down to Belgrade in his Mercedes, right through Partisan-infested territory. His only escorts were his driver and two other enlisted men. Upon arrival at Belgrade, the German garrison commander could not believe that Skorzeny had driven all the way without encountering a single Partisan band.

As soon as he arrived, Skorzeny set about trying to locate Tito's headquarters. Reports quickly began to come in, but they were so contradictory that he realized they were useless. He decided to move into the interior of the country, disguised as a Partisan. He took with him a only small force of men, so as to attract less attention. It took Skorzeny approximately four weeks to locate Tito's secret headquarters—inside a spacious cave, high above a mountain overlooking the town of Drvar and, not accidentally, all avenues of approach.

Skorzeny interrogated a Partisan deserter who gave him complete details about the Partisan defenses at Drvar. The problem of getting at Tito seemed all but impossible: Tito was surrounded by 6,000 seasoned troops, and reinforcements were available from the surrounding countryside. He had a personal escort battalion of 350 hard-core Communist guards, all of them die-hard fanatics. It is no small wonder that, when pitted against SS troops, this escort battalion would fight like wild dogs, giving no quarter to their enemy—wounded or otherwise.

Skorzeny finally decided that the only sure way to get close enough to Tito was to come in disguised as a Partisan unit. He gathered about him a company of his best men from Friedenthal, his main commando school, and developed a plan to personally lead them in. If all went well, he hoped to be able to get close enough to rush the mouth of the cave, surprising Tito's guards and either capturing or killing the Partisan chief.

SS-Hauptsturmführer Adrian von Foelkersam, Skorzeny's chief of staff, was sent to speak with General Loehr in hopes of convincing him that his plan offered the best results of dealing with Tito. Skorzeny had even instructed von Foelkersam to agree to place the commandos under Loehr's command, if that would please the glory-hungry general.

After being made to wait in Loehr's anteroom, von Foelkersam was finally attended to but did not get past the commander of 2nd Panzer Armee, General Rendulić, who told him that Loehr's mind was firmly made up as to how to get Tito. Von Foelkersam pleaded Skorzeny's case, even mentioning that the latter had heard about the planned SS drop from two Yugoslav agents in the field. "If these agents know about the drop," warned von Foelkersam, "it is more than likely that Tito has gotten wind of it too. These paratroopers will be sitting ducks for the partisans."

But neither Loehr nor Rendulić was in a receptive mood: von Foelkersam was perfunctorily dismissed. All Skorzeny could hope for was that General Loehr's sledgehammer tactics would work, when he knew that what was really needed was the skillful use of a surgeon's scalpel.

As early as April 1944, Tito's Partisans had captured the deserter who had helped Skorzeny. After disclosing the information he gave to the Germans, he was tried and shot. Though his presence at Drvar was clearly compromised, Marshal Tito felt secure and decided to remain. He had seven divisions in eastern Bosnia. He would await the German offensive and take any necessary steps then.

The only question that still worried him was the date of this next German offensive. It seemed that it would start any day, especially after German reconnaissance planes began to appear over Drvar, but Tito was not counting on a parachute drop. Amazingly, this piece of information had not reached him. This relaxed atmosphere was also shared by Tito's Chief of Staff, Stephan Jovanović, who had always sworn that a German airborne assault on Drvar was impossible. Jovanović would later eat his words and curse his overconfidence in underestimating the Germans.

CHAPTER THREE

If the trumpet give an uncertain sound, who shall prepare himself to the battle?

The First Epistle of Paul to the Corinthians 14:8

On 20 May 1944, SS-Hauptsturmführer Kurt Rybka was given a small outline of the role that his SS paratroopers would perform in the operation to get Tito. He was told that there were not enough gliders and planes to transport his entire battalion in one drop—even after a concerted effort to bring together more glider units from other parts of Europe. From Nancy, France, came the 3rd Battalion, Air Landing Group 1 with seventeen towing detachments (twelve Henschel-126 towing DFS-230 gliders and five Avias DFS-230s). This unit, under Luftwaffe Hauptmann (Captain) Karl, arrived at Alibunar in Serbia but was then moved to Krusevac on 7 May.

The 4th Squadron/2nd Battalion, Air Landing Group 1 (under Hauptmann Jahnke) was moved from the area of Strasbourg-Polygon-Mannheim to Zagreb, Croatia, in March. This squadron had eight towing detachments (Ju-87 dive bombers acting as towing machines for DFS-230 gliders). In this way, Towing Group 1 was reinforced.

Towing Group 1 consisted of three squadrons, the 1st, 2nd, and 3rd. The first two were composed of HS-126s and Ju-87s towing DFS-230 gliders. The 3rd Squadron contained Heinkel-IIIs towing huge Gotha (Go-242) gliders. Each Go-242 could carry up to twenty-one men, while the smaller DFS-230 could only haul seven to eight men. Thus, Towing Group 1 could only carry about 300 men in gliders. Also on hand was 2nd Battalion, Transport Group 4, with Ju-52 transport planes, which could carry men who would jump by parachute.

Rybka set about organizing his assault force. A total of 654 paratroopers would form the first assault wave. Of these, 314 would jump by parachute while the remaining men, organized into six assault groups, would land by glider. Clearly, the gliders would be overloaded, but Rybka needed as many men in the initial drop as he could squeeze in. The more men he could land, the better his chances for pulling off a successful operation. Each glider group was assigned a name and a specific mission.

- Panther Group—110 men to destroy Tito, his headquarters, and his elite guard (350 strong)
- Greifer Group—40 men to destroy the British military mission
- Sturmer Group—50 men to destroy the Soviet military mission
- Brecher Group—50 men to destroy the U.S. military mission
- Daufganger Group—A composite force of about 50 paratroopers and 20 men of group Svadil (intelligence officers from the Brandenburg Division, Luftwaffe signalers, and interpreters from the ethnic-German 7. SS-Freiwilligen-Gebirgs-Division Prinz Eugen. This group was to destroy the Partisan line and radio communications network, gather intelligence, and evaluate radio techniques plus captured code and signal books.
- Beisser Group—20 men to seize an outpost radio station, after which it was to assist Greifer Group in attacking the British military mission.

If Panther Group was successful, a Swastika flag would be draped over the entrance of the cave. If the assault unit ran into trouble, a red flare would be fired into the air: the signal for Sturmer and Grun Groups to disengage from their own missions and come to the assistance of Panther.

Rybka had further organized the 314 parachutists into three groups: Rot (Red), Grun (Green), and Blau (Blue). The paratroopers would land over Drvar and secure the town.

The glider troops would land on selected areas of the town and outlying key positions, including Tito's mountaintop hideout.

During the night of 22 May, the SS-Fallschirmjäger-Bataillon 500 and its attached units were transported to the three main assembly areas. Dressed in regular Wehrmacht uniforms and driven in German army trucks, platoon after platoon arrived at their destinations. Where it was convenient, trains were used to move each group quickly and quietly into position.

The first wave of paratroopers stationed at Nagy-Betskerek Airdrome was composed of Numbers 2 and 3 companies of the battalion; the battalion headquarters company was also billeted there, as well as a platoon from the 4th (Heavy Weapons) Company. Number 1 Company and the rest of Number 4 Company were placed at Zagreb Airfield. The Brandenburg and Bosnian contingents (Svadil) were also placed at Zagreb.

The second wave of paratroopers would have to await the return of the "Auntie Ju's" (Ju-52 transport plane) and the Italian Savoia-Marchetti transports.These Italian transport planes were piloted by Italians who had remained loyal to the Duce. The remainder of No. 2 Company and the Parachute Training Company were to form the second parachute drop. They were stationed at Banja-Luka and were scheduled to be dropped over Drvar around noontime. A few DFS-230 gliders packed with food and ammunition would also land in conjunction with the second drop.

On the morning of 24 May , all of the tactical and operational officers and commanders met for a detailed briefing of the operation. Included in this meeting were all the branch representatives and even the Ustaschi (Croatian Fascist Guard) and Chetnik group leaders. Under tight security, the plans were handed out and the photo-reconnaissance pictures shown. The glider pilots were instructed to land as close to their targets as possible, even if this meant the irretrievable loss of the gliders due to damage.The drop would be made at 7:00 A.M. the following day, 25 May 1944. It was no coincidence that this date was Tito's birthday.

At 3:30 A.M. on 25 May, the men of the SS Parachute Battalion 500 were awakened at all of the airfields. No sooner had the platoon sergeants turned on the lights in the barracks than they began to harangue their troops, "Mach, dass du weiter kommst!" ("Get going now!"), "Sich fertig machen!" ("Get ready!"), "Sich beeilen, ihr faul bastard!" ("Hurry up, you lazy bastards!") could be heard reverberating throughout the barracks holding Rybka's men.

Many stumbled out of bed without having had a wink of sleep; they had worried all night about that day's battle. Instinctively, they got themselves together and managed to grab a quick breakfast in the base mess hall. Many noticed that the mess sergeant was being especially generous with the rations that morning. Many cynically compared it to a condemned man's last meal, but every plate was cleaned off; everyone seemed especially hungry that morning. On his way to the runway for formation, SS-Sturmmann Schultz wondered where the hell they were going—wherever it was, it seemed like it was going to be hot! Maybe they were going to hell itself to fight the devil. Schultz would remember his thoughts on that Thursday morning with bitter sarcasm, and after it was all over, he would write to his sweetheart that he had done just that!

Standing at attention by the runway, the paratroopers were given a last-minute speech that ended with a warning, "When you land, get rid of your harness quickly and head for the canisters to arm yourselves. Don't stop for any wounded, or you won't leave the field alive." With mixed feelings of excitement and foreboding, the SS paratroopers boarded their gliders and transports.

The planes and gliders began taxiing on the runway, the transport planes and towing machines revving up their engines for takeoff. Fully loaded with men and materiel, the flotilla awaited the "Go" from the ground flight officer. They did not have long to wait,

The SS-Fallschirmjäger-Bataillon 500 is drawn up and prepared for the parachute and glider assault on Drvar, where its headquarters was located in May 1944 (left).

Final preparations before the glider and parachute assault on Drvar (bottom left).

Communications post somewhere at Drvar, during the morning of the assault. Soon after this photo was taken, the post was knocked out by a lucky round from a Partisan mortar (bottom right). Photos: Munin Verlag.

and with a loud roar, the first wave of planes and gliders took off from their airfields. Operation Rösselsprung (Knight's Move) was on! The time was 5:55 A.M.

Flying at an altitude of around 3,000 meters, the men in the transports could see the planes of the Night Fighter Group 7 and the Croatian Air Legion pass them on their way to bomb Drvar. Over the port city of Mokrica whined the air flotilla, flying high across the Sava Valley. At exactly 6:50 A.M., the town of Drvar came into view below the morning fog. At this time, the towing lines were released, and the gliders were on their own. Down below, the German bombers were already making their bombing run.

Tito was having a heavy breakfast of eggs, pork fat, farmer's bread, and an ample supply of goat's milk. While spreading his bread with a thick slab of butter and orange marmalade—a special treat supplied by head of the British Mission, Randolph Churchill—Tito heard the explosions down below and cursed, "Those Nazis can't even let me finish my breakfast! I'm going to make those bastards pay for this!" Jovanović laughed out loud, adding that if anything, he should thank his Fascist opponents for saving him from such a fattening meal. An hour later, both men would be contemplating much simpler things, like whether they would survive that misty Thursday morning of 25 May 1944.

CHAPTER FOUR

Do you know that the war in the Balkans was waged with particular intensity on both sides . . .?

Nuremberg prosecutor

I had no personal insight into the campaign in the Balkans. But from history I know that even before the First World War such excesses did take place in the Balkans.

Waffen-SS General Paul Hausser

The bombardment of Drvar had not ceased when the gliders and transports began their descent. Rybka scanned his mind to see if there was any detail he had forgotten. He couldn't think of anything. All was as it had been planned. He wondered if his plan would be sufficient. His glider began its diving descent, and his attention turned to the steep angle that the fragile glider had taken. The "whoosh" of the wind striking the nose and wings of the glider scared everyone inside who knew how dangerous a glider landing was, especially under fire.

Rybka again thought of the upcoming battle and he felt his heart pumping faster. He was scared, but that was not the reason. "Those damn pills," he thought to himself, "wonderful drugs which quicken your heart and reflexes, but, my God, did they ever give one a thirst!" Rybka now instinctively gripped the handlebars and arched his body back in his seat. From his glider, SS-Mann Fritz Hess saw one of the gliders going down at too steep an angle, almost 90 degrees. He turned away, not wanting to see his comrades diving to their deaths. He later learned that the glider that crashed had contained the leader of Greifer Group and his staff.

Down below, curiosity turned to amazement, and amazement to alarm. A young Partisan stared up into the sky, gawking at the paratroopers who were jumping off and diving at distances of only sixty to seventy-five meters from the ground. Rifle and submachine-gun fire began to fill the sky as the Partisans got over their initial shock.

Even so, the paratroopers continued to jump, their flight taking only twenty seconds—the longest twenty seconds of their lives! An SS-Fallschirmjäger was helpless until he could reach the ground; all he could hope was that a Partisan would not claim him with a lucky bullet. To make matters even worse, the German bombers were still making runs against the town, causing undue casualties among the paratroopers. As the first paratroopers hit the ground and tumbled over in typical fashion, Partisan fire was directed not only at the air but at the ground, and mortar fire began to take its toll. All of the paratroopers reached the ground, but many did not stir once they landed, their parachutes having blown over their bodies by a gust of wind.

Rybka's glider seemed to be almost on the ground. He knew that at any moment his craft could disintegrate into a thousand pieces, but his mind kept racing to what awaited them once they landed. He closed his eyes and gripped the handlebars until his knuckles turned white. He immediately felt the glider racing across the ground, bouncing roughly up and down, left and right, out of control it seemed. Just as suddenly, the fragile craft came to a halt, and the noise of small-arms fire immediately increased in volume. He opened his eyes and saw that his craft had been badly shaken and torn, but he and his men had survived the landing. The question of whether they would get out of the glider alive was one they now faced.

Everywhere stick leaders were running around, pistols in hand, desperately trying to gather their men and head for the canisters to arm themselves. There was no time to stop and mourn a dead friend or help a wounded man. Each man remembered the brief speech made to them before leaving the safety of the airfields, "Don't stop for anyone or any wounded when you land, or you won't leave the field alive!"

At first, this mass of men seemed to be running in every direction; then, just as quickly, they began to take some military order and started to assault the Partisan positions. Many of the paratroopers had witnessed the Greifer glider begin its plummet, watching helplessly as their comrades fell to their deaths. They turned their anger and attention to the Partisans to wipe out the memory. Like madmen, they raced forward in total disregard of the fierce curtain of fire leveled at them. The "pep pills" were certainly working, but so was their rage.

One member of the glider pilots has described his descent into hell.

> *I decided on a dive with the brake parachute. It released itself with hardly a noticeable wrench. The overladen glider with its ten men lost speed quickly and dived at a steep angle, crashing through the mist toward the smoke- and dust-covered goal. The altimeter fell in rapid tempo . . . 1,500 meters . . . 1,000 meters . . . 500 meters. The outline of our objective came in sight. I recognized the walls of the citadel of Drvar and the headquarters building. I saw two Flak [antiaircraft] guns and the wild panic of the Partisans. I then concentrated on the landing. As we flattened out of our dive, I thought to land as close as possible to the walls. Just over the ground I threw away the braking parachute, and, as if hit with a whip, my glider shot forward and, after a few meters' flight, it scraped the ground and landed just before the wall of the citadel.*
>
> *Angry rifle and submachine-gun fire from the citadel met us . . . the paras jumped from all sides over the wall and stormed forward. My Panther group landed in the citadel except for one glider whose pilot had been shot during the dive. The paras jumped from all sides over the wall and began to storm Tito's HQ.*[1]

The men of the Daufganger Group landed so close to Drvar that it took them no time to reach the town and blow up the radio station. Ten paras and the Svadil Group that landed with them attacked the telephone building but became immediately embroiled in a firefight. A heavy enfilade of rifle and submachine-gun fire began to cover the area around the telephone exchange.

A member of Daufganger group described his descent into Dvar.

> *We flew in a southeasterly course toward Drvar. Shortly after the towline was dropped, I noticed that the glider under the command of Lieutenant Breidenbech had already lost a lot of altitude and was probably diving to a fiery crash. It was 7:10 A.M. From an altitude of 800 meters, we started our dive toward the West Cross part of Drvar. My braking parachute failed. I brought the glider out at a 180° turn and flew in again by the landing site picked on the West Cross. While landing, the right wing support tore off of the cabin. Even before we had come to a stop, the doors were thrown open and the SS parachutists jumped out. All during the landing, we were under heavy fire from the eastern slope. No one was wounded, and we crept into a flat gully that led to a house just next to us. Already in this house were some men from the other gliders. The oil containers for our flamethrowers were still in my glider. I rolled once again to get it and brought it back. On my return I found out that pilot Obergefreiter Kielmann had died in his glider.*[1]

In spite of intense fire, the SS paras of Daufganger Group surged forward, oblivious to the fire and their comrades, who seemed to be falling around them, left and right. Three-and-a-half years of bitter fighting in Yugoslavia had taught the Germans that the Partisans gave no quarter. In a war such as this, it was better to take one's life than to fall into the hands of the enemy.

Tito's Partisans fought on like cornered animals. They too realized the depredations their own forces had committed and were not about to be taken captive, lest they pay for the sins of their comrades. They resisted

[1] Quoted from the 1986 Motorbuch Verlag edition of *Die Deutschen Lastensegler-Verbande 1937-1945*, by Georg Schlag, with permission of the author and publisher.

Photo of SS-Hauptsturmführer Siegfried Milius as he appeared while he was still attached to the SS-Division Totenkopf in 1943 (top left). Photo: Gerald Smoak.

Photo of SS-Haupsturmführer Siegfried Milius as he appeared while he served in SS Parachute Battalion 500 in Latvia during the summer of 1944 (bottom left). Photo: Munin Verlag.

with rifles, grenades, pistols, and knives. The fighting turned into a small-scale war in which every room housed a major battle fought to the bitter end. The SS paratroopers had special armor-piercing rounds, which they fired through the ceiling into the upper level of the building to kill the Partisans inside.

Two SS paras entered a back room, shooting from the hip. From the corner of the room a Partisan officer fell back spasmodically, riddled by Schmeisser fire. Before they knew what was happening, both paras were shot in the neck and face by a telephone operator who had hidden underneath a desk. She, in turn, was sent to meet her maker by the blast of a grenade. In this type of fighting, chivalry could end one's life; recognizing this wasn't an option, but a necessity.

Where barricades prevented the paras from entering a room, they used grenades. The loud blast of the grenades would announce the almost simultaneous discharge of thousands of wooden splinters, each seeking a human victim, German or Yugoslav. The dazed Partisans who survived the blast would be shot or bayoneted as expediently as possible. Hand-to-hand fighting was the worst. Bayoneting a man was not given a second thought; after all, it was kill or be killed, and killing a female guerrilla was done with the thought that, given the chance, a "delicate" female was capable of despicable acts against captured comrades, the wounded included.

The few Partisans who had no stomach for this kind of fighting attempted to evade death by jumping out the windows. They were picked off as they jumped by the sniper screen that the paras had left outside to cover any possible escape attempts. Half a dozen satchel charges were enough to send the building flying sky-high. Moving through the town, the SS paras were subjected again and again to the tactics of the retreating guerrillas. Every crevice seemed to harbor a Partisan sniper. Often these were old men and women who would pick off the point man of a German squad, then run like hell before the Germans could exact vengeance.

It is exemplary of these paras that they bothered to take prisoners, even when confronted with the opportunity to avoid such a risky enterprise (many times Partisan "surrenders" were a ruse to get the unsuspecting paratroopers out from their cover). The pictures taken during and after the paratroop assault testify to this. Though weary after such a hard fight, the prisoners showed no physical evidence of abuse by their captors.

Coincidence or suppressed fact? The photo on top is supposed to show Randolph Churchill, Winston Churchill's son, standing in front of Marshal Tito's mountain cave hideout just above Drvar. Randolph Churchill was the head of the British Military Mission to Tito. Now look at the photograph on the right. It is part of a photograph that pictured three "newspaper-men" who had been captured by the SS paratroopers during their initial assault on Drvar. All three men were later able to escape their German captors. The photograph was taken by an SS paratrooper. Note that the facial lines and body habits are strikingly identical in both pictures and that the uniform is identical. Clearly, the individual in question on the left is the same person pictured on the right. Could it be that Churchill's son, Randolph, was for a brief time in the unwitting hands of the Germans? Was this fact so embarrassing to Churchill that it was kept secret all this time?

SS-Obersturmführer Otto Mertely, the one-eyed battalion adjutant (bottom right). Soon after this photograph was taken, Mertely was wounded at Drvar, 25 May 1944. Photos: Munin Verlag.

Unfortunately for men and women on both sides, this was more the exception in the Balkans than the norm.

Two Allied correspondents, Stojan Pribićević of the U.S. press and Mr. Talbot of the British press, were captured by Rybka's men. Pribićević would later escape during the confusing and chaotic fighting. After what seemed like an eternity, the town of Drvar was considered secured; all eyes turned to the hill where Tito's headquarters were located. How were the paras faring there?

The gliders of Panther Group had landed as close to the mouth of the cave as possible and, almost immediately, were fired upon by Tito's escort battalion. The first glider, which landed closest to the mouth of the cave, was completely engulfed by rifle and submachine-gun fire. Nothing could have survived such savage fire, and no one did. The farthest these unfortunate paras got was a few feet from their burning and shattered glider.

The surprise of the glider assault lasted only briefly. As each glider sat down on the ground, it seemed like two more Partisan machine guns would spring to life, riddling the canvas and wood of the glider, almost certainly killing and wounding most of its occupants. The whole hill seemed to be one interlocking field of machine-gun and mortar fire. Several attempts were made to rush the mouth of the cave, but to no avail. The battlefield was littered with the wounded—the dead and dying men of Rybka's battalion.

Just then, the Germans got lucky. One of the last gliders to land had crashed against some boulders near the cave, knocking its passengers unconscious. The Partisans came out from hiding and cheered the destruction of the glider and its paras—which was a mistake. The paratroopers in the glider came to in time to send a hail of bullets in the direction of their enemy, immediately killing more than forty guerrillas, whose twisted, contorted bodies lay all around the shattered glider. Obviously, they had not expected anyone to survive that crash.

With this stroke of luck, the Germans received a surge of newfound strength. They again aggressively attacked the cave, using fire, movement, and what little cover the large boulders offered. Stevan Jovanović knew that, if something was not done, the Germans were sure to overrun the cave. He organized about a hundred men from Tito's personal bodyguard and prepared them for a kamikaze-style counterattack. Jovanović bellowed, "If you fail and the Germans capture Tito, then it will mean the end of our battle for freedom. Are you clear what this means? Tito and the party must continue to live. We can deal with a defeat, but Tito must live!" With one concerted yell came back the reply, "Tito must live!" and off they went, ready to prove to anyone that they were willing to die for their leader. The arrival of about 100 officer cadets from the nearby Partisan Officer Cadet School made Jovanović's plan easier.

The wild counterattack did not push back the SS paras, but it did manage to halt their advance. German air support was out of the question, as the paratroopers were too close to the enemy. The German pilots could not be sure of who they were bombing. This momentary break gave Tito the opportunity to change from his brand-new marshal's uniform and try to find a way out of the trap the Germans had him in.

About the same time, Rybka had been informed that the Panther Group was meeting stiff opposition. Rybka immediately radioed the 2nd German Tank Army headquarters, "Either there is vital documentary material in the cave—or Tito himself! I shall try my utmost to take the cave." Rybka ordered a red flare to be fired into the air, the signal to rally around the headquarters for a counterattack against the citadel.

The time was around 8:00 A.M. Reinforcements arrived in the form of the Greifer, Brecher, and Stürmer groups. Led by Rybka, this new German force began an assault on the cave, with the few mortars they had concentrating their fire in support of this assault. Meanwhile, Tito found a

The cramped quarters of the DFS-230 glider (top). The paratrooper in the center of the picture left his field blouse collar hanging out over his camouflage smock. The right collar tab is not clearly defined but is visible. It appears to be the skull and crossed bones worn by members of the Totenkopf Division.

German reinforcements begin to arrive over Drvar at about noontime on 25 May 1944 (bottom). Photos: Munin Verlag.

hole in the ceiling of the cave, and he and Edvard Kardelj (the Partisan leader in Slovenia) fled.

Once on top, they found Aleksandar Ranković-Marko (the Partisan leader in Serbia) and a score of Partisans. With their aid, Tito managed to escape in an old locomotive that had been kept nearby in readiness for just such an occasion. By then the whole surrounding countryside had been alerted to the plight of Tito and his staff. Numerous guerrilla columns were on their way, all of them converging on the Bosnian town whose name would cover the front pages of newspapers in the coming weeks. The 1st Partisan Brigade from the 6th (Lika) Division was the first major unit to arrive. It immediately began to encircle SS Parachute Battalion 500 at Drvar.

Rybka's assault finally crushed the Partisan defense in and around the cave, but the victory came at a high price for him. Not only had his battalion suffered heavy losses, but a burst from a Partisan grenade had shattered his left arm. All the battalion had managed to capture were some Partisan documents and flags plus Tito's newly abandoned uniform. Failure was a bitter pill to swallow, especially after so many sacrifices. The whole hill overlooking the town of Drvar was a vast cemetery of fallen SS paratroopers. The gods of war are cynical; before the day was over, they would place Rybka and his men inside a real cemetery, fighting for their lives!

The 1st Battalion of the crack 1st Partisan Brigade began to put pressure on Rybka's men on the hilltop. With this new surge of strength, the badly depleted Partisan officer candidates and what remained of Tito's escort battalion managed to push the SS paratroopers from Tito's former headquarters. Rybka gave up this indefensible position at 3:00 P.M. and radioed back to 2nd German Tank Army, "Lying in a totally visible position. Need medical supplies. Enemy is massing his forces."

Hauptsturmführer Obermeier's group of 220 paratroopers had arrived by glider over Drvar at around 11:30 A.M. II/Transport Glider Group 4 had arrived to be met with intense fire from the ground. II/Luftlande-Gruppe 1 (LLG-1) sent off two more gliders, both of them bringing ammunition supplies. Obermeier's reinforcements were the only good news Rybka received that day.

After that, the situation for the SS-Fallschirmjägers worsened by the hour. Toward evening, they began to take heavy fire from every surrounding hillside, partly from mortars. The order was now sent out, "Rally in the town cemetery!" Slowly and in order, the SS paratroopers withdrew. The cemetery lay at the foot of a small hill overlooking the town. For almost half a day, the paras

SS-Mann Jost Schneider a typical member of the 7. SS-Freiwilligen-Gebirgs-jäger-Division Prinz Eugen. Note the Prinz Eugen Divisional insignia on the right collar tab. It was the this Division which came to the rescue of the beleaguered SS Parachute Battalion 500 at Drvar, Yugoslavia. Photo: Jost Schneider.

fought and died for that piece of worthless real estate. It wasn't until about 10:00 P.M. that the main body of the battalion reached the cemetery.

Some paras were already digging foxholes. Many joked that they were digging their own graves, but their jocular words belied their determination and desperation. Everyone knew what awaited if they fell into Partisans' hands.

Friends buddied up and swore to take the other's life if it appeared they were going to be captured. Someone heard one of the doctors say that he would personally blow himself and his wounded up, rather than let the guerrillas get them. With typical German stoicism, each paratrooper saved his last round for himself, and, feeling cornered, vowed to fight like a wild dog. They were determined to make the enemy's victory as costly as possible. Their only hope rested on the advance that the relief groups of the SS Prinz Eugen and 373rd German divisions were making. Both divisions were, in fact, having a bad time of it, especially the Prinz Eugen, which was taking heavy losses against the crack 3rd Proletarian Brigade barring its way.

An order for the Partisans to withdraw had come down straight from Tito himself, "All Partisan units will now disengage from the fighting starting immediately. Strong enemy forces are expected at Drvar within the next twenty-four hours." In spite of this direct order, Tito's field commanders, knowing that Tito was safely away, decided to disobey this withdrawal order so they could exact vengeance for the nerve of these Germans, who had dared to land right in their midst. The Partisans did not need a second lesson. They knew what Rybka's men were capable of, and they did not want this SS battalion to survive to stalk them again. They hoped to wipe out Rybka's battalion to the last man.

As fate would have it, one SS platoon that had been defending a farmhouse on the outskirts of town failed to get the German order to withdraw. It was soon isolated from the rest of its countrymen. All day they had exacted a heavy toll on any Partisan unit that dared to challenge them, but their ammunition was running low, and they had taken heavy casualties. The Partisans vented their anger and frustration on this isolated group of paratroopers; even the Partisan officer cadets joined in the fun.

With each successive attack, even more SS men became casualties. Still, they held their ground against the overwhelming enemy force for many hours. It was around 11:30 P.M. when the last remaining German noncommissioned officer (NCO) took count. Everyone had either been killed or wounded (himself included); ammunition was all but exhausted; he knew they did not have much time left.

The end came just before midnight when the Partisan's final assault of the day succeeded in crushing German resistance. The final seconds in the farmhouse were spent in hand-to-hand fighting, with the bayonet ruling. True to their vows, some of the wounded who had the strength to do so shot themselves, while the German NCO preferred to take a few Partisans with him by blowing himself up in their midst.

Wild-eyed with sadistic joy, these Partisans, the "freedom fighters" the Western world was hearing so much about, systematically decapitated and dismembered the dead and wounded paras, cutting off their sexual organs and extremities in a savage frenzy of butchery.

Back in the cemetery, Rybka's remaining men braced themselves for the worst. Heavy enemy mortars hurled their rounds in, spitting them up with a low thud and sending their hellish message onto the positions on the paratroopers. "Woosh-ka-whump!" and another three or four paratroopers would lie writhing on the ground, dying. The screams of the wounded caused as much anxiety in the paras as did the darkness of the night. The nightmarish dance of death was beginning for Rybka's men.

A corporal of III/LLG-1 described the dramatic experience of this night.

OPERATION RÖSSELSPRUNG (KNIGHT'S MOVE)
25 MAY 1944

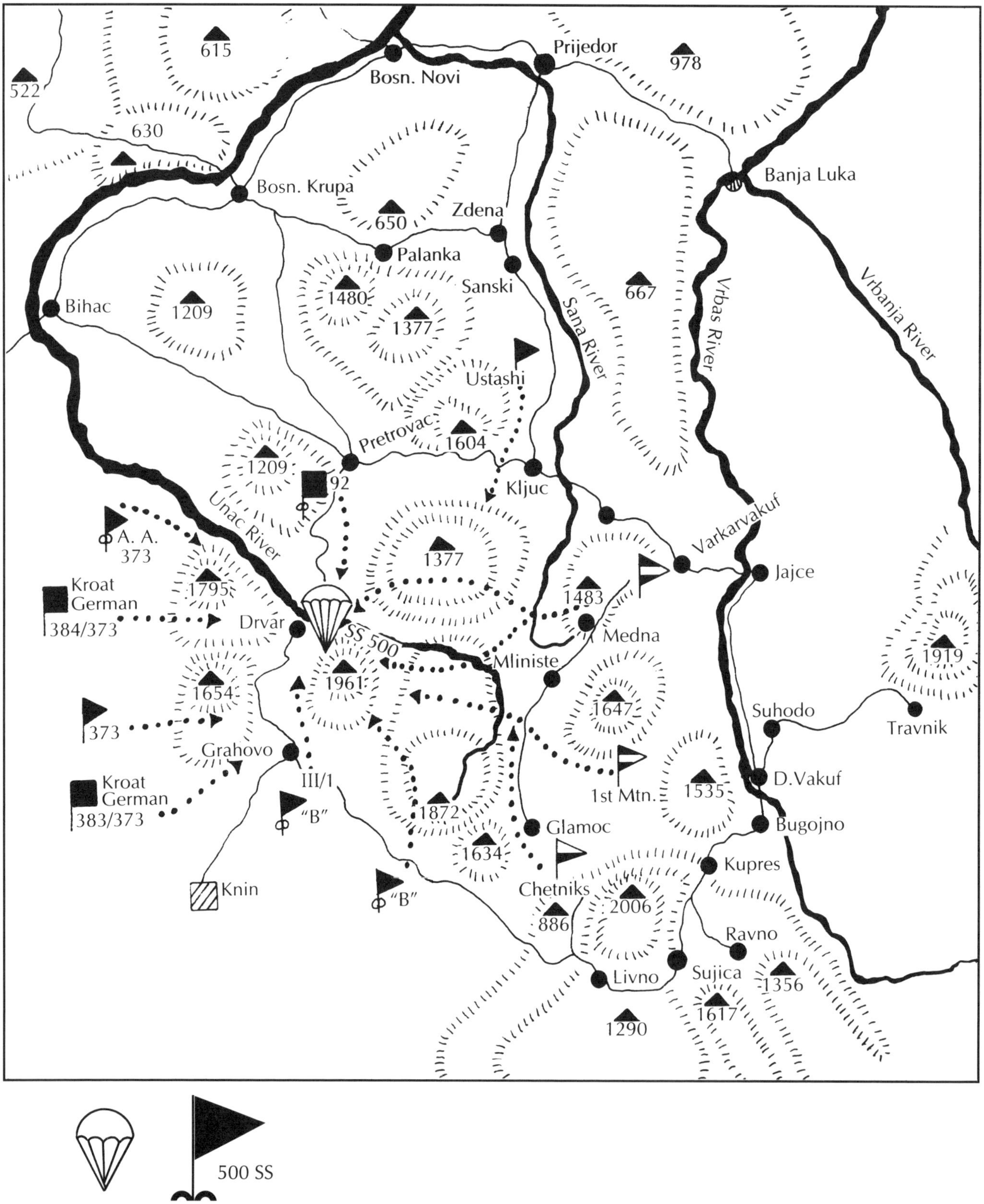

I was able to reach our defense line on the northwest corner of the cemetery where our men were already pressed shoulder to shoulder. A few meters behind me the radio set was set up, and on the southwest corner of the citadel was the wounded collection center. Our radio position was destroyed during the night by a mortar shot. Also, shells fell among the wounded. Many times there was a rain of steel on us.[2]

The battalion doctors, SS-Obersturmführer Dr. Hermann and SS-Obersturmführer Dr. Helmersen, were doing all they could for the wounded, but morphine had long been exhausted. The wounded cried and whimpered all night. A SS-Fallschirmjäger survivor of that night recounted his experience.

I was dug in under a cross in the cemetery. Platoon headquarters was in a tomb from which we had taken off the stone cover and turfed out the coffins. I was never so frightened in my life as I was that night! We had an outer perimeter of positions and a small inner perimeter. After every one of the bandit attacks was driven off, those in the outer perimeter were relieved by a man from the inner line.

That way the burden was not too hard to carry. It was 1:30 in the morning. I was cold, I was sleepy, and the only water was a trickle from a sort of standpipe inside the cemetery, near the chapel. At intervals, Very lights would be fired to see if the bandits were forming up for a new assault. One light had just gone out. Suddenly, without warning, the Reds were climbing over the wall.

A whole mass of our flares burst—all of them white—and in the glare the bandits were silhouetted. We shot them down, but they seemed to be immune to rifle fire and kept coming. Then some of them from behind cover of the wall threw hand grenades and got some light mortars into action.

In that particular attack, the Reds actually got inside the cemetery and held a small bridgehead. Those Partisans in the bridgehead and those outside the wall tried to knock it down in order to form a corridor through which to ferry forward reinforcements of men and weapons. One of our platoons went into a counterattack and wiped out the Reds who had crossed the wall. It was a frightening night.[2]

At moments such as those, the entire fate of the battalion was in the hands of a

[2] Quoted from the 1986 Motorbuch Verlag edition of *Die Deutschen Lastensegler-Verbande 1937-1945*, by Georg Schlag, with permission of the author and publisher.

A. A. 373—Recon battalion of the 373rd Croat/German Infantry Division
384/373—384th Infantry Regiment, 373rd Croat/German Infantry Division
373—373rd Antitank Battalion, 373rd Croat/German Infantry Division
383/373—383rd Infantry Regiment, 373rd Croat/German Infantry Division
III/1 "B"—3rd Battalion, Brandenburg Regiment
II/1 "B"—2nd Battalion, Brandenburg Regiment 1
92—92nd Independent Motorized Infantry Regiment
7 SS—7th SS-Freiwilligen-Gebirgs-Division Prinz Eugen
1 Mtn.—1st Gebirgs Division (German)
USTASHA—200-300 men of the Ustashi Guard (Croat Republic)
CHETNIKS—500 Serbian guerrillas who were loyal to the exiled king of Yugoslavia and who hated Croations, above all. Their next most-hated enemies, respectively, were the Tito Communists, the German, and the Italians. The Chetniks fought their enemies in that order, choosing to side with one group at a time, to deal another enemy a blow.
500-SS—SS-Fallschirmjäger-Bataillon 500

---

few brave paratroopers. Fear and the instinct to survive were the two driving factors behind such a fierce defense, especially after such a grueling daylong battle. Lieutenant Sieg, a glider pilot, summed up the feelings of the beleaguered defenders: "With the coming of light also came the first German reconnaissance plane. The men waved and called to the plane, driving the nightmares of the last night away."

The Partisans were not done with the paratroopers just yet, however. Just a little before dawn, they again attacked the cemetery. This time, they launched their assault from two sides, hoping to break through from one end or the other. Only with the greatest of difficulty was this last attack repelled. The hope that enough time had elapsed for the relief groups to be close inspired the weary defenders to fight on as stubbornly as before. This last assault was broken up just like all the rest.

Later in the morning, the paratroopers heard the distinctive sound of Hitler's sagen (Hitler's saws, a term used by the German forces to denote the particular sound of the MG-42 heavy machine gun). Since the paras knew that the Partisans also had captured stocks of this excellent weapon, they weren't certain whose fire they were hearing and braced themselves for another attack. Then, in the distance, two, then three German schwimmwagens appeared. An agitated member of the battalion jumped up and shouted, sensing that safety was just 1,500 feet away from him, "Das entsetzung gruppe kommend!" ("The relief group is coming!")[3]

As the German reconnaissance vehicles came within shouting distance, the SS paras were greeted with soothing words they would not soon forget, "Wir sind von der Prinz Eugen; wir holen euch raus!" ("We are from the Prinz Eugen; we've rushed to get you!") Victorious shouts and yells erupted within the ranks of the SS paratroopers. The whole cemetery at Drvar seemed to come alive. As more units of the 13. Waffen-Gebirgs-Jäger-Regiment der-SS began to arrive, the paras left their hard-fought positions and began to mingle with their saviors. The reconnaissance battalion of the Prinz Eugen Division had also been the first units to arrive at the scene. The men of the Prinz Eugen had fought a hard-running battle with the Partisans, but they had managed to reach the besieged SS paratroopers in time—and ahead of the Wehrmacht!

Pats on the back and hugs were followed by offers of cigarettes and a sharing of experiences, which the SS-Fallschirmjägers were quite willing to explain in detail, much to the disbelief of their saviors who couldn't comprehend how these men had survived at all!

The attack on Tito's headquarters had been a failure from the point of view of their mission. Tito had not been caught or disposed of, and the battalion was a shambles. It would take several weeks of refitting before the unit could regain its prior strength. But they had survived, and that, apparently, was enough of a victory for them. Anyone who survived the hell that was Drvar could feel proud to have accomplished that much.

The attack was not a complete failure, however. Tito's command had been badly disrupted. After his escape, he had been flown out of Yugoslavia in a U.S. Dakota piloted by a Russian officer. He was sent to Italy under British protection from whence he was moved by sea (in a British destroyer) from the port of Bardi to the Yugoslav island of Vis. He set up his new command headquarters there, but his officers had to make constant trips to and from the island until October 1944, when he again moved his headquarters to the mainland.

This caused problems and delays in the Partisan chain of command and helped the German war effort somewhat. Late that

[3] Quoted from the 1986 Motorbuch Verlag edition of *Die Deutschen Lastensegler-Verbande 1937-1945*, by Georg Schlag, with permission of the author and publisher.

summer the German High Command in the Balkans had discovered that Tito was indeed entrenched on Vis. Plans were drawn up for the SS-Fallschirmjäger Bataillon 500 to be used in an assault on the island, but this plan was dropped for several reasons: one, the British had complete control of the sea and air; two, British troops had been identified on the island; three, the SS battalion was needed elsewhere for a more important mission.

It seemed that the whole Eastern Front was taking another turn for the worse. Finland had begun to show signs of wanting to leave the war, and the Soviet juggernaut was again on the move. For the men of the SS parachute battalion, the Drvar was merely a taste of things to come.

The last actions of the SS-Fallschirmjäger in Yugoslavia occurred in the Partisan-held territory around Petrovac. Later, the battalion was moved to Ljubljana for replenishment and reorganization. It was here that the new commander, SS-Hauptsturmführer Siegfried Milius, took command on 26 June. The battalion had been decimated; the Field Reserve Company was able to put some lightly wounded men right back into the field, but on the whole, the strength of the battalion had been greatly reduced.

Of 1,000 men on 25 May, only 292 were fit for duty on 30 June. Of these, 15 were officers, 81 noncommissioned officers, and 196 enlisted men. Scanning the thinned ranks of his new command, Milius' thoughts centered on what an honor it was to lead such an elite combat unit. He had no idea then that he would be their commanding officer until almost the end of the war. Had he known, he probably would have gotten used to the idea rather quickly. "These were excellent troops," he thought to himself. "A pity so many of them had been lost at Drvar." He wondered then just how many he would lose.

***

For a look at a Danish volunteer with the SS-Fallschirmjägers at Papa, Hungary, read Eric Brorup's story in Chapter 21, page 289.

Die Ausbildung einer Fallschirmjäger-Kompanie der Waffen-SS ist abgeschlossen. In einer Einsatzübung stellt sie ihr Können unter Bewis. Die ersten Springer liegen bereits in Stellung und sichern den Absprung der nachkommenden Kameraden.

Translation:
The training of a parachute company of the Waffen-SS is completed. In a jump training exercise they proved their ability. The first jumpers have already secured the jump area for the remaining comrades.

This photograph was taken on 11 March 1944 by an SS war reporter. The caption and photo originally appeared in a German wartime newspaper. Since parachute training was at the time centered a Madarushka-Banja near Sarajevo. It is obvious that the picture was taken somewhere in the Yugoslavian countryside. Photo: Robert Tyler Collection.

CHAPTER FIVE

I think continually of those who were truly great—the names of those who in their lives fought for life, who wore at their hearts the fire's centre.

Stephen Spender

The reorganization of the SS-Fallschirmjäger Bataillon 500 had to be canceled. The Russians had launched their 1944 summer offensive on 22 June, unleashing a devastating fury on the divisions of Army Group Center. Finland had sued for peace, accepting a baseless and unfair agreement with the Russians that demanded a staggering amount of war reparations, including large quantities of monies and materiel as well as a vast amount of territory that had been Finnish for years. Another demand of the Russians was the withdrawal of German units still fighting in the northern reaches of Finland within a given time period. If the Finns could not accomplish this within the specified timed, the Soviet army would continue offensive actions against Finland.

Although the Finns had held off the Russian bear for more than three years, their country's economy was wrecked. Large amounts of resources had been directed toward the war, and the country was war-weary. To continue a war that was clearly lost did not fit the Finnish character. The Finns had to get the Germans out. If it could not be accomplished by peaceful means or if the Germans refused, the Russians demanded that the Finnish army forcefully eject the Germans. Of course, the time allotted was too short for the Germans to withdraw their 20th Mountain Army from Lappland (in northern Finland). The Russians had done this on purpose, for they wanted the Finns to attack their former ally. That was the way the Russians did business.

The Finns, for their part, had no choice but to comply. There was a slight chance that they could avoid complete Soviet domination if they sought peace before a total collapse. No one could really blame them.

Hitler's rage was uncontrollable. He completely forgot that the Finns had sided with him faithfully during the last three years of the war. It was only after they saw no possible chance for victory that they began to consider their own welfare. He did not see that they were doing what they thought was best for their country, that they really did not have much of a choice. Not realizing the futility of opposing the Finnish request/ultimatum, Hitler schemed and planned. The German armed forces would attack their former ally to gain vital Finnish positions before they were handed over to the Russians.

On 29 June, the SS-Fallschirmjäger Battalion was transferred by train and assigned to Naval Supreme Command Baltic in Gotenhafen (Gdingen) for one such operation. The battalion was to make an assault landing on the Aaland Islands in the Baltic Sea. Before this took place, German units in northern Finland attacked several key bases but met mostly with failure. Now the Finns could justify their actions against the Germans. Hitler's rage had worked against him, and the Russians gloated.

The operation scheduled for the Aaland Islands was canceled. The battalion traveled by train to its new assignment, the 3rd (Germanic) SS-Panzerkorps (tank corps), which formed part of Army Group North at Narva, Estonia. Its posting there was brief, however; by 9 July, transport aircraft waited to take them to Kaunas, Lithuania. This sudden transfer from the northern army group to Army Group Center resulted from the amazing success of the Russian summer offensive.

Mocking the start of the German invasion of Russia exactly three years before, the Russian steamroller had quickly gained momentum, crushing entire divisions of Army Group Center. The Red Army was making a shambles of the army group.

A most striking illustration of the size of the German catastrophe was the number of German generals who, in one way or another, became casualties. Out of forty-seven gener-

A mortar detachment of SS-Fallschirmjäger-Bataillon 500 (right). Photo: Kenneth Nieman.

SS-Hauptsturmführer Siegfried Milius during a conference with the Luftwaffe parachute officer who led the famous paratroop assault on the Belgian fortress of Eben Emael in 1940. This photograph was taken in a Latvian forest during the summer of 1944. Milius is pictured here, second figure from the right (below). Photo: Munin Verlag.

als, only sixteen reached the relative safety of a still haphazard German front. Thirty-one had been killed, captured, or had committed suicide. The German 9th Army alone had lost 70,000 men killed, missing, wounded, or taken prisoner between 22 and 29 June. That averaged exactly 10,000 men per day (the strength of an average division). It was certainly a dark time for the German army.

Transport aircraft took the men of the battalion who were still fit for combat and transported them from Wesenberg (Rakvere) in Estonia to Kaunas, Lithuania. The paratroopers boarded the "Auntie-Ju's" and the Italian-piloted CR-42AS Falcos, seeing familiar faces from their days in Yugoslavia. The Italians avoided asking about particular comrades, for they knew many had fallen at Drvar. Instead, they made small talk and passed out cigarettes in true comradely fashion. The spirit of a New Europe was very much alive and flying over the Baltic state of Lithuania.

The Soviet war machine was swiftly moving westward toward the Polish and East Prussian borders, almost unopposed. By 4 July, the Soviets controlled Minsk and were a mere sixty kilometers from Vilnius. The bulk of the Russian armor turned northwest and headed for Vilnius with its important road and communications network. This city was just 180 miles east of Königsberg, the capital of East Prussia. The Baltic coast was just a hop, skip, and jump away. To keep their offensive rolling, the Russians had to take Vilnius.

Aware of the great danger threatening Army Group North (that of being cut off), every German unit fought with amazing determination. The German corps unit in this area, the 11th, had only a few scratch units in the line. The corps contained parts of the 252nd Infantry Division, 212th Infantry Division, the 201st Rear-Area Security Division, and Armee-Abteilung "D" (containing remnants of infantry divisions 56, 171, 234, 262, and 156). In addition, the 221st Rear-Area Security Division was in reserve just southeast of Vilnius, while parts of 170th Infantry and 5th Panzer divisions covered the southeastern flank.

Several crack panzer divisions had received their marching orders and were on their way to 3rd Panzer Army as reinforcements. Although still in reserve under Army Group South Ukraine, the Grossdeutschland (GD) Panzergrenadier Division began to move out in parts. The I./Panzerregiment GD was the first unit in the division to arrive by rail. Meanwhile, the German 11th Corps had been outflanked and forced to withdraw past Vilnius, leaving behind a garrison in the city. General Hasso von Manteuffel, the commanding officer of 3rd Panzer Army, ordered an immediate counterattack to break the Russian ring around the city and free the trapped German forces.

Colonel Ludendorff, Manteuffel's quartermaster, produced some 300-odd vehicles to follow the German spearhead and bring out the wounded. A battle group was quickly formed from I./Panzerregiment GD and the SS paratroopers. This battle group moved out on 10 July, the same day that the paras had arrived at Kaunas, giving them no rest before being committed. The heaviest fighting for Vilnius occurred between 7 and 14 July. The kampfgruppe attacked under the operational control of 39th Panzer Corps. They immediately became embroiled in heavy fighting. On the first day in action, the SS paratroopers received mention in the *Secret Army High Command War Diary*, winning praise for their actions. The text of the report read as follows:

> *On July 10th, strong Soviet armored forces flanked the German defenses at Vilnius. The Vilnius highway itself was cut off and enemy armored units began to advance toward Kaunas. The stiff resistance put up by the SS Paratroop Battalion 500 and I./GD Panzerregiment, attacking the highway from its flanks, stopped the Soviet advance on Kaunas, placing the enemy armored units in a tight vice. This allowed Kampfgruppe Schmidt to attack in the direction of Vilnius, by Sirvintos, and then by Giedraiciai. By this move the kampfgruppe was able to once again cover the flank of XI. Armeekorps, bringing it some relief.*

The superb tactics of the German forces finally pierced the Soviet armored ring around Vilnius. The tank-mounted SS paratroopers continued to wreak havoc on the units of the Soviet 51st Army, which al-

A group of SS paratroopers moving forward during the summer battles in the Baltic countries, 1944 (top).

SS-Obersturmführer Hermann Wabersich during the breakout from the Vilnius pocket, Latvia, 1944 (top right).

SS paratropper communications post somewhere in the Latvian countryside, summer 1944 (bottom right). Photos: Munin Verlag.

lowed the trapped German forces to withdraw, 39th Panzer Corps covering the retreat. The withdrawing German forces were then subjected to attacks by Soviet armored units. The whole scene might have resembled a wagon train of the old West, with the tank-mounted SS paratroopers acting as the U.S. Cavalry, while the stalking Soviet armored units behaved like marauding "redskins." Time and again these "redskins" threatened to cut off the retreating "wagon train," only to be repulsed each time by the German "cavalrymen."

At one point in the battle, the paratroopers even managed to relieve the last German forces near Vilnius (at the airport). They withdrew in the face of stiff armored pressure, but only after they managed to neutralize it. The paratroopers, having accomplished their mission, withdrew past Kaunas (northwest of Vilnius), following the German retreat between the last week in July and first week of August.

A recently published photohistorical work on the SS paratroopers, *SS Fallschirmjäger der Waffen-SS im Bild*, includes a photograph taken by a former Wehrmacht sergeant. His comments are included here to describe the ferocity of the fighting in those days.

On July 29th, 1944, I took this photo at the bend of the River Memel by Kaunas, of two SS soldiers who passed by my tank after a hard fighting withdrawal. I should think they belonged to a paratroop unit and had resisted until their unit was wiped out. At that time, anyway, we had great respect for these men, which, as far as I am concerned, has not changed.[1]

[1] Quoted from the 1986 Munin Verlag edition of *Fallschirmjäger der Waffen-SS* by Siegfried Milius and Adolf Kunzmann with permission of the authors and publisher.

SS-FALLSCHIRMJÄJAGER-BATAILLON 500 AND THE BALTIC BATTLES
SUMMER 1944

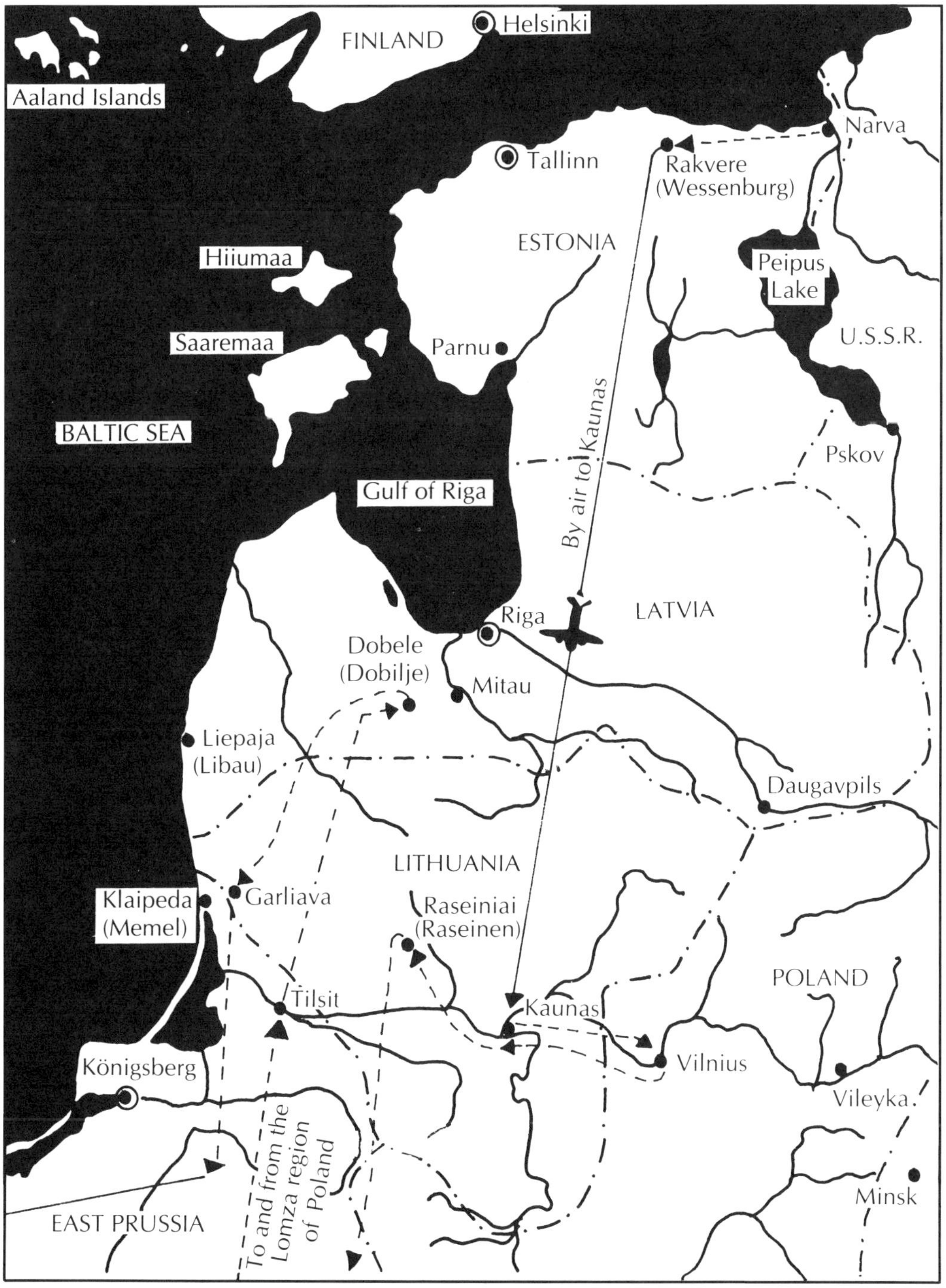

This map show the movements of the SS-Fallschirmjäger-Bataillon 500 in the Baltic countries during the late summer and early fall of 1944.

CHAPTER SIX

He looked upon his people and a tear was in his eye;
He looked upon the traitors, and his glance was stern and high.

Thomas Babington Macaulay

While the remnants of SS-Fallschirmjäger-Bataillon 500 fought desperately to put out the fire in the east, the newly forming SS-Fallschirmjäger-Bataillon 600 was earmarked for yet another dangerous mission. Though not yet completely trained or manned, this new SS parachute unit was to be used in a parachute/glider assault on Budapest, the capital of Hungary. Hungary was an ally of Germany, but it had shown signs of continuing war weariness. Admiral Miklós Horthy, the Hungarian regent, had become more and more recalcitrant. The changing fortunes of war had made him more and more unwilling to follow Germany's ride to Valhalla.

The Hungarian hierarchy had begun to show signs of wavering as early as the winter of 1942–1943. The disaster at Stalingrad was a severe blow to the Hungarian armed forces, and by the following winter, they were desperate to extricate themselves from the Russian cauldron. Admiral Horthy had appointed Miklós Kállay, a man who shared his views, as prime minister of Hungary in the spring of 1942. His appointment was designed to obstruct the German war effort from Hungary's point of view. Still, the Germans had talked the Hungarians into fielding an army in Russia during the summer of 1942. Horthy himself stated that he had agreed only because he feared losing favor with the Germans and, thus, possibly losing his chances of keeping Transylvania from the clutches of the Rumanians.

Hungary's donation to the war in the east had been ten divisions (not counting occupation troops in the Axis rear). These "light" divisions were not prepared to fight a total war on the Russian steppes. They were deficient in training, equipment, and morale. The harshness of the Soviet winter, coupled with the overwhelming strength of the Russian forces facing them, was too much for the Hungarians. Even in the summer when the bulk of the Soviet forces were in full retreat, the Hungarians seemed to lie back, preferring to allow the Russian forces in front of them to withdraw without a fight.

Hungary's ten divisions collapsed under the weight of a Soviet winter offensive launched against them on the Voronezh front in January 1943. A rout ensued. Only two Hungarian formations, the 9th and 13th light divisions, showed any resolve to hold their ground. They thus distinguished themselves from the rest of the Hungarian army but paid the price of being decimated. The Hungarian losses in the field were estimated at about 70,000 men, not including about 50,000 men of the Hungarian Jewish work battalions (most of whom became willing prisoners of the Soviets).

Thereafter, Hungarian intransigence increased, while their value to the war effort shrank. By the beginning of 1944, the Hungarian leadership was flagrant in its desire to remove the country from the conflagration. For example, the Hungarian general staff asked the Germans in February 1944 for permission to withdraw its 18th and 19th light divisions from Russia. Before the Germans had a chance to respond, General Géza Lakatos, the Hungarian commander-in-chief in Russia, withdrew both divisions.

One month later, in March, an SS report on Prime Minister Kállay accused him of outright sabotage. The report described how General Szombathely, the Hungarian chief of staff, had called Colonel Czlenner, head of the Transport Division, and told him that Kállay wished to sabotage all German military transports crossing Hungarian soil. Germany's continuing defeats in the east had caused Hitler to become increasingly concerned about this dubious ally. It was common knowledge that Kállay had let it be known that he would surrender unconditionally to the Allies when they reached his borders.

This statement was not without some foundation, for on 9 September 1943 representatives of the Hungarian government had met secretly with British representatives in Turkey, who had agreed to send into Hungary a British and U.S. group to prepare the way for a surrender. These two groups would be parachuted in at the right time. The time seemed ripe in March 1944, when a three-man U.S. team actually landed in Hungary. The British team did not make it, however. It seems that Hitler had been scheming also. On 18 March, Hitler had received the Hungarian regent and had categorically told him that the Kállay government had to go. An enraged Horthy told him that if the decision had already been made, there was no need for him to remain. To keep the regent from leaving, the Germans staged an elaborate phony air raid, though it fooled no one. Horthy did stay, and when he returned to Budapest on the following day, he was greeted at the train station by a German honor guard drawn from the Brandenburg Division.

The Germans put Operation Margarethe into effect in the early hours of 19 March. From several directions, a German occupation army moved into Hungary. The Hungarians had become aware of this "invasion" in time to put up some opposition, but the Hungarian army refused to take action. For one thing, the majority of the senior Hungarian officers were sympathetic to the Germans: out of twenty-nine top-ranking officers, twenty-one were of German (Swabian) descent. The majority of the Hungarian populace also welcomed the German occupation, greeting the German troops with cheers, much to the chagrin of the Horthy clique. In his memoirs, General Walther Warlimont states, "When their country was occupied, the vast majority of the Hungarians continued to treat the Germans as allies . . ."

Under pressure from Hitler and a strong show of force, Horthy was finally forced to remove the Kállay government and replace it with the right-wing cabinet of Döme Sztójay. On 25 March, the 58th Reserve Panzer Corps sealed off the Hungarian capital and, in a show of force, paraded its Panzer Lehr and 16th SS Panzergrenadier divisions through Budapest. The SS-Fallschirmjäger-Bataillon 500, which at this time had been in training and on anti-Partisan sweeps in Yugoslavia, had been earmarked for the occupation. Had resistance been encountered, they surely would have been committed. As it turned out, it would eventually be the 600th Battalion that would make its appearance felt in Budapest seven months later.

The Horthy clique was not yet finished, however. Enviously, it had watched Italy's desertion in 1943, and the rapid advance of the Russians spurred its haste to make a separate peace with the Allies. When the Allied mission to Hungary failed, Horthy's people took another far more dangerous course. At the beginning of the summer of 1944, Horthy's playboy son, Miklós, Jr., had begun secret negotiations with representatives of the Moscow regime. These representatives were Tito's agents in Budapest who acted as the go-between for the Hungarians and the Russians.

The Soviets had slyly assured the Horthys that their dynasty could survive under Soviet domination. It was typical Bolshevik bullshit, but by now these two jackals were ready to buy any line that offered them continued rule. The Horthys took the Soviet bait, just like Romania had done that August.

The scene was now set for another attempted stab in the back against Germany. The rats were leaving the sinking ship. On 13 September 1944, the Hungarian chief of staff (who was privy to the plot) transmitted an "assurance of loyalty from the regent and the Hungarian nation" to Hitler. This did not deceive Hitler, for he had been receiving disturbing reports from all quarters of the Hungarian government and army. Cabinet ministers and Hungarian generals sympathetic to the German effort had confirmed what his SD (Sicherheitsdienst, the SS Security Service formed in 1932 by Reinhard Heydrich to be the sole Nazi

Party intelligence organization. The headquarters of the Sipo and SD formed the core of the RSHA in 1939. After the discovery that Admiral Canaris, head of the Abwehr or German Secret Service was a traitor, the SD absorbed the Abwehr, becoming Germany's sole security service) agents and members of the right-wing "Arrow Cross" (the Hungarian Fascist Party, headed by Count Ferenc Szalasi, a former Hungarian officer removed from the army for his political beliefs) movement had been saying all along: that Horthy and his clique were now in contact with the Communists. Something had to be done and quickly, before Hungary's ruler was able to throw his nation to the Bolsheviks.

On 23 September, the day the two eastern Hungarian cities of Temesvar and Arad fell to the Soviet juggernaut, Hitler held a last-minute briefing after the regular daily conference. Attending this impromptu session were such court notables as Reichsführer-SS Heinrich Himmler, whose power was just reaching its zenith, and Field Marshal Wilhelm Keitel, Colonel General Alfred Jodl, and Foreign Minister von Ribbentrop, who had lost favor in Hitler's court.

Present at the "Wolf's Lair" (Hitler's East Prussian headquarters) since the second week of September 1944 was Otto Skorzeny, who had stood on the sidelines watching the goings-on at the daily war conferences. Now he was to be the "guest of honor" at this last-minute palaver. That day, Hitler had asked him to stay behind after the conference was over. "Something's up," thought Skorzeny.

Hitler began the meeting with a vicious tirade of accusations and insults at von Ribbentrop for "throwing my one remaining ally to the Russians . . ." Ribbentrop could only stand there, dumb-founded and pale as chalk, while Keitel and Jodl studied the ceiling, hoping to avoid eye contact with their raving Führer. Himmler, whose power and prestige had grown while that of others had dwindled, was the only one in the room who remained unaffected.

Skorzeny studied this scene carefully. It became obvious to him that he would soon be off on another dangerous mission. Hitler continued his ranting speech.

> *Hungary is our last ally. If she falls, my only source of oil, grain, and bauxite will also go. Hungary's collapse will cut off seventy of my best divisions from the main battle front. In the Carpathian Mountains alone, one million German soldiers are standing side by side [with] their Hungarian cousins, defending the Hungarian plains. If Admiral Horthy is allowed to make a separate peace with the Bolshevists, these men will be doomed, and so will our defenses in the southeast. The Red Army will then pour across the Danube plains into Austria. Admiral Horthy is a traitor to his people and to Germany.*

Even after the crippling injuries sustained during the attempt on his life on 20 July, Hitler still had the capacity to see clearly his situation (though he would choose to ignore reality more and more whenever it suited him). Hitler continued, adding that already on 29 August, the pro-German government of Döme Sztójay had been replaced by one headed by Géza Lakatos. A meeting with the chief of the Hungarian general staff, General János Vörös, on 12 September did nothing to allay his suspicions about Hungarian intentions.

Hitler turned his attention directly to Skorzeny, making eye contact with his tall commando extraordinaire. "You, Skorzeny, will deal with this Admiral Horthy!"

The scene was set once again, and the characters chosen. Skorzeny was tasked with capturing Admiral Horthy before he could defect to the enemy. The catch was that Skorzeny had to accomplish his mission without bringing the whole Hungarian armed forces down on him, or worse, having it defect to the Russians. If Skorzeny played his cards wrong (and he didn't have many to play with), the Hungarian army could possibly throw in its lot with the Red camp.

The first plan had been formulated by Hitler, Keitel, and Jodl. As Hitler went on to explained:

> *Four parachute battalions are to stand by in order to lay their hands on the regent. Preparations have also been made for the Hungarian Arrow Cross movement under ex-Major Count Ferenc Szalasi to take over the government.*
>
> *"What paratroop battalions are we speaking of?" queried Skorzeny. Jodl picked up the ball, "You will have a hand-picked Luftwaffe parachute battalion, your own SS-Jagdverband Center, an infantry battalion of officer cadets from Wiener Neustadt acting as glider troops, and the SS-Fallschirmjäger-Bataillon 600."*
>
> *Skorzeny assumed that the officer cadets would perform well since they were veterans, but he had no knowledge of this new SS parachute battalion. "Does this SS parachute unit have any combat veterans?" asked Skorzeny. "What happened to the 500th parachute?"*
>
> *"It's still in the east," answered Jodl.*
>
> *Keitel, sensing that the conversation was safe, joined in, "It's been beat up pretty good, but we can still make some use of them."*

Illustration: Eric Weisman.

Skorzeny stared icily at Keitel. Here was this paper-pushing general, talking about brave men like they were common cannon fodder. He hated Hitler's court lackies. Holding his rage, Skorzeny turned to Hitler and requested that the remnants of the 500th Parachute Battalion be withdrawn to augment the newly formed 600th. If he was to have only four battalions, he would make sure that they were the best. The more veterans in the new unit, the better its chances of performing well.

Jodl imparted in an almost casual manner that the unit was due to be withdrawn soon anyway, because, as he put it, "It has already shot its bolt." Skorzeny had heard enough. He again turned to Hitler and asked about glider support. Hitler, sensing the animosity, replied that two squadrons (twenty-four gliders) were already en route and would be ready for use in a few days.

"Where is Horthy residing?" asked Skorzeny.

"At Castle Hill," came the reply. This palatial fortress was situated in Budapest on the banks of the Danube River.

Skorzeny paused for a minute and then asked, "What kind of [Hungarian] troops are stationed in Budapest?"

Jodl, who had not sensed Skorzeny's anger, was the first to speak, "About three divisions, one of them a Hussar unit."

Four battalions against three divisions. "The odds haven't changed at all," Skorzeny thought sarcastically. "What about the castle itself?" he continued.

"The troops so far identified are listed here," said Keitel, passing a sheet of paper to Skorzeny, who quickly glanced at it. The troops were significant: a reserve flak (antiaircraft) regiment, a tank platoon, two Honved (reserve) infantry battalions, elements of a mountain brigade, plus Admiral Horthy's personal 200-man bodyguard.

"In addition," Jodl added, breaking Skorzeny's thoughts, "a Hungarian parachute regiment has recently made its appearance in Budapest. But intelligence tells us that the commander is a stout anti-Communist, although we have no idea what his reaction will be if Horthy declares that a separate peace treaty has been reached with the Bolsheviks."

That statement met with silence. Seeing the long face on Skorzeny, Hitler broke the silence, "Skorzeny, SS-Obergruppenführer Bach-Zelewsky has brought up a 650mm superheavy mortar, fresh from our victory at Warsaw. He has promised to level Castle Hill and all of Budapest if necessary. In addition, I've given orders that you are to be assisted in carrying out your orders by all German military and civilian personnel. This letter will confirm that order."

Skorzeny was handed a crisp, clean sheet of paper surmounted by the German Eagle and Iron Cross in gilt by the German leader. It was the equivalent of a blank check from Hitler. He could commandeer whatever German troops and armaments were available in the Budapest area. Even so, Skorzeny remained perplexed as to how he could accomplish this difficult task. Compared to this, the Mussolini rescue mission seemed like a Sunday walk in the park.

Hitler continued his sales pitch, "I am giving you a Condor (four-engined bomber) from my Leader's Squadron for your own use. Mind you, Skorzeny, I'm counting on you!"

Skorzeny would have to reconnoiter his objective before unleashing all this force. The Karl-Batterie 428 that Bach-Zelewsky wanted to use so badly was out of the question. Once the Germans began a bombardment like that, the Hungarians would have an excuse to change sides.

There would not be a plentiful number of German troops for this operation, as there had been in March. As a matter of fact, the only division-sized unit earmarked for the operation, Panzergrenadier-Division Feldherrnhalle, which had been in the Budapest area was sent to eastern Hungary on 10 October to fight in the battle for Debrecen. The only unit that could lend Skorzeny's mission some support was the 22nd SS-Freiwilligen-Kavallerie-Division Maria Theresa, made up of ethnic Germans from Hungary.

Within a few days, Skorzeny arrived in Budapest, disguised as a "Dr. Wolff." With typical determination in the face of seemingly insurmountable odds, Skorzeny set about devising a plan that would bring quick results. His hunch had been right. Castle Hill was a veritable fortress stronghold. Barbed wire, mines, antitank ditches, and gun emplacements surrounded it. In addition, an alarming number of light and heavy flak guns pointed up toward the sky.

An air drop or frontal assault was definitely out of the question. In fact, any type of aggressive action would fail. Skorzeny radioed Berlin, "Paratroop drop highly unlikely to be successful. Will try different approach." The parachute battalions were kept on standby however, just in case Skorzeny could find no other way to get to the regent. He had promised a "different approach," but as yet, Skorzeny had no idea what that approach would be. Time was running out.

CHAPTER SEVEN

Christian! Seek not yet repose, Hear thy guardian angel say, "Thou art in the midst of foes—watch and pray."

Charlotte Elliot

The problem of getting at Horthy seemed solved when a report by the SD indicated that Horthy's youngest son, Miklós (the eldest had died in a plane crash in 1942), was to meet with members of Tito's representatives on 15 October. The Gestapo knew exactly where and at what time that meeting would be held. Skorzeny took the opportunity in hand and instructed that Miklós was to be captured and held hostage.

Skorzeny was hoping that his father would think twice about leaving the war if his son's life was at stake. It was a long shot by any means, but it was more than he had to work with so far. The Gestapo rented an apartment on the floor immediately above the one in which the meeting would take place. Skorzeny's aide, von Foelkersam, prepared his SS commandos for possible action, placing them close to the building where Miklós was to meet with Tito's representatives, who, in turn, were acting as go-betweens for the Russians.

The meeting took place as scheduled on 15 October. By 10:00 A.M. all of the plotters were inside the building of the offices of the Danube Steamship Company. Skorzeny's memoirs describe what followed with vivid detail.

October 15th was a bright Sunday. The streets were empty at the time appointed for the rendezvous. My company was in a side street in covered trucks. SS-Hauptsturmführer von Foelkersam kept me in touch with them, as obviously I could not show myself in uniform that day.

If I was to appear on the stage, so to speak, I must be inconspicuous. My driver and another man, both Luftwaffe personnel, were taking the air on a seat in the little garden which occupied most of the square. I drove up in my own car shortly before the meeting began.

When I entered the square, I noticed a Hungarian military lorry and a private car, which was presumably Horthy's, stationed in front of the building of which we had been told. It took me no time to make up my mind and park my own car right in the path of these vehicles, so that they could not get away in a hurry.

The floor above the offices in this building had been occupied the day before by policemen, who were lodged nearby. Others were to enter it from the street about 10:10 A.M. and make the arrests.

Three Honved officers were sitting in the covered lorry but could not be seen from the street. Two others were lounging on benches in the gardens. I was standing by my car, pretending to be fiddling with the engine, when the curtain rose on the drama.

The first German policeman had hardly entered the building when there was a burst of machine pistol fire from the lorry, and the second fell to the ground with a wound in the stomach. The two other Hungarian officers came running out of the gardens, firing their revolvers. I had just time to take cover behind my car when its open door was drilled. Things were getting really hot!

Honved soldiers appeared at the windows and on the balconies of houses. The moment the first shots rang out, my driver and his companion rushed up to me, assuming that I had been hit. The driver was shot through the thigh but could still walk. I gave the agreed signal to my detachment, and we three defended ourselves with our weapons as best we could against the rain of fire from the enemy. It was a most uncomfortable situation, though it only lasted a few minutes.

By then my car was not much more than a sieve. Bullets ricocheting from walls passed unpleasantly near, and we could only put our noses out of cover for long enough to have potshots at the enemy and keep them at least ten to fifteen meters away. Then I heard my men running out of the side street in our direction. Foelkersam had taken the situation in at a glance and posted the first section at the corner of the square, while the others swept through the gardens and began firing at the house fronts. My first assailants now withdrew to the shelter of a nearby house, which was occupied by Hungarians in some strength. I observed that these men were lining up for an assault, and quick thinking inspired us to hurl a number of grenades in the doorway, thereby bringing down the door and some marble slabs, which temporarily blocked the entrance. With that, the fighting ceased. It may have lasted five minutes.

Our policemen came down from the upper floor, bringing four prisoners with them. The two Hungarians, "Mickey Mouse" [Mikki, or Miklós Horthy, Jr.] and his comrade Bornemisza, were bundled into one of our trucks. To conceal their identity, our fellows had tried to roll them up in carpets, with only partial success, I observed, noting the effort required to get the refractory prisoners into the vehicle.

The lorry moved off, and my company withdrew. I was anxious to avoid further scuffles, which were only too likely when the enemy recovered from his surprise.

Fortunately, our retirement passed without further incident.

Some instinct prompted me to follow the truck. Another car and driver were available for me. Barely a hundred meters from the square, under the Erzsébet bridge, I saw three Honved companies approaching at the double. If they got any nearer, they could easily find themselves involved in a mix-up with my men—an eventuality I was determined to prevent at all costs. Time must be gained somehow, but bluff was my only resource. I told the driver to pull up, and ran towards the officer who appeared to be in command. "Halt your men quick!" I yelled, "There's a hell of a mix-up going on up there! No one knows what's happening! You'd better find out for yourself first!"

The trick came off. The troops halted, and the officer seemed undecided what to do. It was lucky for me that he knew some German, as otherwise he might not have understood me. The short pause was vital from my point of view. By now, my own men must have got away in their trucks. "I must get on!" I called to the Hungarian officer, jumped into my car, and made for the aerodrome. When I arrived, the two Hungarians were in a plane, and two minutes later they were on their way to Vienna.

An enraged Admiral Horthy announced over the radio that afternoon that Germany had lost the war, and Hungary had accordingly concluded a preliminary armistice with the Soviet Union. His radio speech, which was carried over the air at exactly 1:00 P.M., had a detrimental effect on his position. It seems that he had failed to notify the cabinet members of the Lakatos government. The cabinet resigned en masse that same afternoon, citing that, since they had not been consulted in Parliament, they could not approve an armistice.

The resignation of the cabinet had an effect on the Hungarian armed forces, which now held their breaths and awaited further direction. None came, however. Horthy was afraid for his son's life. Skorzeny had once again gambled correctly, but he could not lose any time. At any moment, the regent could have a change of heart about the value that he placed on his son's life and order the Hungarian army to surrender to the Reds, or worse still, to fight the Germans.

Whether the Hungarian army would follow such orders was a question Skorzeny did not want to wait and find the answer to.

Hungarian officers conferring after the assault on Castle Hill. Note the arm badge of the Hungarian Fascist "Arrow Cross" movement on their left sleeves. Note also an SS-Fallschirmjäger standing guard in the background (top). Photo: Bundesarchiv.

The tunnel entrance to Castle Hill was used by the men of the SS Parachute Battalion to gain entrance to the palatial fortress (bottom). Photo: Interfoto MTI.

A mad rush for Castle Hill or a paratroop assault would surely cause a battle between Germans and Hungarians, which would spread to the rest of the country—and the Hungarians would then have an excuse for changing sides. It would mean that his mission was a failure, and everything that had been accomplished up to then was for nothing. He had to fight with himself not to take the easy road of attacking. There were enough hotheads like Bach-Zelewsky who wanted to pound the Hungarians into oblivion. He would have to find another way.

In the event that an assault was necessary, the SS-Fallschirmjäger Batallion had been transferred from Hitler's direct control to Skorzeny's command on 14 October. On 15 October, the SS Parachute Battalion and the officer cadets from Wiener Neustadt arrived in Budapest. Slowly, the German forces began to occupy the railway station and other important buildings in the city. German units converged on Castle Hill, laying a ring of troops around the fortress stronghold.

The Hungarians up on the hill had been watching the German movements with much apprehension. Clearly, the Hungarians wished to avoid a fight that was sure to take the lives of many men—men who had survived the war so far and wished to see it end. On Horthy's orders, a Hungarian representative went down to meet with Skorzeny around midnight, requesting a parley. The senior Hungarian officer began the conversation with a veritable river of complaints about all the German activity around the hill. After all, he imparted, peace with Russia did not exclude peace with Germany.

Skorzeny cut short the officer's monologue by injecting some complaints of his own. Why, he asked him, were the German ambassador and his staff confined to Castle Hill (the German embassy was located there)? And why was it that he could not move freely about? "Why can't they come down to the city?" queried Skorzeny. The Hungarian officer was clearly embarrassed. Skorzeny knew that the Hungarians were just stalling for time. He decided to gamble right there and then. "If the mines and barricades leading down from the hill are not cleared from the road by 6:00 A.M.," Skorzeny warned his visitor, "Germany will draw the necessary conclusions!"

Feeling flustered and uncomfortable, the Hungarian promised at least to try and have that wish met. Skorzeny hoped that he had sized up the officer accurately; his plan depended on it. Operation Panzerfaust was on. Skorzeny described the operation most vividly.

Just before 5:30 A.M., when it was beginning to get light, I took my place in my truck at the head of the column. Behind me, I had two Tigers, followed by a platoon of the Goliath company, and the rest of the unit in their trucks. Automatics were set at safety. Most of the men had slumped in their seats and were enjoying a quiet nap. They had the hardened warriors' gift of snatching a bit of sleep when a really tough job lay ahead.

I took the precaution of sending my second-in-command to Corps Headquarters, to ascertain whether there had been any change in the situation, but the answer was in the negative, so zero-hour was adhered to.

In my truck I had Foelkersam and Ostafel, as well as five NCOs who had been in the Gran Sasso show. I considered them my personal assault group. Each was armed with a machine pistol, a few hand grenades, and the new panzerfaust. We were wondering what the Hungarian tanks in the Citadel would do. If necessary, our tanks and panzerfausts would have to look after them.

At one minute before 6 o'clock, I waved my arm as the signal to switch on. Then I stood up in my truck and pointed upwards several times, whereupon we started off, rather slowly, as it was uphill. I could only hope that none of our vehicles struck a mine, which would have blocked our advance and upset our plan. The Vienna Gate emerged out of the half-light—the way was open! A few Hungarian soldiers stared curiously at us. We were soon at the top. "Gradually accelerate," I whispered to my driver.

On our right was a Honved barrcks. "Nasty if we get fired on from the flank," murmured Foelkersam at my side. There were two machine guns behind sandbags in front of the barracks, but nothing happened. No sound could be heard but the rumble of the Tigers behind.

I chose the side street on the right in which the German embassy was situated. We could now travel at a good pace without losing the rest of the column. The tanks were doing a good thirty-five to forty kilometers to the hour, and at length the citadel was not more than a thousand meters away and a substantial part of our task had been accomplished.

Now the great detached mass of the War Ministry appeared to the left, and we heard the distant sound of two heavy explosions. Our men must have forced their way through the tunnel: the critical moment was at hand. We were past the War Ministry and in the square in a flash. Three Hungarian tanks faced us, but as we drew level, the leading one tilted its gun skywards as a signal that they would not fire.

A barricade of stones had been placed in front of the gate of the Citadel. I told my driver to draw aside and signaled to the leading Tiger to charge it. We left our truck and ran behind, while the barricade collapsed under the weight of the thirty-ton monster, which continued its irresistible thrust. Leveling its long gun barrel at the center of the courtyard, it found itself faced with a battery of six anti-tank guns.

We leaped over the debris of the barricade and burst through the shattered gate. A colonel of the guard got out his revolver to stop us, but Foelkersam knocked it out of his hand. On our right was what appeared to be the main entrance, and we took it at the run, almost colliding with a Honved officer, whom I ordered to lead us straight to the Commandant. He immediately complied, and at his side we rushed up the broad staircase, not failing to notice the elegant red carpet.

On reaching the first floor we turned left into a corridor, and I left one of my men behind to cover us. The officer pointed to a door and we went into a small ante-room where a table had been drawn up to the open window and a man was lying on it firing a machine gun into the courtyard. Holzer, a short, stocky NCO, clasped the gun in his arms and flung it out of the window. The gunner was so surprised that he fell off the table.

I saw a door on my right, knocked and walked straight in. A Honved major-general got up and came towards me. "Are you the Commandant?" I asked. "You must surrender the Citadel at once! If you don't, you will be responsible for any bloodshed. You must decide immediately!" As we could hear shots outside, including bursts of machine-gun fire, I added: "You can see that any resistance is hopeless. I have already occupied the Citadel." I was speaking the truth, as I was quite certain that the Mitte Battalion, led by the redoubtable Leutnant Hunke, was just behind me and must have seized all the strategic points.

The Hungarian major-general was not long in making up his mind: "I surrender the Citadel and will order the cease-fire at once." We shook hands and soon arranged that a Hungarian officer and one of ours should inform the troops fighting in the Citadel gardens of the cease-fire. After ten minutes had passed, no noise of battle could be heard.

Accompanied, at my request, by two Hungarian majors to serve as interpreters, I went along the corridor to have a look round. We came to the rooms adjoining the Regent's reception room. I was astonished to find that he was not there and learned that he had left shortly before six o'clock. It transpired later that he had been escorted by Obergruppenführer Pfeffer-Wildenbruch, commander of IX. SS-Gebirgs-Korps, to the latter's residence on the Citadel hill. His family had previously taken refuge with the Papal Nuncio. The presence of Horthy would have made no difference to our plans, which were not concerned with him personally but confined to controlling the seat of government.

While we were looking out of the window over the so-called Blutfeld, a few bullets whistled past. Hunke subsequently explained that it had proved impossible to notify the cease-fire to some of the Hun-

garian posts on the Danube side of the Citadel gardens. Two rounds from a panzerfaust soon convinced them that it would be wiser to abandon resistance.

The whole operation had not taken more than half an hour. Peace returned to the city and the citizens in the vicinity could turn over and go to sleep again. I rang up the Corps Headquarters on a special line and could almost hear the sigh of relief at the other end. Apparently they had considered the success of my coup as somewhat problematic.

Shortly afterwards, the reports came in from the War Ministry (where alone, there had been a short, sharp action), and the Ministry for Home Affairs, and one by one the commanders of the different groups turned up to relate their experiences. Our casualties had been agreeably low, not more than four killed and about twelve wounded. The only serious fighting had been in the gardens. I asked the Commandant about the Hungarian casualties and was told that they amounted to three killed and fifteen wounded.[1]

The SS paratroops had fared rather well. This was not the case with the officer cadets from Wiener Neustadt. They had taken most of the casualties. That very same day, a funeral was arranged for the dead on both sides. This funeral was attended by the Germans and Hungarians. It was important to begin again a peaceful coexistence with the Hungarian army.

Admiral Horthy was escorted out of Budapest that same day by Skorzeny. A German "honor guard" sent him on his way to meet his son, Miklós, who Hitler sarcastically described as "in protective custody." The hyenas had been caught and dealt with. The SS parachute battalion had a few days' rest in Budapest before being relieved. They even had time for a victory parade through the Hungarian capital. They left Budapest almost halfheartedly. For the old veterans of the 500th Battalion, this Hungarian interlude had seemed like a vacation from the rigors to which they were accustomed. Had it not been for Skorzeny's bold plan, they might have left Budapest with a totally different point of view. Siegfried Milius could not thank the gods of war enough.

[1] *Skorzeny's Special Missions*, London: Robert Hale, 1947.

Second from left, SS-Obersturmbannführer Otto Skorzeny after the coup at Castle Hill, 16 October 1944. Skorzeny's chief adjutant, SS-Hauptsturmführer Adrian von Fölkersam, is pictured here, walking to the left and behind Skorzeny. The officer at the extreme right is Obersturmführer Walter Girg (top). Photo: Bundesarchiv.

Below, a funeral ceremony for the fallen Hungarian and German soldiers in the courtyard of the Royal Palace, 16 October 1944. Note the mixture of Hungarian army, Arrow Cross, and German officers. SS-Obersturmbannführer Otto Skorzeny is in the front rank at the extreme right. Photo: Interfoto MTI.

OPERATION PANZERFAUST, 16 OCTOBER 1944

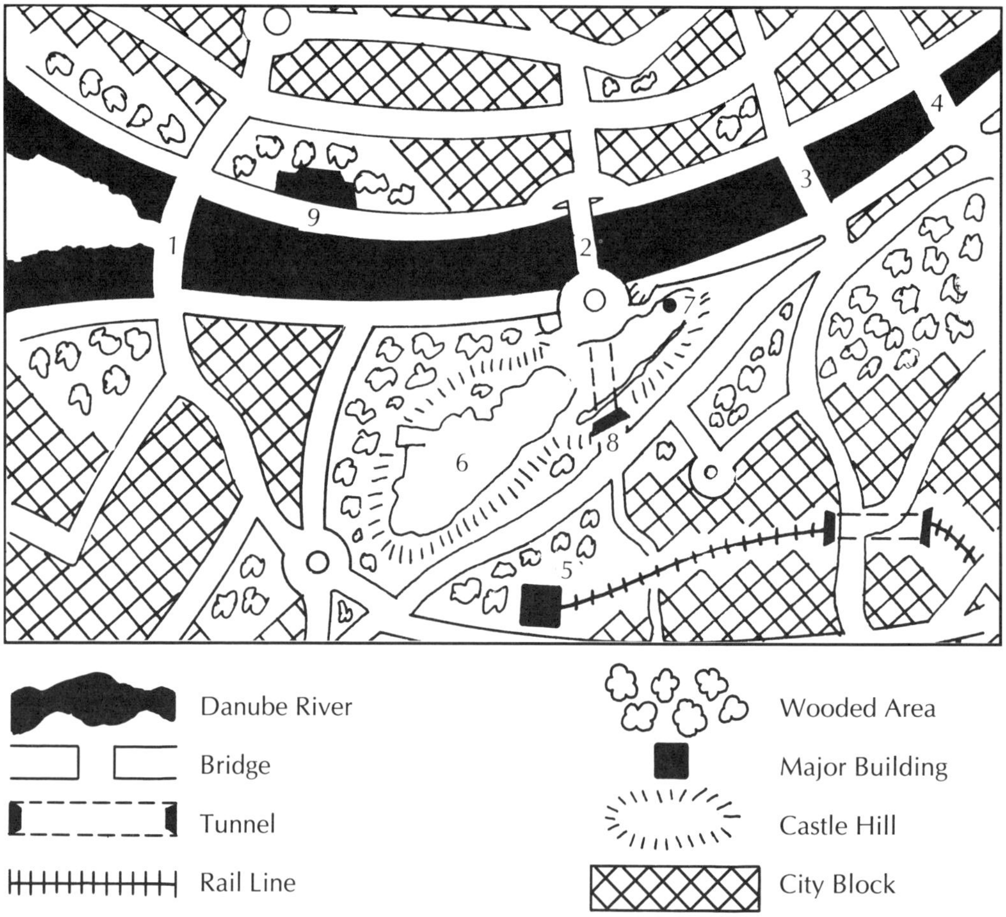

1 – Margaret Bridge
2 – Szechenyi Chain Bridge
3 – Elizabeth Bridge
4 – Szabadsag Bridge
5 – Southern Railway Station
6 – Castle Hill
7 – Royal Palace (on Castle Hill)
8 – Tunnel Entrance to Castle Hill
9 – Parliament Buildings

Men of the 22. SS-Freiwilligen-Kavallerie-Division Maria Theresa. The Hungarian ethnic-German division was in the process of forming into a division when it was called upon to help in Operation Panzerfaust by supplying Skorzeny with men (left and below). Photos: Bundesarchiv.

Parade of the SS-Fallschirmjäger-Bataillon 600 in Budapest shortly after Operation Panzerfaust (right).

Another group of officer candidates from the officer candidate school at Wiener-Neustadt (below). Photos: Bundesarchiv.

CHAPTER EIGHT

Make our faces vizards to our hearts, disguising what they are.

William Shakespeare
Macbeth

The men of the SS-Fallschirmjäger Bataillon were transferred in late October to Neustrelitz, Mecklenburg, where they received more replacements and continued their interrupted training. Winter was fast approaching, and it was bringing Germany's enemies ever so closer. The Third Reich was slowly being strangled by two firm hands, from the east and west. Several stab wounds at the groin (Italy and the Balkans) had left her bleeding. If some miracle did not occur soon, Germany would be defeated come the spring.

A hunter knows that an animal is most dangerous when he is wounded and cornered. Germany's last hopes for survival would soon resemble that very condition. Even during the retreat through France and Poland, Hitler had begun to formulate a plan for an offensive he hoped would change the war. All during the numerous crises of the summer and fall, his mind kept wandering back to the same thought: attack! His military resources were dwindling; he had no hope whatsoever of launching an offensive all along the vast eastern front. Besides, what strategically important objective could he reach with his small reserves? Moscow had long been out of the picture. An attack in the east would simply strip him of his last reserves, making the Russians' job a lot easier.

The situation in the west was somewhat different. The Allies were strung out along a much shorter front. The 1st Canadian, 2nd British, and 9th U.S. armies were in a large salient that stretched from the Dutch coast to the Roer River by Germany's border with Belgium. If a sizable German force could push its way through to the Belgian cities of Brussels and Antwerp, these three Allied armies would be cut off from their supplies and the rest of the Allied force. In turn, these Allied troops could then be attacked while surrounded and destroyed piecemeal.

Hitler hoped that the elimination of such a sizable Allied force would compel the Allies to withdraw their armies back into France and that pressure from the civilians at home would force a complete withdrawal of Allied troops from the continent. Of course, this was a fantasy. As a matter of fact, the German forces available, while substantive, were not large enough to push through to the Belgian coast. Even if they could, they could not have held the Allied units in check. General George S. Patton, the most dashing and charismatic of all the U.S. generals, asked General Dwight D. Eisenhower, Supreme Commander of all Allied forces in Europe, to allow the Germans to get at least that far so he could deal them a more severe blow. As it turned out, that was not needed. Hitler's last reserves were wasted in an offensive that never made it halfway to its objective.

Once again, Skorzeny's commandos were called upon to take part in a desperate operation, and the SS parachute battalion was not spared either, for it would provide two entire companies for a secret mission behind enemy lines. Recruits for the battalion were drawn from reserve units and convalescent companies. The German Navy and Luftwaffe were releasing many volunteers for service in the army and Waffen-SS. The German army even allowed volunteers to join the SS.

A large number of these volunteers were herded together in Schleswig-Holstein, Mecklenburg, and Lower Saxony. These men eventually found themselves in the SS parachute battalion. Though the majority of these men could not receive parachute training due to shortages, they nevertheless became first-class soldiers. With careful guidance from the veterans in the battalion, the new recruits quickly learned their new profession. Morale was quite high; the number of disciplinary cases in the battalion had shrunk to about 20 percent because of attrition and rehabilitation. These men who rushed to join the ranks of the SS-Fallschirmjäger soon proved their mettle in some of the fiercest fighting of the war. Soon, the battalion had once again attained a strength of about 1,000-plus men. The old training company however, had to be incorporated into the battalion.

One day in November, SS-Hauptsturmführer Siegfried Milius received a copy of an order from Skorzeny, who said it had been distributed by the Oberkommandos der Wehrmacht (OKW, or German Army High Command). The order had been signed by Field Marshal Wilhelm Keitel, and read as follows:

> *The Führer has ordered the formation of a special unit of a strength of about two battalions for employment on reconnaissance and special tasks on the western front. The personnel will be assembled from volunteers of all arms of the army and Waffen-SS, who fulfill the following requirements:*
>
> *A. Physically A.1, suitable for special tasks, mentally keen, strong personality*
>
> *B. Fully trained in single combat*
>
> *C. Knowledge of the English language and also the American dialect—especially important is knowledge of military terms. This order is to be made known to all units and headquarters. Volunteers may not be retained on military grounds but are to be sent immediately to Friedenthal near Oranienburg (Headquarters Skorzeny) for a test of suitability, but a value will be put on [their] fighting spirit and temperament. Captured U.S. clothing, equipment, weapons, and vehicles are to be collected and reported for the equipment of the above special troops. Personal wishes of the troops to make use of this kind of captured equipment must take secondary priority. Details will be notified later . . .*

Skorzeny told Milius that he had tried to have the whole operation cancelled shortly after the release of this order. The Allied intelligence sources were sure to pick it up and deduce from its contents just what the Germans were up to. "It's incredible the stupidity which prevails in the German High Command!" Skorzeny said bitterly. "The Führer has heard my argument, but refuses to cancel the order. By the way, I shall have need of your fine troops, if I am to make this mess work!"

Milius just looked in a resigned manner. "You'll have them, Sir," he said, then picked up the phone. "This is Milius, let me speak with Obersturmfuehrer Leifhert of First Company."

Leifhert took less than five minutes to reach Milius' office. He was ushered in and stood at attention in front of his commanding officer.

"Obersturmfuehrer Leifhert reporting as ordered, Sir."

"Stand at ease, Hans. I'll make this short. SS-Sturmbannführer Otto Skorzeny is looking for volunteers for a new unit he is forming. They will be committed on the Western Front and be employed for special tasks of the most dangerous nature. I called you because you are my best company commander, and he will need excellent troops."

"I would like to volunteer, Sir!" came the reply.

Milius looked him squarely in the eye. "We may not see each other again, Hans. Although I cannot divulge too much, I can tell you that the risks will be high."

"I understand, Sir, but I would still like to volunteer."

"Very well. Sturmbannführer Skorzeny will have need of about two companies from our battalion. Will you ask your men if they too wish to volunteer?"

"I will, Sir, but I am sure they will all volunteer."

With that, Milius dismissed Leifhert and called to his orderly, "Decker, send in the other company commander."

Back in Friedenthal, Skorzeny set about testing the abilities of the hundreds of volunteers who flocked to his command. The strength of the brigade he was forming was supposed to field 3,300 men, though he would only be able to muster 2,500 qualified candidates. Out of this number, 500 came from the Waffen-SS, 800 from the Luftwaffe, 1,000 from the Heer, and 200 from the Kriegsmarine (navy). Even after this influx of men, Skorzeny still needed to add entire units to his brigade because the time allotted to form the brigade was insufficient to merge all of the divergent manpower into one tight, homogeneous force.

In this way, the two companies of SS paratroopers from the 600th Battalion found themselves at Friedenthal as whole units. Leifhert's company alone had 180 men. They were at full strength. Other bits and pieces also arrived. Also added to the order of battle were 5th Company of SS-Jagdverbande Mitte and 7th Company, 1st Battalion, 11th Panzer Regiment (of 6th Panzer Division).

The tank crews for the brigade came from 1st Company, Panzerjäger-Abteilung 655; the reconnaissance men from 1st Company, Panzer-Aufklärungs-Abteilung 190 (of the 90th Panzergrenadier Division), and 1st Company, Panzer-Aufklärungs-Abteilung 2 (of the 2nd Panzer Division). The artillerymen came from 1st Battalion/40th Artillery Regiment, while the artillery pieces came from the Führer-Grenadier-Brigade. The brigade staff was formed from Panzer-Brigade 108, while the battalion staffs were formed from Panzer-Brigades 10 and 113.

Skorzeny's brigade was numbered 150 and given the mission of securing the three vital bridges behind the U.S. lines at Huy, Engis, and Amay. For tactical purposes, he would split his brigade up into three forces, "X," "Y," and "Z." From the beginning, Skorzeny's brigade suffered a shortage of equipment. U.S. coats arrived with the "PW" emblem painted brightly on each one—totally useless. They had British equipment, but they couldn't dress in British field jackets if they were to impersonate American soldiers.

The most pressing problem was that of acquiring U.S. vehicles. In the fall, Skorzeny had seen just about every other German officer with a U.S. jeep. Not one could be found in all of Germany. He took matters into his own hands and managed to "liberate" some jeeps. U.S. tanks were more difficult since they were almost nonexistent in Germany. Two Sherman tanks eventually turned up, but one was immediately lost to transmission trouble. They had to resort to disguising German tanks and assault guns with sheet metal to make them look like U.S. vehicles—a bluff that would fool no one. As one U.S. author put it, "These vehicles could only fool untrained American recruits, at a distance, and at dark, maybe." Some U.S armored cars arrived, but not enough. Some British armored cars in good condition were available, but since their presence in a U.S. sector could not be explained, they were eventually returned.

As X-Day approached, Skorzeny frantically prepared his small brigade. Each kampfgruppen (battle group) had a headquarters platoon, three motorized rifle companies (in Ford trucks), two armored infantry platoons, two antitank platoons, two heavy mortar platoons, an engineer platoon, and a signal platoon. The first two kampfgruppen, X and Y, had five Panthers (disguised as number ten tank destroyers), and five assault guns apiece. Kampfgruppe Z had no armor support. Kampfgruppe X was earmarked to work in conjunction with the leading SS Panzer Kampfgruppe, headed by SS-Obersturmbannführer Joachim Peiper, of 1st SS Panzer Division Leibstandarte Adolf Hitler. The commanders of each kampfgruppe were as follows: Kampfgruppe X, SS-Obersturmbannführer Willi Hardieck; Kampfgruppe Y, Hauptmann Scherff; and Kampfgruppe Z, Oberstleutnant Wolf.

Skorzeny's Panzer-Brigade 150 had the mission of securing three vital bridges behind the U.S. lines—at Huy, Engis, and Amay. His force of about 2,500 men was armed with a motley collection of German,

British, and U.S. arms. They lacked sufficient artillery and tanks but had plenty of small arms and automatic weapons. The key to a successful mission lay in the hope that the U.S. front would collapse on the first day of the offensive, 16 December.

On 14 December, Skorzeny set up his headquarters at Schmittheim, where 1st SS-Panzer-Korps had its headquarters. The brigade moved from Wahn to an area near Münstereifel to assemble for the attack. When the leading elements of 1st SS-Panzer-Division Leibstandarte Adolf Hitler began its attack, the 150th Panzer Brigade followed closely behind. The brigade experienced problems from the start. It seems that some knucklehead moved some horse-drawn artillery pieces into the main road leading to the Losheim Gap too soon, bogging down the whole brigade far behind the front. Battle Group Peiper managed to get through by crushing any wagon that did not move out of the way of its Panzers.

To add injury to insult, one of Skorzeny's kampfgruppen commanders, SS-Obersturmbannführer Willi Hardieck, was killed by a land mine as he attempted to bypass a Wehrmacht column. SS-Hauptsturmführer Adrian von Foelkersam, Skorzeny's aide, had to take over command of Kampfgruppe X. By the evening of 17 December, Skorzeny's hopes for a breakthrough were diminishing rapidly.

At a meeting that same evening at 6. Panzer-Armee headquarters, Skorzeny requested permission to employ the brigade as a regular combat unit if a breakthrough was not forthcoming by 18 December. The commander of 6th Panzer-Armee, Dietrich, ordered Skorzeny's brigade to plan an attack against the town of Malmedy, which Kampfgruppe Peiper had bypassed. The expected breakthrough did not materialize, and Skorzeny's command was committed to taking Malmedy.

By 20 December, only two of the three brigade battle groups had managed to reach their assembly positions near Ligneuville. The last battle group, Kampfgruppe Z, was still far behind the front, stuck in a tremendous traffic jam, a major problem encountered by the Germans in the Ardennes offensive. This last battle group was thus earmarked as a reserve for the projected attack—provided that it arrived in time.

The objectives of the attack were the roads from Malmedy to Stavelot, the U.S. artillery positions on the high ground above the road, and Malmedy itself. Skorzeny was well aware of his deficiencies, but he was hopeful of acjieving a quick victory by launching a surprise attack on the defenders. Intelligence reports had stated that the U.S. 9th Air Force had mistakenly bombed its own troops in the town at least three times. He had also gained some valuable information from one of his commando teams, who had inadvertently driven right into the town, thinking it was already in German hands. The three-man team, disguised as U.S. soldiers, quickly got over its shock and left town, but not before seeing enough to give Skorzeny an idea of what he would be up against.

Once again, fate favored the Americans. It seems that a Dutch deserter had been captured on the outskirts of Malmedy on the afternoon of 20 December. This deserter warned the U.S. troops of a German attack to be launched at 3:00 A.M. the following day. When Skorzeny's brigade moved out on the morning of the 21 December, the Americans were waiting for it.

Skorzeny split his attack into two prongs, a right led by Hauptmann Scherff's Y Battle Group and a left led by Hauptsturmführer von Foelkersam's X Battle Group. Kampfgruppe Y followed the road that led from Baugnez to Malmedy and immediately confronted one battalion of the U.S. 120th Infantry Regiment (30th U.S. Division), which was quickly reinforced by another battalion of the same regiment.

Long-range artillery fire, combined with heavy automatic weapons fire, bazookas, and mines, stopped the advance of Scherff's men. Many who were not killed surrendered. The battle group attempted to get through several times afterward, but each assault met with failure.

Von Foelkersam's attack had some success. His battle group moved out from Ligneuville through Bellevaux, attacking along the Route de Falize. He committed a much larger force—two companies of riflemen supported by the Panther tanks disguised to look like M-10 tank destroyers. As they advanced on the town, his battle group was given away by a trip wire set off by a grenadier, immediately lighting up the whole countryside and initiating the battle.

The leading "M-10" tank hit a mine in front of a railway overpass. The other tanks were forced to maneuver around the burning vehicle, while the infantry attempted to set up machine guns by the railway embankment. The 3rd Battalion 120th U.S. Infantry Regiment opened up on the German attackers with everything it had. In spite of this, Skorzeny's men kept on coming. The U.S. artillery units overlooking the road unleashed a terrible barrage on von Foelkersam's men, using a new and deadly weapon called POZIT. The POZIT contained a radar device in each artillery round that caused it to explode a few feet before the shell hit the ground, inflicting severe injuries. This weapon halted the attack dead in its tracks. Many of von Foelkersam's men ran in panic when confronted with this new "secret weapon."

Another column of von Foelkersam's battle group that had made for the Warche River bridge had better luck. A few men of Company K 120th Infantry and some engineers and a platoon of the 823rd Tank Destroyer Battalion guarded this area. The Germans stormed the bridge, eliminating the tank destroyer platoon that was defending the bridge and paper mill nearby. The men of Company K withdrew some distance and decided to make a last stand in the paper mill just north of the bridge. By then, rumors of the "Malmedy Massacre," in which an SS unit near Malmedy had killed hundreds of unarmed Americans on 17 December, had spread. The U.S. soldiers were in no mood to retire, and they vented their anger on Skorzeny's brigade.

When Skorzeny ordered the factory attacked, the Americans were waiting for them with small-arms fire and bazookas. The tanks of 150th Panzer Brigade were brewed up one after the other by the courageous actions of the Americans. The German attack bogged down in the face of this stiff defense. Company K received mention that day in the U.S. 1st Army war diary, and several Americans won high military decorations for their defense of the factory, including two Distinguished Service Crosses and one Medal of Honor.

Personally leading the attack on the factory, Skorzeny had to dash for cover, while the future U.S. Medal of Honor winner, Private First Class Francis Currey, fired his automatic weapon at the tall commando. At the arrival of U.S. reinforcements, Skorzeny withdrew what remained of his force, considering retreat the better part of valor when faced against such a savage resistance. SS-Rottenführer Andreas Meyer was a member of von Foelkersam's headquarters unit. He attests to the ferocity of the U.S. defense on 21 December.

> *I was with von Foelkersam on December 21st, at about 10 A.M. when the message came in that von Foelkersam's adjutant, SS-Untersturmführer Lochner, had been heavily wounded. Foelkersam gave me the order to immediately go and bring Lochner back.*
>
> *It was a dangerous mission and a veritable ticket to heaven, but we managed to reach the front line comparatively fast. Soon we had reached Lochner. Lochner had been wounded at several places. He had been shot in his hand, rib cage, and his stomach. He was lying in an overgrown ditch, which we had to crawl on our bellies to reach, all the time while we received heavy infantry and artillery fire.*
>
> *When we reached Lochner's body, I removed my paratrooper's blouse and*

poncho and used them like a litter. A comrade now came over to my position and gave me some help. We proceeded back to our own line, crawling on our bellies and elbows. The area was a flat, open ground, which was a great help to the enemy. Every two meters or so we had to lie flat because the infantry fire would get very close.

In addition, the fire coming from the paper mill was particularly intense. This fire was coming from our flank, adding to our predicament. I didn't realize it for a few seconds, but my comrade was not moving back with me during a pause in the firing.

I crawled around Lochner and got close to the paratrooper who had been helping me. I turned him over and realized that he was dead. His eyes looked broken up, and blood was dripping from his mouth. In the meantime, four tanks from von Schwerdt's command (panzer IVs) got into action but only for about half an hour. They were all burnt out by phosphorous grenades. Every single crew member was shot as [he] attempted to get away from the burning tanks.

With the help of another comrade, I was finally able to extricate Lochner and return him to our line. I later heard that he died the next day. I can confirm that the concentrated U.S. artillery fire was murder for us. It was only when darkness finally arrived that we could crawl out and retrieve our wounded.

The mood of American soldiers over the tragic incident of December 17th—the road crossing at Baugnez [the "Malmedy Massacre"]—was not known to us. We only found out after the war.

For this and other acts of bravery, Corporal Meyer was promoted to SS-Unterscharführer (sergeant) on 2 January 1945. His hair-raising testimonial described quite fully the fury of the U.S. defense, spurred on by the rumors of a great massacre of defenseless American soldiers by Waffen-SS troops. Only one German tank managed to reach the northern bank of the Ambléve at Malmedy, and a well-placed bazooka round to the engine compartment stopped the tank just about 100 feet from the bridge.

Skorzeny kept up his attack throughout the morning; all the while, U.S. troops opposing his brigade kept up a fierce resistance. As Sergeant Meyer described, the German forces did not know why the Americans were putting up such a stiff defense. By the early afternoon, Skorzeny had counted his losses: if the fighting kept going at this tempo, it would amount to crippling losses. Still, he was not ready to retire. He ordered the attack continued, but no further progress was possible.

By mid-afternoon, Skorzeny called off the attack. He ordered his brigade to withdraw to the crest of the hill south of Malmedy. U.S. artillery batteries were now heavily pounding anything in front of the town. Even Skorzeny's behind-the-lines headquarters was shelled. When he got back to the Hotel du Moulin, where he had set up his command post, Skorzeny radioed 1st SS Panzer Division, "Enemy defense very stiff. Have suffered heavy losses. Request reinforcements if attack is to continue. Skorzeny."

Just he was about to leave for 1st SS Panzer Division headquarters, Skorzeny was caught out in the open by another U.S. artillery barrage on the town. He dove for cover underneath a truck as artillery shells began to explode all around him. The chief of staff's car was blown to bits. After the barrage, Skorzeny ran over to aid the wounded men on the street, when suddenly, three more rounds landed on the street, close to him. Shrapnel flew in all directions, with Skorzeny taking some in the face. He checked his wound and discovered that the deep gash on his forehead had missed his eye by inches. The gash was bleeding profusely. His men took him to an aid station where he was bandaged up.

The doctor who attended to him ordered that he be taken from the battle area, but Skorzeny had other plans. He returned to his headquarters and continued to lead the brigade, although its actions from that point on were all defensive. It had been too badly mauled to offer much more, and as he had predicted, his request for reinforcements had been turned down since there were none to give him.

The brigade was listed as having been withdrawn from the line on 28 December, when it was relieved by the 18th Volksgrenadier Division. German reports intercepted by ULTRA (discussed in Chapter 2) mentioned that Skorzeny's panzer brigade was withdrawing to St. Vith as early as 26 December. It is clear that these reports referred to the order passed down from the German army High Command to disband the unit as of 25 December.

For a brief period, Hans-Joachim (Hajo) Draeger and about 180 other SS paratroopers were attached to Otto Skorzeny's Panzerbrigade 150. The following account by Draeger is enlightening.

> *During the second half of 1943, the opportunity of leaving the SS Signal School in Braunschweig presented itself to me. I had been attending this school and decided to join the SS Cavalry School. The Signal Troops in war did not appear to me to be a good career. They seemed to be a better career in peacetime. At the same time, I thought of the possibility of leaving the Wiking Division from whom I had been temporarily detached in order to attend the signal school.*
>
> *The Wiking Division contained many European volunteers, and the Division was committed to the Russian Front without the slightest possibility of being relieved. Whenever the Das Reich or Leibstandarte divisions were pulled from the front lines for rest and refitting, we would receive their old weapons.*
>
> *After successfully passing the Signal School exam in Braunschweig, I was sent to the SS Cavalry School in Warsaw for weapons training. This course lasted two months. For seven hours a day, we were on horseback. Before noon, five hours were spent riding, while the afternoon gave way to one hour of jumping and one hour of parade riding. At the same time, we learned to drive as many as two and even four horses at once. We were also taught some blacksmith skills.*
>
> *After these two months, I attempted to get into the 2nd SS Cavalry Replacement Squadron, which was a bicycle company. I met a friend there whom I wanted to do duty with. We both looked like experienced officer-recruits but behaved like easygoing Verfügungstruppen soldiers of the old school!*
>
> *In April 1943, an SS-FHA memorandum was circulated, requesting SS officers to volunteer for a newly forming SS unit, a parachute battalion. That evening I asked my comrade, SS-Obersturmführer Fritz Leifheit if he had read the memo. "So, Comrade Draeger, you also read that memorandum?" I answered that I had and that I had volunteered. He told me that he had expected nothing else from me.*
>
> *So in May 1943 we were sent to Sarajevo [Yugoslavia] by way of Vienna and Belgrade. The old Paratroop School No. 3, which had been located at Braunschweig, had now been relocated to Sarajevo to get away from the Allied bombings. We trained there and made our first jumps. I have never had as much leg pain as I had during the first week at the parachute school. Then, in June 1943, we were sent to Papa [Hungary] where we completed our jump training.*
>
> *We came from Sarajevo with our "A" Company to Papa, where we received new NCOs and men. Obersturmführer Leifheit and I had the job of interviewing the new men and deciding which ones were considered fit to serve in an elite unit. The ones we considered unfit were declared to be unacceptable for parachute jumping by the Battalion Doctor, and so were sorted out.*

We were thus able to get an exemplary company. From Papa Obersturmführer Leifheit and I travelled to Budapest to set up weapons, equipment, and vehicles. We were to proceed then directly to the offices of the SS-FHA in Berlin and lay out our requests. This journey took place on July 20th, 1944. Fritz Leifheit and I wondered why the Wehrmacht officer who met us had greeted us with the German greetings [and not the standard "Heil Hitler!"]. About the Army putsch, we heard very few rumors.

After our interview we returned back to Papa. In Papa, the rank and file were trained as paratroopers. In August 1944, we were transferred to Iglau in Czechoslovakia. Here we took part in a parade under the leadership of SS-Obersturmbannführer Droste, for the yearly Harvest Festival. We were all a bit mad that the National Socialist Leader who led the Festival remained wearing his soft cap, while all of us officers and men standing before him had to wear our helmets.

From Iglau the "A" Company was moved to Deutsche-Wagram via Vienna because of actions in Budapest [Skorzeny's coup in October]. At the end of this, we went to Neustrelitz in Mecklenburg, our new residence. In November, 180 men were pulled out of the Battalion and sent under the command of SS-Obersturmführer Leifheit. We traveled by train to Troop Training Ground Grafenwoehr. This was a gathering of seldom-seen people. One noticed someone only when he didn't wear a Knight's Cross! Once again we received intense infantry training. We had to do long marches in the evenings into ready areas. In the mornings we made attack maneuvers over distances, plus close combat. Over and over, we practiced firing at night by flare-lights and moonlight.

I myself was platoon leader of 1st Platoon and was amazed at how well my No. 1 Machine-Gun Team, Siegfried Treu and Bernhard Blanke, fired. After a three-week stay and some hard training, we were loaded on railway cars and moved west. The railway cars were camouflaged with Christmas trees. We unloaded in Köln-Wahn [Cologne].

In the meantime, we had noticed we were getting ready to move out as Americans. We were learning how an American opens his cigarette pack, how he held himself and acted around superiors. Our tanks were dressed in tin so they appeared to be Sherman tanks. We received captured American trucks and began to hear that a great offensive was being planned, which would once again bring German troops to the Channel Coast [of France].

For this, it was important that the Meuse [River] bridges fall into our hands so that the offensive could proceed unhindered. This was definitely a paratroop task, but because we had no air superiority, which was necessary for such an action, someone came up with the idea of a trick in that we would be sent out as American soldiers withdrawing.

It was very clear to us that if caught in American uniforms we would be shot, but who would be shocked by that in the fifth year of the war? From Wahn by Cologne, we were transported by truck across Münstereifel-Blankenheim-Stadtkyll into the initial assembly area.

It was our task to capture Huy, above all to hold the bridges at Huy until the elite divisions Leibstandarte Adolf Hitler and Das Reich caught up. The same thing went for the cities of Luttich and Namur. In Stadtkyll, we turned west and traveled through Hallschlag, Losheim, Bullange, Büttgenbach, further in the direction of Huy.

In Büttgenbach we took heavy artillery fire. We left the vehicles and, as infantrymen, crossed targeted ground and allowed the vehicles to race individually over the streets. As much as I can recall, we suffered no losses. By Waimes we ran into the enemy. We damaged a few American armored cars and captured one of them.

To our great joy, this vehicle was full of K-rations, the American ration packs, and there was even some champagne!

After the battle, we had ourselves a champagne breakfast on the street and then went on in the direction of Malmedy. Due to an artillery barrage, our commanding officer was killed, so SS-Obersturmführer Fritz Leifheit took over the battalion. I was given command of our parachute company. It is funny, but I was totally happy and sad at the same time—to immediately realize the responsibility that one takes on and the cost that a mistake can cost.

We were not held up in Malmedy. We had no heavy weapons and no supplies. It was our task that we should take and hold the bridge at Huy. On the railroad bank before Malmedy, we received heavy fire and lost twelve men. Because of these twelve men, on the following night, I ordered assault group after assault group to retrieve the bodies for a proper burial.

We then laid fast before Malmedy and secured the flanks of the southern group, which was assaulting in the direction of Bastogne near Namur. In our defense positions, we saw little enemy action. Only then and again would an artillery shell come flying in our direction. There are many small things to relate, but none of these were decisive to the battle. I still remember the evening of December 15th, 1944. We were in the battle headquarters of Skorzeny. His last words to us were, "Gentlemen, Germany has worked hard and with difficulty to prepare that which now stands before us. It is now up to you to be brave, conscious, and full of aggressive action to use these weapons."

On the 16th of December, the offensive began and also the first of our troubles! The forward infantry was unable to immediately break through the enemy line, so that those troops held for the breakthrough had to come to the aid of the forward infantry units. As we moved on in stages, again and again, a vehicle would halt due to lack of fuel. Leaders would quickly scrounge for gasoline from every vehicle which could spare it or from abandoned ones.

Smallest motorized discipline went unnoticed. A traveling column was not to be passed; a stationary one, only when the unit's leader had been apprised of the situation and then would move by when security posts had been set up. As it happened, the traveling column was passed up. Vehicles came from the front so the passing unit threaded itself into the other unit and became entangled. When the leading vehicle from the passing unit turned left, the vehicles from the other unit followed!

There was an incurable mess, and from beginning to end, like a still-born child. At least, that's how it appeared to me. We followed the point guard so that we immediately cut out and headed for Huy as Americans, but nothing came from that. A few commando troops changed road signs, cut lines, and attempted to destroy [U.S.] headquarters' units, some even coming out of it alive. Droste was one of the lucky ones.

We then left our vehicles in the shadow of a pair of mountains about fifteen kilometers before Malmedy and set up defense positions in a small wooded area. We laid there until the end of 1944 and then were ordered to retire in the beginning of 1945. During these last few weeks, we had a hard time with the scant supplies that we received.

Draeger had joined the SS-Verfugungstruppen on 30 August 1940. His first posting was to the SS Signals Battalion. He served in that battalion's 2nd Company until being transferred to the newly formed SS Wiking Division. Draeger served in the Wiking Division, being promoted to SS-Rottenführer (Corporal) on 1 September 1942 and SS-Unterscharführer (Sergeant) on 1 November 1942. These promotions came during the heaviest fighting in the Russian Caucasus Region, where the Wiking Division was serving.

During the spring of 1943, Draeger was again sent to school, this time to the SS-Signals School at Braunschweig. Draeger was finally promoted to SS-Untersturmführer (2nd Lieutenant) on 1 March 1944. He had the opportunity to join the SS parachute troops and served with the SS Parachute Battalion 500 and later 600 until he was wounded in 1945. He received the following badges and awards during his military career: Iron Cross, 2nd Class; Iron Cross, lst Class; Eastern Medal; Parachute Jump Badge; Motor Troops Badge Tank Destruction Badge, two awards; Close Combat Clasp; Wound Badge in Silver.)

It wasn't until the beginning of the year that the survivors of the SS parachute battalion who had taken part in the Ardennes offensive were able to return to Neustrelitz. They had served with distinction in Skorzeny's Panzerbrigade 150, and many had medals to prove it. The men who served were retained as one company and used as a whole unit once they were back in their parent battalions. Since the men had seen action together, the German command believed it would be best not to separate them.

The winter of 1945 was bitterly cold in Western Europe, and accompanying this cold spell was another large-scale Russian offensive, launched on 12 January. The dark, ominous clouds that were converging over north central Europe indicated the beginning of the end for the Third Reich. The German nation held its breath and awaited the twilight of war to fall upon it. Germany's Götterdämmerung was about to begin.

An original SS-Fallschirmjäger cuff band worn on the lower left sleeve (right). Photo: Dr. Einar Himma.

(Left) SS-Rottenführer Werner Schrödl, top row, second from left. The photo was taken after Ardennes, when the two SS parachute companies temporarily attached to Skorzeny's 150th Panzer Brigade had retruned to their home base at Neustrelitz. Soon after, they were on their way to the Eastern Front. With the exception of Shrödl, everyone else in the photo was killed. Photo: Werner Schrödl.

CHAPTER NINE

And I looked, and behold a pale horse; and his name that sat on him was Death, and Hell followed with him.

Revelation 6:8

Following the return of the paratroopers who had been assigned to Otto Skorzeny's 150th Panzer Brigade, Milius again reorganized his companies and continued training. Toward the middle of January 1945, the SS-Fallschirmjäger-Bataillon 600 went on alert. The Soviets launched their winter offensive on 12 January, and the German divisions were too few and weak to hold the Red Army drive for more than a few days.

Soon vast gaps in the German lines appeared, through which the mass of the Soviet tank forces poured. Hundreds of thousands of German civilians fled, moving in long, dreary columns toward the safety of the west. All too soon, the Russian tanks caught up with these civilian refugees, crushing the wagons and people with apparent joy. Long columns of motorized Russian infantry would follow, deflowering as many fräuleins as they could find between the ages of eight to eighty and killing anyone left alive.

In only two weeks' time, Red tank columns covered the vast expanse of the Polish plain. Silesia, the last industrial area held by Germany, was just beyond the advance units of the Red Army. Gauleiter Hanke, who was responsible for the Silesian capital, began to prepare for Breslau's defense. He asked Reichsführer SS Heinrich Himmler for some trained parachute troops to, as he put it, "turn Breslau into a Cassino on the Oder." Himmler ordered Milius to prepare a parachute drop on Breslau. Milius gathered about forty to fifty of his best men and prepared to send them to the Silesian capital.

Hanke had wanted at least a battalion, but most of Milius' men now had little if any parachute training. In the whole battalion, he barely had a platoon of men who could jump. Reichsmarschal Hermann Göring finally came to Hanke's rescue, supplying two battalions of paratroopers from the Luftwaffe. Most of these men also had little parachute training, but the Luftwaffe managed to airlift them into Breslau.

The 1st Battalion, 26th Fallschirmjäger Regiment under Hauptmann Trotz arrived first, on 25 February 1945. The 2nd Battalion, Fallschirmjäger Regiment for Special Employment "Schact" under Hauptmann Skau, arrived eight days later, on 5 March.

Milius' battalion was not to be spared, however, and toward the end of January, it was again placed on alert, this time by Skorzeny. The Soviet mechanized spearheads reached out for the Oder River, Germany's last great water barrier before Berlin. Under Hitler's direct instructions, Heinrich Himmler had been placed in charge of a new German army group that was being hastily formed to protect Berlin. Army Group Vistula was a hodge-podge of reserve, convalescent, and Volkssturm formations, all badly armed and equipped units that were entrusted with the task of breaking the Red Army tidal wave before it engulfed Berlin.

Himmler had promised Hitler that "the miracle of Frederick the Great" could be repeated. He would stop the Red hordes threatening the very heart of the Third Reich. Himmler also boasted that he would throw the Bolsheviks back all the way across Poland and over the Vistula River. What he needed most of all was time—time to gather the Third Reich's last available reserves and train and organize them. Himmler solved his dilemma by calling on Otto Skorzeny, whose commandos, Himmler reasoned, were the perfect candidates for holding the Red tide long enough for Army Group Vistula to form. Skorzeny's elite commando battalions would have to hold the

Red Army single-handedly to buy time for Himmler's forces and Germany.

SS-Untersturmführer Hans-Joachim Draeger, whose experiences in the Ardennes offensive were remembered in Chapter 8, wrote of the transfer from the Ardennes and the eventual call up to the Oder River front in January 1945.

> *On the lst of January 1945, we returned to Wahn by Cologne and were there loaded, and a third of our troops had already left Urlaub. As we were to travel to Berlin, more comrades came and asked to be allowed to go on leave, since many lived in and around Berlin.*
>
> *I pulled the rest of the men together from the company and presented those men requesting leave (the Berliners) and asked the men not going on leave if they did not mind working extra, unloading at Neustrelitz, while their Berlin comrades went on leave. No one had anything against it. Later, none of the men that worked at Neustrelitz while some of their comrades went on leave complained.*
>
> *I mentioned that here to show the morale of our troops in the fifth war year. We were already a very close society of great soldiers. We came to Neustrelitz on January 28th, 1945. I was able to promote SS-Oberscharführer Aussenthal to SS-Hauptscharführer on January 30th. Under Leifheit's command that would never have happened, since there was something between the two of them. Aussenthal died while we were crossing the Oder River.*
>
> *During our celebrations on the 30th of January, we received the news that we had to make ourselves ready for action on the Oder River. I was called to Milius, and he told me that I and the rest of my men from the Ardennes offensive were to proceed immediately to the Oder River as a forward group.*
>
> *Someone came to the conclusion that it was best to leave our cadre together [the Ardennes veterans] and not jostle them around, since everyone knew each other by now—sending them forward in small groups. We had been the last to be pulled out of the Ardennes and the first to go out again!*

SS-Unterscharführer Andreas Meyer, a battle clerk from SS-Jagdverband Mitte and a veteran of the famous 150. Panzer Brigade, described his arrival at the Schwedt bridgehead during February 1945. Meyer's Mitte commando battalion had been virtually destroyed while defending Hohensalza Inowroclaw between 18 and 20 January. The 5th Company of this battalion had been attached to the 150. Panzer Brigade and so had escaped the fate of its sister companies at Hohensalza. By mid-February, SS-Jagdverband Mitte had been rebuilt. It then joined SS-Obersturmbannführer der reserve Otto Skorzeny's command at the Schwedt bridgehead. Andreas Meyer, who had remained at Friedenthal during January, reported directly to Skorzeny.

> *I reported in the castle [at Schwedt] to Skorzeny. Late in the morning we drove in an armored half-track into the bridgehead, where Skorzeny inspected the front lines in the Grabow-Nipperwiese area. Three days later I was assigned as clerk to SS-Hauptsturmführer Huncke [the CO of SS-Jagdverband Mitte]. The headquarters was at Niederkränig. I remained here until the evacuation of the [Mitte] Battalion, which the Russians didn't notice till later. In the meantime, Huncke had been wounded, and SS-Hauptsturmführer Fucker took over. We then moved over Angermunde to the Zehden bridgehead.*

Himmler's call to Lieutenant Colonel Otto Skorzeny requesting him to move his commandos to the Oder put the forces into motion that would decide the fate of so many and prove terminal for many of Skorzeny's men. But death was something many had become accustomed to after nearly six years of war.

"Skorzeny," said Himmler, "I want you to establish a bridgehead at Schwedt as soon as possible. I am giving you command of the defenses there, in anticipation of our great counteroffensive."

Skorzeny was dumbfounded. Did Himmler say counteroffensive? With what?

Himmler continued, "You will take every soldier that you can scrape up and move to the Oder at once."

"A front-line command!" thought Skorzeny. So Germany no longer had any need for specialized troops. "The end must be very close indeed," he reflected.

Himmler ended his conversation, "I want you to move out no later than this afternoon. Mind you, Skorzeny, the situation is very critical."

"Well," Skorzeny thought to himself, "the Reich's Heini [a derogatory term used behind Himmler's back, which he detested] knew that much!" He was not given much time. He called Milius and ordered his battalion moved out as soon as possible. He also directed his SS Jagdverband Nordwest and the reforming Mitte commando group to be ready to move out right away. From 5:00 P.M. that same day, Himmler's staff bombarded Skorzeny's headquarters at Friedenthal with intermittent phone calls, inquiring if Skorzeny had left yet. "Have you started yet?" "Why haven't you left yet?" The following day, Skorzeny finally set

SS-FALLSCHIRMJÄGER BATTAILLON 600 AND BATTLES NORTH OF BERLIN JANUARY TO MAY 1945

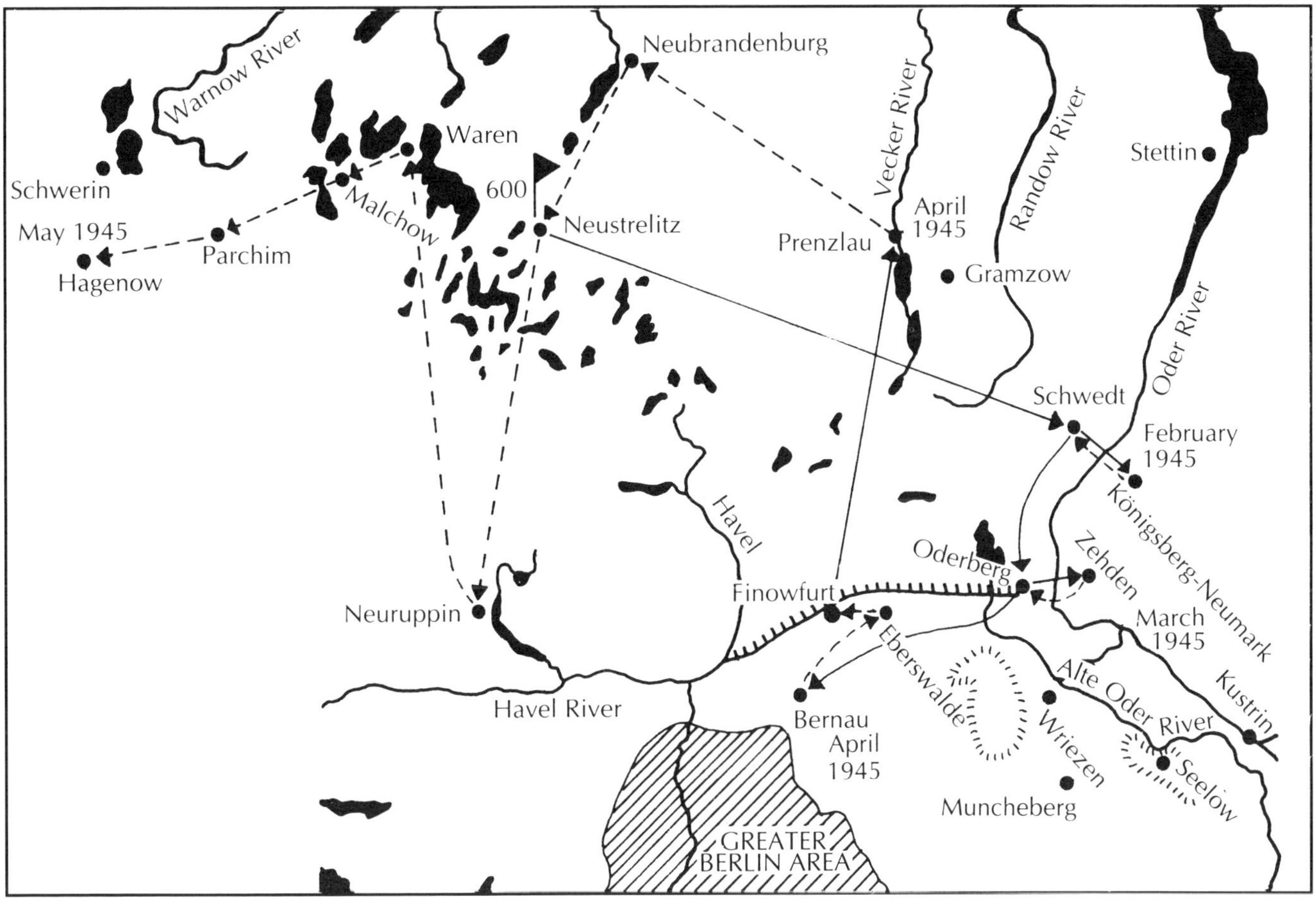

SS-Rottenführer Werner Schrödl on the Oder River heights, January 1945. Snow still covers much of the countryside in this photo (see background). Photo: Werner Schrödl.

out for the Oder, bringing with him the cream of Germany's commando troops. The convoy contained not only Germans but also soldiers from just about every country in Europe. Heading out, his elite force drove east around 5:00 A.M. in the direction of the rising sun.

The advance guard that reached Schwedt-an-der-Oder first was the Ardennes parachute company of Milius' SS Parachute Battalion. The following account was given by SS-Untersturmführer Hans-Joachim Draeger, who was in charge of this advance guard.

> *My first task before leaving was to send telegrams to those comrades on leave, requesting their return. I inquired about the enemy situation at Schwedt. As an answer, I was told nothing definite. It appeared that there were Russian tanks roaming around Schwedt, but more info would have to be brought in by battle reconnaissance.*
>
> *We were to begin our trip at five o'clock in the morning of January 31st. I called the group leaders and the drivers to plan our transport route. I chose the path through Neustrelitz-Strelitz-Alt-Fürstensee-Wokuhl-Lychen-Prenzlau-Gramzow-Schwedt.*
>
> *It was a driving winter day with approximately minus ten to minus fifteen degrees centigrade. All of the drivers had written down the roads to be taken and studied their maps. We then started. Shortly before Alt-Strelitz, I missed the rear truck, so I ordered the column to halt.*
>
> *There was my U.S. jeep, then came the schwimmwagen [amphibious German jeep], then the two motorcycles and sidecars, and one truck. The second truck had disappeared. I sent a sidecar back to look for the missing MTW truck, but it failed to locate it.*

Later I found out that this truck had lost its way at Neustrelitz and had taken the road to the north instead of the one to the south.

Why the driver made such a mistake, I never found out. Now my combat mission was clear. With fifteen men less (the men on the missing truck), I headed for Schwedt. The streets were icy. Shortly after passing Prenzlau, the driver of the remaining truck used his brakes, and immediately the truck swerved to the left and fell off the street, ending up with the right front wheel and the motor block against a tree. The left front wheel was badly bent out. We had a few injured. Immediately, the schwimmwagen, the jeep, and the two motorcycles were made available for the wounded to be brought back to Prenzlau.

This caused a delay of a few short hours, since we had to wait for the vehicles to return from delivering the injured men to the field hospital at Prenzlau. I told those who were lightly wounded to wait on the Chausee Street because the battalion would pass by there in a few hours, and they could be taken care of by the battalion doctor then.

When the vehicles returned, I continued on with our journey to Schwedt. "The combat mission must be accomplished," I kept thinking to myself—just as it had been taught to me at the SS Officer Academy at Braunschweig. In the schwimmwagen, jeep, and two side-cars, we loaded all of the munitions and machine guns, explosives, and Panzerfausts that we were able to.

On the way we ran into individual groups of refugees. Their faces were definitely brightened as they noticed that someone was still heading east. The streets became emptier. At Schwedter Forst, about eight kilometers before Schwedt, a single vehicle came towards us. It stopped, and an air force engineer officer stepped out and asked us where we were going. "To Schwedt," I said. Did I not know that there were Russians there? I said that I knew that, and that was the reason why we were going there.

He then said that behind him there were no more German units. He said he was the last one out from his landing strip. Well, I opined that he could join us, and we could take Schwedt back into our hands. Then he would have some new tasks, but he only wanted to go west. We continued on a little bit more attentive, like a reconnaissance group.

One motorcycle-sidecar would move on ahead of the column, while we waited for any enemy activity against it. Once the clear sign was given, the column would move ahead to the point vehicle. We repeated this procedure until we lost track of the number of times and followed this procedure until we reached our destination.

We reached Schwedt in that way. Schwedt was empty with little sign of life. With caution we crossed the bridge and secured the eastern approach. The crew of one of the motorcycles remained on the west bank to cover our withdrawal, should the occasion arise. It was about noontime when other units began to arrive. Skorzeny now showed up.

Skorzeny had showed up just a few hours after the advance guard. On his way to Schwedt, he had learned that the Soviets were still some miles east of the town. Upon arrival, he managed to find the local commander, an aging officer of the reserve who was better suited for retirement. Skorzeny asked him about his troops and what defenses there were. To Skorzeny's horror, the officer related that, apart from about 150 or so NCOs and officer cadets in an engineering course, he only had about 500 aging men of the Volkssturm and youngsters of the Hitler Youth—and they had not yet been called together!

Himmler's promise to Skorzeny about taking command of Schwedt and "its defenses" was a bit optimistic! There were *no* defenses whatsoever and damn few troops! "I'm sure," Skorzeny was heard to say, "that in the Führer-Bunker, hanging beautifully on a situation map, is the inscription "Fortress Schwedt" carefully written down. What nonsense!"

How would he be able to hold onto Schwedt once the Red Army began to attack in earnest? Skorzeny began to contemplate his options very carefully. He ordered immediate combat status reports from his units and requested that reconnaissance missions be prepared by each battalion. He then sat down and began to ponder his dilemma.

It eventually dawned on him that an almost endless supply of men were available to him; he would gather every straggler who was making his way west through Schwedt. He took five minutes to count the number of men who seemed to be moving west without any apparent purpose but to flee. He counted almost forty. There were his reserves! Soon, his parachute and commando companies had men on the eastern bank of the Oder River bridge on picket duty. These pickets directed any stragglers to Schwedt where Skorzeny's men gathered them up into companies and began the job of retraining them into combat units.

In this way, Skorzeny began to collect his rag-tag army and give the term "Fortress Schwedt" real meaning. His able officers and NCOs turned those defeated dregs into a viable fighting force.

Skorzeny set up his headquarters at the castle in Schwedt and requested that the town Nazi Party chief meet with him. Rousted from bed, the party chief met with the commando leader that same day. Skorzeny got to the point immediately. Every man who could carry a gun was to join the Schwedt Volkssturm. Every woman and old person who could carry a shovel were to be brought out to dig antitank ditches and other earthworks with which to defend their city.

When asked how he should accomplish this, Skorzeny's patience ran out, "Go out yourself with a pick and shovel, and the people will follow you!" The discomposed chief muttered an acceptance of his tasks and hurried his overweight body away from Skorzeny. The best of Germany had met its worst.

Within three days, Skorzeny had managed to scrape up the equivalent of two complete battalions for defending the inner ring defenses, which were slowly taking shape under his guidance. Skorzeny's requests for artillery support were repeatedly denied by Himmler, who argued that what few pieces remained were earmarked for the divisions of Army Group Vistula.

Skorzeny, not the type to accept a fait accompli, sent out patrols into the surrounding countryside. His men found fifteen 75mm antitank guns in a factory some thirty miles south of Schwedt, which the German army High Command had written off when it came within range of the Red Army artillery guns. He also obtained about half a dozen flak guns, which he ingeniously mounted on eight-ton tracked prime movers. In an army dump near Frankfurt-an-der-Oder, his men got as many "Hitler's Sagens" (MG42s) as they could carry.

To match the Red Army's growing attacks and probes, Skorzeny increased his reconnaissance missions. Nightly, the commandos and paratroopers would sally forth from their bridgehead, returning with a few Russian prisoners, most of whom were heavily intoxicated, and some much-needed information about Red Army intentions.

SS-Untersturmführer Hans-Joachim Draeger draws us a good picture of the reconnaissance patrols that Skorzeny ordered during the first week in February.

I reported my strength when the Battalion arrived. I received orders to secure the southern street on the eastern bank and set up my headquarters at Hohenkranig. Skorzeny took me to his headquarters so that I would know where to report.

In Schwedt, to the south, east of the Oder, there were no Russians to be seen. Our Battalion kept arriving so our outlying positions around Hohenkranig were reinforced with men returning from leave. We moved through Niedersaaten and secured positions south and southeast.

On the 4th of February I was called to Skorzeny's headquarters. I was given the task of making reconnaissance patrols from the bridgehead to the southeast because no one had a clear picture of the enemy situation in that direction.

An SS-Untersturmführer of the SS-Jagdverband Nordwest left with me. His task was to reconnoiter in a northeasterly direction. I left on my mission on February 5th, with my jeep and the schwimmwagen. We were only eight men. We traveled by road from Hohenkranig to about three kilometers before Peetzig, where we then took a forest route and found our way by a side street at Peetzig.

We then took the road to Zehden, which we reached via Altenkirchen-Gross Mantel. By Altenkirchen we saw Russian tanks driving at that time on the highway to Zehden. They were apparently T-34 tanks. We counted thirty-six pieces. There were also Russian trucks.

Cautiously, we pulled back to Königsberg-Neumarkt. In Königsberg we saw SS-Obersturmführer Mäcki Markus, whom I greeted and reported my observations. Then we returned to Schwedt to report directly to Skorzeny at his headquarters.

We were ordered that on the following day, we were to make a deeper reconnaissance in the same direction. This recon mission lasted until February 9th. Toward Peetzig, the enemy was nowhere to be seen. We then traveled to Reisdorf, southwest of Königsberg on February 7th and practically stumbled into a Russian armored group.

We stopped, pulled off to the right and to the rear, then pulled to the left and forward so quickly that we were able to get away. Three enemy tanks followed us. Our forward outpost on the Elisen Height [Height Number 63] was in platoon strength as we had left when we went out on patrol.

When we passed

SS-Obersturmführer Droste at the Schwedt Bridgehead February 1945 (left).

SS-Obersturmführer Markus at the Zehden bridgehead March 1945 (below). Photos: Munin Verlag.

SCHWEDT BRIDGEHEAD 1945

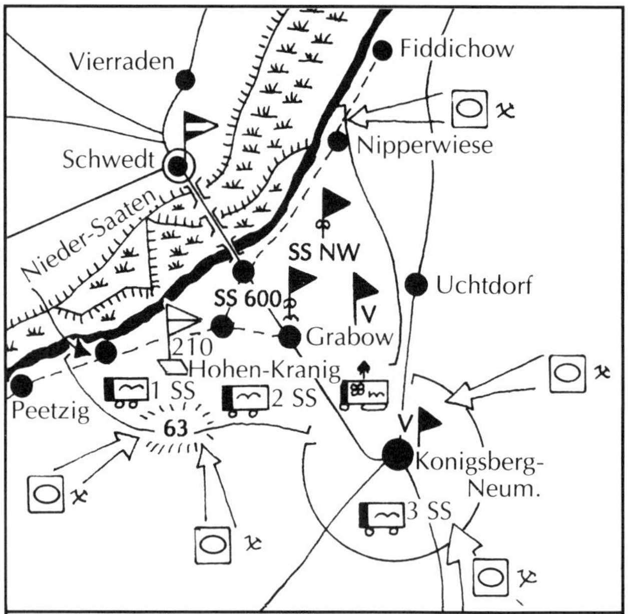

Schwedt/Oder Bridgehead
1 February – 3 March 1945

- Sonderverband Skorzeny
- Motorized Infantry Battalion
- Parachute Unit
- 1st Co./600th Parachute Battalion
- 88mm Flaks Mounted on Eight-Ton Half-Tracks
- 210 Assault-Gun Brigade 210
- Height 63
- Soviet Tank Forces
- V Volkssturm Battalion
- NW SS Nordwest
- SS600 600th Parachute Battalion
- SCHWEDT
- Town/Village
- Marshes
- Man-Made Canal
- Bridge

MAPPING SYMBOLS USED IN THE TEXT

- Tank/*Panzer*
- Mechanized/*Panzergrenadier*
- Assault Gun/*Sturmgeschütz*
- Tank Destroyer/*Jagdpanzer*
- Self-Propelled Antitank
- Flamethrower Tank

Motorized

- Motorized Infantry
- Flak
- HQ Headquarters
- Artillery
- Rocket
- Antitank
- Infantry
- Engineer/*Pionier*
- Cavalry
- Bicycle/*Radfahrjäger*

Unit Size

I I	Battalion
I I I	Regiment
X	Brigade
XX	Division
XXX	Corps
XXXX	Army
XXXXX	Front/Army Group

GENERAL REFERENCE MAP OF SCHWEDT AND ZEHDEN BATTLE AREAS

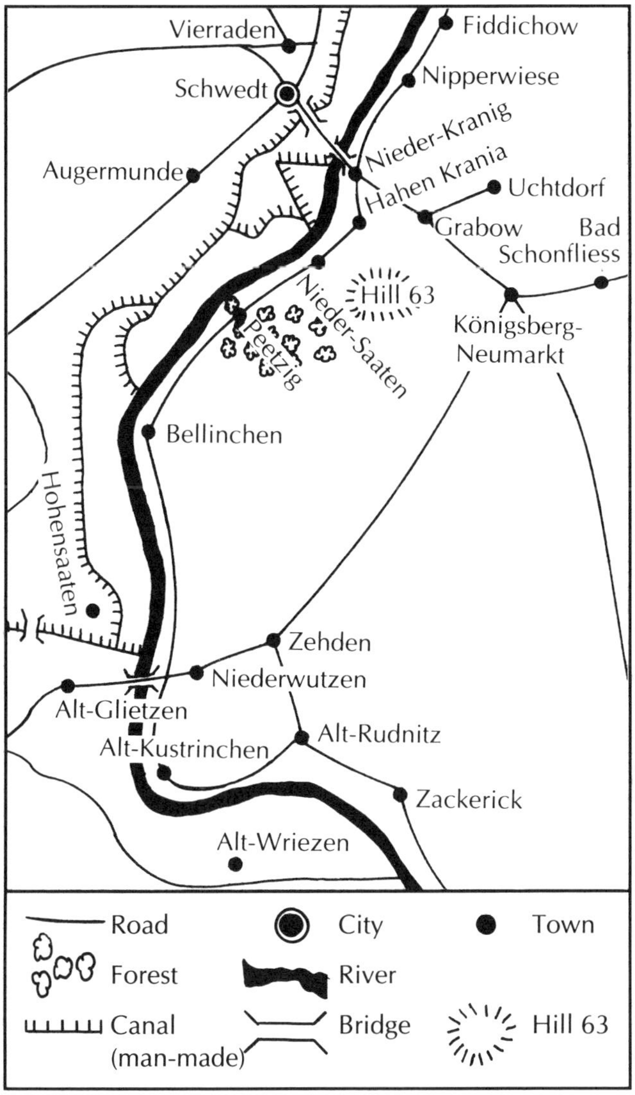

them they told us that three enemy tanks had appeared at the edge of the woods but had then stopped there and pulled back. On Monday, we went out to those woods and managed to find the cottage of the local forester and his wife. We took them with us, much to their relief. The local area inhabitants had left them out in the forest when they had evacuated the area.

After our "rescue mission" in the Peetziger Forest, we ran into an isolated farmhouse where we found many stragglers from shattered units. They were of the opinion that the war was over for them. They only wanted to eat and sleep. We sent them back towards Schwedt.

In the southern part of the Peetziger Forest, we attacked a village on foot that was occupied by the Russians. They discovered us early and fired a few shots from the Maxim-Gorki [machine-gun].

It was not our job to risk getting involved in a firefight where we might have wounded or dead. We were to reconnoiter—to see and to report. All in all, we got the impression that the Russians were trying to cross the Oder River by Zehden, to the south of our Schwedt bridgehead.

Against our bridgehead, they only seemed to be employing light forces. A definite offensive against Schwedt with assembly areas, could not be found. Probably they were located near Zehden.

My reconnaissance activity ended on February 8th, and I returned to my 1st Company, which had temporarily been led by SS-Obersturmführer Scheu. There was also an Unterscharführer Scheu in the company. To differentiate, we called them "Ober-Scheu" and "Unter-Scheu" respectively.

In Königsberg, I heard that SS-Obersturmführer Mäcki Markus had come up before a court-martial and of his execution. On the 6th of February, as we moved on the Königsberg-Schwedt highway, we passed by a long column of female concentration camp

Paratroop insignia for the army and Waffen-SS units.

inmates, being watched by an old Volkssturm man. It made a ghostly impression. Never before then had I seen such a thing.

Skorzeny's command continued to be augmented from various sources. He received a Volkssturm battalion that had been raised in Hamburg. Skorzeny later wrote that this particular battalion, made up of ex-Communist dock workers, "were to fight like devils against the Soviets." Another Volkssturm battalion was raised at Königsberg-Neumarkt for the defense of that town.

Hermann Göring, the Luftwaffe's obese chief, got into the act by calling up Skorzeny and asking, "How's it going?"

"I could do with some more troops," Skorzeny replied. The next day, 600 men of a newly formed battalion of the Hermann Göring Panzergrenadier-Division Number 2 arrived at Schwedt. To the dismay of its youthful, inexperienced commander, Skorzeny disbanded the unit. He used the ex-pilots and ground crewmen-turned-infantrymen as replacements in other, more experienced formations. Being exposed to combat veterans who took them under their wings, the Luftwaffe men soon became first-class fighters.

Other dribs and drabs continued to increase Skorzeny's command. A cavalry officer rode up to Skorzeny one day from out of nowhere and asked, "Can you use us, Colonel?" Skorzeny's reply was a positive yes. He had his own personal cavalry—an entire cavalry company, all of them veterans! Even a Rumanian-SS regiment appeared. Within a little over a week, Skorzeny had amassed a force of some 15,000 men. After his withdrawal from Schwedt, the majority of these men would remain in the area as Infanterie-Division Schwedt. The Rumanian-SS regiment was eventually attached to SS-General Felix Steiner's 3rd (Germanic)-SS Tank Corps, as a corps-unit (Waffen-Grenadier-Regiment der-SS 103 [rumanische Nr. 1]).

The first real armored support Skorzeny received was an assault gun brigade, which he managed to acquire from Himmler. He had pleaded for some armored support at the beginning of February, but it wasn't until 8 February that the assault gun brigade finally appeared at the Schwedt bridgehead.

Major Langel was the commander of the 210th Assault-Gun Brigade. He and the rest of his unit had been without a job because they had lost all of their assault guns to enemy action. On 1 February, he received his new orders.

He was to take his gunless brigade to Angermunde, where they were to await the arrival of their new sturmgeschütz guns. Leaving Stettin on the same day, Major Langel reached Angermunde (about twenty-two kilometers southwest of Schwedt) on 2 February.

On 3 February, the promised assault guns arrived—thirty-one brand-new STGIVs straight from the factory. They had not even been painted. This assault-gun brigade figured prominently in the fierce defensive fighting that marked the beginning of the second week in the

HEADQUARTERS STAFF BATTERY

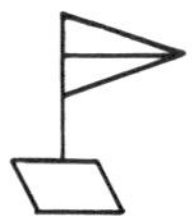

Commanding Officer: Major Langel
Brigade Adjutant: Oberleutnant Althoff
Ordnance Officer: Leutnants Apprich and Ingendaay
Paymaster: Oberzahlmeister Schobe
Troop Doctor: Oberleutnant Dr. Carl
Technical Officer: Oberleutnant Latzel

Platoon Leader:
Oberleutnant Ahrendt

1st Battery
C.O. –1st Lt. Gedeck
P.L. – 1st Lt. Randzio
2nd Lt. Hanstein

2nd Battery
C.O. –1st Lt. Schmeing-Engberding
P.L. – 2nd Lt. Uhlig
2nd Lt. Knechtel

3rd Battery
C.O. –Capt. Vincon
P.L. – 2nd Lt. Kohler
2nd Lt. Naumann

5th Battery
C.O. –1st Lt. Lindemann
P.L. – 1st Lt. Bachmann

Reconnaissance Officer:
1st Lt. Wiloth

Supply Column
210

Schwedt bridgehead. The order of battle for the 210th Assault-Gun Brigade was as illustrated above.

Fully outfitted by 1945 standards, Major Langel received new orders from Third Panzer Army Headquarters to proceed to the town of Schwedt and join Skorzeny's command. Langel departed on 6 February, arriving at Schwedt two days later. Skorzeny immediately deployed Major Langel's assault-gun brigade across the Oder River, spreading his batteries between Niedersaaten and Königsberg-Neumarkt. The armored support could not have arrived at a more convenient time, for on 9 February, the Red Army launched a coordinated attack. The brunt of the assault centered on the towns of Grabow and Hausberg.

Before the Germans could halt the assault, almost the entire SS Parachute Battalion was involved. During this engagement, the 3rd Battery of Langel's brigade, under the command of Hauptmann Vincon, managed to halt a Soviet armored assault in the sector being defended by 2nd Parachute Company. Working in conjunction with the tank-hunter teams of the SS paratroopers, the two platoons of 3rd Battery, led by Lieutenants Kohler and Naumann, managed to destroy thirteen enemy tanks of the Josef Stalin variety. These steel monsters carried a massive 122mm cannon and had 4.1 inches of frontal armor plating, 3.1 inches of side armor, and 2.5 inches of rear plating. In comparison, the average frontal armor of German tanks was only 3.5 inches. .SS-Untersturmführer Draeger was at Grabow when the Soviet attack was launched. He gives us his description of the battle.

The 1st Company still had its headquarters at Hohenkränig. The street to Saathen was secured for about a kilometer. On the 9th of February, we were sent towards Grabow and took up quarters in the empty farmhouses. The

Last barrier before Berlin: the Oder River. SS SMG squad on picket duty. In the background, a destroyed rail bridge. In the foreground, a Panzerfaust antitank rocket lies ready for use. Photo: Bundesarchiv.

Battalion Headquarters was in the basement of the Gutshauses [Good House Inn] at Grabow. There one could find Dr. Hermann and [Hauptsturmführer] Milius."

On the 9th of February, we were able to create a good lunch from the food left in the empty houses, when the Russians suddenly began to fire on Grabow. They hit us with artillery and rocket fire, although they were not using the rockets called "Stalin's Organs." The rockets must have been captured [German] Naval Nebelwerfers as I had never before experienced such a terrible bombardment.

Now we sat in a well-constructed basement and felt secure. We knew that airplane bombs could fall through the roof of a house, artillery shells could explode on the roofs and on the immediate floor. Around three o'clock there was an alarm, and I had to go to the Battalion Headquarters. I ran over to the Gutshauses basement during a lull in the barrage.

Milius told me that the Russians appeared to have broken through and were on the march from Königsberg to Grabow. The lst Company immediately had to get out to the south of Grabow, to the right of the highway going towards Königsberg. There we were to go into positions in order to halt the Russian breakthrough.

Obersturmführer Scheu and I gave our company the orders. We were just 118 men. I don't know for sure if Scheu was already there or whether he came into our positions on the open field. I know I informed my men. The lst Platoon was about 300 meters to the right of the highway to Königsberg. The 2nd Platoon was a further 300 meters from 1st Platoon. I was with the 3rd Platoon immediately behind the 1st, also about 300 meters from the highway.

We met with no enemy units. I think it was here when Obersturmführer Scheu came and ordered us to dig in and hold. Some old Volkssturm men now came and helped us to dig our positions. We spent the night in the open. It was around minus 16° centigrade and a light snow covered the ground, which was frozen to about forty centimeters deep.

We remained in our "positions" the entire night. Out in front and to the left in a forest area, we heard firing during the day on the 11th of February. The unit in the forest was either the 1st or 2nd Motorcycle Company. On February 11th, we were ordered to relieve those men in the forest, on the left side of the highway to Königsberg.

This movement took place around 11:00 P.M., and it certainly took till around midnight since every individual man in the forest had to be relieved foxhole by foxhole. The Russians lay about thirty-five paces in front of our positions in the forest. We could make out some Russian words and the glow of their cigarettes.

It was time to pay attention. No one was allowed to sleep. Someone had given me a large tablet of Pervitin, which I divided among the men. The night remained quiet. On the 12th of February, at the break of morning, the Russians attacked with mortars.

It is uncomfortable to be in a

DEFENSE OF GRABOW BY 1ST COMPANY
9 FEBRUARY 1945

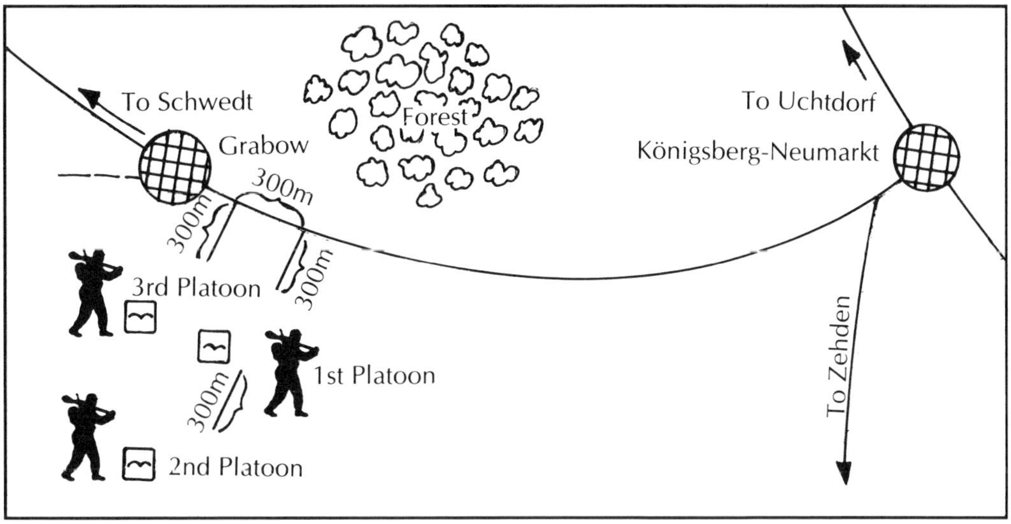

forest during a mortar attack. The grenades all explode in the branches of the trees, sending shrapnel on everyone below. I was glad to have my helmet! In spite of it all, we had many casualties from headwounds. The Russians now attacked with infantry, but we beat them back.

Like Indians, they would creep up to our positions, suddenly rise up and yell their "Hurray!," but by now we were seasoned soldiers. The average number of years of service for most of the men in our company was four and a half years!

It was hard to believe that it was now the sixth year of the war. Russian attacks didn't unnerve us. Never let it be said that we were just 118 men. Each one could fire six well-aimed rounds per minute! That means that every minute we were in a firing position, capable of bringing down 600 Russians. I mean more than a battalion in strength a minute. Even the Red Army couldn't send that many men every minute!

You may even laugh, but we even had occasions to use the Panzerfaust. The stream of fire from the rear of the Panzerfaust in the thick of close combat, gave one ample room. [The backblast of this hand-held antitank weapon could kill you.]

Now the Russians were filling the highway with their tanks. I left our forward positions with my company troop leader. Each of us had a Panzerfaust. We crept up like Indians. Suddenly, I heard the "dumpf!" sound of a Panzerfaust being fired. My company troop leader had fired his weapon. I could see the projectile fly over the tank. "Aha!" I thought, "he aimed for eighty meters, and I imagined it to be sixty meters." I aimed and fired my Panzerfaust. My calculation proved right as my Panzerfaust struck home.

The treads on the side of the turret were damaged. The tank now couldn't move anymore. We crept back and reported to the Battalion. Due to the defensive fighting in the morning, we had incurred heavy losses. From my old lst Platoon of 1st Company, my best machine-gun team of Bernhard Blanke and Siegfried Treu had both fallen in battle.

Blanke, a Mecklenburg farmer with nine years of service with the Verfugungstruppe and Waffen-SS,

Dirlewanger vehicle identification shield (top).

Dirlewanger collar tab (bottom). Illustrations: Antonio Munoz.

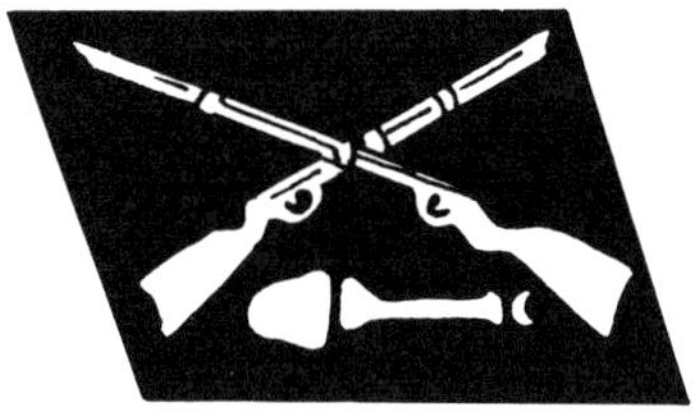

liked to fire his machine gun from the hip better than anyone else could lying down. During the Ardennes offensive he once said to me, "Lieutenant, when things get rotten, come calmly to me. Nothing ever happens to me. I've already been through a lot and stood up to much. I was in the Knochensturm, that lost lot!"

By that he meant he was in the penal unit, which had two crossed bones on their collars [and had] cleared minefields, blown passages.[1]

"Nothing happens to me." Here on the Schwedt bridgehead, a bullet had found its way under his left arm and into his heart. So also, Siegfried Treu gave his life. This man was a tall 1.9 meters in height, red-haired carpenter from Silesia.

From my 1st Platoon many men fell, too many! Walter Birke, one of the best sharpshooters that I have ever known, remained with me until the afternoon of February 12th. The Russians were now coming with a new tank, the "Stalin." We wanted to damage these colussuses and after capturing them, tow them away.

Again and again I attempted with my company troop leader, a Hauptscharführer whose name I don't recall, to come on up on a Stalin tank. This time the Russians were paying attention and took submachine-gun fire, which forced us to withdraw. We went around and attempted once more from the other side of the highway. I took Walter Birke along for security.

We moved into the woods forward, toward the right side of the highway but did not find a good position from which to get a good shot at the tank. I ordered us back. Just before the edge of the woods, in the direction of the flat field, I heard a rifle shot. I turned around. My security had fired from the road ditch with a telescopic rifle at a Russian who was working on a disabled T-34 tank. As I turned my head to see, I noticed how the turret of the Stalin tank was slowly turning in our direction

I next saw a flash from the tank's cannon, heard the explosion, saw fire directly near me and then fell backwards. I suddenly realized that at last I had been hit. I wanted to spring up and look for cover, but it didn't work. My legs would not hold me, so I pulled myself behind a small ground swell and tested my legs with my hands. I found them full of blood and thought, "Shit! Now I'll lose my legs!" The shell, an explosive grenade from the tank, flew 300 paces through the forest and exploded on a tree behind me.

The effect of the wooden splinters was so great that all three of us went flying and landed on our noses, even though we were about forty paces apart! We called out to our company. A

[1] Knochensturm literally translates as "assault bones." There was no SS unit with insignia depicting crossed bones. I believe that the crossed-bones collar tab mentioned here may be the crossed stick grenades emblem of the Dirlewanger-SS penal formation.

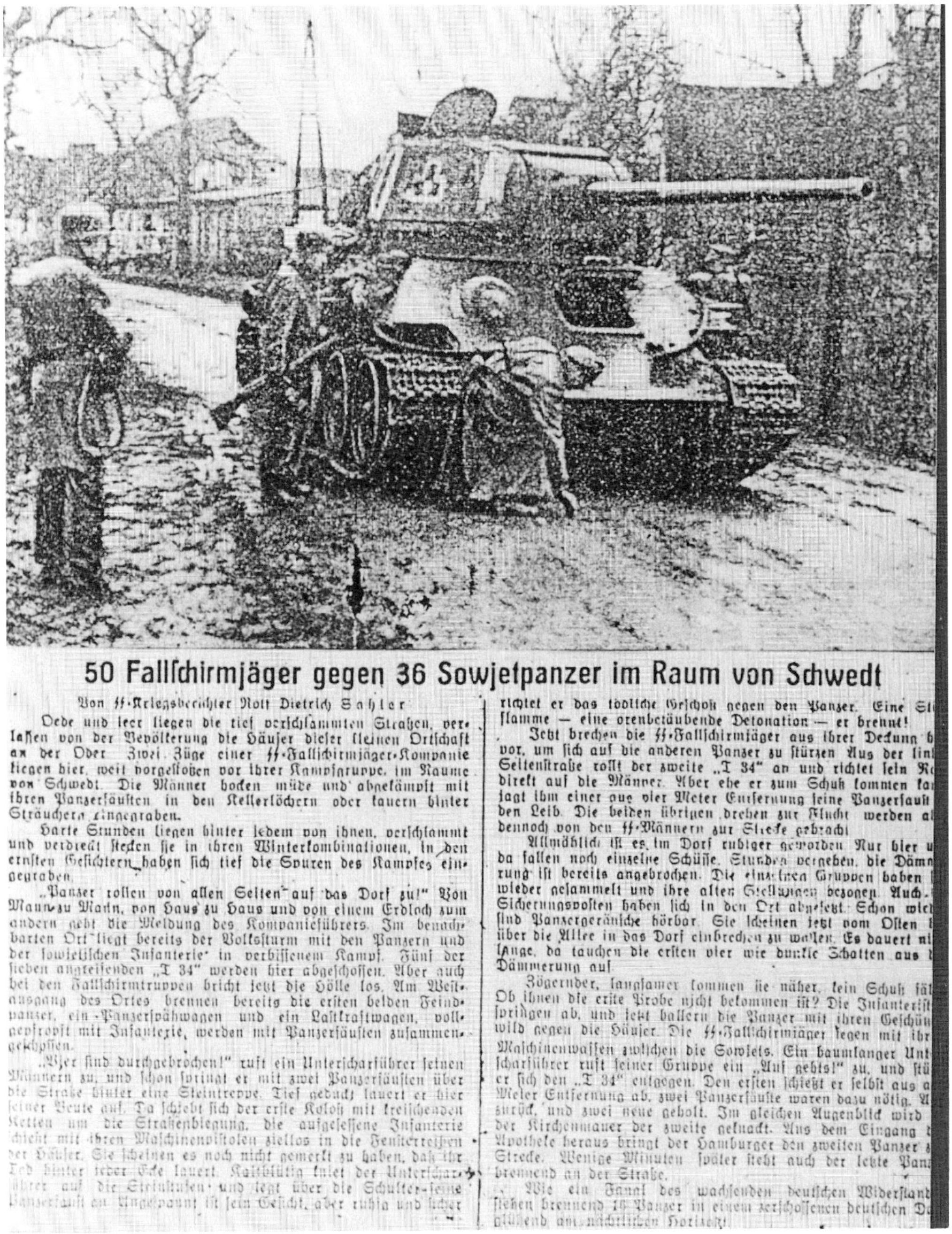
50 Fallschirmjäger gegen 36 Sowjetpanzer im Raum von Schwedt

Von SS-Kriegsberichter Rolf Dietrich Sahler

Oede und leer liegen die tief verschlammten Straßen, verlassen von der Bevölkerung die Häuser dieser kleinen Ortschaft an der Oder. Zwei Züge einer SS-Fallschirmjäger-Kompanie liegen hier, weit vorgestoßen vor ihrer Kampfgruppe, im Raume von Schwedt. Die Männer hocken müde und abgekämpft mit ihren Panzerfäusten in den Kellerlöchern oder lauern hinter Sträuchern eingegraben.

Harte Stunden liegen hinter jedem von ihnen, verschlammt und verdreckt stecken sie in ihren Winterkombinationen, in den ernsten Gesichtern haben sich tief die Spuren des Kampfes eingegraben.

„Panzer rollen von allen Seiten auf das Dorf zu!" Von Mann zu Mann, von Haus zu Haus und von einem Erdloch zum andern geht die Meldung des Kompanieführers. Im benachbarten Ort liegt bereits der Volkssturm mit den Panzern und der sowjetischen Infanterie in verbissenem Kampf. Fünf der sieben angreifenden „T 34" werden hier abgeschossen. Aber auch bei den Fallschirmtruppen bricht jetzt die Hölle los. Am Westausgang des Ortes brennen bereits die ersten beiden Feindpanzer, ein Panzerspähwagen und ein Lastkraftwagen, vollgepfropft mit Infanterie, werden mit Panzerfäusten zusammengeschossen.

„Wir sind durchgebrochen!" ruft ein Unterscharführer seinen Männern zu, und schon springt er mit zwei Panzerfäusten über die Straße hinter eine Steintreppe. Tief geduckt lauert er hier seiner Beute auf. Da schiebt sich der erste Koloß mit kreischenden Ketten um die Straßenbiegung, die aufgesessene Infanterie schießt mit ihren Maschinenpistolen ziellos in die Fensterreihen der Häuser. Sie scheinen es noch nicht gemerkt zu haben, daß ihr Tod hinter jeder Ecke lauert. Kaltblütig kniet der Unterscharführer auf die Steinstufen und legt über die Schulter seine Panzerfaust an. Angespannt ist sein Gesicht, aber ruhig und sicher richtet er das tödliche Geschoß gegen den Panzer. Eine St[illegible] flamme — eine ohrenbetäubende Detonation — er brennt!

Jetzt brechen die SS-Fallschirmjäger aus ihrer Deckung h[illegible] vor, um sich auf die anderen Panzer zu stürzen. Aus der lin[illegible] Seitenstraße rollt der zweite „T 34" an und richtet sein R[illegible] direkt auf die Männer. Aber ehe er zum Schuß kommen ka[illegible] jagt ihm einer aus vier Meter Entfernung seine Panzerfaust [illegible] den Leib. Die beiden übrigen drehen zur Flucht, werden a[illegible] dennoch von den SS-Männern zur Strecke gebracht.

Allmählich ist es im Dorf ruhiger geworden. Nur hier u[illegible] da fallen noch einzelne Schüsse. Stunden vergehen, die Dämm[illegible] rung ist bereits angebrochen. Die einzelnen Gruppen haben [illegible] wieder gesammelt und ihre alten Stellungen bezogen. Auch [illegible] Sicherungsposten haben sich in den Ort abgesetzt. Schon wie[illegible] sind Panzergeräusche hörbar. Sie scheinen jetzt vom Osten [illegible] über die Allee in das Dorf einbrechen zu wollen. Es dauert ni[illegible] lange, da tauchen die ersten vier wie dunkle Schatten aus [illegible] Dämmerung auf.

Zögernder, langsamer kommen sie näher, kein Schuß fäl[illegible] Ob ihnen die erste Probe nicht bekommen ist? Die Infanterist[illegible] springen ab, und jetzt ballern die Panzer mit ihren Geschün[illegible] wild gegen die Häuser. Die SS-Fallschirmjäger legen mit ihr[illegible] Maschinenwaffen zwischen die Sowjets. Ein baumlanger Unt[illegible] scharführer ruft seiner Gruppe ein „Auf gehts!" zu, und stü[illegible] er sich den „T 34" entgegen. Den ersten schießt er selbst aus a[illegible] Meter Entfernung ab, zwei Panzerfäuste waren dazu nötig. [illegible] zurück, und zwei neue geholt. Im gleichen Augenblick wird [illegible] der Kirchenmauer der zweite geknackt. Aus dem Eingang [illegible] Apotheke heraus bringt der Hamburger den zweiten Panzer [illegible] Strecke. Wenige Minuten später steht auch der letzte Pan[illegible] brennend an der Straße.

Wie ein Fanal des wachsenden deutschen Widerstan[illegible] stehen brennend 16 Panzer in einem zerschossenen deutschen D[illegible] glühend am nächtlichen Horizont.

Photograph of a German wartime newspaper article. The story described the battle for Grabow. The large caption reads: "50 Paratroopers Against 36 Soviet Tanks Near Schwedt." The article was written as reported by SS War Correspondent Rolf Dietrich Sahler. The photograph shows three SS paratroopers inspecting a Soviet T34/85, which they had just knocked out. Photo: Antonio Munoz.

rifleman sprang over. I told him to go to the Lieutenant whose platoon of assault guns was nearby [probably a unit from Major Langel's 210th Assault-Gun Brigade] and request that he bring his vehicle over to our position so that we could be laid on the rear carriage of the assault gun and taken back to Battalion Headquarters at Grabow, so that Dr. Hermann could take care of us.

The rifleman came back bringing the message that the Lieutenant had declined my request because he had to remain in his ambush position to await the attack of the Russian tanks.

Now the Battalion was informed of our situation by radio, as the telephone line was always blown apart. They told us that we would be pulled out when darkness fell.

DEFENSE OF GRABOW BY 1ST COMPANY
12 FEBRUARY 1945

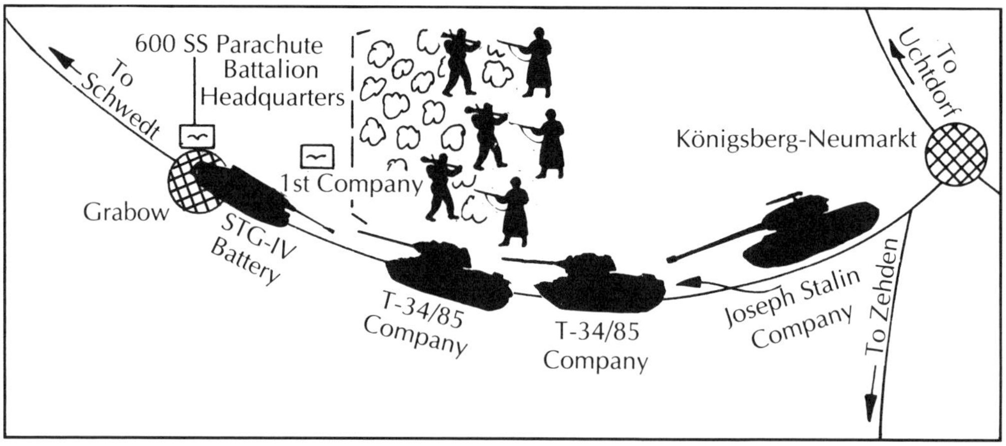

We had been wounded around 2:30 P.M. Until the advent of darkness, about four hours must have passed. I was bleeding from numerous button-hole wounds. It was our misfortune that we were lying on the position which a few days before the Battalion Headquarters had stood the wooden triangles to the right, when one came down the Grabow-Königsberg highway.

A couple of days ago, Jimmy Sauer, an Untersturmführer had been wounded here. This parcel of land was apparently well sighted by the Russian artillerymen. Jimmy Sauer had the habit of doing a one-arm handstand whenever his alcohol intake reached a certain point. There was a story going around that while in the operating room at the hospital where he was having some splinters removed from his lower leg, he woke up from the anesthetic and said, "I shall now show you a real Fallschirmjäger!" Right there and then, it is said that he did his handstand right on the operating table!"

SS-Untersturmführer Draeger's account, though limited to his experience with the lst Company of the SS parachute battalion, nevertheless clearly shows the nature of the fighting at the Schwedt bridgehead. Back on 8 February, the Russians had concentrated numerous artillery units in the village of Johannisgrund. With this artillery assemblage, they began to bombard Grabow. Milius' 2nd Parachute Company and a company of commandos led by Skorzeny himself launched an attack against the Russian batteries.

In spite of serious losses, the paras and commandos moved forward, and the village was reached with the help of Major Langel's assault guns firing as mobile artillery pieces. The company of commandos that Skorzeny employed was apparently made up in part of Russian volunteers. While the Russian batteries withdrew by nightfall, one of them attempted to make a stand at Gutshof. As night arrived, the darkness worked to the Germans' favor.

The Russian infantry defending the artillery battery apparently were fooled into thinking that Skorzeny's Russian volunteers were another group :of Russian infantrymen. Before they realized what was happening, the German commandos and paratroopers were overrunning them. The Russian battery and its accompanying infantrymen were destroyed.

After this engagement, the Russians never again moved their artillery units so close to the front lines, to avoid a repetition of this event.

The following day (9 February) the Russians attempted to break through to Grabow once again. A reconnaissance group from SS-Obersturmführer Markus' 3rd Parachute Company had come under fire when it was ambushed on its way to Bad Schonfliess, a small watering hole just beyond Königsberg-Neumark. In the process, the reconnaissance squad lost two men killed. That morning the squad returned to give their report. Markus contacted Milius, who called Skorzeny. Skorzeny advised launching another reconnaissance immediately and volunteered to come along.

That very afternoon a platoon of paratroopers from 3rd Company accompanied Skorzeny and their company commander (Markus) on another recon mission toward Bad Schonfliess. Leaving their vehicles hidden just at the outskirts of the village, the recon group crept forward. They managed to crawl to the edge of the village, noting the absence of people. Skorzeny ordered the group to slowly make its way toward the other side of town, where the railroad station was located. On their way, they encountered three civilians. The two women seemed to be in a permanent state of shock. The male accompanying them told Skorzeny and Markus that they had been raped numerous times by the Russians who had entered the village two days before.

The man said that more Russians were arriving by rail. He said that he had been spared only because he worked at the rail station. Most of the villagers who could escape had already left, but the ones who had been caught by the Russians had either been raped or killed. Skorzeny ordered the recon group forward toward the railroad station. There they discovered to their horror that the Russians had just unloaded forty to fifty T-34/85s. The monsters were lined up by the railroad tracks in neat rows. The Russians were using the eastern part of the village as a giant staging area for what seemed like an imminent attack on Königsberg. No time could be wasted.

They managed to creep back to the western part of the village without being seen. Skorzeny knew that Königsberg had to be warned and quickly. Before leaving, Markus requested that his men scour the western half of the village for any remaining civilians. Amazingly, they found several villagers who were only too glad to escape. They boarded their vehicles and left the dead village to its fate. They arrived at Königsberg-Neumark a short

SS-Sturmman Werner Schrödl as he appeared in 1945. Schrödl, a member of SS-Fallschirmjäger Bataillon 600, was the recipient of the Iron Cross, 1st Class; the Close Combat Clasp in Bronze; and the Infantry Assault Badge in Silver. Photo: Werner Schrödl.

while later, their eyes full of frustration and hate. The villagers, it seems, did not hesitate to tell their rescuers what they had gone through for two days.

Skorzeny and Markus had time only to partially prepare a welcoming committee for the Soviets when about forty tanks (about two battalions) and three to four infantry battalions fell upon the men of 3rd Company. Skorzeny's men opened up with small arms and submachine-gun fire, the MG-42s cutting deep into the earth-brown ranks of the Red Army infantrymen. The Russians, oblivious to the German fire and most heavily intoxicated, attacked with complete disregard for losses. They moved forward like a pack of hungry wolves with the same battle cry, "Hurray! Hurray, Pobeida!" Their shouts were drowned again and again by the relentless clatter of "Hitler's Saw," mowing them down by the droves.

The Russian tanks entered the action, their guns barking angrily at the success that the German defenders were having against the Red Army infantry. Unterscharführer Erhart Knöfel was a medic with Markus' 3rd Company at Königsberg and had volunteered for the Waffen-SS in the spring of 1941, when he was eighteen years old. His first assignment had been with the Berlin Leibstandarte Adolf Hitler Wachkompanie (Guard Company). During the fall of 1941, Knöfel was transferred to the Leibstandarte Adolf Hitler Division, which was currently fighting in Russia under Army Group South.

In July 1943, he was at a military hospital, after being wounded for the third time. After he fully recuperated, he was sent to medic school and in the fall of 1943 he was posted to SS-Fallschirmjäger Bataillon 500. Knöfel was one of the few SS paratroopers who survived the hell of Drvar in May 1944. His luck failed him slightly when he was wounded for the fourth time at Raseinen, Lithuania, in July 1944. This time his wounds kept him away from his fellow paratroopers between July and October of that year.

He had returned to his battalion and continued to serve as a medic. He was taking care of the men who had been wounded by the Russian assault on Königsberg. He described briefly what transpired next.

> *After our front was pierced on the River Vistula, we were rushed to the Schwedt area and established a bridgehead there. What is there to say? We had some heavy battles around Königsberg-Neumark. When the Russians attacked [on 9 February], we were able to bust up several T-34 tanks, but at a great cost of many dead and wounded paratroopers. We destroyed seven tanks in the first assault. The attack continued, this time the Russians had tied to their tanks women from the area.*

Panzerfausts again began to fly, and tanks began to explode and burn. It was unfortunate about the poor women, but the paratroopers had no choice. The Russians had no qualms about using inhumane methods to complete their mission. Since they had never signed the Geneva Convention Accords on the treatment of captured civilians and soldiers, they felt free to act as they wished.

In spite of a spirited defense by Markus' paratroopers and the local Volkssturm, the German lines were pierced. The Soviets were attacking with too many men and tanks. The small German force at Königsberg just couldn't hold the Red Army tidal wave. Skorzeny ordered a withdrawal to the western half of the town. The battle was concluded when he ordered Markus to withdraw what was left of his 3rd Company and the remnants of the Königsberg Volkssturm to Grabow.

Milius had already been warned, and he had repositioned his remaining companies for the expected follow-up attack. The retreat was conducted in an orderly manner. When they finally withdrew in the early evening, the paratroopers left ten Russian tanks smoldering in and around Königsberg, plus an unspecified number of Red Army infantrymen who would never threaten the Reich again.

Upon arriving at Grabow, Skorzeny found Milius, who had been busy organizing that town's defense while the battle for Königsberg-Neumark had been fought. "You have a very anxious visitor," said Milius. Skorzeny looked up curiously at Milius, who seemed to be doing his best at being mysterious. It didn't take Skorzeny long to understand Milius' behavior.

"I've been waiting here all night, Colonel! I've come to report that all is lost in Königsberg," said the plump uniformed figure, standing rigidly at attention in the safety of the Guteshause basement.

Skorzeny stared hard at the Nazi Party official. "I know," came back Skorzeny's reply. "I was there with your men; where were you?"

As head of the Nazi Party in Königsberg, this man was supposed to lead the town Volkssturm in its defense. Instead, he had fled his command. Skorzeny immediately had him tried and shot.

No sooner had Königsberg been captured than the Russians set out pressing their attacks against Grabow. As early as 9 February, their main effort had been concentrated against Height Number 63, the cornerstone holding Scheu's 2nd and Leifheit's lst Parachute Company. If that hill was taken, the way would be open for a direct assault against Grabow from the south and southwest. Again, it was the timely arrival of Major Langel's 210th Assault-Gun Brigade units that came to the aid of the beleaguered paratroopers. Though ordered to be withdrawn on 10 February for employment at Kustrin, some of its platoons were still in action just east of Grabow on 12 February—a day Untersturmführer Draeger would never forget. Having been wounded at about 2:30 P.M., he and his two assistants, who had also lain wounded for four hours on the battlefield, were finally rescued in the early evening.

Now, while I was lying on the frozen ground, waiting for darkness, another Russian mortar attack fell on our position. I received several more wounds in the upper thigh and my nose. And so it went on until around 6:30 P.M. Then came two motorcycles with sidecars. The sidecars had been converted to carry stretchers, to be used as fast transports for the wounded.

There was only room in the motorcycles for one wounded man a piece, so I told the cyclists to take my men first, and this they did. At the moment you are wounded, you don't feel any pain, probably because of the shock. The real pain begins about one to two hours later. So I was happy to finally get into the hands of our Dr. Wolf Herrmann!

He cut my uniform and underwear from me, wrapped me in bandages and covered me with a blanket. Then I received a shot of morphine and passed out. I was still aware of how I was placed in the ambulance and sent on to the Hauptverband Street in Schwedt. There the Auguste-Viktoria Hospital had been relocated from its original location in Berlin. The service at this hospital was practically as it had been during peacetime. I slumbered during the whole trip to the hospital. As I later heard, they thought I was dead upon our arrival at Schwedt.

I had been carried into the morgue and the battalion had been notified so that I could be picked up the following day for burial. During the night another comrade had died. He was brought into the morgue. It was at that time that, luckily for me I must have murmured, moved, or moaned or somehow made a noise.

Someone noticed me, for one of the orderlies got quite a shock, but had enough sense to call for a doctor. The doctor came, pulled the dead apart from each other, looking for the body which would be warm enough to show signs of life. He finally found me. I was pulled out and sent directly to the operating table. I remember I received a blood transfusion. A strong, middle-aged female doctor had offered a pint or two of her blood.

On the next day, I came to in a room with five wounded. My driver from the Ardennes [offensive] and Schwedt fighting, Pipel Höhne, came into the room. "Untersturmführer, I've come to retrieve your body!" Now, I took that as a great joke, as was usual in the last war years.

It wasn't until after the war that I learned the whole story and that he had been the one selected to pick up my body! For me, the war was over. After eight days at Schwedt,

CHAPTER TEN

Telegram from the High Command of Army Group Weichsel of 28 March 1945:

To
Sturmbannführer Milius
via Pz. A.O., 3

The combat group commanded by Sturmbannführer Milius showed great bravery in the Zehden bridgehead and performed its mission unswervingly, in as much as the circumstances permitted this. I would like to express my special admiration of the bold officers and troops.

High Command Army Group Weichsel
Commander-in-Chief

Signed Heinrici
Colonel-General

I left by train in a cattle-car which was only covered with straw, towards Eberswalde and the 4/532 Field Hospital located therein. It was minus 20° centigrade during the train ride. I had never frozen so much as I did during that train ride. After three weeks at Eberswalde, I went to the Reserve Hospital Saalfield in Saale. Saalfield was bombed by twenty-seven American bombers, even though the rooftops of the large buildings housing the hospital showed the Red Cross sign.

The line held in front of Grabow, though tenuously. Around the second week of February, Skorzeny received a hand-picked company of SS commandos, led by SS-Untersturmführer Schwerdt, who had taken part in the Mussolini rescue mission, as had most of this company. Skorzeny was so glad to see them that he welcomed every man personally. War builds a special bond between comrades that is hard to understand during peacetime.

On 17 February the Reds once again pressed their attack against Skorzeny's bridgehead. This time they launched their typical frontal tank and infantry assault on the village of Nipperwiess, the northernmost defensive position of the Schwedt bridgehead. This village was being defended by units of the SS-Jagdverband Nordwest and a large number of men from the disbanded Luftwaffe battalion that Göring had sent Skorzeny. For the next six days, these men held oft-repeated attacks by the Red Army tank and infantry forces. When Skorzeny, fearing that his men would be totally annihilated, finally ordered a withdrawal from the village, the surrounding area was left littered with the burning hulks of dozens of Soviet tanks.

The German bridgehead felt the full fury of a frustrated enemy. One full-strength Russian tank corps and as many infantrymen threw themselves against the shrinking bridgehead. The Reds even placed "Sovietized" Rumanian troops in the area, apparently aware that somewhere near the bridgehead, the Rumanian-manned Waffen-Grenadier-Regiment der-SS Nr. 1 was operating. Only through a heavy sacrifice in lives was the bridgehead held during the last two weeks of February. Milius' SS Parachute Battalion later departed from the Schwedt area depleted. Losses were especially heavy in platoon and company commanders. The bridgehead was held long enough for the last civilians on the wrong side of the Oder to be evacuated.

An order to withdraw the bridgehead finally arrived, much to the relief of Skorzeny, who—as the situation worsened in the bridgehead—seriously contemplated withdrawing his battered men without orders from higher up. There was no further need for the bridgehead. All that it was doing was draining talented manpower that Germany no longer could afford to squander. The commando and paratroop contingent at Schwedt eventually included all of the following units.

- SS-Jagdverband Mitte under SS-Hauptsturmführer Fucker
- SS-Fallschirmjäger-Bataillon 600, under SS-Sturmbannführer Siegfried Milius
- SS-Sturmkompanie, under SS-Obersturmführer Otto Schwerdt (SS-Obersturmführer Otto Schwerdt was killed in the third week of February, while leading a flank counterattack near Grabow.)
- SS-Scharfschützen-Kompanie, under SS-Obersturmführer Otto Wisler (The Scharfschützen, or sharpshooters company, contained seventy of Skorzeny's best marksmen and was responsible for a large proportion of the Red Army losses at Schwedt.)

SS-Rottenführer Werner Schrödl was an NCO in the lst Parachute Company. He shares his experiences at the Schwedt bridgehead.

We returned to Neustrelitz at the end of January 1945. The SS Parachute Battalion 600 was once again complete. The last units had come back a few days ago—those parts of 1st Company that had been engaged in the Ardennes offensive, were now in Mecklenburg. The Battalion had replenished its weapons stocks and had received a new influx of men from the volunteer reserves.

A new engagement was ahead of us. Somewhere the flames were burning high and, as we had done so often in the past, we were to play "Fire Brigade." The news in those days reported the deep advances of the Red Army into the German territories of East Prussia and Pomerania.

Everywhere, our infantry and tank divisions were fighting for sheer survival—isolated battle groups under heroic leadership, throwing themselves against the Bolshevik flood, allowing countless soldiers and civilians to get away from the murderous Russians.

As the Red Army got closer and closer to the Oder River, the Battalion was transferred to Schwedt an der Oder on January 31st. In the early afternoon, we rolled through the empty, ghostly town. Without rest, we crossed the Oder bridges to the eastern bank. In Grabow, three kilometers east of the river, on the way to Königsberg-Neumark, we occupied the empty houses.

The Battalion battle headquarters was at the Inn. Old trees surrounded the buildings and we put our vehicles underneath them so they wouldn't be detected from the air. Soldiers arriving in the town told us they did not know how far the enemy had advanced or what their positions were. Reconnaissance troops from our battalion cleared things up in the directions of Königsberg, Hohenkrönig, Peetzig, Alt-Kustrinchen and Zehden.

The 3rd Company occupied Königsberg-Neumark and dug in for defense. In those days, nothing much was seen of the Russians. Behind Königsberg-Neumark a few T-34 tanks stalked about, but otherwise it was quiet. All villages were free of the enemy until February 9th.

Apparently, the Russians had done some reconnaissance on February 7th, because paratroopers in Rebdorf, southwest of Königsberg-Neumark, had encountered some very surprised Russians who withdrew after a short fight.

In the following days, it stayed relatively quiet, except more and more civilians arrived at Grabow, fleeing the Russians. They reported horrible stories of crimes against women and children in those occupied villages. The Battalion took advantage of those quiet days to secure their positions in a half-circle around the Oder.

First Company was at Niedersaaten, 2nd Company left and right from the street Grabow-Königsberg-Neumark, 3rd Company around Königsberg-Neumark. The 4th (Heavy) Company was floated around crucial points. In the middle of February, the Russians advanced with strong units towards Grabow. The 3rd Company, after hard battles, withdrew from Königsberg-Neumark to prepared positions next to the 2nd Company. Without heavy antitank weapons, the Battalion succeeded to fight off all of the attacks. In the villages, the paratroopers busted up the T-34 tanks with the Panzerfaust and brought heavy losses on the Russian infantry.

Again and again, tanks tried to get to the Oder bridges, did not succeed, and incurred heavy losses. The paratroopers destroyed in those days over twenty-five T-34s, using the Panzerfaust. The area was not suited for tank warfare. Woods went through heights and valleys, while tanks could only use the roads. Heavy casualties were taken around every village.

At the end of February, the Russians combined strong units and concentrated them around Grabow and the surrounding villages. In order to keep the road to the Oder bridge clear of the enemy in case we needed to withdraw, my company was transferred to this area. One day, as we were about to receive food, looking forward to a warm meal, when someone yelled out, "Russian tanks in our village!"

Forgetting the food and discarding the eating utensils, I ran in a dash behind a wall. There they were! Making a lot of noise as they advanced. I could not see any infantry. They had probably separated from the tanks before the attack. I cursed out loud because I didn't have a panzerfaust in my hands. In my excitement, I never knew there was someone behind me. "Now they can come," I thought, as I fixed my visor.

Detonations and explosions echoed all over the place. "This time they will know; I am in a good position. Anything passing here, I hit from the side," I thought. I just had to be careful not to be seen. The noise of the tanks firing was mixed with that of detonations and hand grenades. Black smoke poured from the farmhouses and burning tanks. In between the humming of the tanks' engines, you could hear the rattle of MGs and MPs. Some of our groups were in a firefight with the enemy infantry.

On the street in front of me, a real big tank rolled from behind a house. I felt the earth shaking underneath my boots. I saw the red flames from the tank's exhaust, as everything happens in a few seconds. I see every detail, just like in a film. Supporting myself against the wall, I slowly followed the movement of the T-34. I aimed a little ahead and with good confidence, pulled the trigger.

I must have hit a tank of diesel fuel because the explosion tore into the engine cover. The tank continued to race ahead, dragging some heavy rope behind him a few meters. Later, someone then finished him up. For hours, we fought from behind house walls, trees, bushes—and gave the Russians a battle to "bend and break." They tried every trick, but we stood our ground. As long as we had ammunition, we never let them get through.

As the night finally approached, we counted seventeen T-34s which had been knocked out with our close-combat weapons. That night, from our dugouts, we heard the grouping of numerous tank divisions of the 2nd Belorussian Front. During the entire night, we listened to the tank motors —fearful rattlings of chains. How far they would go was a question that everyone was thinking, and despite total fatigue, no one closed an eye!

As the Red Army pressed its attacks against the German defenders, the bridgehead became a liability, draining scarce, irreplaceable manpower from the Germans. Someone else at the higher command level had apparently realized this as well. The usefulness of the bridgehead was over. Looking back on the events of the previous four weeks, Skorzeny could hardly believe that he had actually pulled it all off.

The cost had been quite high, however. For more than a month, between 1 February and 3 March, he and his mixed bag of men had held off the best that the Red Army could throw at them. The Russians had been temporarily halted, and Army Group Vistula had been given a grace period that it badly needed to organize. As he later related in his memoirs, Skorzeny and his European Division had formed the bulwark of the defense along the Oder River by Schwedt. "My European Division," he wrote, "since I had men from just about every nation in Europe!"

Sturmbannführer Siegfried Milius was standing on the west bank of the Oder River by the bridge at Schwedt. Scanning the eastern bank with his field glasses, he could see the German rearguard troops fighting off the advance guard of a large Red Army motorized force, trying to cut them off from the bridge. He himself had crossed only minutes before, taking personal charge of the bridge demolition teams.

The anxiety of getting his rear guard across the bridge in time was running through his mind, causing his head to throb at the temple. Though a cold and merciless wind was blowing full blast, he felt hot and was developing a cold sweat. Exhaustion was quickly overtaking him. He knew his men felt no better. The last platoon was running across the expanse of the bridge, the men turning every few seconds to let off a couple of rounds of rifle and submachine-gun fire.

Many did not make it across the bridge. Their bodies lay lifeless as the blood oozed out of big gashes in their uniforms and quickly dried in the wind. No sooner had the last soldier cleared the steel and concrete structure than Milius ordered the sappers to detonate the charges. In a series of thundering explosions, the last bridge near Schwedt blew, sending blocks of concrete and twisted steel girders flying up and landing all over the river and its banks.

It was all over. Another bloody battle had reached its climactic conclusion. The war, however, was far from over. The battle at Schwedt bridgehead had merely been the prelude to the last act of this bloody carnage. No sooner had Milius collected his troops than an order came from headquarters in Berlin, requesting that he supply forty trained paratroopers to the German capital for its defense. Milius complied, and the men loaded on two waiting lorries for the trip south.

This forty-man platoon was under the command of SS-Obersturmführer Schürmann. It was directed to Friedenthal at the last minute. What became of that platoon has been lost in time. As for Milius and the SS paratroopers, they went to Oderberg for much-needed rest. Though the motorized complement of the battalion could carry up to 1,000 men, many trucks left Schwedt empty.

As they departed on that overcast and wintry day, Milius wondered if, after all they had been through, it was humanly possible to go through much worse. He received his answer that same month on a small and otherwise insignificant parcel of land that the men of the SS Parachute Battalion would come to know as the Zehden bridgehead.

The SS Parachute Battalion was stationed at Oderberg at the beginning of March and was scheduled to receive some badly needed reinforcements. The man-made Finow Canal ran in an east-west direction, immediately below Oderberg. At the eastern end, the Finow Canal connects with the Oder River.

The area from the Finow Canal running north to just before the city of Schwedt was manned by a conglomeration of hastily put together forces under the control of Divisional Staff for Special Employment Number 610 (Division z.b.V. 610). Milius' battalion was attached to this division temporarily. The 610th Division had two SS-police regiments, the 8th and 50th, plus a composite regiment made up of two Volkssturm battalions. Lieutenant-General Lendl, the former commander of the German 213 Security Division, was in charge of the 610th.

Lendl ordered Milius to move his battalion to positions just east of Oderberg, by the Oder. That area had been held by a newly formed naval infantry regiment. One paratrooper related just how relieved those naval infantrymen—sailors, until recently—had been to see Milius' men arrive.

The battalion stayed in that area very briefly, for by 9 March, the paratroopers found themselves once again on the east

GERMAN POSITIONS 5 MARCH – 23 APRIL 1945

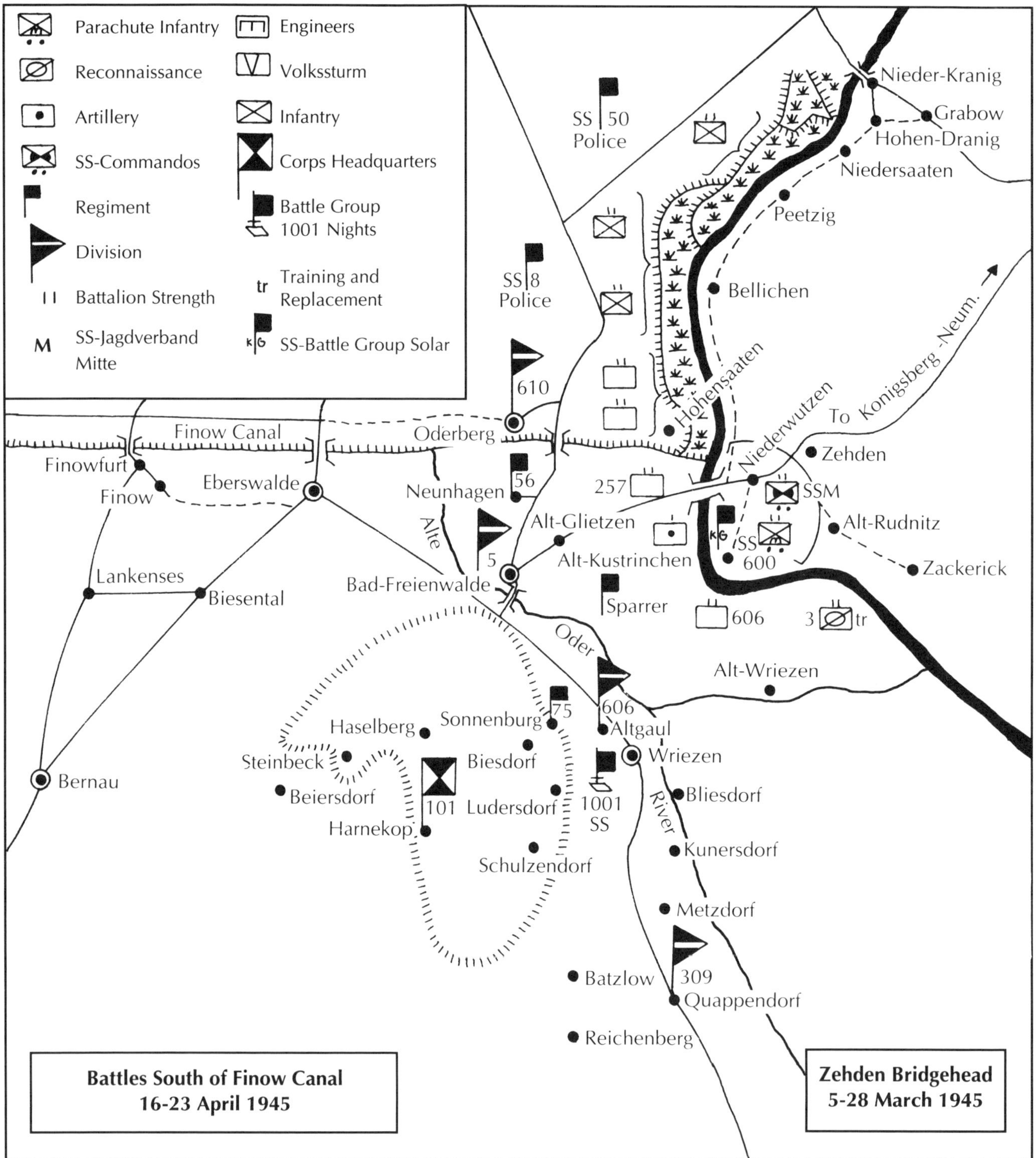

This situation map describes the positions of the German forces between 5 March and 16 April 1945. By 23 April, all German units shown here were pushed back across the Finow Canal or south and west of this area. The 610th Division z.b.V. was transferred to the Stettin area to the north during the first week in April. By 16 April, the Germans had brought up the 1st Marine Division to cover the positions vacated by Division 610 z.b.V.

bank of the Oder River. All along the Oder River front, the Germans were scraping together the last available forces that Germany possessed. A small bridgehead had been established on the east bank of the Oder.

While bridgeheads like these did serve to distract the Red Army somewhat, they were unnecessary affairs. The Germans, however, believed that the Russians would feel compelled to wipe out these bridgeheads before any serious attempt could be made against Berlin. In other words, they thought that no major offensive could be launched by the Red Army while a threat remained on their flanks. As for stopping the Russian probing attacks, it was wiser and cheaper to do it behind the natural barrier of the Oder, instead of giving the Red Army an "open target."

Milius and other officers must have wondered what good would be accomplished by holding onto an area no larger than four square kilometers, an area that was sure to be exposed to intense enemy artillery fire.

Orders were orders, however, so like all professional soldiers, they obeyed. Milius still had some doubts, and his only solace was the bridge that connected the bridgehead to the western bank of the Oder River. If the bridgehead had to be withdrawn, the Germans would have to retreat across that bridge. As his battalion crossed the bridge, he studied the humble but stout structure, a tribute to German engineering. Milius figured that even if the Russians tried to blast the bridge into oblivion, the steel and concrete structure would hold itself together long enough for the German units to withdraw. What he didn't take into consideration was the one factor that always seems to be the decisive point in most plans—human error.

Earlier, during the first week in March, Skorzeny had been given the task of organizing a battle group to occupy the bridgehead by the town of Zehden and hold it for as long as was necessary. His earlier success at the Schwedt bridgehead had worked against him, in that the German army High Command believed Skorzeny could pull off another defense. On paper, all seemed the same: the same planner (Skorzeny) and the same type of units (Milius' battalion and SS-Jagdverband Mitte). However, the two situations were not identical.

Milius was given command of the mixed paratrooper/commando force. The name of the "kampfgruppe" was to be called Solar SS-Jagdverband. Mitte was headed by SS-Hauptsturmführer Fucker. SS-Obersturmführer Ftitz Leifheit, who had once commanded the SS Parachute Training Company and led that unit under Skorzeny's Panzerbrigade 150, was charged with command of the 600th SS Parachute Battalion. Milius' command was augmented by a platoon of armored

Photograph of SS-Sturmann Walter Redecker. Photo: Walter Redecker.

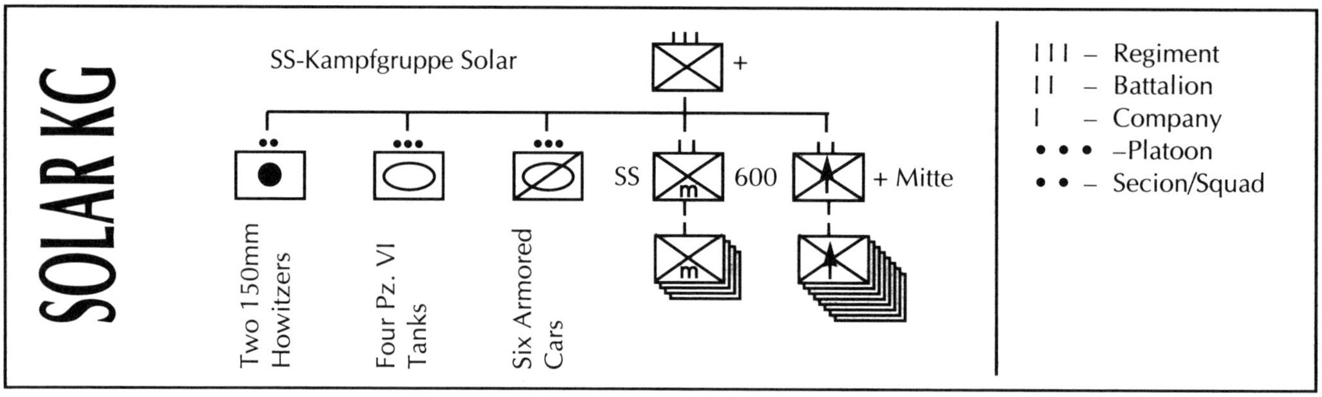

cars from the 3rd Armored Car Training and Replacement Battalion, plus two infantry howitzers from a reserve infantry gun company. SS-Obersturmführer Otto Schwerdt was given command of the reconnaissance platoon and an additional armored support in the form of four aging Panzer Mark IV tanks that happened to be in the area.

At this time, the SS Parachute Battalion could count on about 800 men. SS-Jagdverband Mitte had about the same number. On paper, the battle group looked like the illustration above.

On 7 March, two days before the bulk of SS-Kampfgruppe Solar crossed into the bridgehead, the Russians had launched their first serious assault against the German enclave on the east bank of the Oder. That attack was halted only with the greatest difficulty. Another coordinated and determined Red Army assault might wipe out the bridgehead before the Germans could have time to make a proper defense of it. Milius figured that the Reds wouldn't allow him the time to dig in his battle group, so he decided to throw them off balance by going on the offensive.

Under orders to expand the bridgehead and hoping to throw the Red Army units on the defensive, Milius ordered his parachute battalion, supported by some of his meager armor, on the attack. On Saturday 10 March, just a day after moving into their positions west and southwest of Alt-Rudnitz, the SS paratroopers launched their attack. The paras managed to surprise the Russians and gained some ground, though the assault failed in its goal of capturing Alt-Rudnitz.

The Russians, who were numerous, had regrouped after their initial surprise at this unexpected German offensive and had finally stopped the paratroopers. The attack's success was more psychological than military. The Russian command had been shocked by the audacity of the Germans. They decided to call off any immediate attack plans until they could establish just what new German forces were facing them.

Enemy activity thereafter regressed to heavy and indiscriminate bombing of the German enclave. The Reds seemed to have an endless supply of artillery rounds, so the bombardment lasted for days on end. Eventually, the Russians discovered just what forces were opposing them. They decided that the better part of valor was to soften that new German resistance up before resorting to any further frontal assaults.

Fate stepped in to favor the Russians. The Germans were preparing to demolish the only escape route from the bridgehead, the bridge over the Oder River, when the time came to retreat behind it and blow it up. The officer and troops charged with the demolition were from the army's 257th Engineer Battalion. The officer in charge, a reservist, failed to take the precaution of temporarily disarming the large explosive charges that his men had set on the bridge.

During a heavy artillery barrage directed against the bridge itself, a lucky round hit a main charge, setting off all of the other explosive charges on the bridge. The whole structure went up like a gigantic fireworks display. Milius' men realized instantly what had probably happened. For his negligence, the officer in charge was court-martialed and never seen again. However severe his punishment, it did nothing to correct the terrible predicament of Kampfgruppe Solar.

It took the Russians less than a week to resume their attacks against the small, cut-off German bridgehead. By Friday 16 March, the Reds were ready to strike. That day, the troops of SS-Jagdverband Mitte were the ones to receive the brunt of the assault. In hard, costly fighting, the Russians managed to gain five to six kilometers, but the SS commandos halted the advance and managed to regain some of the ground that had been captured by the Reds. The Red Army attack had just been halted right in front of Niederwutzen.

Once again the German bridgehead was subjected to an intense and murderous artillery barrage. Against this seemingly infinite number of enemy guns and rounds, Milius' small artillery contingent was a speck of dust—more of a spiritual defiance than a physical threat to the Red Army forces besieging the group. To make matters worse, there was a severe shortage of artillery guns and rounds on the entire German battle front. What few German batteries there were lying on the western bank of the Oder near the bridgehead were forbidden from using any rounds in protecting the doomed bridgehead. So critical was the German artillery shortage that when the Russians launched their attack on Berlin, some batteries quickly ran out of live rounds and began to fire training rounds marked: "BLANKS! For Instructional Purposes Only!" The German gunners simply had nothing else to throw at the Russians.

The number of German casualties unable to be evacuated began to rise. The added burden caused confusion and apprehension within the German bridgehead. The men wondered if they would get out of this kessel (pocket). By far, the greatest cause of casualties was the abundant artillery. On Sunday 25 March, the Russians launched another major attack. This assault was the beginning of the end for the German bridgehead. Erhard Knöfel, the SS paratrooper/medic in Milius' battalion, describes briefly the events and feelings of the men trapped in the bridgehead.

The whole bridgehead area had been no larger than four kilometers by four kilometers, and a partial enlargement was successful. Later this changed. We had no heavy weapons. The Soviets began to attack steady from 25 to 27 March. They were using well over 500 cannons of all calibers, one tank brigade, two infantry brigades, plus one cavalry division!

The Red Army troops began to make headway by sheer numbers, pushing the German defenders closer and closer to the east bank of the Oder River. Milius described those last days in March as the worst that he and his men had yet experienced. His shrinking bridgehead was covered by thousands of craters, from the massive Russian artillery barrages against the haggard German troops. "The whole area," wrote Milius, "was like a cratered lunar landscape!" With ammunition almost exhausted, Milius knew that the time had come to withdraw from the bridgehead.

The thought that his command would be completely wiped out weighed heavily on his mind. He had already taken over fifty casualties. On 26 March, the Russian attack had carried to the outskirts of Alt-Kustrinchen, where Milius had set up his battle group headquarters. Russian katyusha rockets screamed overhead by the hundreds, slamming hard on the already pulverized town. The basement of the local church was acting as an aid station to the scores of wounded men, as well as being the headquarters of the battle group.

The scene in the basement bunker was apocalyptic—row upon row of dirty, wounded men, bandages bloodied and infected, whose owners just moaned out loud or whimpered incessantly. Their twisted and agonized bodies disturbed some of the orderlies so much, that they risked being blown up outside the bunker just to get away from that terrible scene for a brief time. The stench of the emaciated men and the infected wounds permeated through the whole bunker and even reached the surrounding area of the church. The dust that the Russian bombardment kicked up seemed to remain suspended in midair.

An old and rickety table served as the operating area for Dr. Herrmann. More than once, the dilapidated table had to be steadied for fear that it would fall apart. Above, the depleted parachute companies were nearing exhaustion. While Milius' 2nd Company was situated at Alt-Kustrinchen, the lst and 3rd companies were strung out in the surrounding countryside. The 3rd Company was the linchpin of the defense, connecting Milius' paratroopers with the commandos of SS-Jagdverband Mitte. Only with the greatest difficulty had 3rd Company been able to keep the connection intact, and it had been achieved at a great cost in lives. Many more lives would be lost before this battle would be over.

It was 1:00 P.M. on 26 March. The 3rd Company had already thrown back a total of four enemy attacks. Incredibly, the men had actually cherished them, for it was the one time when the Red Army artillerymen would hold their fire for fear of hitting their own advancing infantry. The supply of Panzerfausts was all but exhausted. At the rate they were going, another attack would completely use up the remaining antitank weapons. The thought ran through everyone's mind: what would they use against the Russian tanks once the Panzerfausts and satchel charges were exhausted?

The Soviets didn't give them much time to ponder that question. From their foxholes, the paratroopers could see the silhouettes of a dozen T-34 tanks coming over the ridge, some 1,000 yards away. Moments later, they saw the long rows of brown uniforms, heralding the arrival of two-full strength Russian infantry battalions, following close behind the thirty-five-ton tanks. Already, their katyusha rockets and 152mm cannons were slamming their rounds into the positions of 3rd Company.

Everyone hid in their foxholes and waited for the Russian troops to get closer and their artillery to cease before sticking their heads out of their positions. The company commander ordered his men to hold off firing their antitank weapons until the enemy armor was within half the required distance for an effective shot. He did so, knowing that every round would have to count. This meant, however, that the paratroopers would have to wait until the armored beasts would be almost on top of their positions. Despite the strain this placed on the paratroopers, no one broke and left his trench. Experienced veterans, they all knew that it was far better to have a T-34 run over your trench than to get caught out in the open. You couldn't outrun a T-34, and Russian crews took special pleasure in crushing any hapless German who tried.

Small-arms fire had already begun. The SS machine gunners concentrated their fire at the advancing Russian infantry, between the armored vehicles. It was useless to fire at tanks. It only caught the attention of the tank crews who made a point to run over the foxhole and traverse the tank's tracks over it, back and forth and sideways, until the ground gave in under the weight of the tank, crushing the foolhardy soldier who had fired his rifle at the vehicle.

Panzerfausts began to fly, hitting the steel monsters with such force that they were pulled from the ground a few feet by the sheer impact. Turrets blew off like giant firecrackers, killing their four-man tank crews, and more often than not, the brave paratrooper who had fired so close.

Half a dozen burning wrecks littered the battlefield, but the remaining Russian tanks got through the positions of 3rd Company. There was nothing that 3rd Company could do but radio Milius and inform him of the penetration. The Red Army infantrymen were already swarming over the foxholes and trenches, in spite of heavy casualties. Hand-to-hand combat broke out along the whole line. Milius' 3rd Company, already badly depleted, was cut to pieces. What remained of the company withdrew southwesterly, only to be caught in the open by the enemy tanks that had overrun them earlier. At his headquarters, Milius knew that the time had come to withdraw from Alt-Kustrinchen.

The seriously wounded were sure to die if they were moved, yet death was just as certain if they remained to be caught by the Reds. Milius ordered every platoon to construct makeshift stretchers. A platoon under SS Obersturmführer Wabersich volunteered to cover the withdrawal, almost a death sentence for those who chose to stay behind. SS-Unterscharführer Erhardt Knöfel gives his description here of the last few days in the Zehden bridgehead.

When I think of the bridgehead at Zehden, I see it as an inferno! By noon on March 27th, most of our weapons had been shot up. Only in the night of the 28th did we make our escape. The Oder dams were destroyed, raising the level of water in the Oder River.

When we crossed, we had to navigate 2,000 meters of water! The Oder Bridge was destroyed, so we could not cross. Myself, I took over two and one-half hours to swim to the other side of the river. I placed my submachine gun, medical equipment, shoes, belt, utensils, etc. in a bag which I had to drag along with me during the crossing. I shall never forget the cold waters of the Oder River!

When we finally reached the far shore, helpers put us into beds with warm blankets to help us from freezing. We had many dead comrades. After the war, I visited the local cemetery in the area and recognized many names.

SS-Rottenführer Werner Schrödl had survived the battles at the bridgehead near Schwedt. His testimony below confirms the experiences of other paratroopers.

At the beginning of March, everything at the Schwedt bridgehead was evacuated, after very heavy losses of platoon and company leaders. We withdrew from the bridgehead on March 2nd, 1945. After withdrawing the troops and equipment, the bridges were destroyed by higher authority.

On March 5th, the Battalion, along with SS-Hauptsturmführer Fucker's Jagdverband Mitte, [was] transferred as a battle group to the Zehden bridgehead. The battle group commander was Sturmbannführer Milius, the leader of our SS Parachute Battalion 600. The battle group was under the control of Division 610 z.b.V. of General Lendle.

The Russians got to the Oder everywhere. Besides Zehden, the only other bridgeheads of resistance were at Kustrin, Frankfurt-an-der-Oder, and Langenberg. In the Zehden bridgehead, the SS Parachute Battalion 600 replaced a marine infantry unit which never had training and equipment to combat a Russian attack. I never saw troops clear out as fast as those marine soldiers, who had never seen a Russian infantryman.

Also engaged and under the battle group was a heavy-infantry gun company (150mm) under SS-Hauptsturmführer Reiche. He had two cannons at the bridgehead, four on the western bank of the Oder. There was also an armored car platoon and four Panzer IVs from a battle unit. The companies prepared for defense. Construction units dug trenches into the sands on the heights of Alt-Kustrinchen.

It was quiet for a while. The bitter cold in the last few months and the first days of March gave way to sunny spring weather. The ice on the Oder and Oderbruch rivers melted. Grenade explosions and the appearance of Russian reconnaissance groups signaled the end of our rest. At night, the paratroopers would listen as the Russians dug in.

In the moonlight, they were digging no further than 100 meters from our positions. Flares illuminated the scene, and machine guns rattled. The battle of Zehden had begun! Often the Russians would try and penetrate the Battalion area, but just as often they would be thrown back. The paratroopers didn't give a foot!

Day by day, the increasing artillery and mortar fire showed us that the Russians were being reinforced. In the evening, we hear marching music. Over the loud speaker a traitor from the "National Kommitee Freies Deutschland" (National Committee for a Free Germany) asked us to put down our weapons. They promised us good care, food, and much more. We know too well what to think about those Russian promises and answered with a burst from our machine pistols.

"Traitor!" one yells back, "Miserable pack of rascals." A short click and the speaker is turned off. On March 22nd, we heard in the distance a steady hum. The air began to vibrate, and we began to imagine just what exactly was happening in those hours above the Oder. The hum turned out to be the multithousands of detonations from the artillery bombardment of 1,000 cannons, laying a barrage over Kustrin and its environs.

Russian assault troops of the 8th Guard Army succeeded in breaking through and establishing a bridgehead west of Kustrin. German infantry and tank counterattacks were not successful, mainly because our infantry could not keep up with the tanks since the Russians overpowered our men by sheer weight of numbers.

At Zehden, we did not know what had happened further south, though noise and heavy detonations did not look good to us. The companies were ready. Would all this be a new offensive of the Russians? We had no time to think. Whenever the Russians reached our trenches, it was always a matter of shovels, guns, and the Kapp knife.

Just defeating them, and once again we could hear a renewed attack with their shouting "'Urrah!"

The casualties from mortar fire and the constant attacks by the enemy were substantial. Despite it all, our companies held onto all our dugouts. At the positions of our 3rd Company, the Russians succeeded in advancing up to the trenches. Only with the aid of their rifles, machine pistols, shovels, and Kapp knifes were our comrades able to defend their trenches by a hairline.

Daily we were now seeing Ivan increasing in strength. Bigger units were attacking, supported by hundreds of cannons and mortars. Despite it all, we did not retreat.

On March 25th the Russians gathered against the Battalion an entire tank brigade, cavalry division and two rifle brigades. March 25th brought something never seen before. With 500 can-

nons, Stalin Organs, and medium and heavy mortars, we were subjected to three days of continuous fire.

The bombardment destroyed our trenches and dugouts. The barbed wire and our advanced machine-gun posts were also destroyed. I was relieved at 4:00 A.M. and was just about to go to sleep as the "magic" began. At the first bursts, I jumped out of the bunker because the construction of it was much too weak to withstand such heavy calibers.

Outside it was hell. Round after [round], each shell came crashing down on the sandy ground. The air was full of dust and the smell of gunpowder. I sat somewhere a little below ground and only could wait and hope—hope that I wouldn't get a direct hit and I would once again get out of the mess. In one hour, I was almost deaf. I tried to shut out the bombardment, not to think about anything. My mouth was dry, and my eyes were burning.

All around us was the constant noise of the shells, whining past us and crashing into the ground, bursting. In between the artillery attacks came more Russian attacks. Despite many casualties, we were still able to defend, many times man to man. In the morning of March 27th a fearful howling tore us out of our tiredness. I saw in the distance countless smoke streaks rising towards the sky and threw myself flat down. Stalin organs was the thought that went through my head, as the earth shook from the blasts. Fifty to sixty meters ahead of us rose a mountain of dirt. In a carpet bombing, the next detonations were behind us. Dirt and dust was in the air and all around us. Smoke was rising up from everywhere. We were lucky, terribly lucky. As far as I possibly could see, none of our trenches got hit, but our quarters was destroyed.

Because of heavy losses, the commander of our battle group, Sturmbannführer Milius, ordered a withdrawal over the Oder in the night of 27-28th March 1945. Only now did we find out that the only bridge over the Oder had been destroyed by the Russians. Between 10–11 P.M. the individual companies started to swim through the ice-cold water, under a full moon, only partly taking their weapons.

Non-swimmers found beams and boards to hold on to. For the lightly wounded there were large barn doors. The heavily wounded crossed the river by boat with the engineers. The crossing went very disciplined. The Russians only found out at the very end of it. The 4th Company lost all its guns and mortars.

The following days, the battalion recruited in Oderberg, and was supplemented with volunteers, and rested in quiet. From the original 1,000 paratroopers at the beginning of the battles at Schwedt and at Zehden bridgeheads, on 28 and 29 March, only a little over 200 reported back to Oderberg. The rest were missing, wounded, or dead. The battle fought by the battalion was honored by the High Command of the Wehrmacht. It reported:

> *27 March 1945—The units at the Zehden and Langenberg bridgeheads again held their positions against the Bolshevist attacks.*
>
> *28 March 1945—Our bridgeheads at Zehden and Langenberg held.*
>
> *29 March 1945—Heavy battles were fought today at our Zehden and Langenberg bridgeheads.*

Milius notified his companies and the SS commandos. The commandos still held Niederwutzen, though their positions there were precarious to say the least. This was the situation facing Kampfgruppe Solar on March 26. That Monday, at the height of this fierce defensive fighting, Dr. Josef Goebbels (Hitler's Minister of Propaganda) was to write the following: "Enemy reconnaissance activity against our bridgehead at Zehden and Politz increased."

It was a good thing that the men of the SS Parachute Battalion and SS commandos never found out about Minister Goebbel's comment, for it belittled the sacrifices that they were making.

By 27 March, Milius' command was dangerously close to being overrun. On that day, the new commander of Army Group Vistula, Infantry General Gotthard Heinrici, made a final attempt to relieve the 9,000 trapped German defenders at Kustrin. Once again, due to superior enemy forces, the German counterattack failed.

That same afternoon Generaloberst Heinz Guderian notified Führer Headquarters of the failed attempt. Hitler flew into a fury, lashing out at his "defeatist" generals. A special conference was ordered for the following day, to "find the culprits" of this new failure. Isolated from their own lines and apparently forgotten by their leaders in Berlin, the men of Kampfgruppe Solar continued to die by the droves. The time had come to save as many of them as possible.

By the evening of 27 March, the Zehden bridgehead had shrunk considerably. The average depth was a mere 500 yards from the bank of the Oder River. The wounded who had not been transported to the western bank of the river were lying by the sandy bank on the eastern side. The bank gave very little cover to the Russian artillery barrages, however. A Russian 152mm round would land with such force that it could kill anything within seventy-five feet, burying many with the pounds of dirt that the impact of the shell would kick up. Having done everything possible to hold up the Red Army, Milius' command began to cross the Oder. Thus, they crossed, using every imaginable thing that would float. The remnants of his battle group moved accross the icy waters.

A select number of volunteers again chose to cover the withdrawal. And just as at Alt-Kustrinchen, this mission was a veritable death sentence to the brave men who chose to stay behind. Their chances of escaping were practically nil, but many chose to step forward and volunteer. There was a full moon that Tuesday night of 27 March 1945, which worked against the retreating Germans. From their positions, the Red Army tankers and machine gunners had an unobstructed view of the Germans' last withdrawal. On barn doors, wooden planks, rubber tires, and anything that would float, the Germans attempted to negotiate the river.

Exposed as they were, Milius' men suffered heavy losses while crossing the river. Many men just proved too tired and weak to negotiate the treacherous currents and simply drowned. Others were blown apart in the icy cold water by a Russian tank or artillery round. A total of eighty men died crossing the Oder River on that night. SS Unterscharführer Andreas Meyer remembered that terrible night with vivid clarity. He was attached to the battle group's headquarters company, though he was "on loan" from SS-Jagdverband Mitte.

> *The headquarters of SS-Hauptsturmführer Fucker was located at Alt-Kustrinchen.*[1] *It transpired that the bridge over the Oder River, connecting the bridgehead from the west bank received a direct artillery hit, which landed on the demolition charges which had not yet been prepared. It was in this way that we found ourselves practically on an island. After numerous, repeated attacks against our weakly held posi-*

[1] The headquarters of SS-Fallschirmjäger Bataillon 600, SS-Jagdverband Mitte, and SS-Kampfgruppe Solar were all located at Alt-Kustrinchen.

tions, the Russians began their great attack on March 25th.

The relationship of the enemy forces versus our own stood at about fifteen to one in the Russians' favor. Since the morning of the 25th, a hardly imaginable artillery and mortar barrage had come down on our positions. This fire continued in its intensity for the next two days.

Our positions by then looked like the cratered Moon. Our manpower was decimated. Our lines were gradually pushed back as we awaited at any moment for the enemy storm to finally break us.

At ten o'clock on the evening of March 27th, a Hungarian 1st Lieutenant appeared at the headquarters. He was the commanding officer of a Hungarian police unit. He said that he and his men had lost contact with the front to the north and wanted to pull back his men.

The commanding officer, Siegfried Milius, gave me the task of showing the Lieutenant and his troops where the northern area of the front was, so I left the headquarters with the officer and headed for the trenches. It was a moonlit night. There was a factory in the neighborhood of the destroyed bridge. The factory had been destroyed and set on fire by Russian artillery fire, so we could clearly see how the Russians, like ants, were pouring out of their positions, forming up for yet another attack.

We assumed from this that our positions in the north were no longer held. While we were still marching north, I heard someone loudly call my name. I answered and the runner called to me that I should return immediately to the headquarters.

As I came into the headquarters' bunker, it was already evacuated. The petroleum lamp was still burning and an MG-42 [machine gun] was leaning against the wall. The War Diary lay on the table. I took the diary and the machine gun and made my way to the bank of the Oder, about 100-150 meters away. At this moment, the Russians stormed from the south with a loud "Hurra!" and a terribly loud shell landed into the river bank where our unit was dug in.

That the enemy had attacked yelling, "Hurra!" and not the usual "Hurray!" made us suspect that they were German Seydlitz troops.[2] *Upon reaching the river bank, I asked about Milius. It was said that he had already crossed the Oder River in an assault boat. The senior officer of the battle group present, Korvetten-Kapitan von Behr, was of the opinion that we could not hold out against such an overwhelming force and ordered every man for himself.*[3]

We took off our clothes and jumped into the Oder, which only three weeks before had floating ice! Inspite of that, we did not notice the cold. In the meantime, the Russians were concentrating their artillery and mortar fire on the Oder River itself. Because of this, many comrades were wounded, killed, or drowned.

The river by Oderbruch was noticeably wide, and at the time, I was

[2] The Red Army raised and sponsored a Russian- and German-officered "Anti-Fascist" Division, raised from turncoat German prisoners, who now "professed" their belief in Communism. It was led by German General Seydlitz, thus its name, Seydlitz Division.

[3] This title is a Kriegsmarine (navy) rank and stands for "Corvette Captain." It is most likely that von Behr was either a specialist with the SS commandos or simply a recent officer-recruit from the then-useless navy.

not a very good swimmer, so I would not have had a chance to survive. I thus drove myself further northward, hugging the eastern bank of the river. I then ran into an Untersturmführer, Unterscharführer, and a Rottenführer from our unit—all in the same predicament! By the bank, we again took up arms which we found lying around. These weapons included machine pistols and Panzerfausts.

The four of us now made our way northward, dressed only in our underwear. We had already heard of the horrible acts committed by the Russians during their march through East Prussia. We all had our illusions as to what awaited us members of the Waffen-SS in case we were taken prisoner, so the seriously wounded which we encountered begged us for the "mercy shot."

In the meantime, the Russians in the northern area had stormed forward with only light forces to the banks of the Oder. We, however, were able to overcome these relatively small groups and about 1:45 in the morning [of the twenty-eighth] reached the destroyed bridge.

There we found a machine-gun team, still holding the open on-ramp to the bridge. The engineers had constructed a footbridge across the bridge pilings, way above the water. The machine-gun team leader asked us if they could expect more stragglers. In spite of our negative reply, he said he would still hold the ramp open for another half hour, then pull back his men.

We were thus able to cross here and make our way to the west bank of the river. On the west bank, I ran into Obersturmführer Schwerdt to whom we could give our thanks, in that he had practically saved our lives. His armored vehicle had a quadruple 20mm Flak gun mounted on the chassis, which was used to fire over our heads with tracer rounds at the Russian infantry troops.

I had already known Obersturmführer Schwerdt quite well from our shared actions at Malmedy. On the west bank, I looked for a pair of shoes in an abandoned house, then headed further west on foot, towards Neuenhagen. Meanwhile, the Russian artillery had moved their fire towards the west bank of the Oder River. Around 4:30 in the morning, I reached the supply column. There I met my sergeant-major and my other comrades from the Headquarters' Company.

In the early evening [of the twenty-seventh], they had crossed over, on a large barn door, acting as a raft. Our joy was great! I received a woolen blanket, warm drinks, and a sandwich. It wasn't until five days later, that I was in a proper military attire. In the meantime, I had worn an S.A. [Stormtrooper] uniform.

We then remained approximately ten days in the Oderberg-Bralitz area. The Battalion Headquarters was located at Oderberg, in an inn called Haus Oase [Oasis House Inn]. From here we then moved north to the area between Oderberg and Angermunde. There we received reinforcements from SS-Panzergrenadier-Regiment Nr. 7, which had been pulled back across the East Sea from Hela. In the meantime, Skorzeny's SS-Jagdverband Mitte had been ordered back into the Alpine Fortress [in Austria].[4]

[4] Mr. Meyer stated in a letter to me that only two companies from SS-Jagdverband Mitte ever left the Berlin area for Austria. Skorzeny and his small force of some 250 men had to travel through Czechoslovakia on their way south to avoid the advancing Allied forces. The remnants of the commando group that remained on the Oder River front remained attached to Kampfgruppe Solar until the last days of the war.

SS-Obersturnführer Fritz Leifheit, temporary commander of the 600th SS Parachute Battalion, had this to say about the last battles on the Oder in a letter to Eric Brorup 14 July 1989.

I returned to the SS Parachute Battalion 600 with my company in January 1945. The Battalion had in the meantime been activated for the eastern front. With still another unit we formed from now on a battle group under the leadership of Milius, while I took over the 600th Parachute Battalion.[5]

We were engaged in heavy battles with heavy casualties, especially on two Oder River bridgeheads. On the last one, the Russians cut us off from the bridge, and we had to swim across on 29.03.45 at 9 P.M. The Oder had melted water and was overflowing and very rough. This happened in a very disciplined and orderly manner—no panic, therefore, we had very few casualties.

SS-Hauptscharführer Aussenthal was listed as missing after that. I spent over one hour, under fire, in the water, got to the other shore totally frozen and without strength. I only had my shorts on.

Yes, those were bad times, but wtill it was not the end of it. Newly equipped, we then fought the Russians for another six weeks, northeast of Berlin (Eberswalde, Prenzlau, Neubrandenburg). After maneuvering between the Amis [Americans] and the Russians, we finally dissolved (destroying our weapons and equipment) and disappeared in small groups into the woods.

SS-Sturmbannführer Siegfried Milius had remained by the west bank of the Oder River, once he and his staff had crossed. His efforts had not ceased upon reaching the relative safety of the other side of the river. He had been attending to the wounded during the early morning hours of 28 March 1945. Milius' command had suffered 500 wounded, many of whom belonged to his own battalion. He stayed with them, making sure that they were moved to the rear as quickly as possible. The sudden shift of the Red Army artillery from the east bank of the Oder River to the west bank caught his attention.

The quiet of the eastern bank also confirmed his fear that all was now lost on that side. He knew that it was all over for the German rear-guard troops who had chosen to cover the withdrawal. The sun began to rise. Milius now realized that he hadn't slept in almost forty-eight hours. The weariness and fatigue of several days' fierce fighting suddenly landed on him like a rock. He found it more and more difficult to gather his thoughts. He stumbled a few times as though he were drunk. "No time to pass out," he thought to himself. There were still more things to be done. He forced his body to conform to his will, knowing that his men felt the same exhaustion.

The vehicles which were to take him and his men to Oderberg for several days of rest had by then arrived. He slowly gathered all of the able-bodied men together, for he seemed to be stuck in slow motion. Little by little, the SS paratroopers and commandos fell into some type of order.

It was about then that a telegram ar-

[5] This letter proves that Leifheit took over command of the SS Parachute Battalion while Milius commanded the Solar Battle Group. It does not mention that Milius was wounded in the ankle at Bernau, and that SS-Obersturmbannführer Schrage-Bobet took command of Solar. Leifheit also mentions that the battle group was fomed shortly after he and his 1st Company (formerly the Field Training Company of the parachute battalion) were transferred back to the battalion (which implies January 1945, even though SS-Jagdverband Mitte served alongside Milius' battalion at Schwedt bridgehead in January and at Zehden bridgehead in March, when it was part of the battle group).

rived from the headquarters of Army Group Vistula, via 3rd Panzer Army and Division z.b.V. 610. Milius casually opened the telegram and read the message enclosed, all the while, trying to fight a deep sleep that seemed to be overtaking him. The message seemed to wake him from his trance. His men even noticed the change on his face.

"Men of Kampfgruppe Solar, attention!" Like robots, the ranks stiffened to his call. "I will read to you a message from General Heinrici, the commander of our army group." All ears and mental processes still functioning strained to hear to what Milius had to say as he read the telegram (see telegram at top of next page).

Milius looked across the ranks and noticed that the tired, worn-out bodies were straighter than before. They too had received renewed energy from the telegram. Milius felt very proud of them. It was good that they received this much recognition. He folded the telegram and placed it in his pocket.

SS-Hauptsturmführer Fucker had already organized SS-Jagdverband Mitte's remnants into a marching column. Milius then ordered the battleworthy troops that remained of his parachute battalion onto the waiting vehicles. It didn't take the paratroopers long to mount the waiting lorries. In a flash, all thirty-six of them clambered on board. *Thirty-six!* With his men mounted, Milius signaled the driver forward. He sank down between two SS paratroopers who were fast asleep. He had opted to drive to Oderberg with his men. Oblivious to the noise of the truck engine and the bumpy road, Milius closed his eyes and quickly fell into a comalike sleep.

One of Goebbel's diary entries for 28 March 1945 included this little piece of iniquitous information: "*Stong enemy attacks on our bridgehead at Zehden failed, apart from a minor break-in.*"

What is so amazing about this is that it was not meant for the ears of the German nation. It came from the *private* diary of Hitler's propaganda minister. Either Goebbels had been misinformed about the actual situation at Zehden bridgehead or the great propagandist was fooling himself.

Milius and the survivors of the battle for the Zehden bridgehead spent the first ten days of April recuperating in Oderberg. For all intents and purposes, the battalion had been shattered. Milius had gone into the bridgehead with 800-plus men. When he finally pulled out, he had a total of only thirty-six effectives. His worst fears had come true. Even with the addition of the lightly wounded cases who were beginning to rejoin the battalion, he was hard-pressed to rebuild his command.

After only a few days at Oderberg, the replacements began to trickle in. A special call for volunteers throughout the Waffen-SS had to be instituted to reform the 600th Battalion. The Waffen-SS NCO and officer academies and training schools particularly were bombarded with urgent appeals for volunteers. Many men responded to the call. It was this influx of trained and seasoned veterans and youthful recruits that eventually brought the SS parachute battalion back up to combat efficiency. There were not enough SS cadets and NCOs, however, to completely fill up the empty ranks of the battalion. Men from the Luftwaffe and Kriegsmarine were now absorbed into the battalion.

Milius had little time in which to train the raw recruits, but he was luckier than most commanders since his own battalion had combat-experienced cadre to shear the rough edges off the recruits. During the third week in April, the battle group was moved to an area between Oderberg and Angermunde. There they received additional reinforcements from the 4th SS-Polizei-Panzergrenadier-Division's 7th Armored Infantry Regiment. Before the start of the Red

March 28th

To Sturmbannführer Milius via Pz. A.O.K.3

The combat group commanded by Sturmbannführer Milius showed great bravery in the Zehden Bridgehead and performed its mission unswervingly, inasmuch as the circumstances permitted this. I would like to express my special admiration of the bold officers and troops.

High Command Army
Group Vistula
Commander-in-Chief

Signed Heinrici
Colonel-General[6]

Army offensive on 16 April, Milius' battalion had again reached a fairly strong level of manpower, numbering 800 to 850. The unit, however, was virtually a new formation—only a handful had parachute experience.

With the Russian storm about to break once again, Army Group Vistula desperately sought to give every sector of the front at least a small, mobile force that it hoped would hold off any breaks in the lines. The German situation on the Oder River front was dismal. Colonel-General Heinrici's strongest mobile division possessed no more than seventy-nine armored vehicles. His weakest had only two tanks. Plans were quickly implemented, and "fire brigades" were formed behind the front. Milius' Battle Group Solar contained the 600th Battalion plus the newly reformed SS-Jagverband Ost (which had absorbed, or was absorbed by, the remnant Mitte Jagdverband that remained behind when Skorzeny was ordered to Austria).

Another kampfgruppe formed was Kampfgruppe 1001 Nächte (Battle Group 1001 Nights). Milius was obliged to give up one of his parachute companies for this newly formed battle group. Reluctantly, Milius gave the order for one of his companies to join 1001 Nächte. With their new semiautomatic MP-43s and 44s slung over their shoulders and burdened by the added weight of the Panzerfaust rocket, the men moved onto the waiting trucks and left Oderberg for the area just north of Wriezen, where the new battle group was forming. For six days after the start of the Red Army offensive, this company was separated from the rest of the parachute battalion. It rejoined its parent unit for the defense of Finofurt.

Kampfgruppe 1001 Nächte was commanded by Wehrmacht Major Blancbois, who had seen service with Infantry Regiment Number 5 in Stettin. He had seen service in France and in Russia, where he had led a tank battalion of the 12th Panzer Division and was wounded. He had been awarded the much-coveted Knight's Cross for his actions while leading this tank battalion. From Russia, he had been sent for a period of convalescence, but the crushing Russian offensive in January 1945 had catapulted him from convalescence to the post of Army Group Adjutant for Army Group Vistula.

He was handed the position of kampfgruppe commander, whose armored reserve was to protect the front covered

[6] Quoted from the 1986 Munin Verlag edition of *Fallschirmjäger der Waffen-SS Im Bild* by Siegfried Milius and Adolf Kunzmann, with permission of the authors and the publisher.

by the German 101st Army Corps. Blancbois' command consisted of a mixed bag of men and machines typical of the rag-tag army facing the Russian army. His men were from the Heer, Waffen-SS, and RAD (Reich-Arbeits-Diens, or National Labor Service). The main units were SS-Jagdpanzer-Abteilung 560, Panzeraufklarungs-Abteilung Speer, and the SS parachute company.

SS-Jagdpanzer-Abteilung 560 (SS Tank Destroyer Battalion 560) had recently seen service under the direct disposal of the Reichsführer-SS Headquarters' Staff (Kommandostab Reichsführer-SS) before being posted to 1001 Nächte. Its armored strength comprised eight Sturmgeschütz III assault guns and forty-four Jagdpanzer 38-T Hetzer (Bushwacker) tank destroyers. On 7 April, the 560th SS Tank Destroyer Battalion had the above-named vehicles in various stages of combat readiness.

	Combat Ready	Short Repair	Medium Repair	Long	Total
STG-III	6	–	2	–	8
Hetzer	37	1	6	–	44
Total	43	1	8	–	52

The SS Tank Destroyer Battalion was code-named "Suleika." The first commander was SS-Sturmbannführer Wost; the last commander was Hauptsturmführer Markowz, who took over when SS-Wost was killed. The unit had the following companies: one headquarters, one SS parachute "escort," three tank-destroyer, one assault gun, and one supply.

Headquarters command had a few armored recovery vehicles. The one drawback to SS-Jagdpanzer-Bataillon 560 was the relatively weak armor with which it had to work. The tank destroyer companies were composed of the Hetzer (Bushwhacker) tank destroyer, a slow-moving vehicle with a top speed of only forty-two kilometers per hour.

Although the vehicle's gun, a 75mm PAK 39 L/48, had already been superceded by the far-superior L/70 cannon, it was still useful against most Russian armored vehicles. The one advantage of the Hetzer was its low silhouette, making it hard to hit. compared to the heavy Russian tanks, however, these vehicles were next to useless. We need only look at the Hetzer's tank gun versus that of the Soviet "Joseph Stalin" tank to see.

The assault gun company was made up of the Sturmgeschütz III vehicle, with the 75mm PAK 39 L/48 gun. This vehicle had also been superceded by the STG-IV assault gun, but it could hold its own against most medium Russian vehicles. Since it was turretless, it was at a disadvantage in the attack, but by this stage in the game, the Germans were not attacking, but defending. Although the Russian T-34 could run rings around the slow assault gun, speed was not essential in the defense.

Panzeraufklarung-Abteilung Speer was formed from the militarized members of the RAD. Organized as combat troops, they made a two-company escort battalion to protect the German Armaments Minister, Albert Speer. It contained one motorcycle company made up almost ex-

SS-Unterscharführer Andreas Meyer as he appeared in January 1945. Photo: Andreas Meyer.

BATTLE GROUP 1001 NIGHTS

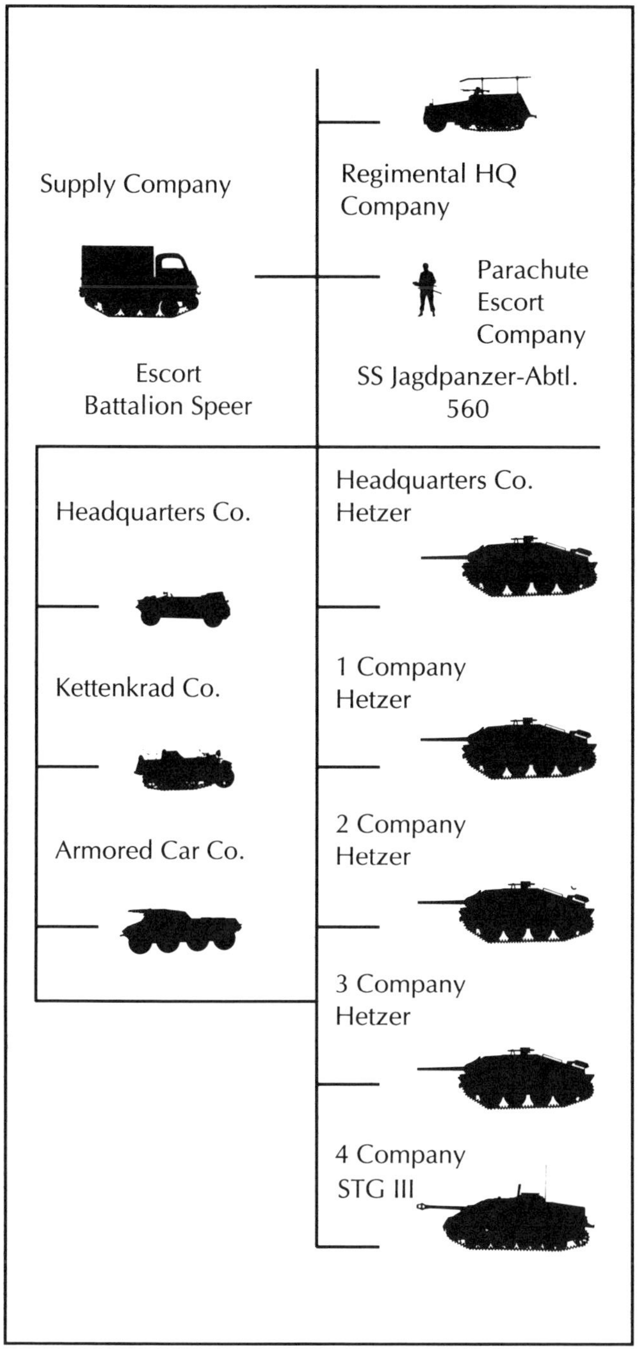

signifying its elite status within the kampfgruppe.

Facing Army Group Vistula on 16 April 1945 was a formidable and sizable Russian force. Opposite the German 9th Army's so-called "divisions," the 1st Belorussian Front had amassed 3,155 tanks and self-propelled guns, distributed mainly in the seven tank and mechanized corps of its front. Seventy-seven Russian rifle divisions comprised the bulk of the 1st Belorussian Front. These figures do not include the dozens of artillery divisions and rocket-launcher regiments that gave this "front" a total of 16,934 barrels of various calibers.

Third Panzer Army was in even worse shape. Against the better judgment of his senior army commanders, Hitler had considerably weakened Army Group Vistula when he ordered the transfer of the 39th Panzer Corps, with its 10th SS-Panzer-Division Frundsberg from the Oder River front on 30 March. He further depleted the army group when he ordered the additional transfer of the 25th Panzer-Division and Führer Grenadier Division on 2 and 3 April. The 2nd Belorussian Front, facing 3rd Panzer Army, could count on 951 tanks and assault guns, plus 6,642 artillery pieces of various calibers. The 25th Panzer Division alone had accounted for sixty-five tanks and forty-three armored personnel carriers (as of 30 March).

When the Red Army finally launched its attack on Berlin, those troops were dearly missed. The battle for Berlin had

clusively of the semi-tracked Kettenkrad motorcycle and another armored car unit, whose vehicles carried the quadruple (four-barreled) 20mm flak gun, as well as half-tracks carrying the short-barreled 75mm L/24 cannon. The SS parachute company, as mentioned earlier, formed the escort company of the battle group,

Right: Comparison of the Jagdpanzer 38-T and Joseph Stalin II.

GUN	PEN	AB-F	AB-S	AB-R	TYPE
75mm L/48	5.0"	2.3"	0.7"	0.7"	JgdPz 38-T
122mm L/43	5.7"	4.1"	3.6"	2.5"	JS-II

GUN: Cannon type
PEN: Penetration of the gun at 500 meters
AB: Armor basis. This is the armor thickness times the function of the slope
AB-F: Frontal armor
AB-S: Side armor
AB-R Rear armor

been decided before it had begun. The battle, after two and one-half brief but furious days, turned into a running withdrawal, with the German forces desperately trying to escape Soviet captivity while the Russian units competed with each other to see who was to reach Berlin first. The German capital attracted Russian units like a magnet.

It is against this backdrop of Germany's Götterdamerung that the final chapter of the history of the SS Parachute Battalion is written. Of the handful of men who served in the parachute battalion and managed to live through the war, their most common statement was that the battles on the Oder River were the toughest they experienced in the war. Though many knew that the war was lost, their sense of duty kept them going. Russian atrocities and the U.S. plan to divide Germany gave the Germans reason for continuing their futile resistance.

Reserve/Strategic

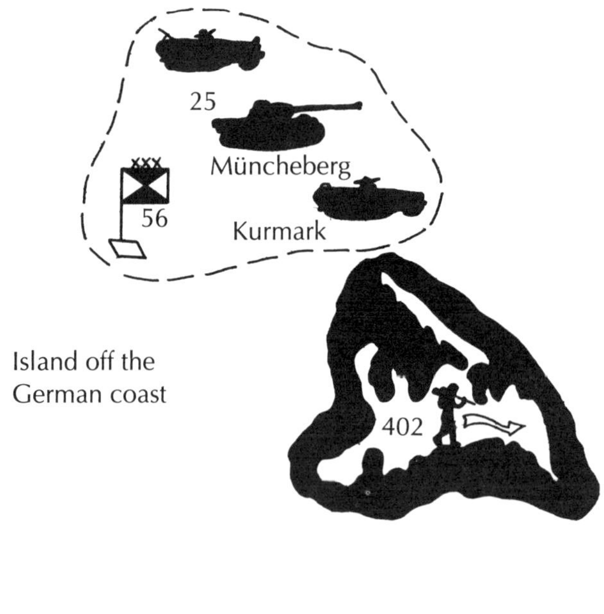

ARMY GROUP VISTULA THE GERMAN SITUATION: APRIL 1945

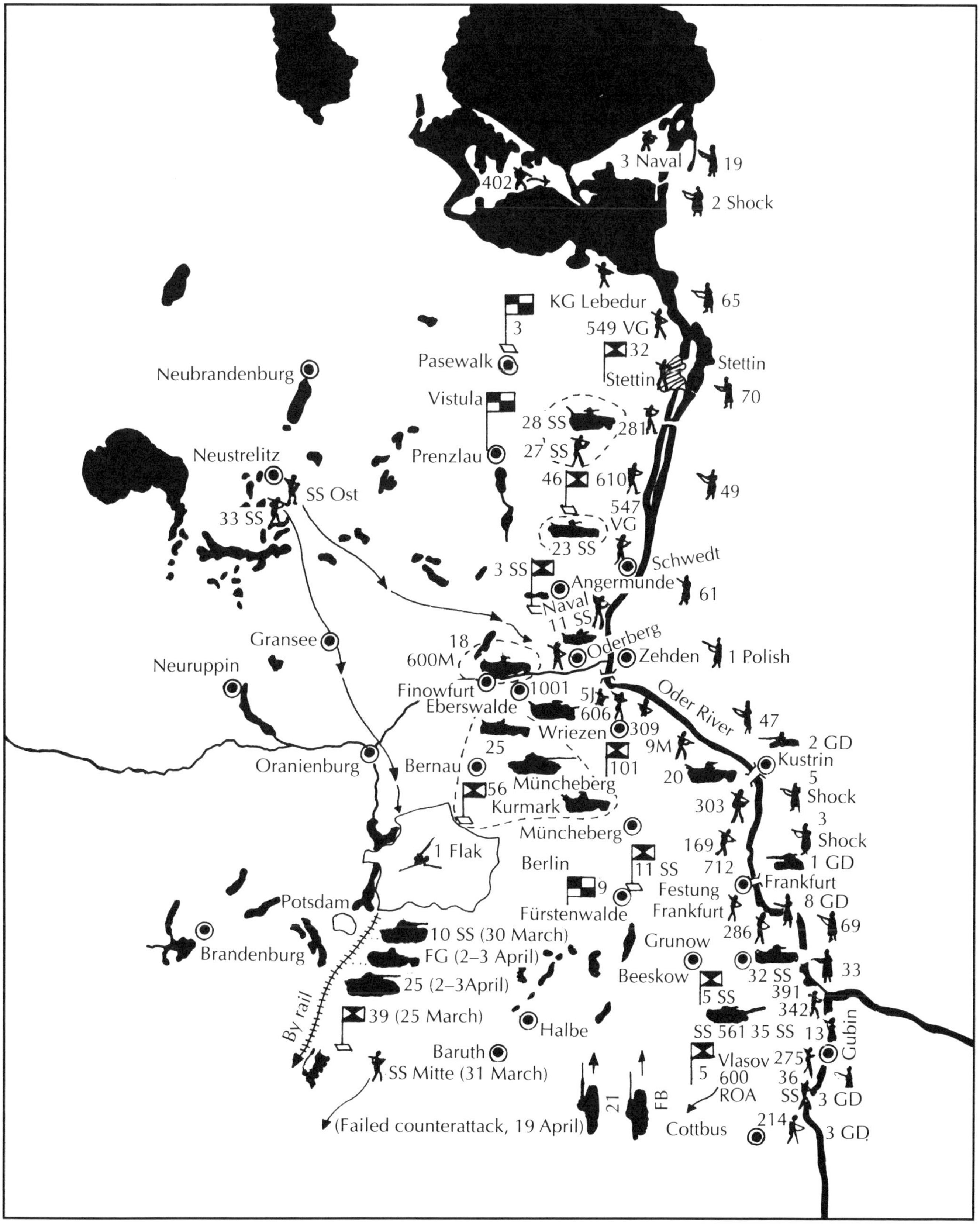

EXPLANATION

Festung Frankfurt:	Fortress garrison of Frankfurt. The German commander at Frankfurt had a total of eighteen infantry battalions for the defense of the city. These battalions were formed from Volkssturm, Hitler Youth, SS, SD, and police units.
Müncheberg:	Müncheberg Panzer Division.
KG Lebedur:	Battle Group von Lebedur. About a brigade in strength, made up of mixed units.
Stettin:	Infantry Division Stettin. Raised and formed in the city of the same name.
SS Wallonien:	28. SS Freiwilligen-Panzergrenadier-Division Wallonien.
SS Langemarck:	27. SS-Freiwilligen-Infanterie-Division Langemarck.
3 Marine Naval:	3. Naval Infantry Division.
549 VG:	549. Volksgrenadier-Division.
SS Ost:	SS-Jagdverband Ost.
600:	SS-Fallschirmjäger-Bataillon 600.
33 SS:	33. Waffen-Grenadier Division der SS Charlemagne. This unit, like the 27th and 28th SS divisions, was a battle group with the strength of a regiment.
1001 Nächte:	Battle Group One Thousand and One Nights. This unit contained SS-Panzerjäger-Abteilung Speer, plus one compnay from SS Para Battalion 600.
600 Vlasov ROA:	Panzergrenadier Division No. 600 (Russian No. 1). Part of the German-sponsored Vlasov army. Withdrew from the area of 275th German Infantry Division in spite of pleas by the commander of the German division.
281:	281. Infanterie-Division. Redesignation of the 281st Security Division.
610:	Division Number 610. Controlled a number of Volkssturm and police units, including Police regiments 50 and 8.
27 SS:	27. SS-Freiwilligen-Grenadier-Division Langemarck.
28 SS:	28. SS-Freiwilligen-Panzergrenadier-Division Nederland.
1 Naval/1Marine:	Referred to by either title, it was formed from German naval personnel.
11 SS:	11. SS-Freiwilligen-Panzergrenadier-Division Nordland.
18:	18. Panzergrenadier-Division.
5J:	5. Jäger-Division. 5th Light Division.
606:	Division Number 606. Controlled numerous Volkssturm, police, and reserve units.
309:	309. Infanterie-Division Gross Berlin.
9:	9. Fallschirmjäger-Division. 9th Parachute Division.
20:	20. Panzergrenadier-Division.
25:	25. Panzergrenadier-Division.
303:	303. Infanterie-Division.
169:	169. Infanterie-Division.
712:	712. Infanterie-Division.
286:	286. Infanterie-Division. Redesignation of the 286th Security Division.
32 SS:	32. SS-Freiwilligen-Panzergrenadier-Division 30 Januar.
391:	391. Infanterie-Division.

342:	342. Infanterie-Division.
275:	275. Infanterie-Division.
214:	214. Infanterie-Division.
402:	Division Number 402. A static formation made up of various stationary units. This division took over for the 3rd Marine Division, which in late April was transferred from its position by the German coast to just north of the Finow Canal, northwest of Eberswalde.
Göring:	Flak-Regiment Göring.
Hermann Göring:	Panzerjagd-Verband Hermann Göring. Tank Destroyer Brigade Hermann Göring.
30. Januar SS:	32. Freiwilligen-Panzergrenadier-Division 30. Januar.
X 10 Volkssturm:	Ten battalions made up of elderly homeguardsmen.
SS-Mohnke:	SS Battalion Mohnke made up of Hitler's bunker guard unit.
15. SS:	SS-Fusilier-Batallion 15. This unit belonged to the shattered 15. Waffen-Grenadier-Division der-SS (lettische Nr. 1). It entered Berlin from the north, just before the Red Army ring of steel completely encircled the German capital.
SS Mitte:	SS-Jagdverband Mittet, which belonged to Otto Skorzeny's SS commando force. Left the Berlin battlefront 31 March for Austria with Skorzeny himself.
Führer Escort:	Panzerdivision Führer Begleit.
Führer Grenadier:	Panzergrenadier-Division Führer Grenadier.
36 SS:	36. Waffen-Grenadier-Division der-SS Dirlewanger. In reality, not more than a reinforced regiment/brigade made up of penal convicts, homosexuals, poachers, Communists, degenerates, etc., from the German prison system. Very unreliable unit.
35 SS:	35. SS-Polizei-Grenadier-Division. Made up of police school units and several combat police regiments.
von Gaudecker:	Divisionsgruppe von Gaudecker. Divisional formation von Gaudecker.

Illustration: Eric Weisman.

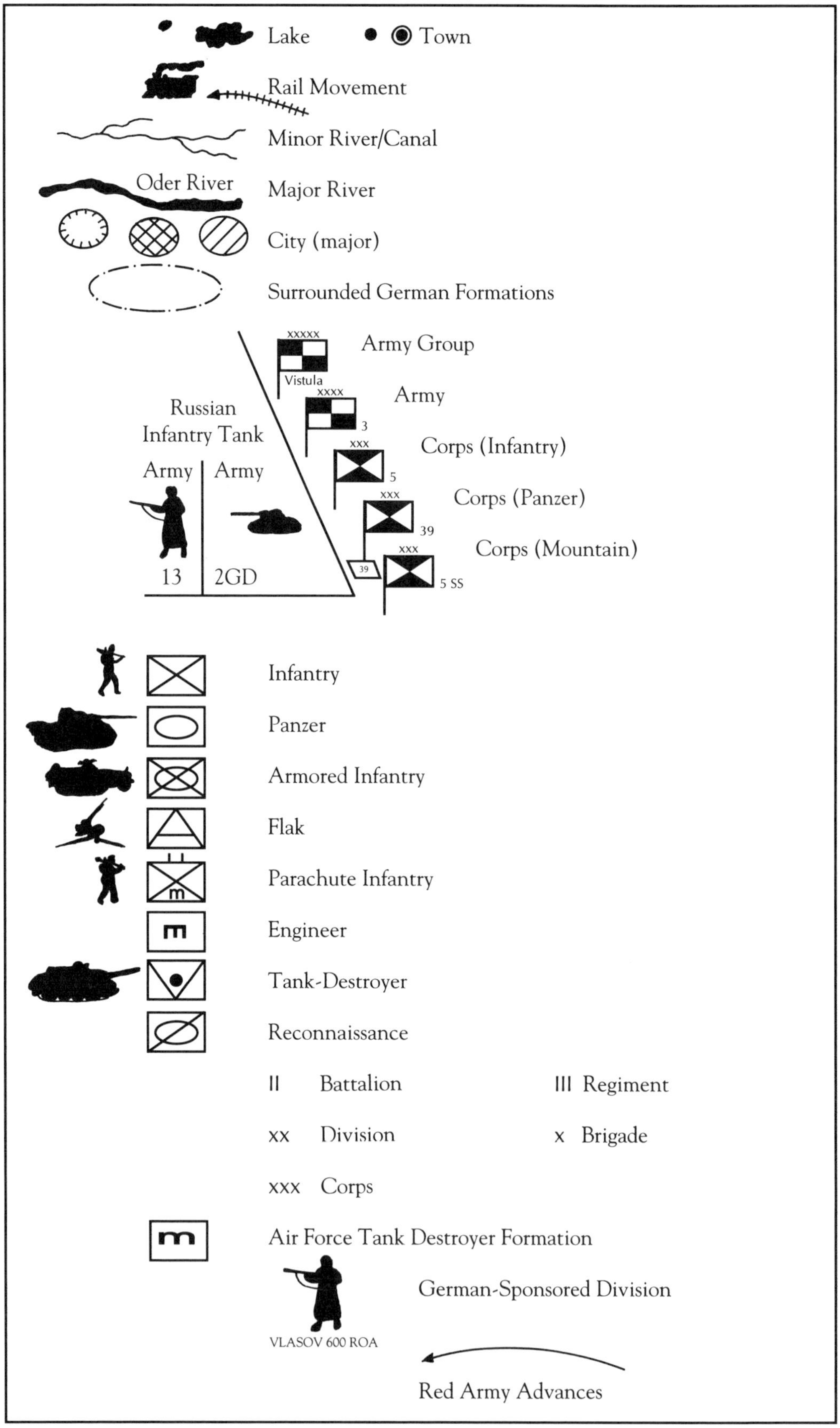
Lake
Town
Rail Movement
Minor River/Canal
Oder River
Major River
City (major)
Surrounded German Formations
Vistula
Army Group
Army
Corps (Infantry)
Corps (Panzer)
Corps (Mountain)
Russian
Infantry Tank
Army
Army
13
2GD
Infantry
Panzer
Armored Infantry
Flak
Parachute Infantry
Engineer
Tank-Destroyer
Reconnaissance
II Battalion
III Regiment
xx Division
x Brigade
xxx Corps
Air Force Tank Destroyer Formation
German-Sponsored Division
VLASOV 600 ROA
Red Army Advances

ARMORED STRENGTH OF ARMY GROUP VISTULA 8 APRIL 1945

Obkdo. der Heeresgruppe Weichsel — H.Qu., 8.4.45
O.Qu./V(Pz)Br.B. Nr.70/45 gKdos.
6 Ausfertigungen — 2. Ausfertigung: I d

Stand der Panzer und Sturmgeschütze am 4 July 1945

Einheit	Art	Pz.Kpfw.,Stug.,Jagdpanzer				
AOK 9			a	b	c	d
Pz.Gren.Div.Kurmark						
Pz.Jg.Abt.Kurmark	Stug.	III	12	–	–	–
	Pz.	IV	2	1	–	–
	Pz.	V	1	1	–	–
	Pz.	VI	–	–	–	–
	Jagdpz.	38	15	1	1	–
Pz.Abt.Brandenburg I/26	Pz.	V	28	1	–	–
Pz.Div.Müncheberg	Pz.	III	1	–	–	–
I.Pz.Abt.29	Pz.	IV	1	–	1	–
	Pz. (L)A	IV	1	–	–	–
	Pz.	V	18	3	–	1
	Pz.	VI	8	–	5	–
	Jagdpz.	IV	1	–	–	–
20.Pz.Grn.Div.						
Pz.Abt.8	Pz.	IV	15	–	1	–
	Pz. (Flak)	IV	3	–	–	–
	Pz. (L)A	IV	16	–	–	–
25.Pz.Grn.Div.						
Pz.Abt.5	Pz.	IV	1	–	–	–
	Pz. (Flak)	IV	2	–	–	–
	Pz. (L)A	IV	6	–	3	–
	Pz.	V	28	3	3	–
Pz.Jg.Abt.25	Stug.	III	30	–	–	–
	Pz. (L)A	IV	12	–	–	–
	Jagdpz.	IV	–	1	–	–
5.Jäger-Div.						
Pz.Jg.Kp.1005	Pz. (L)A	IV	10	–	–	–
	Stu.	III	1	–	1	–
SS-Pz.Abt.502	Pz.	VI	27	2	1	–
Pz.Jg.Abt.Berlin	Stu.	III	8	2	2	–
Pz.Jg.Abt.Döberitz	Stu.	III	17	–	1	–
	Pz.(L)	IV	7	–	2	–
SS-Gren.Div.30.Januar						
SS-Pz.Jg.Abt.32	Stu.	III	20	–	2	–
	Stu.H.	42	8	–	1	–
Fkl.Pz.Abt.303	Stu.	III	2	6	–	–
Sich.Div.391	Pz.	III	2	–	–	–
	Pz.	IV	–	–	1	–

Einheit	Art		Pz.	Kpfw.	Stug.	Jagdpanzer
			a	b	c	d
169.Inf.Div.						
Jagdpz.Kp.1230	Jagdpz.	38	10	-	-	-
Stu.Lehr.Brig.111	Stu.	III	33	-	-	-
	Stu.H	42	9	-	-	-
	Pz.(L)A	IV	5	1	-	-
Jagdpz.Komp.1129	Jagdpz.	38	10	-	2	-
Pz.Jagd-Abt.2	Japdpz.	38	24	-	1	-
Pz.Jg.Komp.15	Pz.	IV	1	-	-	-
	Jagdpz.	38	15	1	-	-
SS-Pz.Jagd-Abt.560	Stu.	III	6	-	2	-
(Kampfgr.1001 Nächte)	Jagdpz.	38	37	1	6	-
9.Fallschirm.Div.						
Pz.Jg.Abt.	Stu.	III	-	-	1	-
	Jagdpz.	38	8	-	-	-
600.Inf.Div.(russ.)	Jagdpz.	38	8	-	2	-
	T	34	-	9	-	-
SS-Stu.Gesch.Abt.105	Stu.ital.L	6	10	1	-	-
SS-Pz.Komp.105	Stu.ital.L	6	-	-	1	-
	Pz.m.13/40		10	-	5	-
AOK 9 insgesamt:			489	34	46	1

einsatzbereit	=	489	=	86%
kurze Instandsetzung	=	34	=	6%
lange Instandsetzung	=	46	=	8%
		569		
im Zuführung	=	1		

Pz.AOK 3

Pz.Jagd.-Abt.1	Stu.	III	4	-	-	-
	Stu.	IV	11	-	1	-
	Jagdpz.	38	5	-	-	-
Pz.Jagd.-Abt.281	Jagdpz.	38	10	-	-	-
18.Pz.Gren.Div.						
Pz.Abt.Schlesien	Pz.	IV	25	-	1	2
Pz.Ausb.Verb.Ostsee	Pz.	V	2	-	-	-
	Jagdpz.	IV	2	-	-	-

Einheit	Art		Pz.	Kpfw.	Stug.	Jagdpanzer
			a	b	c	d
SS-Freiw.Div.Nordland						
Pz.Abt.11	Stu.	III	24	2	-	-
	Pz.(L)V.	IV	10	-	-	-
SS-Pz.Abt.503	Pz.	VI	10	-	2	-
	Pz.(Flak)	IV	7	1	-	-
Stu.Gesch.Brig.184	Stu.	III	23	-	1	-
	Stu.H	42	8	-	-	-
SS-Freiw.Div.Nederland	Stu.	III	1	-	-	-
SS-Pz.Jg.Abt.23	Stu.H	42	1	-	-	-
Pz.Jagd.Abt.6	Pz.	III	-	-	-	1
	Jagdpz.	38	15	-	4	-
Stu.Gesch.Brig.210	Stug.	III	12	-	1	-
	Stu.H	42	13	-	2	-
	Pz.(L)V	IV	11	-	2	1
Festung Stettin	Pz.	IV	-	-	1	-
	T	34	-	-	1	-
	Pz.AOK 3 ingesamt:		220	4	20	4

einsatzbereit	=	220	=	90%
kurze Instandsetzung	=	4	=	2%
lange Instandsetzung	=	20	=	8%
		244		
im Zuführung		4		

Vert.Bereich Swinemünde						
163.Inf.Div.						
Jagd.Pz.Komp.1234	Jagdpz.	38	10	-	-	-

Zusammenstellung				
AOK 9	489	34	46	1
Pz.AOK 3	220	4	20	4
Vert.Bereich Swinemünde	10	-	-	-
Heersgruppe Weichsel (Vistula) ingesamt:	719	38	66	5

einsatzbereit	=	719	=	87%
kurze Instandsetzung	=	38	=	5%
lange Instandsetzung	=	66	=	8%
		823		
im Zufügrung	=	5		

Erläuterung: a = eisatzbereit, b = kurze Instadsetzung, c = lange Instandsetzung, d = Zuführung

This armored strength list is divided by combat unit as well as armored vehicle type and combat readiness in each formation. It also shows percentages. The types of armored vehicles ranged from superheavy Jagdtiger tank destroyers to obsolete, worthless Italian M13/40 tanks. They were a hodge-podge of vehicles that were supposed to stop the 4,100-plus tanks and assault guns of the 1st and 2nd Belorussian Front.

CHAPTER ELEVEN

On 16 April 1945 the Soviets unleashed their demons on Army Group Vistula. The German 9th Army had to cover a 130-kilometer-wide front with only 235,000 men, 833 tanks and assault guns, and about 4,000 artillery pieces and mortars. Third Panzer Army was worse off, having to cover a front of 150 kilometers with only 105,000 men, 223 tanks and assault guns, and 1,850 cannons and mortars, including 300 to 400 flak guns acting as artillery pieces.

Against this pitiful force, the Red Army unleashed its massive artillery barrage. Thousands of cannons of various calibers erupted into action, pulverizing the whole first-line German defense. Entire Russian Katyusha rocket regiments screamed in unison, sending their terrifying payloads into the German hinterland. The noise of the barrage was deafening and horrifying all at once.

Battle Group 1001 Nachte was quickly moved and deployed around Wriezen, about thirty-seven kilometers east-northeast of Berlin, ready to counterattack the inevitable breakthrough that would occur. The Russians moved forward hoards of tanks and assault guns. Some of these steel monsters carried the massive 100mm and 122mm cannons. The SU-152 assault gun, the heaviest Russian vehicle, carried a huge 152mm cannon. Against this tidal wave of Russian armor, the small kampfgruppe was outgunned as well as outnumbered.

During the evening of 16 April, the kampfgruppe had received reports from front-line units of a Russian armored breakthrough that afternoon. Patrols were dispatched through the Wriezen highway, via Thoringswerden and Alt-Lewin. As daylight faded away and night arrived, the patrols encountered more and more enemy tanks and infantrymen. Clearly, the reports had not been wrong. SS-Untersturmführer Tiso, platoon commander in 1001 Nächte, received orders to make a reconnaissance in force north of Wriezen. He was to be supported by the battle group's escort company of SS paratroopers.

The reconnaissance mission turned into a full-fledged firefight when it came up against one of the Russian spearheads. The advance element of a strong Russian force, comprising four heavy tank battalions (about eighty heavy tanks) and supporting infantry, walked right into the path of Tiso's recon force. The SS paratroopers quickly dismounted the Hetzer tank destroyers and hugged the ground, preparing to give the tank destroyers some infantry support.

Panzerfausts were drawn and prepared to be fired. Protected by darkness, the Red Army tanks moved forward, only to be surprised by the loud, dull sound of hollow-charged rounds from several Panzerfausts, firing all at once. The leading enemy tanks were stopped dead in their tracks, while Tiso's tank destroyers barked, sending 75mm antitank shells flying toward the Russians. Just as fast as the Panzerfaust was fired, the paratrooper instinctively dropped the empty casing to the ground before the tube turned red hot, scorching the inexperienced user who was foolhardy enough to hold it. The backblast of the rocket leaving the casing could kill anyone lying immediately behind a Panzerfaust being fired.

Tiso's platoon was deployed on the right side of the highway just north of Wriezen, in a wooded area. The paratroopers began to creep forward, camouflaged by the darkness of the night and the thickness of the foliage. The burning Russian tanks worked against them by illuminating the area. The Red Army tankers concentrated their fire on the wooded area where Tiso's Hetzer platoon had been concealed, blasting the trees to smithereens with their Joseph Stalin 122mm shells.

Another Panzerfaust hit home, this time on the side armor of a Joseph Stalin-II tank. The paratrooper who fired the round was close to the armored beast, allowing the hollow-charged round to easily pierce the tank's 90mm armor plating. A loud explosion in the turret was instantly followed by an even louder explosion that sent the turret flying skyward. The noise was a rapid one, "Whump! Ka-Boom!" Another Red Army tank was destroyed.

Other paratroopers had similar results, sending the Russian armor scurrying for cover on the opposite side of the highway. Soviet infantry moved forward. Against them, the paratroopers held their own, but not without suffering some losses. Precious men and ammunition were used up that night.

By then the entire Hetzer Battalion had been alerted and was on its way, with the rest of the battle group following close behind. More Red Army tank units appeared and attempted to bypass the stiff German defense north of Wriezen. The battle group held up the enemy advance for a total of two and one-half days. Heavily outnumbered, the German defenses were soon overwhelmed at several places, and the other units were outflanked. The battle group was forced to withdraw before it too was surrounded.

Wriezen fell on 19 April, and Kampfgruppe 1001 Nächte made a fighting withdrawal through Kunersdorf, Vevais, Ludersdorf, Biesdorf, and Haselberg. At Haselberg, Major Blancbois reorganized his command to try once again to stem the Red flood that seemed to be engulfing everything in its path. First Company, SS-Jagdpanzer-Abteilung 560, and the SS Paratrooper Escort Company were detailed to hold up the enemy armored advance between Haselberg and Harnekop to give the battle group time to reorganize. A battery of German assault guns from Sturmgeschütz-Brigade III, which happened to be in the area, lent whatever support it could.

The date was 20 April 1945, Hitler's birthday. With the aid of the paratroopers and the few assault guns from Brigade III, the remaining four Hetzers of 1st Company met the Red Army tanks. As before, the German armored crews were outnumbered and outgunned. They nevertheless met their end fatalistically. "Graaf!" shouted SS-Hauptscharführer Nicolai, "This is our last stand!"

Outside these future coffins, the paras were still fighting the Red Army tanks, stalking the vehicles with the calm demeanor of a professional plying his trade. They handled the Panzerfausts and satchel charges as easily and efficiently as a surgeon handled a scalpel. The bigger the tank knocked out, the tougher the soldier appeared in the eyes of his companions.

Photograph of SS-Obersturmbannführer Willi Hardieck, the first commander of Kampfgruppe Z/Panzerbrigade 150. Hardieck gave this picture to SS-Unterscharführer Andreas Meyer on 17 December 1944 – just three days before Hardieck was killed in the Battle of the Bulge. Photo: Andreas Meyer.

Hauptscharführer Nicolai's words proved prophetic, but not before he and his tank crew blew up thirteen enemy armored vehicles. Nicolai's tank destroyer was finally destroyed by a lucky hit from a Russian artillery round. Eventually, the rest of the battle group was committed to the fighting around Haselberg and was effectively put out of action during these battles, though remnants of the kampfgruppe would continue to serve independently until the end of the war. Between 17 and 19 April, the 1st Company, SS-Jagdpanzer-Abteilung 560 alone had lost SS-Obersturmführer Link, the company chief, plus all of its platoon leaders. The roster of officers killed in the other companies included the following men:

Obersturmführer Dittmann, Company Chief, 2nd Co.; killed on 18 April
Untersturmführer Leier, Company Leader, 2nd Co.; killed on 19 April
Obersturmführer Sprenger, Company Chief, 3rd Co.; killed on 19 April
Hauptsturmführer Leihs, Company Chief, 4th Co.; killed on 19 April

---

On 19 April, while Kampfgruppe 1001 Nächte fought it out near Haselberg, Sturmbannführer Siegfried Milius received orders to move out from his reserve position north of Oderberg. He had been receiving as many survivors from SS-Jagdverband Mitte as possible. This numbered about 200 men but did not include 250 men of that commando formation who were sent to Austria with Otto Skorzeny at the end of March. His command was still referred to as Kampfgruppe Solar, the name used at the Zehden bridgehead. Soon, his kampfgruppe absorbed the hastily reformed SS Jagdverband Ost, which had been destroyed in January at Hohensalza.

Milius was then ordered to link up with 1001 Nächte and attack the flanks of the Red Army armored thrust reaching Bernau, about twenty to twenty-one kilometers west-southwest of Haselberg. Bernau was just a hop, skip, and a jump from Berlin. Luckily, his battle group still possessed enough gasoline and motor vehicles to make the trip. Corporal Werner Schrödl remembered those desperate two days when he and his fellow paratroopers, supported by a few armored vehicles, tried to hold up the Red Army.

> *The Russians opened the offensive against Berlin on 16 April 1945, at 3 A.M., with 40,000 cannons and rocket launchers. For two days the soldiers of 9 Army defeated them under heavy losses, mainly at Seelow Heights/Müncheberg, on the road to Berlin and the Seelow/Wriezen-Bad Frienwalde road.*
>
> *On the third day, the Russians broke through. Endless columns of tanks and infantry units stormed towards Berlin. On March 19 the SS-Fallschirmjäger-Bataillon 600 was transferred to Bernau in order to cut the Russian tanks off. Out of some woods, in the early morning of April 20th, the men saw it [was] impossible to attack with just a few tanks against overwhelming masses. The battalion then proceeded in the direction of Finowfurt, where they confronted Russian units coming from Eberswalde.*

Meanwhile, 1001 Nächte had been pushed back to the Haselberg-Steinbeck area after some very hard fighting. The battle group had gotten some much-needed rest in a forest near Melchow. It was there that the last Hetzers were refueled and ammuniton distributed. Every round would have to count more than ever, for there was no more ammunition available. Blancbois quickly organized what was left of his command and readied his battle group once again.

General chaos began to overtake the entire German front. On 19 April, Red Army armored spearheads took Harnekop, the site of 101st Army Corps

9TH ARMY FIGHTS FOR ITS LIFE, 16-19 APRIL 1945

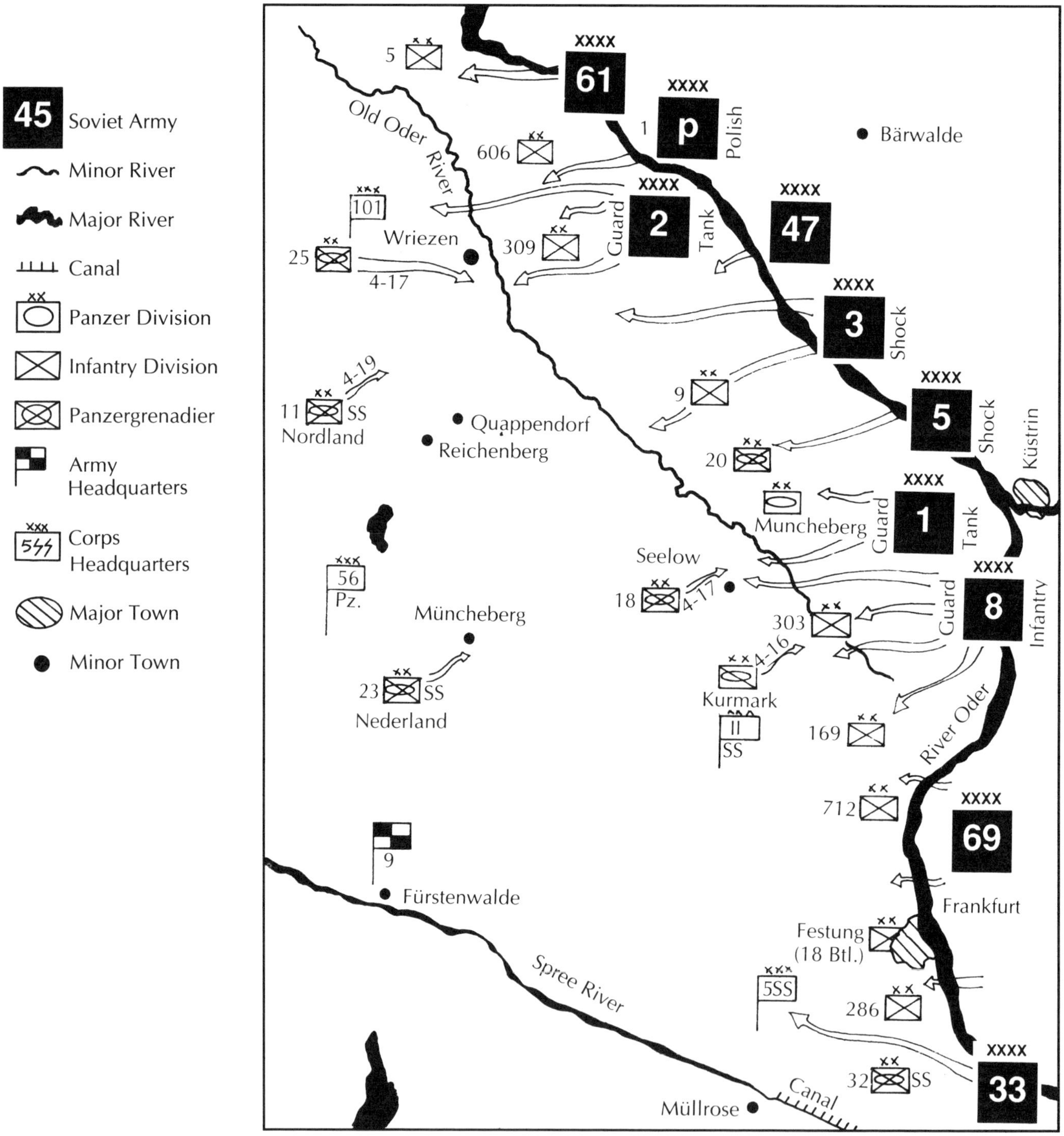

Headquarters. To the east and southeast, Schulzendorf, Metzdorf, Batzlow, and Reichenberg had already fallen. This area had been held by the weak Division Staff 606 z.b.V. and the 309th Infantry Division Greater Berlin, but both of these units had been shattered.

Further north, the 5th Jäger-Division's right flank was dangerously exposed. This unit was the southernmost division of the 101st Army Corps. The commander of the 5th Jäger-Division,

THE GERMAN SITUATION
24-28 APRIL 1945

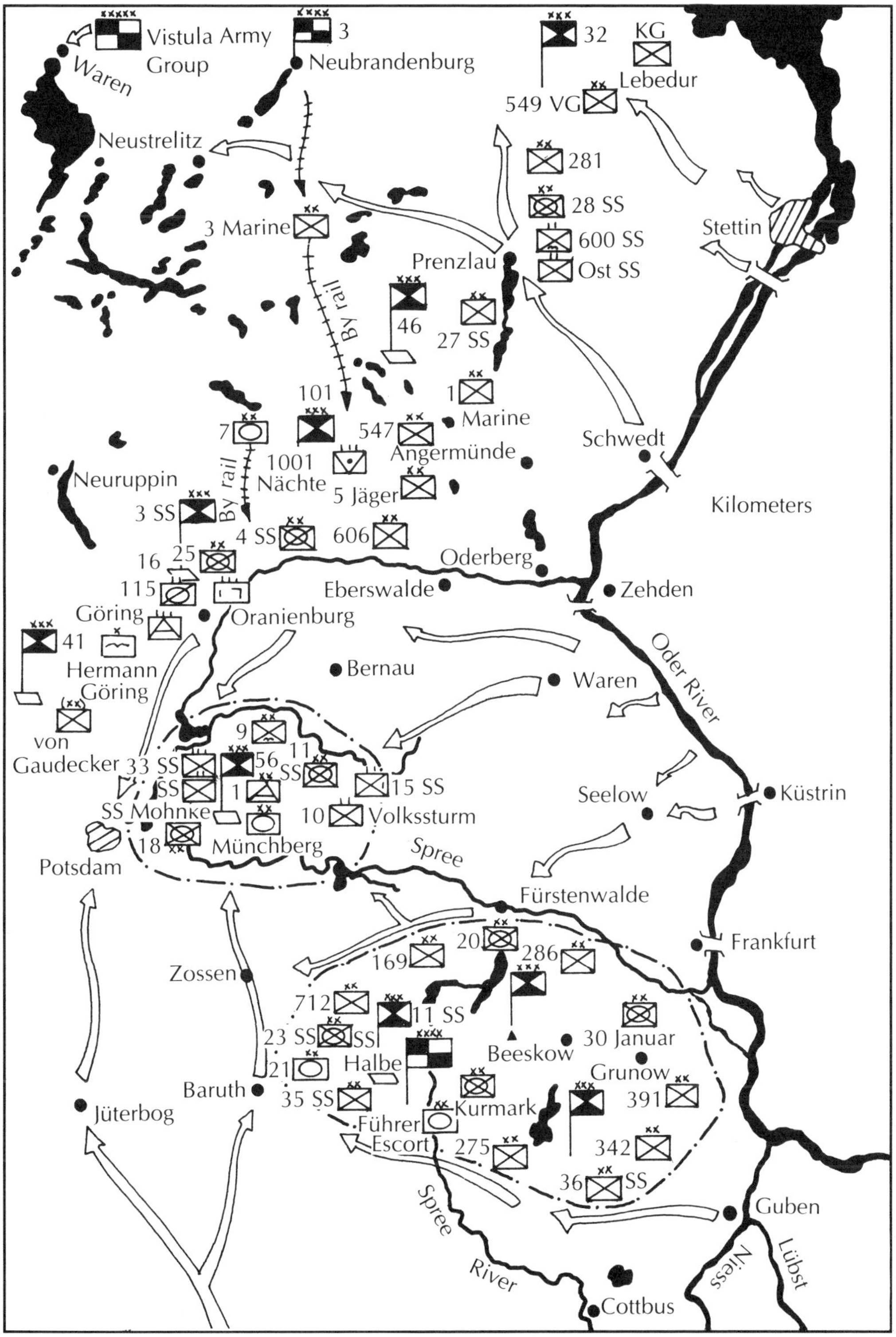

The German situation between 24 and 28 April 1945. The Battle for Berlin began in earnest on 24 April and lasted until 8 May. Most German units attempted to make their way toward the U.S. and British lines. The failed counterattack at Prenzlau was the last German offensive.

General-Leutnant Thumm, ordered a withdrawal. With this order, the fate of the German defenses between the Finow Canal and as far south as Strausberg was sealed. In any case it was irrelevant since a wide enough gap had already been made through which the entire Red Army could go. Any bigger hole would not make much difference.

SURROUNDED GERMAN UNITS IN THE GREATER BERLIN AREA

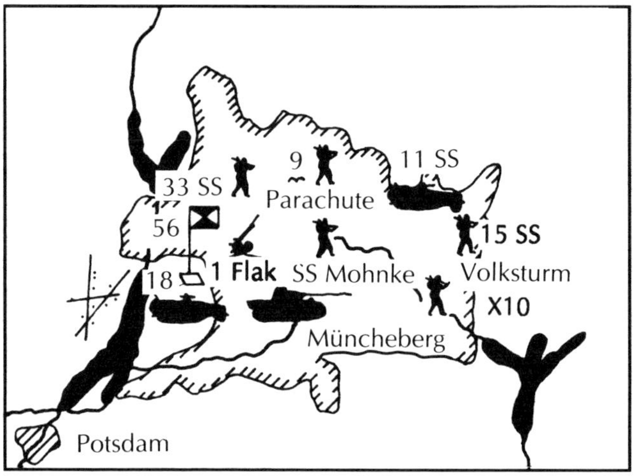

The scene was set for the closing chapter of the history of the SS paratroopers. With their advance against enemy forces at Bernau behind them, the battle group headed toward Finowfurt, where they met the advanced enemy formations coming from Eberswalde. Sturmbannführer Milius had hurt his ankle dring the brief fight at Bernau.

SS-Obersturmbannführer Schrage-Bobet took over command of the battle group at Nieder-Finow. Other units that pushed northward toward Finowfurt included the 25th Panzergrenadier-Division, Sturmgeschütz-Brigade 111 and 184, and SS-Panzergrenadier Regiment 7, from which Milius' battle group had already drawn replacements.

These units crossed the Finow Canal, the SS paratroopers and commandos providing cover for the withdrawal. The right flank of 3rd Panzer Army was attacked by the 2nd Belorussian Front and pierced on 20 April between Stettin and Schwedt. While these events were happening further north, SS General Felix Steiner's 3rd SS (Germanic) Tank Corps (which had been without combat troops since 17 April) was ordered to the Finow Canal area to organize a southern defense for 3rd Panzer Army.

Plans were also in the works in Berlin that envisioned the use of Steiner's tank corps as the component formation for a drive from the Finow Canal, in the direction of the German capital. Fantasy was again overtaking the men inside the Führer bunker. Steiner's command was virtually nonexistent. At first, his forces had included the following units: Engineer Regiment Staff No. 16, 630th Engineer Battalion, and the 115th Reconnaissance Battalion.

Other reinforcements were on their way to Steiner, but their grandiose-sounding names bore no relation to their actual combat strength. The 4th SS-Polizei-Panzergrenadier-Division, which had been virtually destroyed during the fighting around Danzig, was en route to Steiner after its remnant cadre had been evacuated by sea. Already, one of its regiments, the 7th, had been fighting under the command of 3rd Panzerarmee. The 4th SS Police Division reached Gamzow, where it was rebuilt partially with the help of a reserve unit, SS-Ausbildungs-Brigade 103. This replacement and forming brigade was under the command of SS-Obersturmbannführer Lang.

On 21 April, Steiner's command was elevated to the rank of Armeeabteilung (Army Detachment), not quite an army but stronger than a corps-size unit. Again, Hitler was dealing in fantasy. Not only was Steiner to block any enemy

THE GERMAN SITUATION (NORTH OF BERLIN)
28 APRIL 1945

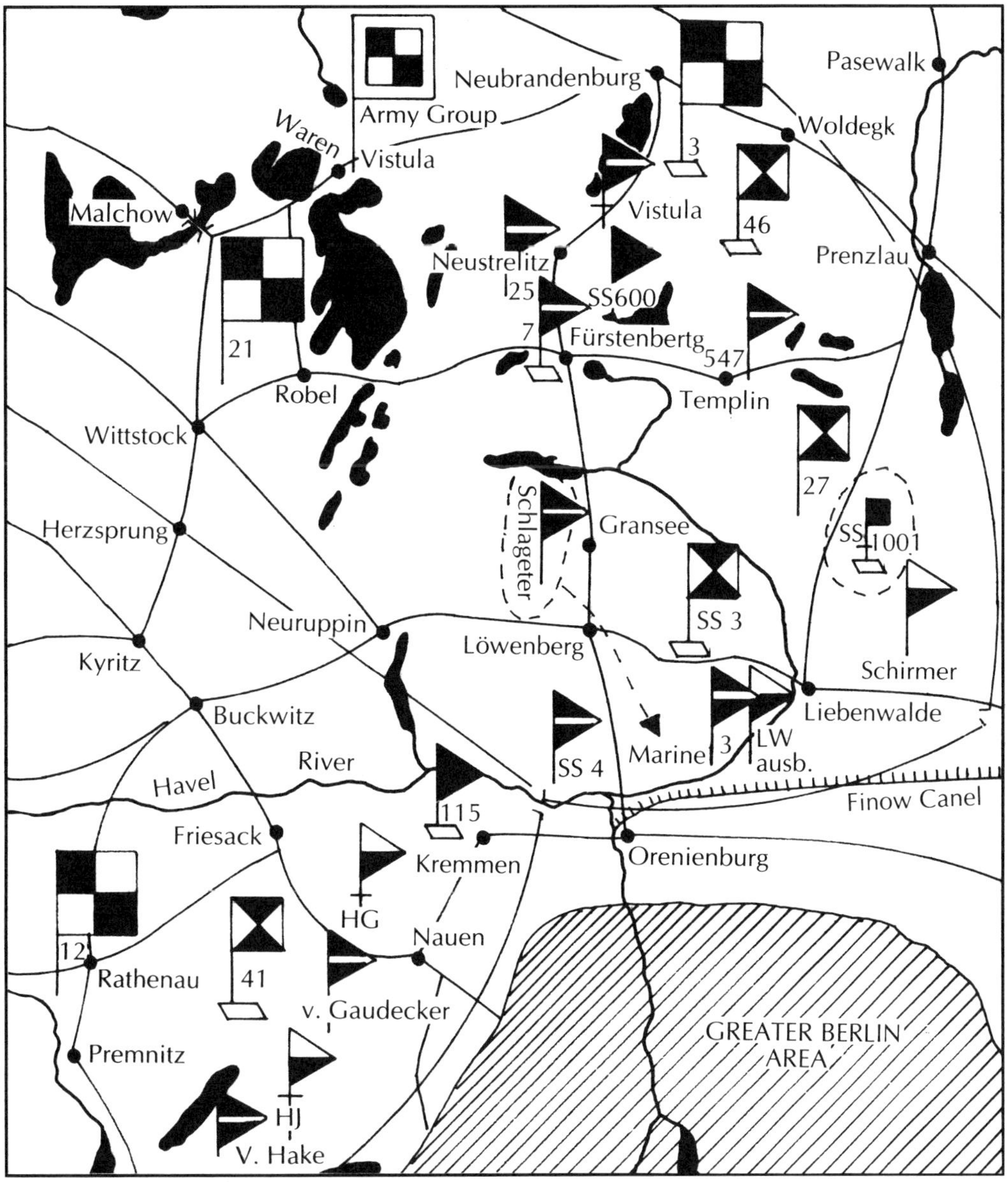

Armored
547 - 547th Volksgrenadier Division
Schirmer - Brigade Group von Schirmer
LW ausb. - Air Force Training & Replacement Brigade
Schlageter - Schlageter Division Battle Group
115 - Armored Reconnaissance Battalion 115
HG - Tank Destroyer Brigade Hermann Göring
v.Gaudecker - V-Waffen-Division Oberst von Gaudecker
HJ - Tank Destroyer Brigade Hitler Jugend
V. Hake - Division Group von Hake

In reserve or moving up

Army Group

Army

Corps

drive along the line of Opandau-Oranienburg-Finow-Finowfurt, but he was to prepare an attack from the Finowfurt area against the right flank of the advancing Red Army forces reaching for Berlin. This attack was to coincide with one that was supposed to be launched by the German 12th Army, aimed at attacking the Russian forces south of Berlin. Both attacks would end with the relief of Berlin, breaking the encirclement that was taking form.

It was an incredible order, since Steiner possessed neither the manpower nor the armored strength to accomplish such a mission. He would be hard-pressed to defend the right flank of the 3rd Panzer Army, let alone go on the attack. Further units were allocated to Steiner's 3rd (Germanic) SS Tank Corps, although, as before, their great-sounding names bore no relationship to their actual condition: Panzer-Jagd-Division Weichsel (Vistula), Panzer-Jagd-Brigade Krampnitz, Pionir-Bataillon 968, Fortress-Machine-Gun Battalion 116, etc.

Later in the day, Hitler turned over command of three divisions, the 5th Jäger-Division, 25th Panzergrenadier-Division, and the long-promised 4th SS-Polizei-Division to Steiner's SS corps. The German bridgehead at Finowfurt, supposed to be used for the projected attack, was itself attacked by the Russians, forcing all of the German units still south of the canal to withdraw. Milius' battle group covered the withdrawal. Werner Schrödl again enlightens us with his revelations.

> *On orders from Corps Headquarters, the Battalion was ordered in a hurry to Prenzlau, to participate as corps reserve against General Rokossofsky's 2nd Beylorussian Front which had crossed the Oder on April 20th, between Stettin and Schwedt. Fighting, the Battalion retreated via Burg Stargard to Neubrandenburg and finally via Waren-Malchow-Parchim and the area of Ludwigslust.*
>
> *Parts of the Battalion surrendered to the Americans, which crossed the Elbe on May 2nd and penetrated into the area Wismar-Schwerin-Ludwigslust-Dömitz. Other parts of the Battalion withdrew through the woods and surrendered to the Americans near Grabow on May 4th, 1945.*

The paratroopers and commandos were sent to Prenzlau from their defensive positions on the Finow Canal. They had to travel about fifty kilometers to Prenzlau. On 26 April, the Russians finally broke through at Prenzlau. The combat group, under the command of Schrage-Bobet, had the task of delaying the Red Army advance, as corps reserve, giving larger division-size units the chance to assemble for a counterattack. The 3rd Panzer Army was diverting the 25th Panzergrenadier Division, 547th Volksgrenadier Division, and the newly arriving 7th Panzer Division for this attack. Had the SS paratroopers and commandos known that one of the divisions earmarked for the attack, the 7th Panzer, had arrived without a single tank or assault gun, they would have realized that the counterattack was doomed to fail. They were going to be sacrificed for a battle that was already lost. The German attack, as expected, petered out just as quickly as it began.

The mixed paratrooper/commando battle group fought stiff rear-guard actions as it withdrew from Prenzlau, toward Neubrandenburg. There, it defended that city in a battle that lasted all night. The next battle took place at Neustrelitz. In this engagement the German 46th Panzer Corps tried, once again, to hold off the main brunt of the 2nd Beylorussian Front. The German units involved included the 7th Panzer, 25th Panzergrenadier, Panzer-

Jagdverband Weichsel, 547th Volksgrenadier, and the 1st Marine Division, by far the strongest unit, containing 12,000 men—although its ex-sailors were inexperienced.[1]

The attack was launched on 28 April, and as most predicted, it failed to stop the Red Army advance. Organized resistance then fell apart. The Commander of 3rd Panzer Army, General Hasso von Manteuffel, ordered a general retreat and "every unit for themselves" in an attempt to reach the U.S. lines. His only concern was to save his men from what awaited them under Soviet captivity.

With this order in hand, the paratroopers and commandos separated and attempted to make their way west. The paratroopers tried to reach the German 12th Army lines southwest of Neustrelitz. The Soviet forces were making quick headway, so much so that the battalion was cut off at Neuruppin, about forty-four kilometers south-southwest of Neustrelitz. No one there wanted to make a stand. In the end, the burden of defense thus fell on the paratroopers. The army units could not withdraw and had to help in the defense, but the bulk of the fighting was done by the paratroopers.

The parachute companies still had about eighty to one hundred men apiece. The final agony of the SS-Fallschirmjäger was about to begin. On 30 April, the Russians attacked Neuruppin with no less than two tank corps and an equal number of cavalry divisions. Armed only with Panzerfausts, satchel charges, and small arms, the paratroopers held off the enemy thrusts during the entire day.

Again and again, the enemy tanks would lead off the attacks, only to be repulsed. In turn, the Cossack cavalry regiments would signal their charge with a loud "Hurrah!"—their horses galloping at a mad pace, sabers twirling in menacing circles high above their heads. The seasoned paratroopers had to draw on every available ounce of courage to keep from running away from such a scene. Machine-gun fire met the Soviet charge, spewing deadly steel at the rows upon rows of cavalry regiments.

Horses fell by the dozen, whining as they hit the ground. Soon the field was littered with burning tanks and dead or dying men and horses, their contorted bodies adding to the grisly scene. The odor of burning flesh and steel nauseated both sides.

By the end of the second attack, an entire Russian tank battalion (about twenty vehicles) and hundreds of horses and men covered the battlefield. SS-Obersturmbannführer Schrage-Bobet had followed the events of the day's fighting from the trenches. The time was 1:30 P.M. The second assault had already used up the remaining Panzerfausts. More "armored fists" could be had at battle-group headquarters, though the supply was not inexhaustible. Schrage-Bobet quickly ordered runners to be sent from each company to bring back a fresh supply. In groups of twos and threes, they scurried off toward the command post. Not half an hour had passed before the enemy launched its third attack of the day. Schrage-Bobet had to deal with this assault with only satchel charges and grenades.

Once again the Red Army troops flung themselves at the paratroopers. As in countless other times, the paras allowed the enemy tanks to overrun their positions. Quickly springing from their one- and two-

[1] The head of the German navy, Grand Admiral Karl Doenitz, had released 50,000 sailors for duty as ground combat troops in 1945. In this way, four marine infantry divisions were formed (numbers 1 through 4). All these men were physically fit, but they had no experience as infantrymen. Many paid for this lack of training with their lives, including Vice Admiral Scheurlen, the commander of the 2nd Naval Division.

man foxholes, the tank hunters went to work on the steel fortresses.

They taped grenades together and flung them at the tracks of the T-34s. The grenades went off with a loud bang, causing the tracks to disengage from the steel wheels. Some paras tried to climb onto the beasts themselves, only to be cut down by the accompanying Russian infantry. By far the most effective way to knock out a tank was for a para to run up to the vehicle from the side, jump in front, and fall to the ground on his back—with the prayer that the wide tracks of the tank would pass on either side. The paratrooper had just a few seconds to attach a magnetic mine or satchel charge to the belly of the beast. There was supposedly sufficient time for the sapper to escape to safety, though many times the trip underneath the tanks was the last.

The paratroopers held the lines once again, but it cost the battalion dearly. At exactly 3:15 P.M. the Russians launched their fourth attack. Again, as before, the terrifying Katyusha rockets were heard screaming overhead and slamming into the positions held by the SS paratroopers. The wooded heath that had shielded them began to burn uncontrollably, scorching its occupants and polluting their lungs with thick, black smoke. Once again, they armed and fired the Panzerfausts, and once again they repulsed the Red Army. By the time dusk arrived, the paratroopers had thrown back nine major enemy attacks.

The last of those attacks had broken the back of the parachute companies. Number 3 Company, for example, which had begun the day with eighty-four men, could only muster thirty. The other parachute companies were in similar states. With the cover of darkness, Schrage-Bobet ordered his men to withdraw. Forming up the remaining men on a dark, deserted back road just west of Neuruppin, Schrage-Bobet took count. He had started the day with about 400 to 440 men. He now counted about 180, barely one full-strength company. Further resistance was useless. He ordered his men to group into assault platoons and move off.

By traveling at night, they managed to escape the Russian encirclement and evade Red Army units. They traveled to Waren, some sixty-four kilometers north of Neuruppin. From there, his small force turned west and passed Malchow and Parchim. They then marched to Hagenau, southwest of Schwerin, where they surrendered to elements of the U.S. 9th Army.

It was finally over! The German capital had fallen, and Hitler had committed suicide. The fear of being turned over to the "tender mercy" of the Russians engulfed the survivors of the SS Parachute Battalion. Their fear stemmed from the fact that the Allies were handing over German units that had fought mainly on the Eastern Front but had surrendered to the western Allies over to Russian forces.

Luckily for the paratroopers, the Americans simply placed them in a U.S. work camp for prisoners. Two survivors of that parachute force once again take us back to those days when no one wanted to be the last soldier to die in a war that was almost over. SS-Untersturmführer Draeger, you will recall, had been wounded in February at the Schwedt bridgehead. We again take up his story.

> *In June the town was taken over by the Russians. Also in that month, the Russians handed over the provinces of Thuringia and Saxony to the Americans.*
>
> *After eight days of Russian occupation, the Americans arrived with trucks and ambulances and took us to Königstein in Taunus. Here all*

AN SS VETERAN'S STORY

Hans Becker was a volunteer in the SS-Fallschirmjäger-Bataillon from the very beginning till the end. He fought with the battalion during the final battles on the Oder River. He received a serious head wound on 9 May 1945 while attempting to escape from the Soviet tentacles that stretched to the areas on the Elbe River. Becker was shot just east of that river, by the town of Wittenberge.

Wounded as he was, he was handled roughly but eventually ended up in a Russian military hospital, where he received minimum of aid. Amazingly he survived, but mentally and physically he was a broken man. In December 1949, he and a score of other German prisoners were collectively tried and sentenced to death. His death sentence was commuted to hard labor in a Siberian gulag (a death sentence in itself).

Hans Becker surprised his captors once again by living through the hell of the Soviet gulag. He lived in these camps until 1955 when the last of the German prisoners were released and sent home. Becker's home had been in West Prussia. When he arrived home, he quickly found that his homeland no longer existed. Prussia had been partitioned between Poland and the Soviet Union.

Worse still, he could not find any of his family. His numerous attempts at trying to locate them met with indifference on the part of the Communists in charge of the new Polish territories. Suffering from his physical disabilities and depression at not being able to locate any of his family members, Hans Becker opted to relocate to the Western zone in Germany. He lives a quiet life there today. Though decades have passed, Hans Becker still suffers from mental and physical handicaps.

(Becker's story was told to me in a letter he sent me dated 11 January 1990.)

SS men were pulled out and assembled after about three weeks, and sent to concentration camps.[2]

Then I was sent to Bad Mergentheim, where German Army doctors and nurses were said to have been told that they would be home by Christmas, if they mistreated the SS men. You could not believe how it was!

After five and a half years of the best comradeship with the German Army, this demand was easily followed. Now we were the "black sheep."

Erhard Knöfel, another survivor, had similar experiences, and his comments show the bitterness he still feels today over the behavior of some of his countrymen.

At Hagenau we were only 180 men. End of the war—now we became prisoners of the Americans, but one hour later I and two comrades managed to escape. I almost broke my neck in the attempt. We headed in the direction of the Elbe.

One adventure followed another. We were totally destroyed, pursued, and betrayed. We were now the scum of the nation. Notwithstanding all of this, we were not taken prisoner. This time period would make another book, but nobody in this country [Germany] would care!

The story of Hans Becker (above), another survivor, is a fitting end to our story on these brave paratroopers.

[2] Draeger's use of the term "concentration camp" seems a bit harsh, since they were merely work camps. True, life in these camps was hard, but in no way could it be equated with what we today consider a concentration camp.

PART TWO

Illustration: Eric Weisman.

SS-PANZER-BRIGADE GROSS

SS-FESTUNGS-GRENADIER-REGIMENT BESSLEIN

OSTTÜRKISCHEN-WAFFEN-VERBAND DER-SS

SPECIAL FORCES OF THE WAFFEN-SS

CHAPTER TWELVE

. . . the race is not to the swift, nor the battle to the strong. . .

Ecclesiastes 9:11

As the Soviet summer offensive of 1944 swept through Belorussia, it soon reached the borders of Poland and the Baltic countries. A dangerous bulge was forming between the German armies of Heeresgruppe Nord (Army Group Nord) and Heeresgruppe Mitte (Army Group Mitte). If the Germans continued to let the situation develop, this Soviet thrust would eventually make its way to the Baltic coast and sever land communication between both army groups. In effect, Army Group Nord would be cut off and thus vulnerable to being destroyed piecemeal—annihilating another thirty-four German divisions and making the Soviet's advance into Germany that much easier.

By the end of July 1944, the German situation was critical. Reinforcements to the badly mauled and almost completely destroyed Army Group Mitte had already drained the small reserves of the German Armies East and German Army High Command. In the end, the few available reserves in the east consisted of various army and Waffen-SS training centers and schools that had been set up in the conquered territories. Other odds and ends were hurriedly sent to 3rd Panzer Army (responsible for holding the line connecting both army groups), but time was against the Germans. The precariousness of the German front in this sector prompted a decision to mobilize all rear-area installations and units for combat.

On 1 August 1944, the 3rd Soviet Mechanized Corps of the 51st Army broke through the weakly held German defense lines between Autz and Mitau (in Latvia). It quickly overcame scattered German delaying units and reached the town of Tuckum, just a few kilometers away from the Gulf of Riga. Army Group North was cut off from the rest of the German armies fighting in the east. If something was not done soon to correct the situation, the Soviet ring around Army Group Nord would be too strong to break.

At this critical moment in the summer of 1944, Generaloberst Ferdinand Schorner, the new commander of Army Group Nord (as of 25 July 1944), began scraping together emergency combat units for a counterattack that he hoped would restore land contact between both army groups. One of these emergency units was SS-Panzer-Brigade Gross, formed from the SS Armored Training and Replacement Regiment located at Dondangen, Latvia. Armored Troop Training Regiment Seelager had been hastily transferred to SS Troop Training Ground Windau (Ventspils) on the Baltic Sea coast in Courland, Latvia. As was common practice in the Waffen-SS, this emergency unit was named after its commander, SS-Obersturmbannführer Martin Gross.

Martin Gross was born on 15 April in Frankfurt-am-Main. He began his military career as an enlisted man, joining the ranks of the Leibstandarte SS Adolf Hitler Regiment on 10 May 1933, the year of Hitler's ascension to power. He was selected for officer candidate school and was commissioned an SS-Untersturmführer (2nd Lieutenant) in 1935. Gross was promoted to SS-Obersturmführer (1st Lieutenant) two years later, in 1937. During the Polish campaign he served as commanding officer of a platoon in 1st Company/1st Battalion LSSAH, winning the Iron Cross, 2nd Class, which was awarded on 1 October 1939.

Obersturmführer Martin Gross was given command of 1st Company during the fall of 1939 and promoted to SS-Hauptsturmführer (Captain) on 30 January 1940. He served as commander of 1st Company during the western campaign, being wounded on

26 May, during the last battles in France. On 20 July 1940, he received the Iron Cross, lst Class, for his performance. Gross continued in the Leibstandarte during the Balkan campaign and was transferred to the 5th (Heavy Weapons) Company/1st Battalion LSSAH during the early part of the Russian campaign on 19 November 1941.

Soon after Gross took command of 5th Company, the Soviets launched their first of many winter offensives, and Gross became a casualty during the heavy defensive fighting near Rostov. The wounds were serious enough to keep him out of the war for most of 1942. When his wounds got better, he was attached to the main SS office in Berlin. Later in the year, he was put in command of the SS Panzer Replacement Battalion. The decision had already been made to upgrade the Leibstandarte formation into a full-size division. A new Panzer Abteilung was thus ordered formed and began taking shape on 1 February at Wildflecken Training Ground. Later on in the year, this battalion was expanded to a regiment of two battalions, and Gross was selected to lead its 2nd Battalion. He was promoted to Sturmbannführer on 30 January 1943.

Soon after, the newly reorganized Leibstandarte Division was sent to the Eastern Front and participated in the battle to retake Kharkov. Gross distinguished himself by being on the forefront of the attack. For his actions, he earned the German Cross in Gold (more correctly, War Order of the German Cross), the next-highest decoration to the much-coveted Knight's Cross, receiving it on 28 March 1943.

On 4 July 1943 the Germans launched Operation Zitadelle, which turned out to be the greatest tank battle in history. Sturmbannführer Gross was nominated for the Knight's Cross for his actions during the height of the battle, on 12 July 1943. The height of the offensive was reached on that day, when more than 600 German and 900 Soviet armored fighting vehicles clashed near the town of Prokhorovka. Both forces' tanks and assault guns fought at point-blank range. Gross' battalion alone destroyed more than ninety enemy T-34 tanks in more than three hours of fierce fighting. His Knight's Cross was awarded ten days later, on 22 July 1943.

After the failure of the Kursk offensive, the Leibstandarte Division was transferred to Northern Italy to re-form. Gross was then transferred to SS Armored Training and Replacement Regiment Seelager in the spring of 1944. He was commander of the Panzer training regiment there, and when his unit was mobilized for frontline service, he led it.

Panzer Commander, SS-Panzer-Brigade Gross. Illustration: Eric Weisman.

The "brigade" began forming between Talsi and Sabile (see map on page 140), having an initial strength of only ten outdated Panzer III and IV tanks formed into two weak tank companies (five vehicles apiece). Later, the tank strength increased to fifteen vehicles, and by mid-August, an entire tank company from the 103rd SS Heavy Tank Battalion (Tiger tanks) had been attached. This would increase Brigade Gross' tank strength to some thirty vehicles. Eventually, a Panzer battalion of four companies was formed.

To this small tank force were added two Panzergrenadier battalions, each with four companies. The 1st SS Armored Reconnaissance Training and Replacement Battalion also joined the brigade. The commander of this three-company battalion was SS-Hauptsturmführer Walter Runge. Formation of these initial elements of the brigade took only a few days, beginning on 3 August 1944.

By 8 August, Brigade Gross had to be committed against an entire Soviet mechanized cavalry corps south of Libau, Latvia. The Soviet cavalry corps was held back by the combined efforts of Gross and some Wehrmacht units. At one point in the battle, Gross even counterattacked but was too weak to follow through and had to retire quickly, as the 52nd German Security Division covered its withdrawal.

The brigade continued to receive reinforcements by dribs and drabs, so that by 16 August it had attained the strength of about 2,500 men. On that day the brigade was subordinated to 39th Panzer Corps, 3rd Panzer Army, and was attached to an ad hoc Panzer division titled Strachwitz. Gross became an integral part of this division on 18 August. Strachwitz had been formed from various splinter groups; aside from Gross, it contained 101st Panzer Brigade, 337th Signal Battalion, 3rd Battalion/19th SS Artillery Regiment, and the divisional staff of the 337th Infantry Division.

Nineteen days after the Soviets had severed land communication between Army Groups Nord and Mitte, the Germans had finally amassed a sizable force that they hoped would be able to restore the German front between both army groups by fighting their way through the Soviet defenses. The date of attack was set for 20 August. A divisional order from Panzer Division Strachwitz outlined the plan of attack.

> ***Advance Phase:*** Panzer Brigade 101 with attached SS Panzer Grenadier Battalion from Brigade Gross to open the area near the railway station at Berzupe.
>
> SS-Brigade Gross (minus one Panzergrenadier and recon battalion) to follow Panzer Brigade 101 upon leaving the wooded area north of Berzupe and to be deployed as follows—right flank, Panzer Brigade 101; left flank, SS Panzer Brigade Gross.
>
> ***Dividing Line:*** Jaunorigi (101), Judni (Gross), Jaunzemiji (Gross), Bukuli (101)
>
> ***Task:*** Panzer Brigade 101 to hold themselves ready along the Remte-Biksti road in the area of Jurgi to attack the small but deep enemy lines by Biksti and break through and get over the Berze area by the railway station at Berzupe. After this breakout, the SS Panzer Grenadier Battalion 1 Gross—after crossing by Berzupe—will leave weak security forces in Dzukste, east of the road Kocas-Tilmani-Spirgus, and along the road Veckrogi-Lielpicas-Silapicas-Kirpji-Silisi, to attack Tuckum and take the western section from the south by storm.
>
> ***Follow-Up:*** Both brigades should hold ready to continue their advances over Kemmern, or along

the coast road with forces alongside Schlock to meet with Corps Command Kleffel (holding the Riga sector).

At 4:00 A.M. on 20 August, Panzer Division Strachwitz attacked, totally surprising the Sóviet forces. By 5:00 P.M. that same day, the unit reached Tuckum. It reported back to corps headquarters that it found forty-eight knocked out T-34 tanks inside the town, thanks to the heavy guns of the cruiser *Prinz Eugen*, which had been firing offshore in the Gulf of Riga. The bad news was that units of Brigade Gross and 101 had counted about sixty enemy tanks and 500 trucks filled with Red Army infantrymen between Mitau and Tuckum. According to plan, a few armored half-tracks from SS-Panzergrenadier-Bataillon 1 Gross were left guarding the town of Tuckum, while the main component units were sent elsewhere.

The rest of SS Panzer Brigade Gross continued the attack with Panzer Division Strachwitz. At the same time, Korps Kleffel attacked from the Riga area with its 81st and 93rd Infantry Divisions. Supporting the attack of these two units were Sturmgeschütz brigades 202, 909, and 912. In addition, Panzerjäger Abteilung 731 and 753 helped. The initial success of the attack propelled the German forces to Dobele, almost forty kilometers south of Tuckum. Fighting for Dobele began on 23 August, but another Soviet armored thrust during this time threatened to split Army Group Nord in half. This new thrust was steadily advancing further north in Estonia. A dangerously exposed sector of front between the Wirz Sea and Lake Peipus was the target of this new Soviet armored assault.

The German command reacted as quickly as possible. SS Panzer Brigade Gross was transferred north and directed to the Dorpat area, under the control of the 18th German Army. The brigade arrived at Elva (just southwest of Dorpat) on 25 August. Since the brigade had been transferred hastily, some elements were still in the Dobele (Doblen) area at this time. On that day, they were located on the Dobele-Autz road. The leading elements of the brigade remained at Elva, holding Soviet probing attacks, while waiting for the rest of the brigade to catch up. Any further advance with just the small leading elements, or even half of the brigade, would have invited disaster with so many enemy armored units in the area.

Three days later the brigade, once again whole, was ordered to move to an area northeast of Joesuu to block a Soviet drive there. The brigade had arrived too late to help prevent the fall of Dorpat (25 August), but it moved north of the town where elements of 18th German Army were strengthening newly dug positions.

The beginning of September 1944 found

SS-Sturmann SS-Panzer-Brigade Gross. Illustration: Eric Weisman.

SS-PANZERBRIGADE GROSS, 8 AUGUST 1944

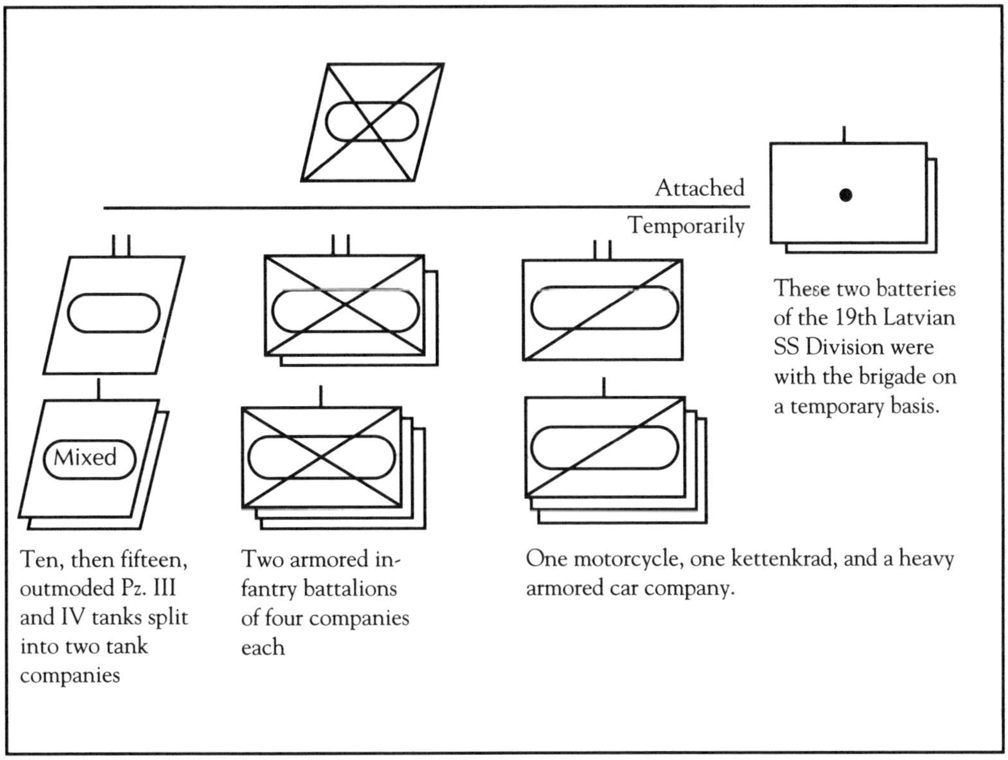

BY 15 AUGUST 1944, THE FOLLOWING UNITS HAD BEEN ADDED:

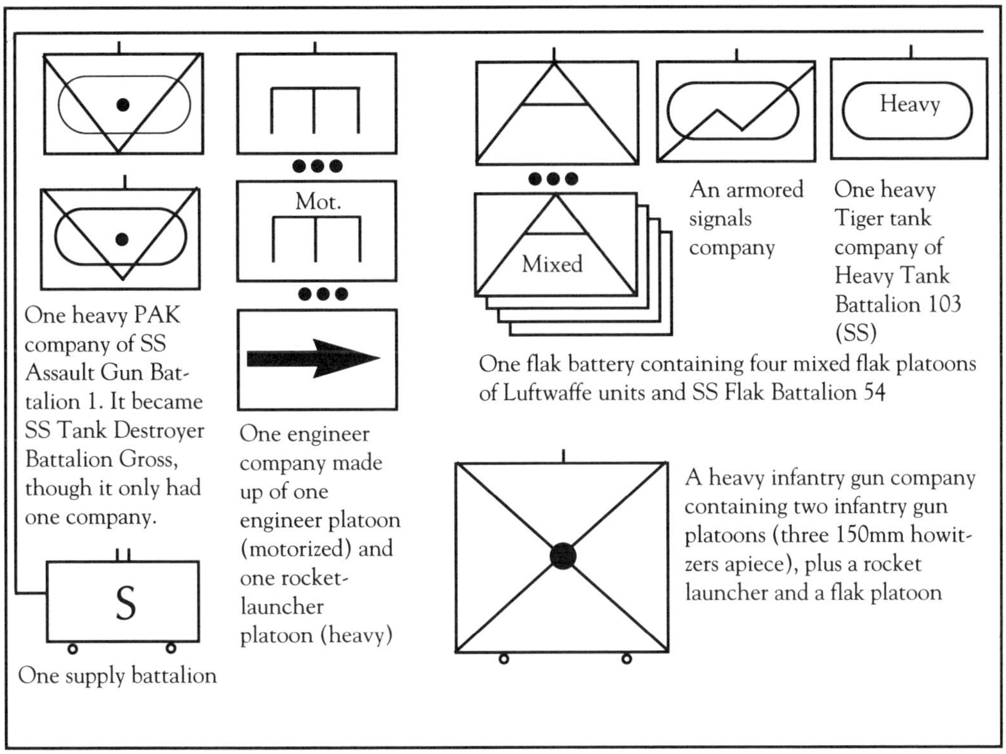

SS-PANZER-BRIGADE GROSS AND THE BALTIC BATTLES JUNE–SEPTEMBER 1944

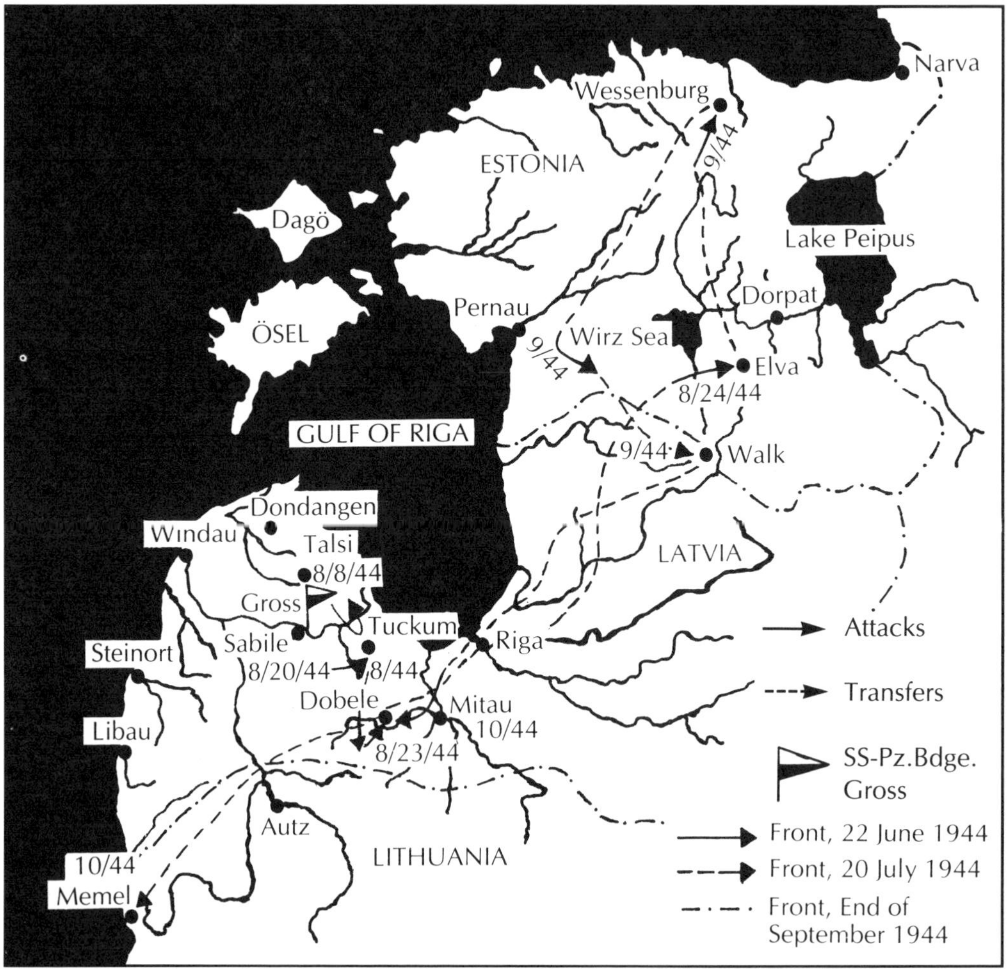

SS-Panzer-Brigade Gross and the battles for the Baltic countries, June – September 1944

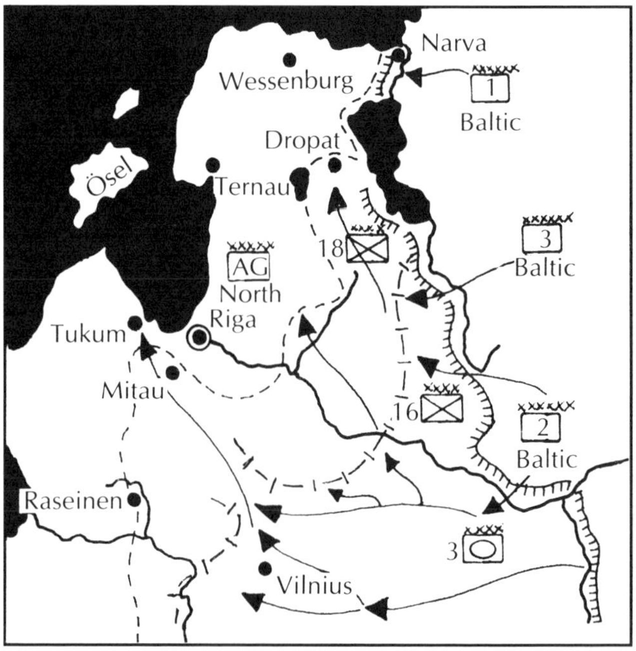

The situation of Army Group North, 22 June – September 1944

SS-Obersturmbannführer Martin Gross (right) as he appeared in April 1945. To his left is SS-Standartenführer Joachim Peiper of the 1st SS-Panzer-Division Leibstandarte Adolf Hitler. At this time, Martin Gross headed SS Railroad Group Gross. This was the codeword used by the 12th SS Tank Regiment to deceive enemy intelligence. Photo: Munin Verlag.

SS Panzer Brigade Gross defending a front line with other German units between the Wirz Sea and Lake Peipus. On 18 September, the Soviets again tried to outflank the German defenses at Narva by striking at the northeast corner of Lake Peipus. This time they hit the Division z.b.V. 300 (Division for Special Employment 300). The Estonian rifle regiments of this newly formed division (2nd, 3rd, 4th, and 6th) consisted of volunteers, many with previous service in either the Estonian armed forces or the German-sponsored Schuma (police) frontier battalions—or both. The men in these four regiments did not lack bravery, but they did lack any serious antitank weapons. Faced with continous armored attacks, it's a wonder they held out as long as they did. But the line finally gave way, and the German units fighting on the Narva front had to withdraw.

SS Panzer Brigade Gross covered the withdrawal of Army Detachment Narva, blocking temporarily any Soviet stabs at the German lines around Wessenberg (Rakvere).

Following the German withdrawal of Army Group Nord from Estonia, Gross retreated south, past Pernau in September, and participated in the defense of Walk on the borders of Estonia and Latvia. It then withdrew past Riga and further south-southwest by Dobele. In October 1944, it found itself fighting stiff rear-guard actions near the port city of Memel. In November the brigade (or what was left of it) was transferred to SS Troop Training Ground Sennelager near Danzig in East Prussia. There it met other parts of the brigade that had come by sea from Windau to Danzig.

The brigade then traveled to Westphalia (SS Military District XII) by train. It arrived at SS Troop Training Ground Steinhagen. On 9 November, Martin Gross had been promoted to an SS-Obersturmbannführer (lieutenant colonel) and transferred to the 12th SS Panzer Division HJ. At this time SS-Sturmbannführer Fritz Herzig, who had fought with the brigade, was promoted to SS-Obersturmbannführer and given command of the brigade.

The decision to disband the brigade and use its remnants to replenish the badly depleted SS Panzer divisions in the west had already been made. These worn-out units were attached to the newly-formed 6th Panzer Armee preparing for the Ardennes offensive. To cover this transfer of personnel from the disbanding SS Panzer Brigade Gross, the German High Command decided to deceive Allied intelligence into thinking that Gross was the nucleus of a new SS Panzer regiment, SS-Panzerregiment 26 Reichsmarschall. Martin Gross was then referred to as the commander of the new SS Panzer-regiment.

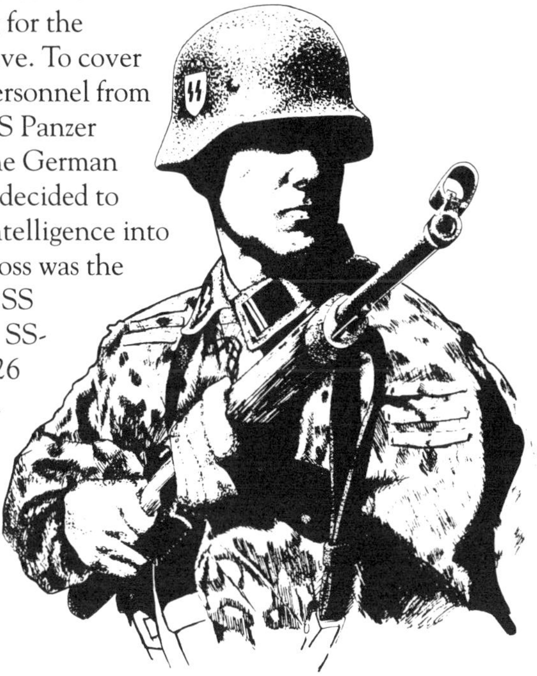

SS-Rottenführer SS-Panzer-Brigade Gross. Illustration: Eric Weisman.

Actually, the formation of such a new SS panzer division had been discussed in the summer of 1944. A lack of manpower, however, forced the idea to be dropped. An entry in the Field Command Staff Files of the Reichsführer-SS on 16 July 1944 noted that the formation of this division had been withdrawn. In the end, this never-formed SS Panzer division served as a perfect cover for the disbanding Brigade Gross. Gross was last listed as being stationed in the town of Padone in early December 1944. The brigade was officially disbanded that month and was not carried on in the roll of active Waffen-SS combat units by January 1945.

Three members of the brigade won the German Cross in Gold while serving under Gross: SS-Sturmbannführer Alfred Arnold, the former commander of SS Panzer Aufklarungs Ausbildungs Abteilung 1 (incorporated into Brigade Gross); SS-Hauptsturmführer Fritz Herzig, whose Tiger company from103rd SS Heavy Tiger Tank Battalion fought under Gross; and SS-Oberscharführer Friedrich Schwatz, a noncommissioned officer.

SS-Oberscharführer SS-Panzer-Brigade Gross. Illustration: Eric Weisman.

Sturmbannführer Arnold was killed in action on 9 October 1944 while covering the withdrawal of Brigade Gross in Latvia. He was posthumously awarded the German Cross in Gold on 6 January 1945. Hauptsturmführer Herzig went on to lead the SS Heavy Tiger Tank Battalion 503 as an SS-Sturmbannführer. He received his German Cross in Gold on the same day as Arnold. (Herzig would later win the Knight's Cross in April 1945.) Oberscharführer Schwatz received his award on 7 January 1945.

As for Martin Gross, he went on to lead SS Panzer Regiment 12 Hitler Jugend until the end of the war. Having been wounded at least three times during

FIELD POST NUMBERS OF THE SS-PANZER-BRIGADE GROSS

FP No.	Unit
34 951	SS-Kampfgruppe Gross 27 August 1944 to 4 September 1944
34 951	SS-Panzer-Brigade Gross 4 September to January 1945
26 871	Staff plus 1.-3. Companies, Mixed Tank Battalion Gross
44 989	Staff plus 1.-3. Companies, Armd. Recon Battalion Gross
25 581	Staff Co. plus 1.-4. Cos., Armd. Infantry Battalion 1 Gross
46 794	Staff Co. plus 1.-4. Cos., Armd. Infantry Battalion 2 Gross

Commanding Officer:
SS-Obersturmbannführer Martin Gross (Brigade Leader)
Commanding Officer:
SS-Obersturmbannführer Fritz Herzig (Tank Battalion)
Commanding Officer:
SS-Sturmbannführer Alfred Arnold (Armd. Recon Battalion.1)
Platoon Commander:
SS-Untersturmführer Josef Scheibler (Armd. Infantry Battalion.2)

Recipients of the German Cross in Gold SS-Panzer-Brigade Gross:

SS-Sturmbannführer Alfred Arnold
SS-Obersturmbannführer Fritz Herzig
SS-Oberscharführer Friedrich Schwartz

RECOLLECTIONS FROM A BRIGADE GROSS GUNNER

Who can still remember the hard batttes which were fought in Latvia during that fateful year of 1944? I was a member of the newly formed SS-Panzer-Brigade Gross, which was raised and organized in the area surrounding Tuckum, Latvia. I can recall the problems we had at that time, trying to get our tanks combat ready when they had only been used for training purposes!

During that time, I was the gunner of the only Tiger tank in the brigade. Directly on the left side of my gunner's seat inside the Tiger tank was a sign which read, "THIS TANK IS NOT COMBAT READY NOR WATERTIGHT." Worse still was the fact that after every five or six shots fired from the gun barrel, the gun would read "FIRE-STOP." The feeling that you get when you are inside a tank and cannot fire the cannon while your tank comes under fire from enemy guns is very frightening.

Still, we were considered the "Fire-Brigade" of the front. During our trip to the staging area we encountered our first problem. As we reached a small river, one of us got out to direct the tank driver [over] a narrow wooden bridge. The wooden planks were quite thick, but would the bridge hold? The bridge was checked out and slowly the Tiger tank was directed onto the span. So narrow was the bridge that the driver did not wish to look left or right, but kept his sights right in front. Right in the middle of the wooden bridge it happened!

With a very loud crack and an equally heavy knock, we fell into the stream! Besides a few scrapes and bruises, plus the shock of the fall, we were all right. Members from the Grossdeutschland were able to pull our tank from the stream so that we could continue our trip to the staging area near Tuckum.

I can also recollect that our brigade tanks contained some old Pz. III, IVs, and one Pz. V—all of them very old and not combat ready. The first day at the staging area was a particularly hard one for us. We had already lost our tank commander from a shot to the head earlier in the day when our second tank commander died in the afternoon in the same manner. Our third tank commander was an SS-Unterscharführer who had recently joined our group from the Luftwaffe.

We were not even together for fifteen minutes when the order came to move out. We had been ordered to advance in the direction of Tuckum, without infantry support, in order to observe the effects of the heavy shelling on that town by the German heavy cruiser *Prinz Eugen*. To this day, I cannot understand that order. After crossing slightly fallen ground, we had to rise again among some huge shell holes, before reaching the city. In the process, we were able to knock out a few antitank positions.

Suddenly, we found ourselves in the middle of a huge lumberyard. Cut wooden planks were stacked up high for as far as the eye could see. We just as quickly began to receive small arms and antitank rifle fire from all directions. We could only see directly in front of us through the gun sights and machine gun slits. Just vaguely, covered up by smoke and piled wooden planks, we recognized the turret of a T-34 tank!

Our reaction time was faster than the enemy's (much to our relief!). One round was all it took, and the enemy tank smoldered where it stood lying in wait, damaged. We then proceeded to clear a path free of the enemy infantry by firing high-explosive shells and bearing machine-gun fire on the enemy infantry hiding all around us. Moving forward a few hundred meters, we moved through a narrow, winding alley. At that moment another T-34 came into our gun sights. The T-34 was so close that [its] 76mm gun was now within effective range and capable of knocking out our own tank.

Sweat was pouring from all of us as the situation reached its zenith. We quickly loaded the first round and just as quickly fired the armor-piercing shell. "A miss! Shit! Load again before they burn us." The first shot had knocked out a telephone pole next to the T-34. Luckily for us, the sun was directly in their eyes. Our next shot blew off the turret of the T-34 with a loud bang, sending it skyward about twenty feet.

Our mission was practically over by then, as most of the town had been cleared of the enemy. We were given a few days' rest but that ended just as quickly as it had begun. Another enemy breakthrough by Elisendorf (Molkerei) with enemy armor threatened to cut off Tuckum once again.

Within a short while we were able to knock out an additional six T-34s, and soon the enemy penetration was destroyed. I had been wounded during those last few battles, which ended my service with SS-Panzer-Brigade Gross. I was initially sent to a military hospital at Windau (Ventspils).[1]

[1] Klaus Gerson, *Der Freiwillige*.

SS-PANZER-BRIGADE GROSS

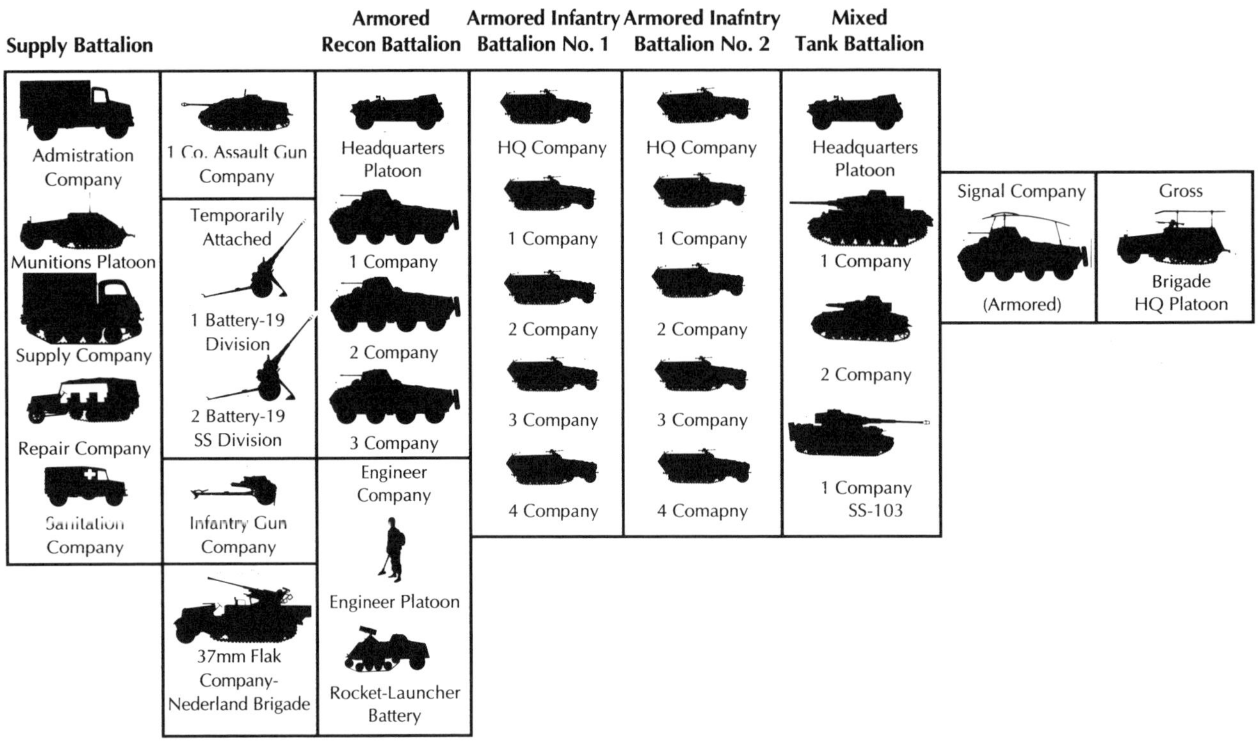

On pages 145 through 147 are photographs of the original recommendation in German for the awarding of the German Cross in Gold to SS-Sturmbannführer Fritz Herzig of SS-Panzer-Brigade Gross. Page two gives Herzig's name, birthplace (Wiener-Neustadt), rank, combat unit, and previous awards (Iron Cross, 2nd; Iron Cross, 1st, Wound Badge in Black; and Eastern Medal). Page three details the actions for which Herzig was nominated: Tuckum 4 August 1944, Irlava 14 August, Tuckum-Kemmern 17-19 August, and Jekava 15 September. The proposal was approved 16 November 1944. Certificate: Michael J. Belles Collection.

his career, he again became a casualty in the spring of 1945, earning the award of the Wound Badge in Gold. On 8 May, SS-Obersturmbannführer Gross was one of only 328 officers of the Hitler Jugend Division to survive the war. As one author put it, "Martin Gross was one of the outstanding tank commanders of the Waffen-SS."

The ability of SS Panzer Brigade Gross to perform as well as it did, considering the rapidity with which it was formed and outfitted, is a credit to its officers and men. The brigade was in action for almost four continuous months, having taken less than four days to form initially. Though Gross is belittled by the more famous SS Panzer divisions, its importance to the German war effort during a critical time on the Eastern Front cannot be overlooked. It was instrumental in reestablishing (for some time) land communication with the besieged divisions of Army Group Nord, and it later covered the withdrawal of this army group from Estonia to Latvia. Even after it was disbanded, SS-Panzer-Brigade Gross served a purpose by supplying its remaining personnel to the depleted 1st, 2nd, 9th, and 12th SS divisions during the late fall of 1944. It can be said that this Panzer brigade was one of the most obscure and underrated units of the Waffen-SS.

SS-Panzer-Brigade "G r o s s"

Reichsführer-SS Persönl. Stab
Eingang 16. Nov. 1944
Tgb.Nr. 1234/44

V O R S C H L A G

für die Verleihung

des Deutschen Kreuzes in Gold

408

O.U., den 3. Oktober 1944

SS-Sturmbannführer u.
Brigade - Kommandeur

Verliehen: 6.1.45

An
OKH/PA/P5 1.Staffel
a.d.D.

Name	Rufname	Geb.-Tag Geb.-Ort	Dienstgrad Dienststellung Dienstverhältnis	Truppenteil	Heimatanschrift
Hersig	Fritz	18.7.1915 Wiener-Neustadt	SS-Stubaf. Ia SS-Pz.Brig. Gross aktiv	SS-Pz.Brig. "Gross"	Ellwangen/ Jagst, Mühlgraben 6

Bisher verliehene Kriegsauszeichnungen mit Angabe der Verleihungsdaten		Bereits genannt im Ehrenblatt des deutschen Heeres	Anerkennungsschreiben des OKH. erhalten?
EK II	10.7.41	nein	nein
EK I	10.9.41		
Verw.Abz.schw.	17.3.43		
Ostmedaille	4.11.43		

Hat bereits ein Vorschlag zur Verleihung des Ritterkreuzes vorgelegen?

nein

Bisherige Kriegsverwendung seit 1939

1.9.1939	-	1.12.1939	Ord.Offz.SS-A.Rgt.
1.12.1939	-	1.11.1940	Kp.Fhr.3./Kradschtz.E.Kp.
1.10.1940	-	10.6.1942	Kp.Chef 3./SS-A.A.5
11.6.1942	-	1.11.1942	Hpt.im Stabe SS-Pz.Rgt.2
1.11.1942	-	30.1.1943	Sstabskp.I./SS-Pz.Rgt.2
31.1.1943	-	30.4.1943	Kp.Chef 8.(s.) SS-Pz.Rgt.2
1.5.1943	-	31.10.1943	Kp.Chef 1./SS-Pz.Abt.
1.11.1943	-	29.2.1944	Fhr.II./SS-Pz.A.u.E.Rgt.
1.3.1944	-	4.8.1944	Kdr.SS-Pz.Unterf.Lehr-Abt.
seit		4.8.1944	Ia SS-Panzer-Brig."Gross"

SS-Sturmbannführer Herzig gehört der Brigade als Ia an und hat sich als solcher auf das höchste bewährt.

1.) Am 4.8. beim Angriff auf Tuckum blieb die rechts eingesetzte Panzergruppe infolge starker Abwehr liegen.
H. fuhr ungeachtet des schweren Feuers im Pkw. nach dort und setzte die Gruppe im persönlichen Einsatz neu an, wodurch das 1. Angriffsziel erreicht werden konnte.

2.) Am 14.8. gelang dem Feind ein Einbruch im Raum von Irlava, der zur Bildung eines feindlichen Brückenkopfes führte.
H. fasste sofort alle verfügbaren Reserven zusammen und zerschlug im energischen Zupacken den feindlichen Brückenkopf.

3.) Beim 2. Angriff auf Tuckum am 17.8. hat sich Herzig insofern auf das Höchste bewährt, als er für den Zusammenhalt der Kampfgruppe mit seiner ganzen Person sich einsetzte und so die geschlossene Stosskraft gewährleistete.

4.) Beim weiteren Durchstoss der Brigade am 19.8. auf Kemmern war durch Feindeinwirkung die Führung durch Funk verlorengegangen.
Herzig, der in meinem Befehls-SPW. fuhr, bootete ungeachtet des schweren Infanterie-Feuers in dem Waldgebiet zwischen Tuckum und Kemmern mehrmals aus und überbrachte den Einheitsführern meine Befehle unmittelbar.

5.) Bei den Absetzbewegungen im Raume Jekava hat er sich durch persönlichen Einsatz erneut bewährt.
Als am 15.9. die Brigade vom Feind bereits überflügelt war, durch Ausfall der Geräte eine Verbindung zu den vorne eingesetzten Teilen nicht mehr bestand, brachte er diesen im schwersten Feuer den Befehl zum Absetzen.

6.) Am 17. gelang dem Gegner im Abschnitt des im Brennpunkt eingesetzten Batl. ein örtlicher Einbruch.
H., der sich zu dieser Zeit dort befand, fasste aus eigenem Entschluss alle dort stehenden Kräfte zusammen und führte persönlich den Gegenstoss, welcher zu vollem Erfolg führte.

H. ist ein alter verdienter SS-Führer, der sich in allen Feldzügen voll bewährt hat.
An Auszeichnungen wurden bisher verliehen:

EK.I, E.K.II, Verw.Abz.(schw.), Sturmabzeichen (Silber), Ostmedaille, Ostmark-Sud.Ld.-Erinnerungs-Med. Gold. HJ.-Ehrenzeichen.

Ich halte Herzig für die Verleihung des Deutschen Kreuzes in Gold für würdig.

SS-Sturmbannführer u.
Brigade-Kommandeur

CHAPTER THIRTEEN

Set ye Uriah in the forefront of the hottest battle.
II Samuel 11:15

During the night of 11 January 1945, the long period of relative quiet on the Eastern Front was suddenly broken by a massive and all-consuming Russian artillery barrage that lasted for hours on end. It proved to be the heaviest barrage experienced so far in the course of the war. This devastating hail of death-dealing shrapnel was followed by an equally huge Russian armored flood. Numerous armored spearheads easily penetrated the bloodied and disintegrated German front, where in most places, the withdrawal of the German forces was nothing more than a rout.

Endless columns of Russian motorized infantry and military hardware followed in the wake of these enemy spearheads, and by the evening of 16 January, Soviet tank columns were already approaching Cracow. So far, Army Group Center (formerly, Army Group "A") had been unable to halt the advance of Marshal Koniev's lst Ukrainian Front. The situation was getting desperate. If the Russians were not halted soon, their armies would overrun Silesia in two weeks, and with it would go Germany's remaining coal mines and arms factories.

On 17 January, as the Russian juggernaut neared Breslau, capital city of Silesia, Gualeiter Hanke gave the codeword "Gneisenau," which immediately placed into motion the activation of all reserve, replacement, and training units in and around the city. Men on leave, convalescing, even old men and boys from the Hitler Youth were all herded together and formed into Volkssturm (People's Assault) battalions.

Thus, four regiments were formed immediately for Breslau's defense. Later in February, a fifth regiment (Regiment Möhr) was added to the city's defenses. Each regiment was named. The Waffen-SS contribution to the city's defense was SS-Festungs-Grenadier-Regiment Besslein, so named after its commanding officer, SS-Obersturmbannführer Georg-Robert Besslein. The then youthful thirty-four-year-old Besslein gathered his command from the following training and replacement units within Breslau and the surrounding countryside:

- SS-Infantry-Gun-(Howitzer)-Training and Replacement-Battalion-l (located in Deutsch-Lissa)
- SS-Armored-Infantry-Training and Replacement Battalion-l (in Breslau)
- SS-NCO-Administrator's School (also in Breslau)
- Army-NCO School Striegau
- Army-Driving-Training and Replacement-Battalion-28 (in Schweidnitz)
- Army-Veterinary-Training and Replacement-Battalion-8 (also in Schweidnitz)

Initially, Lt.Col. Besslein could count on the following weapons: six heavy infantry guns, twelve light infantry guns, sixteen heavy 120mm mortars, and eight 88mm FLAK guns. Four days later on 21 January, the garrison of Breslau received a boost in the arm by way of 100 light field howitzers taken from the nearby Borsig factory before it was blown up so it wouldn't be captured by the Reds. In this way, Besslein's artillery strength was augmented by an additional twelve light howitzers, making a total barrel strength of fifty-four tubes—quite a formidable artillery force. Besides its own artillery, Regiment Besslein received support on many occasions from other artillery units within the surrounded city garrison.

SS-Obersturmbannführer Georg-Robert Besslein, the commander of SS-Festungs-Regiment Besslein, as he appeared in 1945. Photo: Robert Tyler.

Other units that later worked hand-in-glove with the SS were Klose (Nr. 41), and Stephan (Nr. 42), whose ranks Besslein had helped train and supply with his own NCOs, etc. The order of battle for the regiment yielded fifteen companies split into four battalions commanded as listed at the bottom of the page.

Not all of the battalions would fight together however. The 2nd SS-Festung Regiment Besslein would soon be attached to the German 609th Infantry Division. Later, the 3rd SS-Festung Regiment Besslein was also withdrawn from the regiment to help bolster another unit.

It wasn't until 19 January, when the Russian tank armies were already swarming closer to the city, that Gauleiter Hanke gave permission for the evacuation of all noncombatants from the city. No preparations or steps had been taken for the orderly and safe conduct of the city's civilian population.

Although Hanke had plenty of time to plan the city's evacuation, he had avoided taking any steps that might be misconstrued as defeatism by the gentlemen in Berlin. He also wished to allay any fears of the Silesian people, who had heard terrifying stories from Minister Goebbel's propaganda ministry about how the Russians had behaved when they entered the first German villages in East Prussia. These stories were true, and they elevated the fears and anxieties of the civilians in Silesia,who knew how close the Eastern Front had come.

So thousands of half-starved and scared women and children, even the old folks, began a trek westward in the middle of the worst winter Europe had ever

Commanding Officer of the Regiment:
SS-Obersturmbannführer Georg-Robert Besslein

Adjutant:
Hauptmann Eberhad Seiffert

Commanding Officers of Each Battalion:
1. Festungs-Grenadier-Regiment Besslein - SS-Hauptsturmführer Greger
2. Festungs-Grenadier-Regiment Besslein - SS-Hauptsturmführer Zielske
3. Festungs-Grenadier-Regiment Besslein - Hauptmann Sommer
4. Festungs-Grenadier-Regiment Besslein - SS-Obersturmbannführer Scharpwinkel

experienced, with absolutely no plan or, worse yet, no provisions for food and lodging. The civilians in the towns and villages they crossed ran the gamut of human feelings. Many turned a blind eye to the sufferings of these refugees, while others gave more than they could afford. The children suffered worst.

Exemplifying this all too well is the case of a young Silesian woman whose husband worked at the city railroad station. Told by her husband to take their six-month-old daughter, Gabby, and flee the city before the Russians could surround it, she left in such haste that she was ill-prepared for the hardships awaiting her and her daughter. Her husband remained behind to join the Volkssturm.

She made it to her parents' village near Spremberg, more than 190 kilometers from Breslau, but not before her little baby died of malnutrition and frostbite and she herself was the object of repeated rapings by columns of Russian troops who constantly overran the poor refugees. The lucky ones survived. More common was the sight of fifty or more wagons, whose crushed occupants, regardless of age or sex, lay prostrate—frozen and lifeless pieces covering the countryside as a reminder of the inhumanity of war.

Back in Breslau, the German defenders resolved to defend their city—their piece of Germany, if you will—to the end. There was no need to harangue these men into action. They needed no further description from Minister Goebbel's propaganda department of what would befall them and the civilians they were defending should they lose. Every man from commanding officer to lowly private knew that it was a battle to the last, with no quarter given nor taken. After each Russian assault was repulsed, the brutality of the Russians on their prisoners would be felt all the more. This and the low German casualties throughout the battle continually instilled anger and renewed ferocity among the city's defenders, inspiring their fighting spirit to new heights.

That Breslau was not overrun immediately was the miracle of the grenadiers of the German 269th Infantry Division, whose bravery gave the city's defenders time in which to prepare an adequate defense. For seven days, from 21 through 28 January, the grenadiers threw themselves at the advancing Russian tank formations. Using teller mines, Panzerfausts, plastic charges, and even hand grenades, they were able to destroy seventy-six enemy tanks within three days' time.

By 26 January, the Russian advance guard had reached Marsdorf, north of the city, and crossed the Oder River, establishing a bridgehead. The bridge there had already been blown, but the frozen river was thick enough to support the heaviest of Russian tanks. Three days later, the Russians had established another bridgehead at Peikewitz, south of the city. Meanwhile, a ferocious counterattack had been taking place between German forces trying to dislodge the determined Russian defenders holding the bridgehead north of the city.

Finally, the German units under 269th Infantry Division managed to destroy the Russian position. On 29 January, the 269th Infantry Division was transferred south of the city, while the new 609th Infantry Division took over its positions.

The 609th was not a regular front-line unit, having only been raised in Dresden a few weeks before, with its troops thrown together from schools, detachments, stragglers, plus a regiment of ground staff from the Luftwaffe, Luftwaffe Officer Candidates, and, of course, the SS-Fortress-Grenadier-Regiment Besslein. The division was commanded by General Ruff, while the Volkssturm battalions (thirty-eight in all) were commanded by S.A. Obergruppenführer Herzog.

Thirty-two artillery batteries defended the city and supported the 609th Infantry Division. Unfortunately, the

SS-FESTUNGS-GRENADIR-REGIMENT BESSLEIN
ILLUSTRATIVE DIAGRAM

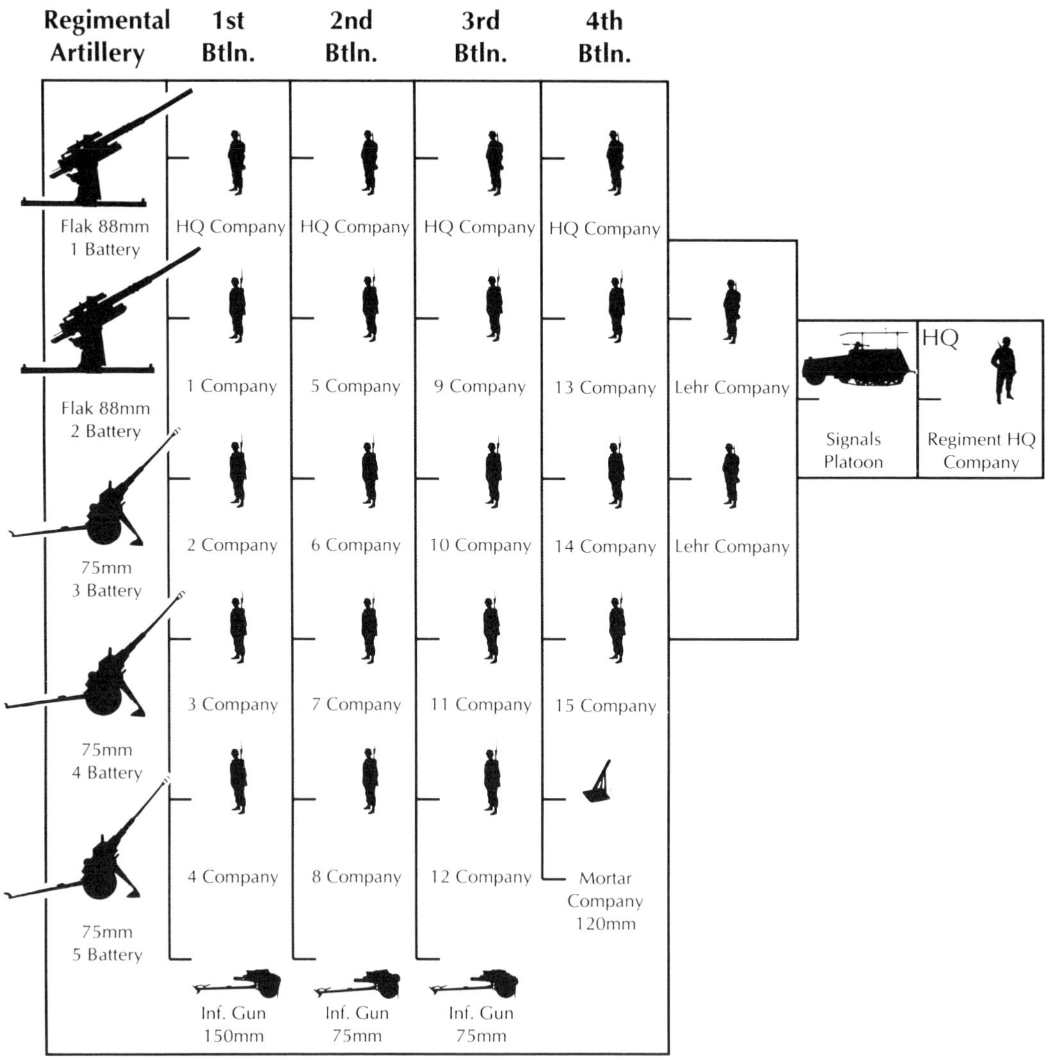

types of guns in these batteries were a hodgepodge of different types and nationalities. The German artillery crews were astonished at having to learn to fire not only German, but Italian, Soviet, Polish, and even Yugoslavian field pieces. Except for fifteen assault guns, the city was tankless. When the Russians finally surrounded the city, 35,000 troops stood ready to defend the capital of Breslau, supported by 15,000 Volkssturm. Thus a total of some 50,000 men defended the city, and of this number, 29,000 (more than 50 percent) became casualties.

The first real combat action for Regiment Besslein came on the night of 1 February, when the regimental adjutant, Hauptmann Seiffert, led a company made up of the SS-NCO Administrator's School in a counterattack south of Breslau, near Treschen. This successful counterattack was the first of many such actions by the SS regiment. It was also an object of curiosity as well, since a female civilian doctor had followed the SS company into the field and treated the wounded men there in the most primitive of conditions. For her courage, she had the singular honor (though many women would later follow her example) of winning the first Iron Cross handed out to a woman from Breslau.

Though the Russian bridgehead north of the city had been destroyed, the one south of the city, by Peikewitz (also spelled Peisterwitz) still remained doggedly in the hands of the Soviets. Repeated German attempts to dislodge the Russians failed until 18 February, when Besslein's SS-regiment began to earn its reputation as a tough fighting unit.

Earlier, before the regiment was fully operational (this occurred on 8 February) the units under Besslein's command had only experienced minor skirmishes with the Russian besiegers. The only other notable encounter had occured on the night of 3, February when Hauptmann Seiffert led another raiding party into the positions in front of Besslein's regiment. He and Oberfeldwebel Schulze led a platoon of engineers equipped with flamethrowers into the enemy-held town of Waserbon. Seiffert had declined the offer by Besslein to use two companies of the 609th Infantry Division (whose men had been hastily put together from stragglers, men on leave, etc.). He correctly deduced that they would be unable to perform well because they could not function as a whole, integrated unit.

Armed with flamethrowers, machine guns, and grenades, Seiffert's assault troops stormed forward. They caught the Russians by surprise, overcoming them by the use of flamethrowers. The sudden and savage streams of red and white flames, licking and smoldering anyone and everything in their paths, broke the morale of the Russians, who panicked and ran from their positions. Clouds of smoke rose up from Waserbon, while jet streams of flames chased the defenders down the streets.

Those who chose to fight the German assault troops were either burned out or blown out. A Russian firing pit showed itself at the instant it opened up against the German troops. "Flamethrowers forward!" came the cry. "First squad, give covering fire to the roasting party."

The Russian fire seemed to have been suppresed temporarily by the German return fire. Just as the Russians recovered from the return fire, the German roasting party beganits job. "Burn the Bolshevick bastards out!" shouted Oberfeldwebel Schulze. Slowly, other men in the squad moved toward the house involved, using as much cover as possible. "Cease firing, flamethrowers back! Assault troop forward! Blast those cockroaches out." But this was not accomplished before Oberfeldwebel Schulze was killed by a lucky shot. The Germans took the town, but Schulze was dead—a great victory and an equally great loss.

Time and again, just as the Russians believed that Breslau's defenders were on their last leg, the beleaguered German units would regroup and counterattack—managing to throw the bewildered

Illustration: Eric Weisman.

THE SIEGE OF BRESLAU: THE APPROACH TO THE CITY

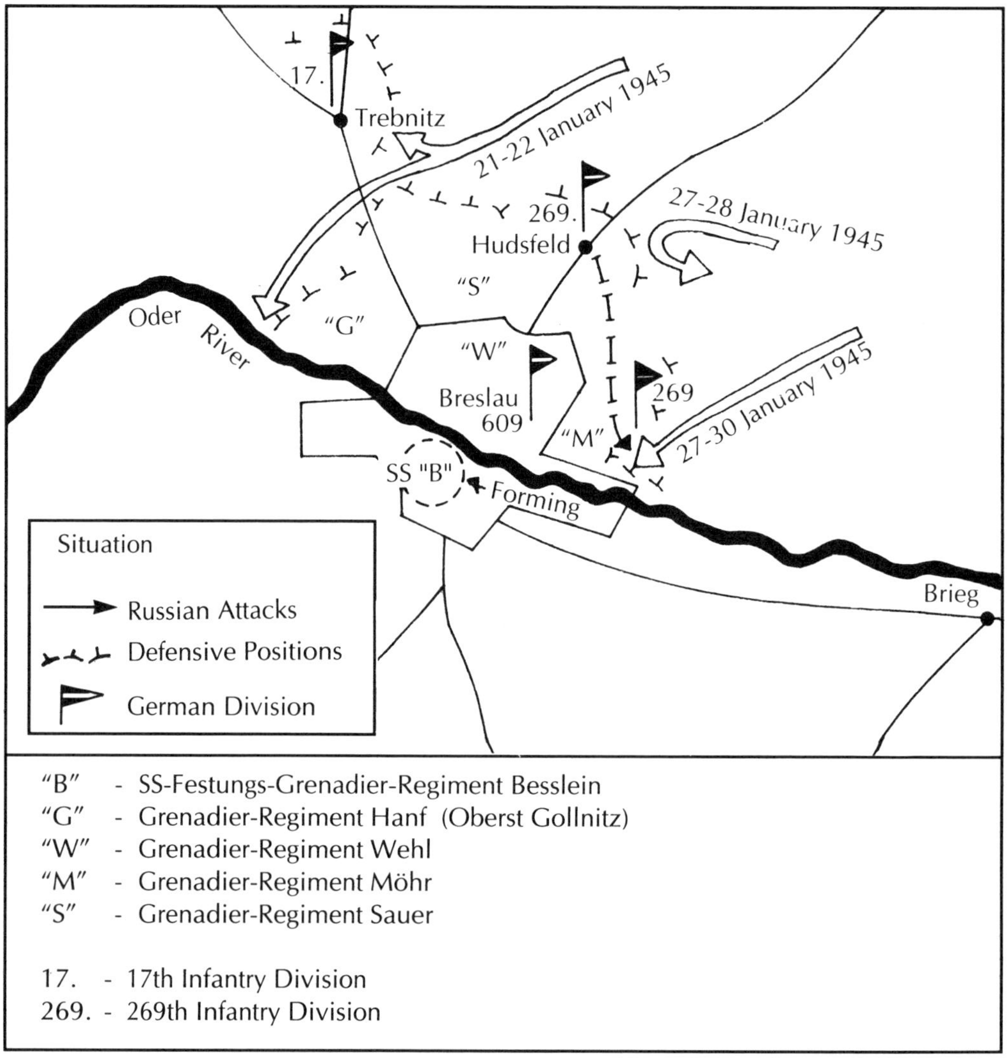

Russians out of a city block, single building, or section of rubble. The Russians couldn't understand Breslau's stubbornness. It only inflamed their passions to points of dastardly acts, which, in turn, fueled the hatred that the Germans had against the Russians.

The Russian peasant had a simplistic idea of his German adversary—only what his propaganda had told him. He saw this dogged German resistance as another sign of Nazi fanaticism. "Don't those Nazi pigs know the war is lost?" They remembered reaching the outskirts of Breslau only to find fanatical Hitler Youth firing at them from every house. "EVERY HOUSE A FORTRESS" and "SURRENDER? NEVER!" were scrawled or painted on every building. It's no wonder that the Russian soldier hated this stubborn eleventh-hour defense, just as peace and victory lay so near their grasp.

A Russian soldier was heard yelling to his adversary in the German trench lines, "Hey you Nazi lout! The war is over; give it up and save yourself!"

The German replied, "I'll see you in hell first, you murdering bastard! We'll never surrender!"

"Don't they know the war is over?"

THE SITUATION AT
THE END OF JANUARY 1945

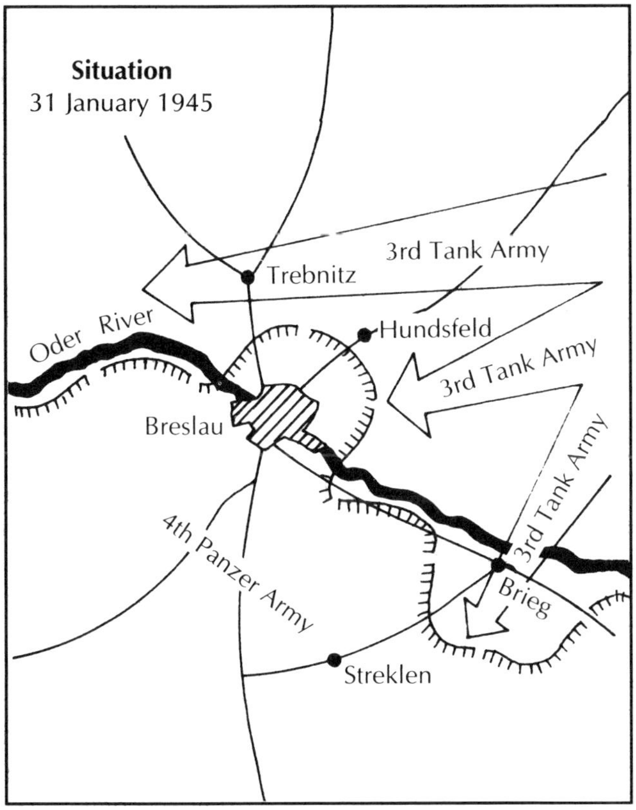

tion as the toughest and best regiment in the garrison.

Besslein's SS-regiment was supported not only by his artillery strength and that of the garrison, but by two platoons of Goliath tanks. These tanks were armored, tracked, remote-controlled vehicles no taller than one or two feet and capable of delivering a seventy-five kilogram explosive in all sorts of terrain. They were ideal for streetfighting.

The two platoons arrived by rail before the Russians finally encircled Breslau on 1 February. They were used for the first time in Breslau's streets and suburbs, the first of many surprises for the Russians.

Dawn on Sunday 18 February 1945 arrived quite early, 1:45 A.M. to be exact. At least that's what the meteorologists said, for the dust and smoke clouds from the constant bombardment were so great that they remained suspended above the city for weeks on end, making it hard to tell night from day. On this day, the gray dawn was disturbed by the sounds of German artillery fire. For what seemed too brief a time, Besslein's men heard their guns pounding the Russian defensive positions around the Peikewitz bridgehead, then they went on the attack, bayonets fixed, hearts pounding, and feet scrambling over the rubble and debris of a dying city.

For many days, Besslein's SS regiment had been fighting south of the city in the area of Deutsche-Lissa. Bitter house-to-house fighting proved successful against what German intelligence said were four separate Russian rifle divisions. Mines and bombs were set off by Besslein's sappers inside houses and buildings that the Russians had spent all morning or afternoon taking.

was repeated many times on the Russian side. This question was beyond the understanding of a simple German housewife or her six-month-old baby who died before she could learn to speak. But it was for those two simple human beings that Breslau's defenders were fighting—those two and the countless thousands like them who suffered and died. Two million German civilians died in World War II, 75 percent of them at the hands of the Russians during the last eight months of the war. The bodies of the dead civilians seemed to be saying the same thing, "Look at the blood-spattered plains around you. Look at our dead children and womenfolk you've raped, and you will see why Breslau shall never surrender!"

The battle for the Peikewitz bridgehead marked a milestone for the fledgling SS-regiment. It was this and future battles that earned Besslein's unit the reputa-

THE ENCIRCLEMENT

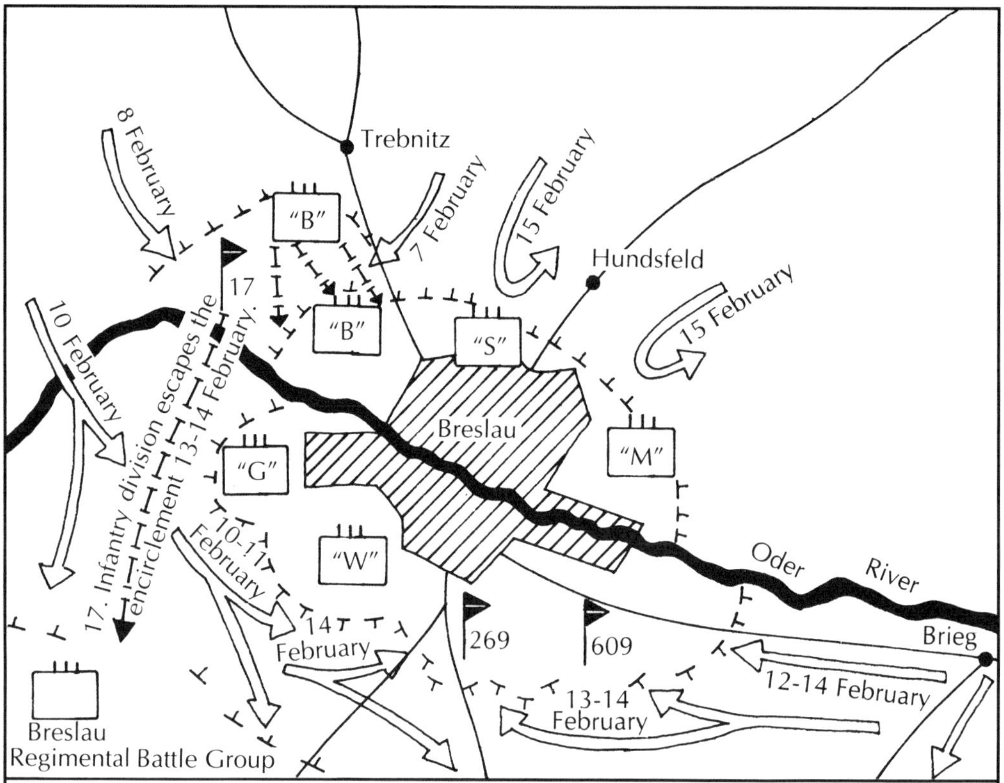

"B" - SS-Festungs-Grenadier-Regiment Besslein
"G" - Grenadier-Regiment Hanf (Oberst Gollnitz)
"W" - Grenadier-Regiment Wehl
"M" - Grenadier-Regiment Möhr
"S" - Grenadier-Regiment Sauer

17. - 17th Infantry Division
269. - 269th Infantry Division

Thereafter, the platoon or company would go into a counterattack and push what remained of the Russian force out of burning rubble that once was a building.

From floor to floor, street to street, and in tunnels and sewers, the fighting continued, sometimes burying friend and foe during a bomb explosion. The Hitler Youth and Volkssturm units in the adjacent sectors on either flanks of SS-Festungs-Regiment Besslein worked hard to build barricades and tank traps. Whole sections of city blocks were blown up at night to create fire-zones in which to knock out enemy armor.

Besslein's men got the opportunity to prove their worth by counterattacking a well-defended Russian bridgehead. To support this assault, the Breslau garrison commander, General Major von Ahlfen, supplied Besslein with his only armor: fifteen assault guns of the newly formed Panzerjaeger-Abteilung Breslau, under the command of Oberleutnant Wente. Leutnant Leo Hartmann, a member of the unit, recalls how it had been formerly Sturmgeschuetz Brigade 311 (the Lion Brigade) and attached to the 17th Infantry Division, until handed over to defend the city. The assault guns were too few

THE SIEGE

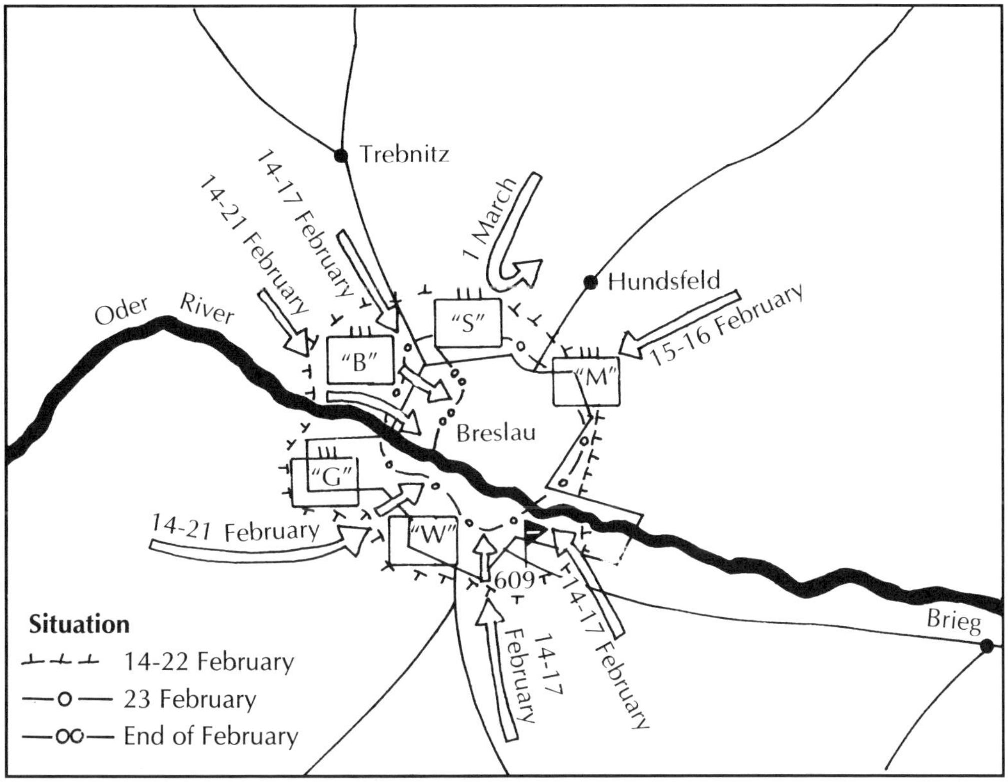

and precious to be wasted in the initial assault, so they followed Besslein's infantry, who in turn followed the Goliath tanks into the fray.

The last few German preparatory artillery rounds fired were smoke-producing shells, which it was hoped would cover the advance of three Goliath tanks, each packed full of explosives and ready to go. At exactly 6:00 A.M., the little tanks advanced, unobserved by the Russian troops around the besieged bridge at Reichstrasse 5. Before the enemy knew what was happening, the first armored midget tank had reached the second span of the bridge.

Lieutenant Kearn, who had been maneuvering his little armored Goliaths through the thick smoke screen, stopped the second vehicle in a pile of debris in the middle of the bridge, while he placed the third on the first span. Without losing much time, he ignited all three explosive packages. The blast came off in unison, causing the ground to shudder. People heard the blast for miles around and saw the thick clouds of dark smoke rise to the sky, unveiling to friend and foe the destruction that it caused.

Both spans of the bridge and its pillar lay in the cold waters of the Oder. Contorted pieces of smoldering steel, whose girders were grotesquely twisted in all directions, seemed to be pointing an accusing finger at everyone.

"Forward!" came the order, and the men of SS-Festungs-Grenadier-Regiment Besslein surged forward, bayonets fixed and ready. The wild charge quickly brought the regiment within the first line of trenches. The Russians fought tenaciously, but the fury of Besslein's men was unstoppable. The men fought like "berserkers," their adrenalin flowing within them like a gushing water faucet.

One of these was SS-Unterscharführer Kraus, a member of SS-Untersturmführer Franz Budka's platoon. Both men belonged to the lst Company, 1st

RESISTANCE TO THE BITTER END

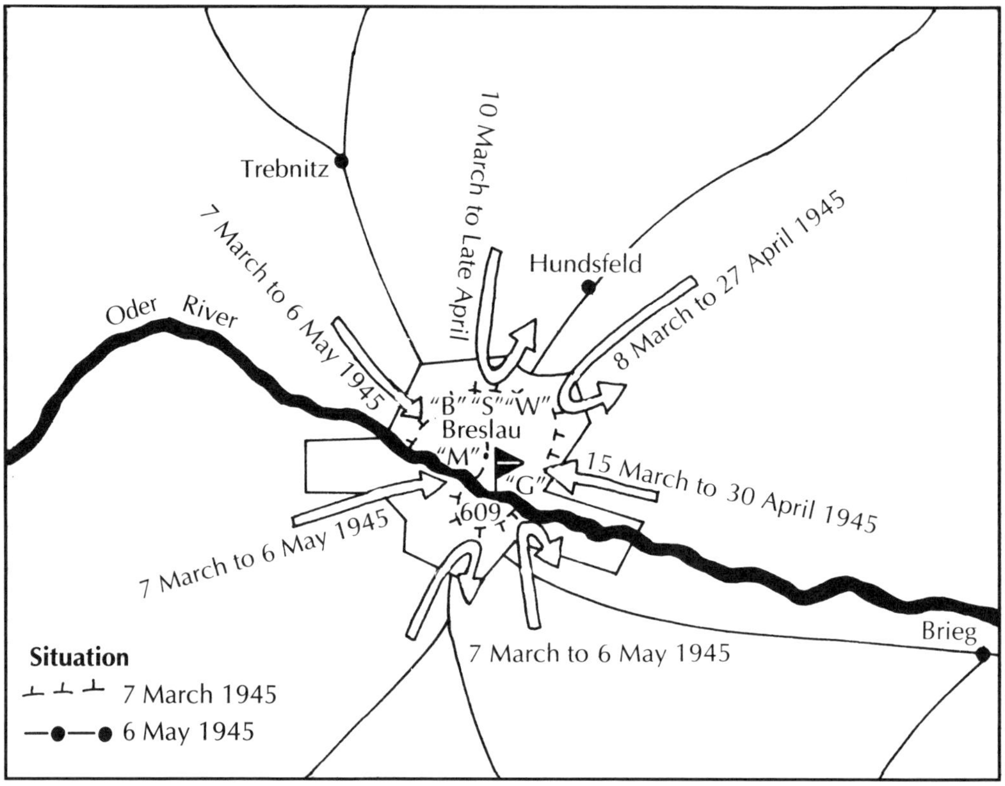

Battalion SS-Regiment Besslein and figured prominently in the combat record of their regiment, as well as in the history of the city's defense. Kraus carved out a little piece of history for himself when he jumped a Russian trench system from one side and flushed the enemy from there by rolling up their flank using grenades and a bayonet.

Besslein soon received the news: "The enemy bridgehead is burned out!" Such bravery deserved mention, and Kraus' actions won mention in the OKW war communique.

Besslein's SS-regiment moved south to meet a new Russian threat. Legnetz had fallen to the enemy on 19 February. A German counterattack aimed at raising the siege of Breslau had been defeated. One of the relief divisions had been the 17th Infantry Division, led by Major General Sachsenheiter. The 17th had made the long trek from the Vistula to the Oder River. For two weeks, it had fought hard and stubbornly in the area of Malt-Neumark, thereby holding off the city's encirclement from the west. It had surrendered much of its heavy weapons to the fortress troops before breaking out to the west on the night of 14 February, and yet five days later and much weakened, it had participated in the relief attempt.

Besslein's companies moved south through the ruins of a once beautiful city. Their morale was high as they marched, singing battle songs. One company had adopted black top hats while another company had chosen stiff hats called "korks." They wore them behind the lines, and the hats were considered prized possesions among the grenadiers. The defense of Peikewitz and Deutsche-Lissa lay behind them.

At Deutsche-Lissa they had fought elements of four Russian divisions. At Gandau, the regiment had just as stubbornly defended the airfield, which was the last supply point for the garrison.

When it finally fell, an airstrip was hastily constructed from one of the major avenues of the city. It was possible then to bring in more supplies and reinforcements, like two parachute battalions that arrived by plane on 25 February and 5 March. The latter also marked the arrival of the newly appointed garrison commander, General Niehoff, who had been chosen by the new commander of Army Group Center, Field Marshal Ferdinand Schürner.

By 20 February, the Russians again had made some headway in the southern part of the city. They had advanced a total of two kilometers, entering the bus depot. Regiment Möhr operated in this sector of the city. It contained many officers and NCOs from Besslein's command. Hauptmann Seiffert had personally taken command of Volkssturm Battalion 55 when its commander was killed in action.

Late in the afternoon of 20 February, Seiffert launched an attack against the bus depot and the surrounding city blocks. Russian antitank guns and dozens of T-34 and Stalin tanks fired point-blank at the attacking Germans. Seiffert's men fought heroically, the Hitler Youth in particular going wild. Soon more than a hundred of them lay dead on the battle-scarred streets, but the remainder were able to push the enemy back. Panzerfausts, the poor-man's flak, were plentiful and destroyed more than a dozen tanks. Goliath tanks detonated what few live Russians escaped from inside the depot and attempted to flee southward, bleeding and shell-shocked.

The battle raged on for days. On 26 February, the Russians again came at the defenders, who then included Regiment Besslein with a mass of tanks. Again, as before, the attackers were thrown back. Fighting grew exceptionally gruesome at the beginning of March. Regiment Möhr alone was responsible for disposing of 100 enemy tanks and antitank guns since 1 March. In the area of Besslein's command, his men had set up large banners facing the Russian positions that read: "HERE FIGHTS SS-REGIMENT BESSLEIN." As the Russian units began to get acquainted with Besslein's troops, they attempted to avoid that regiment as often as possible.

The beginning of March saw the commander of 2nd SS-Festungs-Regiment Besslein replaced by SS-Obersturmführer Willy Zizmann. SS-Hauptsturmführer Zielske had been killed on the Maria Hofchen Strasse. Zizmann, who had been fighting as commander of 10th Company SS-Festungs-Regiment Besslein, took over. He, too, was later replaced, by SS-Obersturmführer Alfred Rogge, who led 1st Battalion after SS-Hauptsturmführer Gregor was wounded.

On 10 March, twenty-two-year-old SS-Unterscharführer Grunne, from Regiment Möhr, received the Knight's Cross for his bravery in action as squad leader in the regiment. SS-Untersturmführer Franz Budka also won the much-coveted Knight's Cross when on 29 March, in the burning ruins of August Strasse, he rallied his 1st Company and held position, even as the buildings they occupied went up in flames.

The Russians, in the insurance building across the street, had set fire to the whole block. Budka directed the fire of his company—naked from the waist up and choking from black smoke—against the advancing Russian troops, using handmade catapults to hurl the grenades at great distances. He had learned this trick from the Hitler Youth. At the same time, Budka rallied some men in an attempt to put out the fire raging out of control in the building.

The German army war communique for 30 March described the heroic fighting at Breslau, with special mention of Budka:

> *. . . in the fortress of Breslau the lst Company of SS-Regiment Besslein under the leadership of SS-Untersturmführer Budka withstood with fa-*

natical will all breakthrough attempts, fighting from the cellars of a burning houseblock, causing great casualties to the enemy.

Budka received his Knight's Cross on 19 April 1945 at the age of twenty-five. A friend of Budka's, Hermann Heberle who was in the 1st Platoon, recalls that his friend died of wounds inflicted on him on 6 May 1945, as he attempted to break out of the city. Three days later, SS-Obersturmführer Alfred Rogge also received his Knight's Cross. He, too, would only wear it for a short time; three days later, as a prisoner of war, he died of wounds received in battle—wounds the Russians neglected to treat. Rogge would have been twenty-two on 1 August 1945.

The wounded SS-Obersturmführer Willy Zizmann, former commander of 2nd SS-Festungs-Regiment Besslein, received the very rare Honor Roll Clasp, while SS-Obersturmbannführer Georg-Robert Besslein received his Knight's Cross on 30 April, ten days before Breslau's capitulation, after seventy-seven days of fighting a larger enemy force.

The fighting in April continued to be as bloody as during the two previous months. Slowly, the Russians were able to push the determined and tenacious defenders towards the city center. By 2 May, resistance was useless. General Niehoff radioed Schörner for permission to surrender. Schörner, known for his fanatical determination, refused. Hanke, the Gauleiter who had been partly responsible for the grievous civilian losses for not evacuating the city sooner, escaped Breslau in Niehoff's own plane. While Hanke chose to flee, the garrison commander chose to stay and share the fate of his men and that of the civilians. Soon, even such staunch believers in a German victory as Field Marshal Schörner left their commands, hoping to avoid capture by the Russians. Schörner, dressed in civilian clothes, was found by the Americans.

The capture of most of Silesia had taken from the middle of January,

Illustration: Eric Weisman.

when the Soviets began their winter offensive, until the end of March, when the whole of Upper Silesia was taken. The Russian advance into Silesia had put millions of people on the move. Of an estimated 4.7 million inhabitants living in Silesia at the beginning of 1944, only 620,000 remained by 16 April 1945. Of the Breslau garrison of 35,000 troops and 15,000 Volkssturm, 29,000 were killed or wounded—more than 50 percent. The civilians still living in the city at the beginning of the siege also suffered grievous losses. Postwar estimates put the number of civilian deaths at 40,000.

An especially gruesome captivity awaited the German survivors of the battle of Breslau. Of the thousands of soldiers sent to Siberian work camps, few returned to tell about it. Georg-Robert Besslein himself, would not return until 9 October 1955. His SS Fortress Grenadier Regiment Besslein made some crucial contributions to the city's defense. Had it not been for them, Breslau would have surely fallen sooner. It is, therefore, a tribute to the fighting spirit of the men who served in the regiment that the city commander was overheard to say that the Besslein Regiment was the best unit in his command.

This was a remarkable accomplishment for men who were considered reararea or garrison troops at best. The regiment existed for only a few months, yet its men won numerous Knight's Crosses. Of the fifty or so independent Waffen-SS formations from battalions to whole divisions, Regiment Besslein was twenty-third highest on the list to have won the most Knight's Crosses. Not a bad record considering its short lifespan and its small numbers—3,000 to 3,500 men.

CHAPTER FOURTEEN

Do hate the traitor, though they love the treason.
Samuel Daniel
"Tragedy of Cleopatra"

Toward the end of 1943, the project of raising a Turkic Moslem SS division began to take form within the SS-FHA (SS High Command). For more than two years the SS-FHA had been working with the RSHA (the Reich Security Main Office) on a plan to raise such a division. Reichsführer-SS Heinrich Himmler had been informed about the tenacious fighting qualities of the Turkic volunteers. He had learned how three Turkic battalions had fought to the last man at Stalingrad and how other Turkic battalions had been destroyed but not routed during the fighting withdrawal from the Caucasus.

He had been impressed when told how another Turkic battalion had broken out of a pocket near Kharkov and reentered it just to rescue the body of their beloved German commander. This German officer had been an enlightened commander who had treated his Turkestani volunteers with respect. The men of his battalion had repaid him with great devotion to duty.

One of the chief architects of the Turkic-SS program was Dr. Rainer Olzscha, a physician whose interest in the East had been aroused by his own professional investigation of epidemics and who later co-authored a book on Turkestan. At a time when the Waffen-SS was grasping for more manpower, Dr. Olzscha was trying to sell his pan-Turkic ideas. The one uniting element tying all of the men who were pushing for a Turkic-SS unit was the desire to use the innate hatred for Stalinism that Turkestanis felt.

In 1924 Turkestan disappeared from Soviet maps. The Soviet Union banned all languages except Russian, suppressed religion, and imposed collectivization methods. The Turkestanis hated these Communist attempts to destroy their personal and ethnic freedoms. It's no small wonder, then, that more than 50,000 Turkestanis lost their lives fighting on the German side and an additional 12,000 were wounded. This belittles the estimated 3,000 Turkestanis who deserted the German camp during the war. Of these 3,000, only about 300 to 500 joined the Communist partisans, while the rest deserted to anti-German/anti-Communist partisan forces.

Himmler's mind was finally made up when a German army officer by the name of Andreas Meyer-Mader convinced him of the possibility of using such a Turkic formation. Major Meyer-Mader had contacted Himmler through the SS-FHA, offering to help raise and command such a Turkic SS unit. Himmler was impressed with Meyer-Mader's credentials. Not only had Meyer-Mader been commanding a Turkic battalion for quite some time, but he had also seen foreign service in China while attached to the German embassy before the war. As such, Major Meyer-Mader was considered an expert on "Eastern peoples."

In November 1943, Major Meyer-Mader had his meeting with Himmler. One month later Himmler requested Gottlob Berger (the head of the SS-FHA) to arrange a meeting between the Grand Mufti of Jerusalem, Major Meyer-Mader, and a representative of the SS-FHA. The meeting was to be arranged to get the approval of the Grand Mufti to raise a Turkic-Muslim SS division whose political and military goal was the fight against Bolshevism. Himmler and Meyer-Mader had agreed that the presence and the "blessing" of the Grand Mufti would be needed so that his "spiritual leadership" (religious and political influence among Muslims) would help facilitate the setting up of a division.

Meanwhile, Himmler decided to appoint Major Meyer-Mader as head of the pro-

An Eastern Turkick volunteer reading a newspaper produced expressly for the Moslem volunteers of Turkistan. Note the mixture of Army/Waffen-SS clothing and insignia: the Wehrmacht national eagle on the soft cover and the left collar tab (blank) of the Waffen-SS (top right).

Another view of the newpaper Neu Turkistan *(New Turkestan), plus a handbook titled* Muslim Turkistan *(bottom right). The Muslim volunteers were allowed to practice their religion. Each battalion also had a religious Ihman (clergyman). Photos: Bundesarchiv.*

posed "Muselmanischen SS-Division Neu-Turkistan." Meyer-Mader was transferred into the ranks of the Waffen-SS and promoted to the rank of Lieutenant-Colonel (SS-Obersturmbannführer), effective 1 January 1944. On 14 December 1943, another meeting was held in Berlin; present at that meeting were the Grand Mufti of Jerusalem, Major Meyer-Mader, and three Turkestani officers from Mader's 480th Turkic Battalion. At the end of this meeting, they all declared themselves ready to support the formation of a Turkic SS unit.

Three weeks later on 4 January 1944, another meeting was attended by the same men and SS-Sturmbannführer Schulte, who was deputy head of the SS-FHA, and therefore Gottlob Berger's immediate subordinate. They finalized details of raising the proposed Turkic SS division. They decided to disband the German Army 450th, 480th, and I/94 Turkic battalions and transfer the men to the Waffen-SS. These volunteers formed the regimental cadre for the division. Among other decisions made, all men agreed that the unit would begin assembling in the SS Work Camp at Poniatowa, Poland.

Approximately 30,000 potential Turkic recruits existed in the form of prisoners of war (POWs) and men working in Germany. The divisional formation staff was headquartered at Lublin, also in the General Government (the part of Poland run by the German army and not annexed by Germany). The new division used the SS Troop Training Ground at Trawniki, Poland, for field exercises.

By 6 December 1943, the I/94 Turkic Battalion had been disbanded and its men sent to the Waffen-SS. On 18 December, the men of the 450th Turkic Battalion were transferred, and the 480th Turkic Battalion soon followed. By the middle of January 1944, the Turkic SS Division was taking form. SS-Obersturmbannführer Andreas Meyer-Mader began making the rounds of German POW camps, recruiting men for his division. He recruited not only Turkestanis, but also Azerbaijani, Kirghi, Uzbek, and Tadzhik volunteers.

It is not known how many Muslims volunteered from a direct appeal by the Grand Mufti of Jerusalem to their religious or political beliefs. Certainly many volunteered because of him, but many also joined to escape the spartan life that the SS work camps offered them. They had been, after all, Russian soldiers who had been forced to work. Soon enough, Meyer-Mader had assembled enough men to form three infantry battalions into the lst Eastern Muslim SS Regiment (1st Ostmuselmanische SS-Regiment).

Typical of these Turkic volunteers was Waffen-Untersturmführer der-SS Alichan Kuliew (also spelled "Kuliev"). Kuliew was born in Turkestan on 3 June 1921 and conscripted into the Red Army in 1939. He attended Flight Leader School but was taken prisoner at Brest-Litovsk on 24 June 1941. He was an early volunteer in the Turkestani Legion formed by the German army and spent three months in Warsaw undergoing an officers course. Later, Kuliew helped to raise two companies of Turkestani volunteers. He was attached to the 3rd Company, 783rd Turkic Battalion as a platoon leader but was later transferred to the battalion staff as ordnance officer.

Wounded at Rostov, he was taken to a field hospital in Dnepropetrovsk, but he later returned to the 783rd Battalion, which had been reconstituted to a strength of about 500 men. The 783rd participated in the fierce fighting around Stalino in 1943 and later fought Russian partisans twelve kilometers east of Bryansk. The 783rd then went to the Nikolai bridgehead, where 300 of the battalion's 500 men lost their lives. While he and the rest of the survivors of the 783rd Battalion were in southern France for rest and refitting, Kuliev heard the request for Turkestani volunteers and offered himself. He was soon on his way to Poland via Stettin, Germany.

By the end of January 1944, Meyer-Mader had brought a force of some 3,000 men into the regiment. The formation of the division began to slow, mainly because of a shortage of equipment of every type, including such elementary items as uniforms and boots. Photographs aptly illustrate the extent of these shortages and the plight of the soldiers. The pictures show men of the Ostmuselmanische SS-Regiment in an assortment of mixed German Army and Waffen-SS uniforms.

In the beginning of February 1944, the regiment was transferred to Belorussia for further training and for antipartisan duty. Meanwhile, the SS-FHA made an attempt to introduce a sizable German cadre into the proposed division, while recruitment of volunteers continued. In late January 1944, the SS-FHA had requested the German army to release Captain Billig from duty with the Wehrmacht Welfare Office in Rostock for assignment with the Waffen-SS. This occurred on 5 February. Unknown to the SS-FHA, Billig had a serious drinking problem, which would adversely affect his performance in the Ostmuselmanische SS-Regiment.

During the first half of March 1944, the Ostmuselmanische SS-Regiment under SS-Obersturmführer Meyer-Mader was conducting antipartisan operations from their base at Yuratishki (also spelled "Juracizki") near Minsk when Meyer-Mader was killed at the hands of a lucky partisan sniper. The newly appointed SS-Hauptsturmführer Billig took command of the regiment. On 28 March 1944, this unfortunate turn of events shattered the morale of the regiment. Transferred into the Waffen-SS, perhaps against his wishes, it is quite possible that Billig resented his new duties, which took him far away from his comfortable post at Rostock.

January 1944 movements are shown in the map on page 166.

Meyer-Mader's death unleashed the unreliable elements in the regiment, who

This map depicts the movements of the Ostmüselmanische SS-Regiment/Osttürkischen Waffen-Verbande der-SS between January and October 1944. When the regiment returned from Russia in July 1944, it was temporarily attached to the Dirlewanger Penal SS Brigade and participated in crushing the uprising in Warsaw by the Polish Home Army.

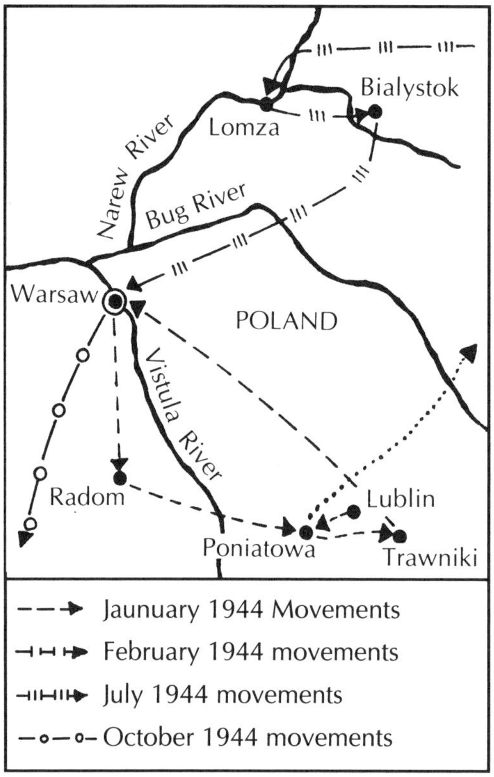

began to foment trouble. Though unsubstantiated, there is speculation that the Ostmuselmanische SS-Regiment had been infiltrated by Red agents for the express purpose of destroying the morale of the unit and, if possible, the desertion of its men. One example of this involved a unit of the German Brandenburg commandos operating in Russia that had trouble with its native contingent. A well-known anti-Communist in the battalion identified a recent recruit as the culprit in all of the unrest. This newcomer, a Ukrainian, staunchly denied his guilt. Still, the German officers decided to keep an eye on the Ukrainian. When the battalion later mutinied, killing most of the Germans, the real instigator was revealed to be the so-called "anti-Communist" all along. Of course, the falsely accused Ukrainian had been killed as well.

It was into this environment of resentment and mistrust that SS-Hauptsturmführer Billig made his appearance at Yuratishki on 28 March 1944, and he exacerbated the situation by his inept handling of his first encounter with the troublesome elements in the regiment. Completely drunk, he removed the problem by shooting seventy-eight suspected mutineers. Though he probably thought that this would rid the regiment of its mutinous mood, it only had the effect of alienating the volunteers and inspiring a hatred for this "newcomer."

After this incident, Billig's command was short-lived. He was transferred on 6 April, but his replacement did not show up until 27 April. Even though the regiment had no official commanding officer, volunteers continued to arrive, a large batch arriving between 7 and 12 April. On 27 April, SS-Hauptsturmführer Hermann arrived to take control of the demoralized unit.

Hermann's first step was to take stock of the situation. He soon found out that, as a result of Billig's actions, hundreds of volunteers had deserted the regiment. He immediately began to defuse the situation and soon regained control of the regiment by his sensible command. On 2 May 1944, just a week after his arrival, Hermann was removed from command. He was not transferred, however, but remained on the regimental staff until the SS-FHA could find another commander. The unit would again be without a commanding officer until June 1944.

In July 1944, the Ostmuselmanische SS-Regiment was ordered back to Poland and out of the control of the Higher SS and Police Leader Minsk. That month, it was located in Lomza, then Bialystok. Toward the end of July, it was attached to the SS Dirlewanger Penal Brigade, which helped quell the uprising in Warsaw. In the meantime, the search for a suitable commander continued. One such officer was presumed found when the name of Major Harun-el-Raschid was presented to Himmler in July 1944.

Born in Austria as Wilhelm Hintersatz on 26 May 1886, Raschid had served as a colonel in the Austro-Hungarian army in

AREA OF OPERATIONS OF THE OSTMUSELMANISCHE SS-REGIMENT IN RUSSIA

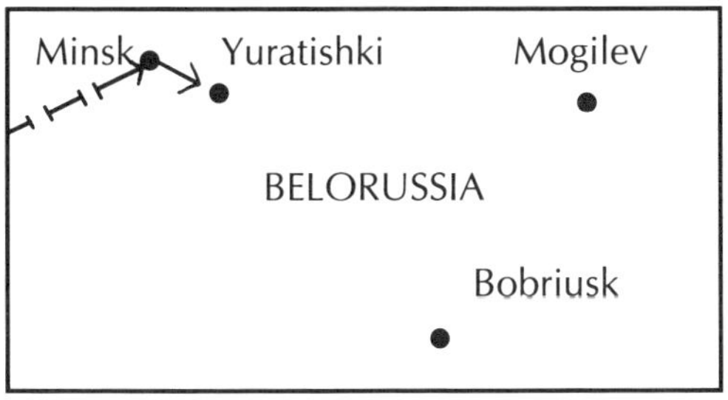

World War I. He had been attached to the Turkish general staff, working with the famous Enver Pasha during most of the war. During his stint in Turkey, he converted to the Islamic faith and took the name of Harun-el-Raschid Bey, which had been officially given to him by the sultan of Turkey. His most recent position before being transferred to command the Turkic SS unit was that of liaison officer between the RSHA and the Grand Mufti of Jerusalem. He had recently come over from the army, receiving the rank of Sturmbannführer (major) on 24 August. On 1 September, just eight days later, he was promoted, this time to lieutenant colonel (Obersturmbannführer). Thirty days later, on 1 October, he was once again promoted and became an SS-Standartenführer. It was hoped that this new commander would improve the unit's morale.

The formation of the Ostmuselmanische SS-Division Neu-Turkistan was on the mind of Reichsführer-SS Heinrich Himmler throughout the summer and fall of 1944. On 2 May a directive by him had stated implicitly that "the formation of an eastern Muslim division, from members of the Turkish race, is to be carried out this year." By the fall of 1944, that prospect had disappeared: not enough recruits were forthcoming to raise the proposed Turkic-SS division.

Thus, on 1 October, Himmler implemented the changes described in the memorandum on page 168.

The Ostmuselmanische SS-Regiment remained under the control of the SS Direlwanger Penal Brigade during the rest of the summer of 1944. SS-Hauptsturmführer Hermann had been killed in action on 11 July 1944. Apparently, after the fighting in Warsaw, the Dirlewanger Brigade had been sent to Radom for rest and refitting. It is not known whether the Ostmuselmanische SS-Regiment followed the Dirlewanger Brigade, but on 12 October on the advice of Obergruppenführer Gottlob Berger, the Dirlewanger Brigade was transferred to Slovakia.

Per Himmler's order written on 20 October (but backdated to 1 October), the formation of the Osttürkischen Waffen-Verbande der-SS would take place in Slovakia. If indeed the Turkic-SS Regiment was detached from the Dirlewanger Brigade before that unit's transfer to Slovakia, it nevertheless found itself in that country by the end of October 1944.

The three battalions of the 1st Ostmuselmanische SS-Regiment were divided on racial lines. There were battalions of Turkestanis, Azerbaijanis, and Tatars from the Idel and Ural Mountains. Each battalion had a staff company headquarters plus five rifle companies. In addition, the regiment had a regimental staff company.

Political events outside the Ostmuselmanische SS-Regiment began to take a turn for the worse. While in the process of forming into a brigade, elements in the RSHA began to make overtures about the eventual transfer of the brigade to General Vlasov's Russian Army of Liberation. This was contradictory to the Turkestani National Committee's aims for an independent state. This committee, sponsored by the Ost-Ministerium and headed by Veli Kayum Khan, head of the Turkestani "government in exile," had been in charge of the political and national leadership of the Turkestani volunteers. It had successfully raised the morale of the Turkestani volunteers by supporting independence for Turkestan and, with the aid of Veli Kayum Khan and the SS-FHA, setting up

To the SS-FHA
From the RFSS

1. As of 1st October, I order the formation of the Osttürkischen Waffen-Verbande der SS.

2. The Osttürkischen Waffen-Verbande der-SS will gather all usable eastern turks (Turkistanis, Volga and Ural Tatars, Crimean Turks, Aserbaijanis, etc.) for the military and political training of these volunteers to make the formation a purposeful fighting unit . . .

3. Make up:
   A. The staff of the Osttürkischen Waffen-Verbande der-SS is to consist of the following units, as per the accompanying order of battle. The responsibility of the formation will be with the commander, Waffen-Standartenführer der-SS Harun-el-Raschid.
   B. The formation of the fighting units will be found in the special orders from the SS-FHA.

4. Formation:
   The Osttürkischen Waffen-Verbande der-SS is to be subordinated to the SS-FHA in all matters of formation, training, supply, as well as leadership, directly to Amt-II. The SS-FHA is responsible for all political and cultural questions, also for any propaganda tasks.

5. Formation to take place in Slovakia, particulars to follow.

6. Personnel:
   A. The German officers in the Osttürkischen Waffen-Verbande der-SS will be briefed by the commander, in direct agreement with the SS-FHA. Amt-ii, Abteilung Ie.
   B. Necessary NCOs and other ranks will be supplied by the SS-FHA, Amt-V, Abteilung IIa.

7. Material:
   Information about weapons, equipment, and vehicles will come from the SS-FHA, Abteilung Ib.

Signed

H. Himmler
RFSS

schools at Dresden and Göttingen to train religious imams for the military units.

The Dresden school mainly trained priests for the Turkestani Waffen-SS, while the school at Göttingen taught imams for the Turkic units in the Wehrmacht. The teachings of both schools centered on pan-Turkic/pan-Islamic beliefs. Slowly, the news of the infighting within the SS about the fate of the Turkestani men filtered down through the ranks of the forming brigade. They resented the thought of being transferred to a Russian-controlled organization, and, eventually, this resentment would lead to a mutiny within the ranks of the men.

On 2 November 1944 (and again on 2 December), the forming Osttürkischen Waffen-Verbande der-SS was listed under the control of the Higher-SS and Police Leader Slovakia. Other formations under this command included the 178th Reserve Panzer Division, SS-Brigade Dirlewanger, the 14. Waffen-Grenadier-Division der-SS, 18. SS-Freiwilligen-Panzergrenadier-Division Horst Wessel, and the SS-Kampfgruppe Schill.

Events in December proved disastrous for the morale of the Turkestanis. The brigade had been reorganized into three regiments: Turkistan, Azerbaijan, and Idel-Ural. A Christmas party had been organized by the German and Turkestani men of the Turkestani regiment. Though they were Muslims, the Turkestanis went along with their German comrades, if only because free gifts were offered. It was during this Christmas party that the men of the Turkestani 1st Battalion, led by a Waffen-Obersturnführer Gulam Alimov (or Alim), deserted the brigade.

Waffen-Unterstrumführer Alichan Kuliew, who was a witness to the mutiny, later testified as to what occurred on the night of 24 December 1944.

Waffen-Obsturnführer Alimov, Waffen-Untersturmführer Dr. Dschaparow, Waffen-Hauptscharführer Safajew, Obscharführer Füttermeister, and Waffen-Untersturmführer Asatpalvan were in Alimov's room for the Christmas gathering. Five minutes later, SS-Untersturmführer Lampa and SS-Oberscharführer Flamek arrived.

Both Germans had numerous Christmas packages that contained cigarette packs, candy, and even games, Alimov gave them a very cheerful welcome.

The Germans then presented the gifts to the men present. Thereafter, soup and cognac were drunk at the table. The cognac was brought by Waffen-Untersturmführer Asatpalvan, the company commander of the 3rd Company. Asatpalvan proceeded to distribute the cigarettes that were in the packages. After this, Asatpalvan turned to the Germans, telling them that their superior [SS-Hauptsturmführer Fürst] was very bad in nature and by race. He also said to them that he had no doubt that Fürst thought bad of him (Asatpalvan), and accused him by noting that two other Turkestani officers had been sent to the officer's course at Myjave instead of him.

Alimov interrupted the conversation, telling the Germans that he did not understand why Fürst had not first consulted with him about the transfer of the

This map shows the movement of the Ostmuselmanische SS-Regiment to Myjava, Slovakia, during the third week in October 1944. SS-Standartenführer Harun-el-Raschid then took control of the forming Osttürkischen Waffen-Verbande der-SS.

two officers. SS-Untersturmführer Lampa replied that he did not understand all of this. Then the Germans were roughly dismissed from the room. Alimov then gave the order to leave the battalion, calling for the horses in order to withdraw. They went over to the area of the Azerbaijani Battalion, but its men would not leave with him. He then took his men into the forest, shooting myself and SS-Untersturmführer Spiese, who happened to be coming back from Myjava.

The order to prevent his escape come too late. Standartenführer Harun-el-Raschid Bey received news of the mutiny by Waffen-Unterscharführer Alim. Bey took the report and pounded himself on the shoulder. He them proceeded to the high mullah's room and there slept the night.

That same evening Bey awoke and telegraphed Pressburg [Bratislava] for an aircraft to take off and drop leaflets in the wooded area where Alimov was supposed to be. The leaflets asked Alimov to return and that all would be forgiven. Waffen-Obersturmführer Alimov disregarded the offer and remained in the town of Propat. A part of the men who went with him by force then force-marched toward Nove Mesto and returned to the barracks.

Alimov was arrested by Slovaks who were friendly to the partisans.

A private in the Ostmuselmanische SS-Regiment. The Oriental features are highly recognizable. Note also the double blank collar tabs signifying that this volunteer was an SS-Schütz, or private (right).

Men of the Ostturkischen Waffen Verband der-SS (below right). Again, several volunteers of Asiatic descent are clearly visible. Two men in the photograph are wearing the Turkistani arm patch, including the only German in the photograph (the officer). Photos: Bundesarchiv.

Men of the Ostmuselmanische SS-Regiment. Note the easily recognizable features of four of the seven volunteers left. The photograph below left depicts five volunteers all of Asiatic descent. Note also the mixed Wehrmacht/Waffen-SS insignia. Photos: Bundesarchiv.

Approximately 450 to 500 men of the 1st Battalion deserted that Christmas night; about 300 eventually returned to the battalion. No doubt the reason given by Alimov and Asatpalvan was an excuse. The mistrust spawned over RSHA's handling of the proposed transfer of the brigade was the real cause for the desertion. Furious, Reichsführer Himmler, through the offices of the SS-FHA, ordered the Osttürischen Waffen-Verbande der-SS reorganized on 30 December (but he backdated the order to 15 December). (A copy of the order appears on page 173.)

The SS-FHA pointed the accusing finger for the Christmas Eve desertions at the brigade's commanding officer, Harun-el-Raschid Bey. A report dated

Eastern Muslim volunteers roasting a side of beef (right). At least one volunteer in this photograph is wearing a camouflaged cover over his uniform. The volunteer in the German helmet (second from left) is wearing the arm patch of the Turkoman volunteers.

Photograph showing a Turkick volunteer still in the service of the Wehrmacht. Note the insignia arm patch of the Eastern Turkick volunteers clearly displayed on the right sleeve, about six inches below the shoulder. When these volunteers were grouped into the Ostmuselmanische-SS Regiment, they continued to wear the army-style arm patch (below). Photos: Bundesarchiv.

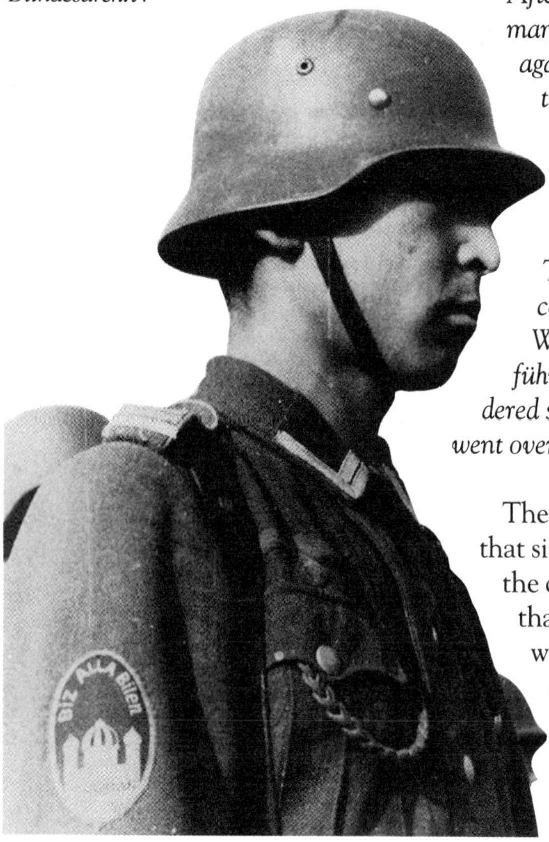

around the middle of January 1945 from Amtsgruppe D of the SS-FHA stated the following:

> *After the present commander unilaterally and against the orders of the SS-FHA placed command in the hands of untrustworthy men, a group of the Turkestani under the command of Waffen-Obersturmführer Alimov murdered some NCOs and went over to the Partisans.*

The report went on to say that since the behavior of the commander showed that he couldn't or wouldn't take the necessary steps, he should be relieved of his command and SS-Hauptsturmführer Fürst given the leadership of the brigade, especially in regard to the Waffengruppe Turkistan.

The removal of Harun-el-Raschid Bey was by then a foregone conclusion. Someone had to take the fall for the mutiny, and Harun-el-Raschid—guilty or not—was the most likely candidate. He left the Osttürkischne Waffen-Verbande der-SS in February 1945, and Fürst took command. Fürst was promoted to SS-Sturmbannführer one month later.

The command of the 1st Battalion Waffengruppe Turkistan had been severely depleted. Gulam Alimov had deserted with all of the company commanders and most of the 1st Battalion. The ordnance officer of the battalion, Waffen-Untersturmführer Dr. Dschaparow, was acquitted of any involvement in the mutiny, as was the company chief of the battalion, Waffen-Untersturmführer der-SS Safajew. Replacing the company commanders was not easy, but it is known that Waffen-Oberstrumführer der-SS Nasarow, who had been attached to the SS-FHA, was transferred to the 1st Battalion, possibly taking over command of the battalion.

Reorganization of the Osttürkischen

To the Osttürkischen Waffen-Verbande der-SS

From the SS-FHA

1. With effect from 15 December 1944, the Osttürkischen Waffen-Verbande der-SS is to be reorganized as follows:
    A. Staff, Osttürkischen Waffen-Verbande der SS
    B. Staff Waffengruppe Idel-Ural, Staff Waffengruppe Turkestan, Staff Waffengruppe Krim
    C. Fighting units as "B," formed as battalions
2. In the framework of the personnel at its disposal, the following units are to be formed first:
    A. Staffs for the three Waffengruppen (abridged regimental staffs to be worked out in the Osttürkischen Waffen-Verbande der-SS and approved by the SS-FHA, Amt-II, Organization Abteilung Ia.
    B. Each Waffengruppe will have:
        aa. Two infantry battalions
        bb. Five infantry companies per battalion
        cc. Formation of other units will be notified by the SS-FHA
3. Formation to take place around Miava, Slovakia.
4. The commander is to be responsible for the formation of the fighting units.
5. The Ostmuselmanische SS-Regiment forms part of the Osttürkischen Waffen-Verbande der-SS and is thereby considered dissolved.
6. Any Azerbaijanis are to be separated out of the Osttürkischen Waffen-Verbande der SS and delivered directly to the Kaukasicher Waffen-Verbande der-SS.

Chief of the SS-FHA

Signed Juttner
SS-Obergruppenführer and General der Waffen-SS

Waffen-Verbande der-SS began in January. Conforming to the SS-FHA orders of 30 December, the Azerberjani regiment in the brigade was withdrawn and transferred to the Kaukasicher Waffen-Verbande der-SS, which was then forming in northern Italy. The 3rd Battalion of the old Ostmuselmanische SS-Regiment formed the basis of the Waffengruppe Idel-Ural, while the 1st and 2nd battalions of the regiment formed into Waffengruppe Turkistan.

The men who formed Waffengruppe Krim came from another Waffen-SS formation that had been disbanded on 31 December 1944, the Waffen-Gebirgs-Brigade der-SS (Tatar Nr. 1). The origins of the Tatar brigade date back to the summer of 1944 when all of the Crimean Tatar Schüma battalions were gathered together and formed into a volunteer mountain regiment of the Waffen-SS.

In May 1944, the regiment began to take form at Moorlager Training Camp.

Out of all the Waffen-SS cuff bands, only that of the Osttürkischer Waffen-Verband der-SS was in colors other than black and white. Because this unit was composed primarily of Muslims, its cuff band sported the traditional Muslim colors of green and white.

Osttürkischer Waffen-Verband der SS

At that time only a battalion had been formed, but by June a sufficient number of men had been gathered to form a regiment or a small brigade. SS-FHA decided in late June to expand the unit into a brigade and issued an order outlining such an expansion on 8 July. In July, the forming brigade left Moorlager for Hungary for training under the Replacement Army and also for garrison duty.

SS-Standartenführer Fortenbacher received command of the brigade. Unfortunately for the unit, Fortenbacher believed that his post was a sort of demotion, since the unit was not considered a first-class formation. Because of this, he took little interest in his command, and the brigade made little headway. Another reason for the brigade's stilted development was that weapons and equipment were slow to arrive because the Waffen-SS' higher priority was arming the primary divisions. The order to disband the forming brigade finally came in December 1944, and the men were ordered to join the Osttürkischen Waffen-Verbande der-SS. The brigade had at this time a sizable number of men: 11 officers, 191 NCOs, 2,219 men, plus 1,097 ethnic Germans—in all some 3,518 men.

The German Field Post numbers shown on page 175 illustrate the transfer of Tatar personnel into the Osttürskischen Waffen-Verbande der-SS.

It wasn't until February 1945 that the Tatar personnel from the disbanded Gebirgsjäger Brigade joined the Osttürkischen Waffen-Verbande der-SS. The Tatar 2nd Battalion as described above contained four infantry companies (5th through 8th). This means that its 1st Battalion had 1st through 4th companies. The order of battle of the Krim Waffengruppe differed from the other two Waffengruppen in that its battalions had four rifle companies instead of five. Another difference between it and the other two was that it had two independent companies not under the control of either the 1st or 2nd Battalion, but rather directly controlled by the Waffengruppe company headquarters.

This was the last listing of the brigade in the German situation maps (Lage-Ost). I cannot confirm this movement. Certainly it sounds suspect, since the German army was in the process of withdrawing from northern Italy, beginning in April 1945. One of the last and few remaining documents from the brigade apparently was written from Austria. This leads me to believe that the Osttürkischen Waffen-Verbande der-SS was moved into Austria sometime in April. If indeed it had never been ordered to northern Italy, it is almost certain that it never reached there.

The end of the war in early May 1945 put an end to the brigade. Whether it was committed during those last days of the war is not known. My belief is that it continued to train but was never used in combat. Had the war lasted another six months, it is highly likely that it would have been used, but its effectiveness would still have rated it as a second- or third-class fighting formation. German military reversals had sapped the morale of the brigade.

One final note should be made about this unit. The brigade was one of the more exotic Waffen-SS formations, and its distinctive cuffband insignia marked it as such. Most of the Waffen-SS forma-

REGULATION PLACEMENT FOR ARM SHIELDS, BADGES, AND CUFFBANDS

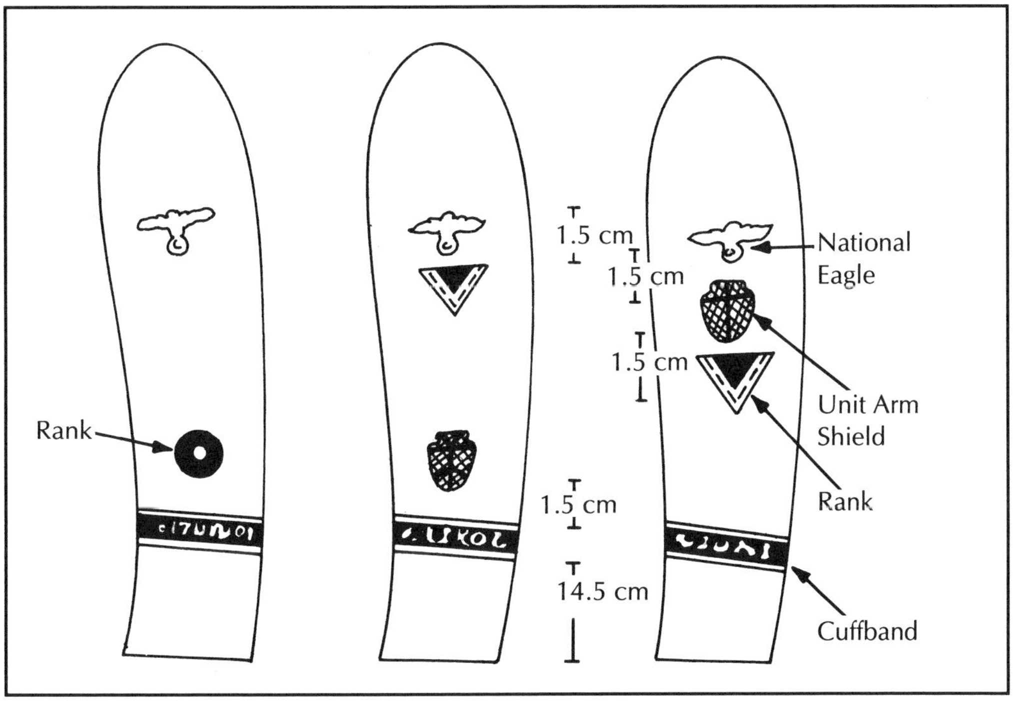

Field Post Number	Date	Personnel
21 581	25 August 1944	Regiment staff plus units, Waffen-Gebirgsjäger-Regiment der-SS (Tatar Nr. 1)
21 581	27 February 1945	Staff plus units, Osttürkischen Waffen-Verbande der-SS
21 581	5 March 1945	Staff plus units, Waffengruppe der-SS Krim
33 948	25 August 1944	Staff of the 2nd Battalion, plus 5th through 8th companies, Waffen-Gebirgsjäger-Regiment der-SS (Tatar No. 1)
33 048	27 February 1945	Staff of the 2nd Battalion, plus 5th through 8th companies, Waffengruppe der-SS Krim

tions had personalized insignia, among them the unit cuffband worn on the lower left sleeve of the tunic. All of the other formations had black cuffbands with white lettering spelling out the unit's name; the cuffband of the Osttürkischen Waffen-Verbande der-SS had the Islamic colors of white on a green background.

Date	Higher Command	Area of Operations
4 January 1944	SS-FHA	Lublin, Poniatowa, Trawniki, Warsaw, Radom
26 February 1944	Higher-SS and Police Leader	Yuratishki, Minsk, Central Russia
11 July 1944	Higher-SS and Police Leader "General Government"	Lomza, Bialystok, Poland
29 July 1944	SS-Dirlewanger Penal Brigade	Warsaw, Poland
15 September 1944	Higher-SS and Police Leader "General Government"	Radom, Poland
30 October 1944	Higher-SS and Police Leader Myjava "Slovakia"	Myjava, Slovakia
1 March 1945	48. Volksgrenadier-Division 8th Army reserve	Myjava, Slovakia
31 March 1945	8th Army reserve	Northwest of Modrovka and west of Backov
15 April 1945	Unknown	Possibly in Austria

Commanding Officer	
SS-Obersturmbannführer Andreas Meyer-Mader	January – March 1944
SS-Hauptsturmführer Billig	28 March – 5 April 1944
None	6 April – 26 April 1944
SS-Hauptsturmführer Hermann	27 April – 2 May 1944
SS-Sturmbannführer der Reserve Franz Liebermann	June – 11 July 1944
SS-Sturmbannführer der Reserve Franz Liebermann	29 July – 30 August 1944
SS-Hauptsturmführer Reiner Olzscha	September – October 1944
SS Standartenführer Harun-el-Raschid Bey	October – December 1944
SS-Hauptsturmführer Fürst	January – May 1945

Date	Activity
4 January 1944	Activation, forming and training of Muslim volunteers
26 February 1944	Training, security duty, antipartisan operations
11 July 1944	Withdrawal from Russia, refitting
29 July 1944	Defense of Warsaw
15 September 1944	Refitting, training
30 October 1944	Security duty, reorganizing, training
15 Novermber 1944	Security duty, reorganizing, training
30 December 1944	Reorganizing, training, security duty
15 January 1945	Reorganizing
27 February 1945	Reorganizing, reforming
1 March 1945	Security duty, reforming
31 March 1945	Reserve duty, training
15 April 1945	Reserve duty, training

Unit Strengths (Approximations)	Date
3,000	January – February 1944
3,500	March – May 1944
4,000	June – July 1944
4,000	August – September 1944
5,000	October – December 1944
5,000	January 1945
8,500	February – May 1945

FORMATION OF THE OSTTÜRKISCHEN WAFFEN-VERBANDE DER-SS
FROM THE OSTMUSELMANISCHE SS-REGIMENT

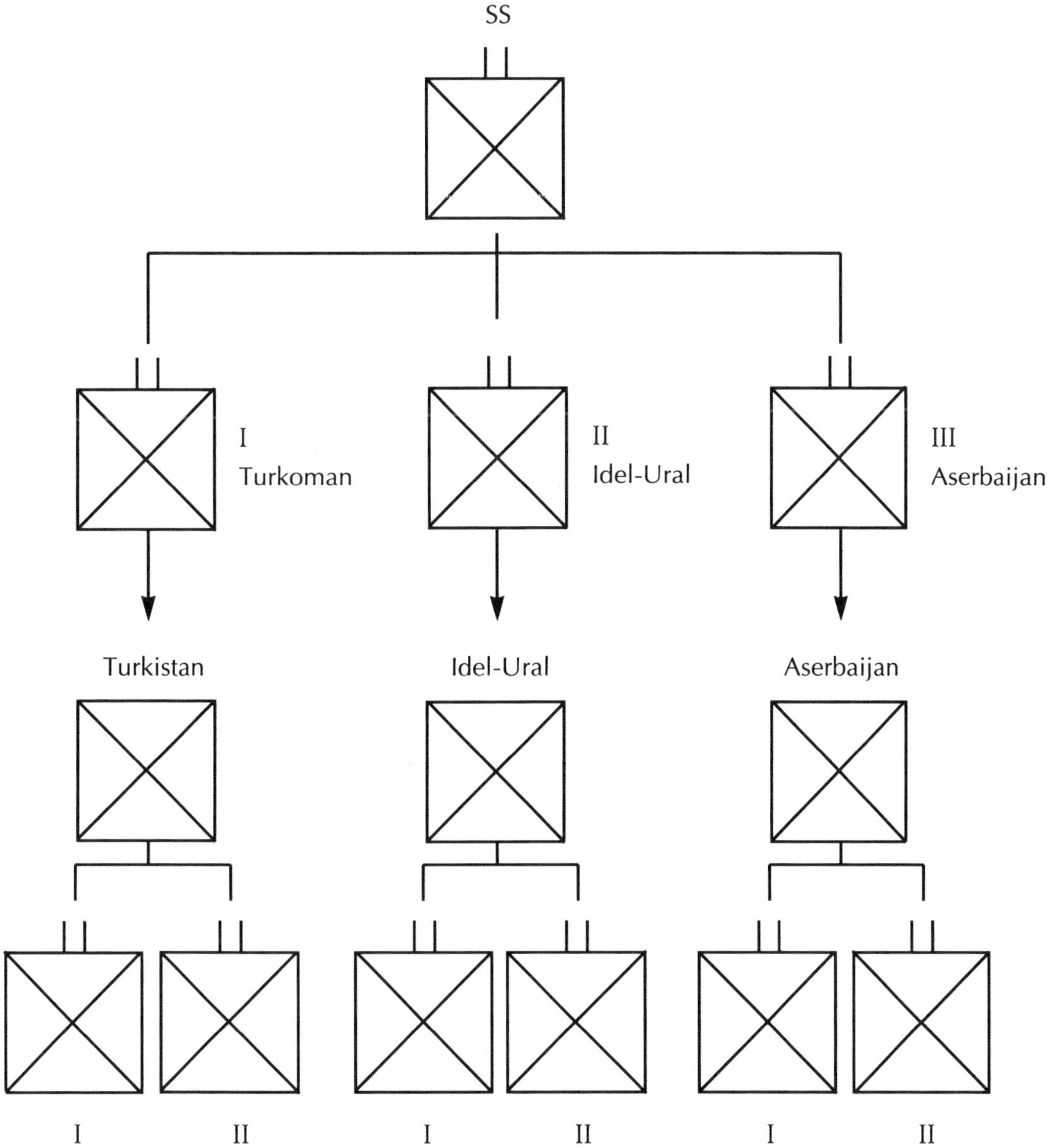

ORDER OF BATTLE OF THE OSTTÜKISCHEN WAFFEN-VERBANDE DER-SS NOVEMBER 1944

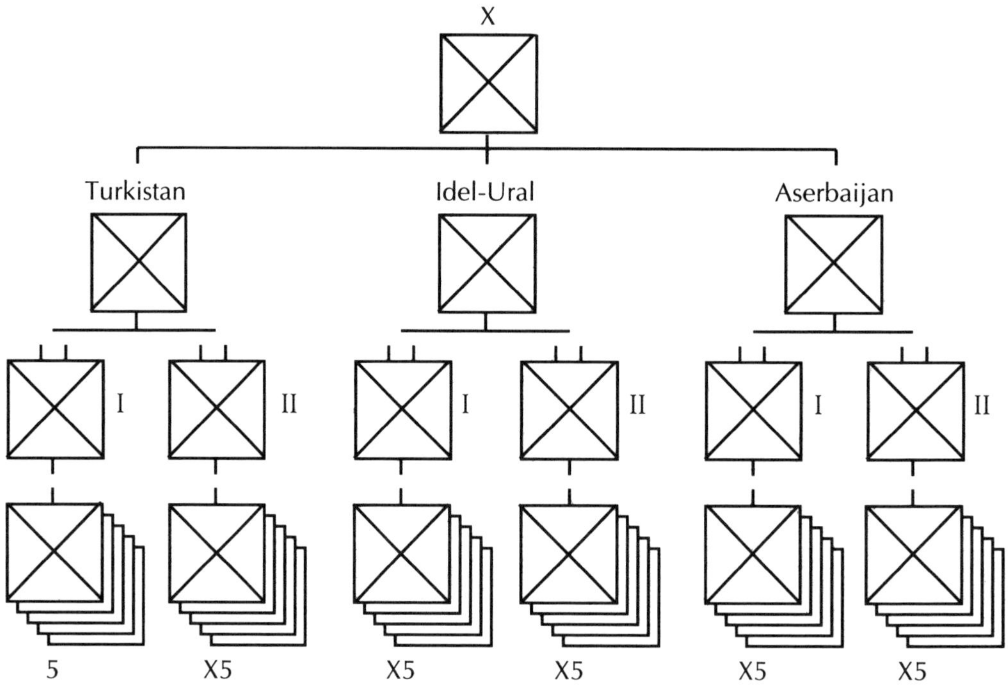

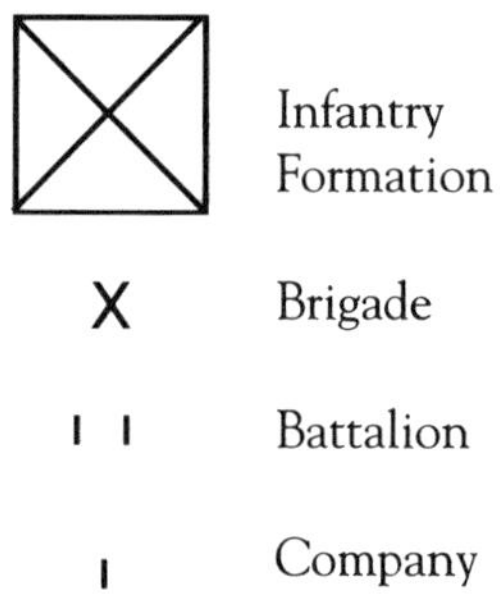

The brigade had a headquarters company. Each Waffengruppe and battalion in the gruppe also had a headquarters company.

The Aserbaijan Waffengruppe was withdrawn from the brigade in an order dated 15 December 1944. Its place was taken by Waffengruppe der-SS Krim, a formation made up of Crimean-Tatar personnel.

OSTTÜRKISCHEN WAFFEN-VERBANDE DER-SS, MARCH 1945
SCHEMATIC ORDER OF BATTLE

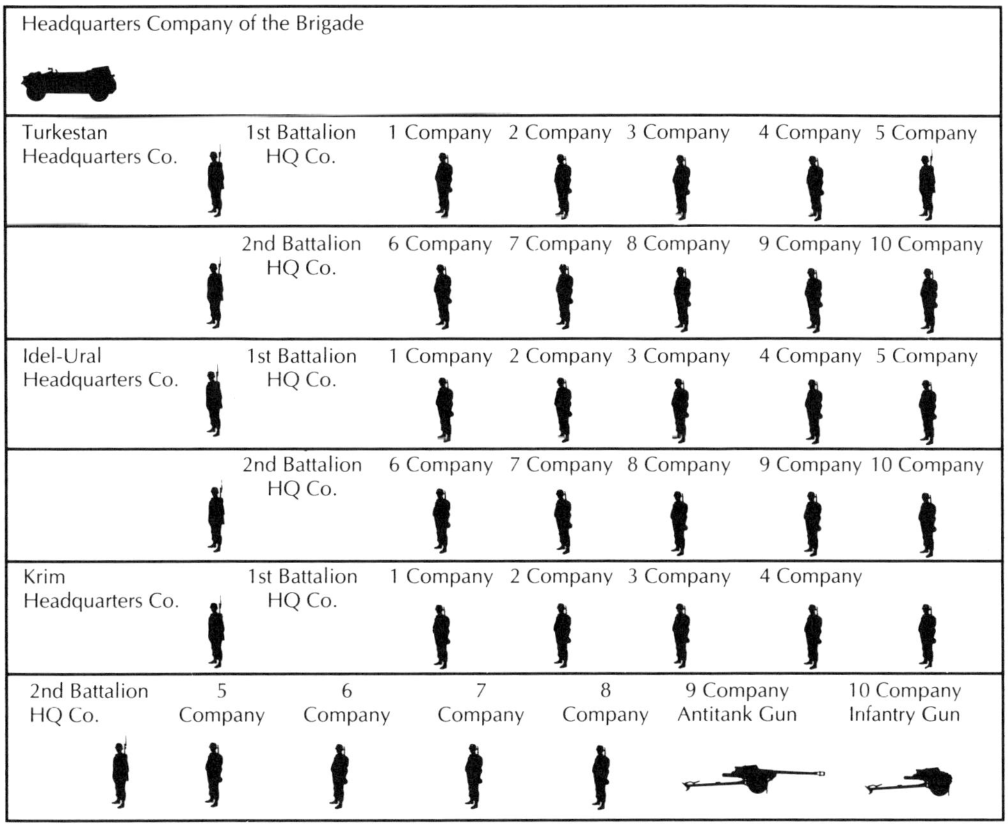

THE OSTMUSELMANISCHE SS-REGIMENT ORDER OF BATTLE, JULY 1944

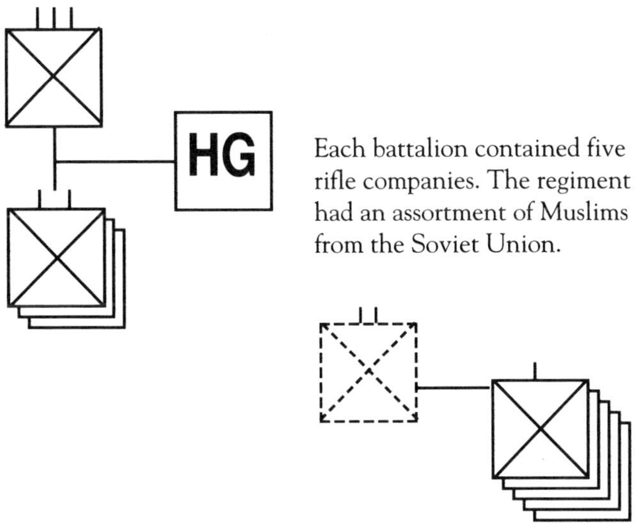

Each battalion contained five rifle companies. The regiment had an assortment of Muslims from the Soviet Union.

THE OSTTÜRKISCHEN WAFFEN-VERBANDE DER-SS ORDER OF BATTLE, MARCH 1945

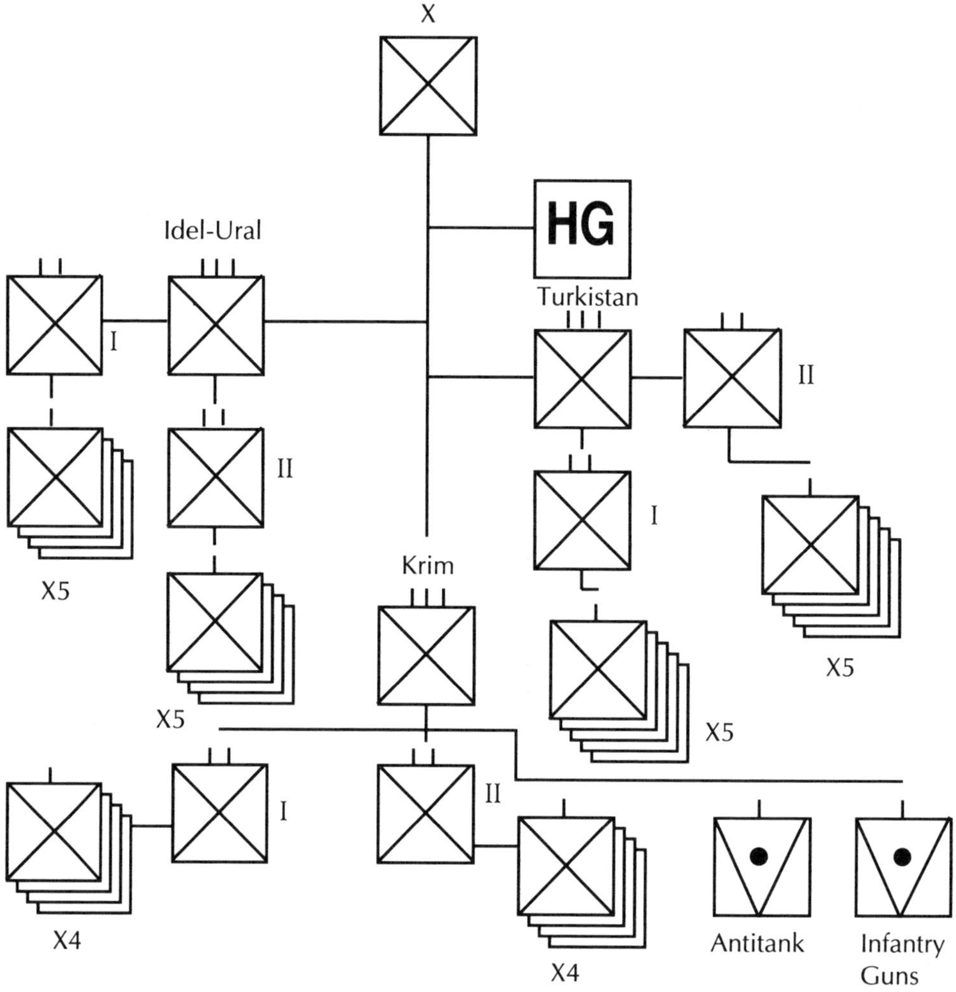

CHAPTER FIFTEEN

Shall man into the mystery of breath
From his quick beating pulse a pathway spy?
Or learn the secret of the shrouded death,
By lifting up the lid of a white eye?
Cleave thou thy way with fathering
Desire of fire to reach to fire.

George Meredith
Hymn to Colour

The SS-Jäger Battaillon 500 was formed originally from the 2nd Battalion, SS Field Recruit Depot Bobruisk in central Russia. The purpose of this depot with its three recruit-training battalions was to supply recruits to the front-line Waffen-SS divisions and to perform security duty behind the rear of the German front, a duty which eventually led all three battalions into antipartisan operations.

SS Field Recruit Depot Bobruisk was established in February 1942 and received Field Post Number 25 357 on 15 February 1942. As the 2nd Battalion, it was responsible for the 5th, 6th, 7th, and 8th Field Recruit companies. Enlistment in the Waffen-SS was still voluntary until the middle of 1943. The frontline units of the Waffen-SS suffered from a shortage of replacements resulting from high losses in battle and an insufficient number of volunteers. It was common knowledge that Waffen-SS units always received the most difficult missions and thus were prone to the worst losses. This fact was the primary reason potential recruits avoided the Waffen-SS.

As a result, the head of the Waffen-SS Recruiting Office, Gottlob Berger, had been forced since 1940 to look outside the Reich national borders for additional recruits. At first, the recruiting office sought foreign volunteers from Western European countries. Another, much larger, group of volunteers came from the various ethnic German colonies all over Europe. These volunteers came from every country that Germany had either overrun or been allied with. The racial makeup of the volunteers in the 2nd Battalion was primarily ethnic Germans from Romania, and there was a sprinkling of Reich Germans in each of the three recruit battalions, mainly in the officer and NCO ranks.

Since the recruit depot was stationed in Belorussia (White Russia), it was used by the commander of rear-area security as a source of manpower for his antipartisan drives. The first drive in which the 2nd Battalion was used was Operation Maikafer (Cockchafer).

Maikafer began on 30 June 1942. The depot's 1st and 2nd battalions served together to clear the area north of the Bobruisk-Mogiliev road. This was part of a larger operation carried out by the German 221st Security Division.

Maikafer was quickly followed on 20 July by Operation Adler (Eagle), in which the German 286th Security Division and local militia participated. The 2nd Battalion and its sister unit, the 1st, only played minor roles in this drive. Other operations soon followed that involved the 1st, 2nd, and 3rd battalions of the SS field recruit depot individually or together. Operations Panther, Falcon, Owl, and Griffin soon followed in 1942.

The beginning of 1943 found the 2nd Battalion participating in Operation Schneehase (White Hare), launched on 28 January 1943, but the only two operations of note were concurrent in July and August 1943. The first, in July, was fought against regular Soviet forces attacking the Parichi bridgehead, which was being defended by the 2nd Battalion and units of the German 36th Infantry Division. During this time, the 2nd Battalion was directly attached to the 36th Infantry Division.

The following month the 2nd Battalion, temporarily named SS-Kampfgruppe Beilhack (named after the 2nd Battalion Commander SS-Sturmbannführer Beilhack), was attached to the German 203rd Security Division. The battalion fought in and around the town of Sluzk until September, when it was withdrawn for rest and refitting. On 15 September, it was redesignated an SS-Jäger-Bataillon (in this sense, it would translate into SS Hunter Battalion).

The 1st and 3rd battalions also became hunter battalions, and the SS Field Recruit Depot Bobruisk took the name SS Supply Command Central Russia. At this time, the three hunter battalions had yet to receive any differentiating numerical designation. Strangely enough, the rifle companies in each of the three battalions underwent a reshuffling. The 2nd Battalion, which up until then contained the 5th through 8th companies, received the lst through 4th companies. The 1st and 3rd battalions' rifle companies were similarly changed. The field post numbers of the three battalions remained the same.

In the early part of 1944, military leaders reached a tentative decision to form the 1st and 2nd battalions into an SS assault regiment and ordered a staff for it to be formed. The initial staging area for the staff was at Debica, Poland. The commanding officer for this new regiment was SS-Standartenführer Rudolf Pannier, the recently appointed commander of SS Supply Command Central Russia. He had previously commanded SS-Truppenübungsplatze Moorlager (SS Troop Training Grounds Moorlager) in Bereza-Kartuska, near Brest, Poland.

Pannier had come under the scrutiny of Reichsführer-SS Heinrich Himmler through several derogatory letters, all anonymous. It seems that Pannier had earned the dislike of some of his subordinates at Moorlager. The thrust of the complaints against Pannier was that hewas a harsh, sadistic disciplinarian who reveled in punishing his men. A member of the 14th SS (Ukrainian) Division, Waffen Obersturmführer der-SS Krokhmaliuk had the following to say about Pannier.

> *Standartenführer Pannier was a hardnosed individual who tolerated no nonsense whatsoever. I can recall arriving late for duty one day, whereupon he made me run several laps around the office building in deep snow. No one that I knew resented his brand of discipline since he was fair about it.*

Though Pannier was a disciplinarian, he apparently was not a sadist. Thus Himmler ignored the anonymous complaints against him as the usual resentment of a few disgruntled subordinates; this aspect of Pannier's character may even have worked in his favor, for Himmler was constantly on the lookout for just such men to lead his frontline units. On 28 April 1944, Pannier received the German Cross in Gold, a prerequisite steppingstone to gaining the next highest decoration—the much-coveted Knight's Cross.

Other officers on the staff included: SS-Sturmbannführer Dr. Jakob Nagel (medical officer), SS-Hauptsturmführer Schaeffer (regimental weapons master), and SS-Hauptsturmführer Kurt Kanzy (regimental administrative officer). Dr. Nagel was reported wounded in action during the first part of 1944 while serving with the 2nd Battalion at Bereska-Kartuska. At this time, the 2nd Battalion was thought to be under the operational control of the 292nd Infantry Division. Employment of the 1st Battalion (redesignated SS-Jäger-Bataillon 501 on 8 February 1944) and the 2nd Battalion together as one regiment proved impossible because of the military situations at the front and rear lines. As a result, the idea had to be scrapped and the staff disbanded.

It wasn't until 15 April 1944 that the 2nd Battalion was numbered, becoming SS-Jäger-Bataillon 500, its final designation. This full title did not appear in the German Field Post list until 14 June 1944, however. During the spring and summer of 1944, SS-Jäger-Bataillon 500 continued to serve in Belorrusia, principally in the rear of the German 9th

Strength of SS-Jäger Bataillon 500
30 June 1944

Commanding Officer:
SS-Sturmbannführer Beilhack

Officers	NCOs	Men	Total
14	120	1,137	1,271

Army/Army Group Center. Up until the beginning of July 1944, it performed rear-area security duty or participated in antipartisan drives.

Second Battalion withdrew westward along with the rest of Army Group Center under the onslaught of the Soviet summer offensive (begun on 22 June 1944). It participated in the fierce crushing of the Warsaw uprising in late August/early September, which devastated most of the city and eventually forced the Polish Home Army to surrender. To this day, the Soviets have been criticized for not coming to the aid of the Poles. They certainly had the capability and opportunity to do so, since their forces were sitting on the banks of the Vistula beside the city. Many have speculated that the Soviets permitted the Polish Home Army to be destroyed so it would be easier to install their own Communist-sponsored regime after the war.

A typical battalion casualty at Warsaw was that of radioman SS-Sturmmann Johann Klein, an ethnic German from the town of Mechiasch, Romania. Born on 19 January 1925, the nineteen-year-old private was listed as missing in action from his 3rd Company in September. He was probably killed and buried under tons of rubble in Warsaw. In October the battalion was finally pulled from the lines and withdrawn to West Prussia. The unit was disbanded that month, and the survivors were transferred to SS-Panzergrenadier-Ausbildungs-Und-Ersatz-Bataillon 35 at Bruss (near Konitz).

SS-JÄGER-BATAILLON 501

The history of SS-Jäger-Bataillon 501 largely parallels that of Bataillon 500, except its development had a few more twists and turns. Formed originally as the 1st Battalion, SS Field Recruit Depot Bobruisk on 15 February 1942, the battalion controlled four companies (1st through 4th). It received Field Post Number 24 697 later that month and kept this number until disbanding.

First Battalion took part in Operation Maikafer along with the depot's 2nd Battalion. This antipartisan drive, begun on 30 June 1942, was the first of many such operations. Much like the 2nd Battalion, the 1st Battalion spent the rest of 1942 pursuing the Soviet partisan bands in Central Russia. First Battalion took part in all of the major operations, including Panther, Falcon, Owl, and Griffin.

Operation Griffin began on 16 August 1942 and involved no less than eleven battalion-level units of the German Rear Area Command, plus two rifle battalions from the German 59th Army Corps. A document written by the Operations Section/Army Group Center dated 17 August 1942 describes the partisan situation and the actions each unit took in Griffin.

Part five of that document dealt with forces under the control of the SS and Police Leader in Central Russia. Operation Order Number 36 detailed the objective of the 1st, 2nd, and 3rd battalions as that of clearing the region on both sides of the Dnieper with Stary Bykhov as the area's central starting point. The commander of SS Field Recruit Depot Bobruisk was ordered to have all three battalions in place no later than 3 September, with actual operations to begin on the following day. All three battalions—while under the overall control of the SS and Police Leader Central Russia—were nominally under the command of the German 286th Security Division.

SPECIAL REPORT

The partisan detachments in the Smolensk-Witebsk-Orscha area continuously reinforced by partisan march battalions prevented for weeks almost the entire highway and railway traffic on Orscha-Witebsk line. The Greif (Griffin) operation hold against this group resulted in its full annihilation.

The units subordinate to the commander of Security Division Number 286 moved in two weeks of combat, despite great difficulties in action in marshy forest land, reached the principal strong point of the partisans, destroying twenty camps and inflicting on the enemy great losses. The partisan commander Blochin was killed in the fighting. Another commander and the chief of staff were captured.

Enemy losses:	796 dead
	599 POW
On the whole:	1,395

Booty:
- 5 guns
- 7 antitank cannons
- 3 antiaircraft guns
- 2 serviceable tanks
- 17 machine guns
- 6 mortars

Other objectives of Operation Griffin as outlined in the operations orders included eliminating enemy groups in the Cherikov-Propoisk-Krasnopolye region; clearing the area to the north and south of the Borisov-Orsha railway line; combing the area around Chaussy; clearing the region and appeasing the populace on both sides of the Orsha-Witebsk road up to the border of 286th Security Division's area of operation; clearing the region northeast of Smolensk; and pacifying the people in the forest regions near Gusino (region Golynki-Ossjuki northward from Krasny Nikolino). The document was signed "for the Commander in Chief of the Group Armies, Chief of the General Staff, Woehler."

A special report by the Chief of Operational Rear, Commander of Security Troops, Army Group Center, dated 30 September 1942, gave an overall assessment of Griffin's success.

The 1st Battalion spent the first half of 1943 in a similar manner, launching antipartisan drives under the control of the local German security division, which happened to be in the area. In September, a major reorganization of the battalion took place. Whereas before, the 1st Battalion had contained the 1st through 4th companies, those numerical designations were given to the 2nd Battalion (now redesignated SS Supply Command Central Russia).

Second Battalion received 1st, 2nd, 3rd, and 4th companies. In turn, 1st Battalion contained companies 8th through 12th.Third Battalion, which up until then had contained companies 9th through 12th, was also redesignated and got companies 5th through 7th. Notice that the 3rd Battalion lost one company in this process, while the 1st Battalion gained one company. This reorganization took place officially on 15 September 1943. The new order of battle for the three SS battalions was as follows:

1st Battalion
8th Rifle Company
9th Rifle Company
10th Rifle Company
11th Rifle Company
12th Rifle Company

2nd Battalion
1st Rifle Company
2nd Rifle Company
3rd Rifle Company
4th Rifle Company

3rd Battalion
5th Rifle Company
6th Rifle Company
7th Rifle Company

In addition to the above-mentioned forces, SS-Supply Commander Central Russia had an additional battalion. This "fourth" battalion was titled Arbeits-Abteilung der Waffen-SS Russland Mitte (roughly translated as: Construction/Work Battalion of the Waffen-SS Central Russia). Its Field Post Number, 22 825, remained the same throughout its existence, from 31 July 1942 to 10 February 1945 when it was stricken from the field post roll. Though SS Field Recruit Depot Bobruisk came into existence officially on 15 February 1942, it is quite possible that this "fourth" battalion may have represented the permanent Waffen-SS personnel at the depot, including guards, instructors, construction crewmen, cooks, officers, and so forth. Little else is known about this unit or why it was kept on the active field-post rolls long after the camp at Bobruisk had been evacuated. Perhaps it was used independently as a combat group until February 1945; perhaps it was just an oversight on the part of the German Military Postal Service (though this seems very unlikely).

Designation of the 1st Battalion as SS-Jäger-Bataillon 501 seems to have occurred before the 2nd Battalion was officially listed as SS-Jäger-Bataillon 500. An

A typical SS-Grenadier. Photo: Munin Verlag.

entry in the German Field Post numbers as early as 8 February 1944 refers to the 1st Battalion as SS-Jäger-Bataillon 501. While the 2nd Battalion is simply mentioned as an SS-Jäger-Bataillon. The 3rd Battalion did not even carry the title Jäger-Bataillon at this time. This is interesting since we know the 2nd Battalion, or SS-Jäger Bataillon 500, had been reorganized and ready once again for combat on 15 April 1944. It had been reorganized at SS Troop Training Grounds Moorlager (Bereza-Kartuska, by Brest Litovsk). The 3rd Battalion, however, had also been listed as reorganized at Moorlager a month later on 15 May.

Stranger still is the fact that on 14 June 1944 the 1st Battalion (SS-Jäger-Bataillon 501) was disbanded. When the old 3rd Battalion was reformed and pronounced combat-ready once again, it was unofficially-officially referred to as SS-Jäger-Bataillon 501. This seems quite confusing, I must admit, but—though twisted—the story is actually quite clear. This is what happened.

While the 2nd and 3rd battalions were regrouping at Moorlager, the 1st

Battalion (SS-Jäger-Bataillon 501) was in combat. In fact, the commanding officer of the 1st Battalion, SS-Sturmbannführer Hans Beyerlein, received the German Cross in Gold on 4 June for his actions and those of his battalion in that fighting. We do not know the state of his battalion at the end of the fighting, but apparently the decision to disband his command had been made in May of that year.

Thus on 14 June 1944, the old 1st Battalion, which had briefly carried the title SS-Jäger-Bataillon 501, was disbanded. The 3rd Battalion (Field Post No. 26 133) was renamed SS-Jäger-Bataillon 501, having been SS-Hauptsturmführer der Reserve Rudolf Bachmann. With this new designation, the old 3rd Battalion (now 501) had its rifle companies reorganized. SS-Jäger-Bataillon 500 received an additional rifle company, as did SS-Jäger-Bataillon 501. An order of battle for 14 June 1944 shows the changes illustrated at the bottom of the page.

The strength of SS-Jäger-Bataillon 501 on 30 June 1944 is shown at the top of this page.

SS-Jäger-Bataillon 501
30 June 1944

Commanding Officer:
SS-Hauptsturmführer der Reserve
Rudolf Bachmann

Officers	NCOs	Men	Total
9	113	820	942

Bear in mind that these figures are for 30 June 1944, by which time the battalion had already been committed to combat on the front lines (remember that the Soviets launched their 1944 summer offensive on 22 June). This means that almost certainly the battalion had already seen extensive service—thus the relatively small number of men still active by 30 June. This also suggests that the 500th Battalion may have met less resistance or fared better than its sister unit, the 501st.

The battalions spent the rest of the summer of 1944 fighting almost exclusively in 3rd Panzer Army's area of operations. Accounts mention both the 500th and 501st battalions as fighting under the control of Battle Group von Gottberg in July and August to keep open a line of communications for the withdrawing formations of Army Group Center.

The SS-Jäger-Bataillon 501 was eventually committed to the fighting at Warsaw, being temporarily attached to the notorious SS Penal Regiment Dirlewanger in August and September. In early October, the battalion left the front lines for Bruss, West Prussia, where it was dis-

SS-Jäger-Bataillon 501 (old 1st Battalion)
- Disbanded

SS-Jäger-Bataillon 500 (old 2nd Battalion)
- 1st Rifle Company
- 2nd Rifle Company
- 3rd Rifle Company
- 4th Rifle Company
- 5th Rifle Company (this unit added from the old 1st Battalion)

SS-Jäger-Bataillon 501 (old 3rd Battalion)
- 1st Rifle Company (formerly the 5th Rifle Company)
- 2nd Rifle Company (formerly the 6th Rifle Company)
- 3rd Rifle Company (formerly the 7th Rifle Company)
- 4th Rifle Company (this unit added from the old 1st Battalion)

banded and its remaining personnel absorbed into SS-Panzergrenadier-Ausbildungs-und-Ersatz-Bataillon 35 (SS Armored Infantry Training and Reserve Battalion 35). This training battalion supplied trained replacements for the combat units of the Kommando Stab Reichsführer-SS, and the SS Assault Brigade Wallonie. Since the Kommando Stab Reichsführer SS furnished men for the German SS and police units on antipartisan security duty, it is likely that some, if not many, ex-501st Battalion members may have found themselves in such units after October 1944.

SS-SONDERVERBANDE z.b.V. FRIEDENTHAL

The first Waffen-SS commando unit was formed as a direct result of several causes. First of all was Hitler's desire to employ elite combat troops, much like the British had been doing since the beginning of the war. The British commando units had been formed to strike at German targets at a time when the British army was unable to fully engage the German army on the European continent. What began as harassment evolved into a well-organized organization, leading elite commandos on surgical strikes against vital points in the enemy hinterland. Hitler wanted to copy this tactic.

Second, there had been a war going on between the German Secret Service (the Abwehr) and the SD (Sicherheitsdienst, or SS Secret Service) for control of the Third Reich's intelligence operations. Reichsführer Himmler, who had been running the RSHA since SS General Reinhard Heydrich's assassination, was more than willing when Hitler brought up the proposal of forming an SS commando force. The Abwehr had been using its Brandenburg commandos since the start of the war. Here was Himmler's chance to "get into the business," as it were, by forming his own commandos.

During the summer of 1943, a camp was established north of Berlin, near the town of Friedenthal, and recruits began to show up. One of the earliest volunteers was an Austrian major who had seen service in Russia under the banner of the Leibstandarte SS Adolf Hitler Division, SS-Sturmbannführer Otto Skorzeny, whose name will be linked forever to the SS commando force. In fact, he is considered the founding father of that elite group of men, much like SS General Paul Hausser is considered the father of the Waffen-SS.

SS-Jäger Engelbert Stryczek, a Hungarian-born member of SS-Jäger-Bataillon 501. Born in Ratibor in 1926, this ethnic German volunteer was listed as missing in action in December 1944 (top). Photo: Munin Verlag.

SS-Unterscharführer Erhart Knöfel, shown here holding his daughter while on leave in 1943. On his right is an SS-Fallschirmjäger commando (above). Photo: Erhart Knofel.

The size of this elite unit quickly grew to one, two, and eventually three companies of volunteers. A headquarters company was formed to administer these men, and the unit was named on 5 August 1943—the official formation date. Admiral Canaris, a staunch opponent (albeit a secret one) of Hitler and the Third Reich, reluctantly supplied specialists from his Abwehr organization to train the Waffen-SS and SD volunteers in the art of espionage and unorthodox warfare.

Numerous specialty schools at the camp taught such necessary arts as radio communications, demolitions, hand-to-

Hitler's commando extraordinaire: Lieutenant-Colonel Otto Skorzeny. Photo: Bundesarchiv.

hand fighting, tracking, and every possible facet of espionage and guerrilla warfare. The early batch of volunteers had to have at least one specialty, such as a foreign language. Germans who had served abroad and especially those who had lived for a long time in a foreign country were especially sought. Recruits also went through parachute and glider training, though this was on a smaller scale.

Skorzeny's rank and elan quickly earned him command of this new formation. He had been wounded in Russia, almost certainly because of his aggressiveness and complete disregard for danger, and that boldness would lead him to eventual fame in the coming years. He quickly began to organize his companies into small platoon-sized assault groups (sturmgruppen). The men were not given any warning as to when the alarm bell would sound, requiring them to pack their gear and weapons and stand by to go into action at a moment's notice. With practice, he got them trained to react within fifteen minutes from the moment the alarm sounded. By the end of the summer, Skorzeny declared his men ready; they needed a mission to prove their mettle.

That opportunity developed during the latter part of the summer of 1943. As described earlier in Chapter 8, the King of Italy, with the tacit approval of Mussolini's Fascist Grand Council, had placed that country's Fascist leader, Benito Mussolini, under house arrest high atop a mountain fortress. Hitler knew that Italy was planning on suing for peace and, if possible, joining the Allies. He also knew that he must stop this. Hitler knew that Italy had a strong Fascist element that was still willing to fight on the German side till the bitter end, but he needed Mussolini to rally them.

Skorzeny and his commandoes received the mission to locate and rescue Mussolini. Against great odds, them were able to pull this off. When Skorzeny arrived at Hitler's headquarters (the Wolf's Lair) with Mussolini, he wore a Knight's Cross officer around his neck expropriated from another—per Hitler's direct orders. The resounding success of the commandos' first mission earned them—and Skorzeny, especially—a reputation that ensured they would be used in other seemingly impossible missions, including a raid in Yugoslavia designed to locate Tito's secret hideaway and capture or assassinate the Partisan leader. (For a full description of the Mussolini rescue, see Chapter 1; for more information on the Tito operation, see Chapter 2.)

SS JÄGER BATAILLON 502

In mid-April the Friedenthal Sonderverband became the SS-Jäger-Bataillon 502 on 17 April 1944. Although the Field Post Number for the unit remained the same throughout its existence, its name and composition changed many times, as shown on page 190.

Notice that for a few months at the end of 1943, the Friedenthal title did not really apply, but the Jäger-Bataillon 502 title was not yet official. In fact, when Skorzeny wrote his first memoir, he men-

EVOLUTION OF THE FIRST WAFFEN-SS COMMANDO FORMATION, 1943-44

Formed officially on 5 August 1943

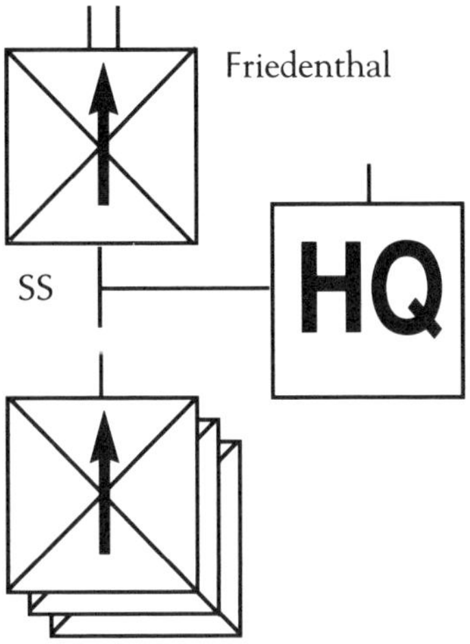

SS-Jagdverbande z.b.V. Friedenthal
One headquarters plus three German rifle companies

Redesignation of Friedenthal to 502 on 17 April 1944

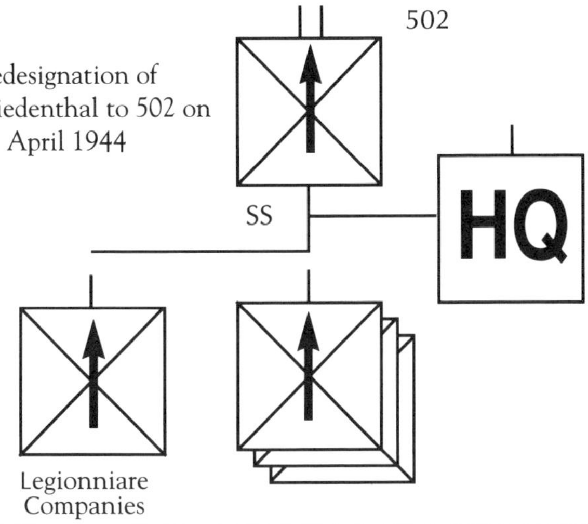

SS-Jäger Bataillon 502
Same composition as above but also included a Legionnaire company made up of foreign personnel

Redesignation of 502 to Mitte on 10 November 1944

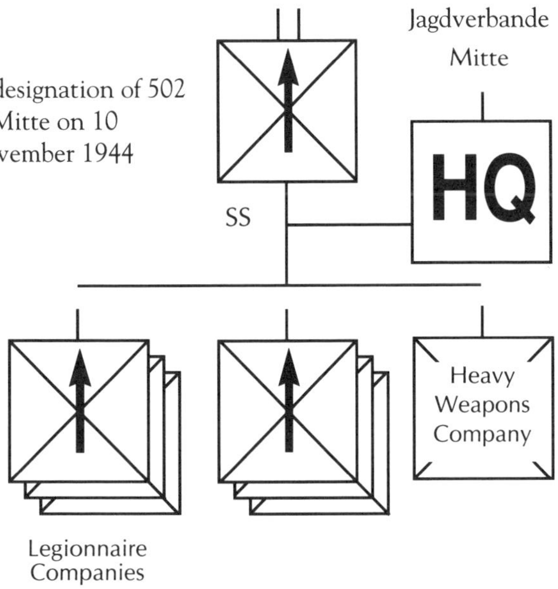

SS-Jagdverbande Mitte
Had one headquarters, three German rifle, one heavy weapons, and four Legionnaire rifle companies. The first commando operation performed by the men of 1st Company SS-Jagdverbande z.b.V. Friedenthal, under the personal command of Otto Skorzeny and SS-Untersturmführer Otto Schwerdt, was the rescue of Benito Mussolini on 12 September 1943.

F.P. No.	Title	Date
48 312	SS-Sonderkommando z.b.V. Oranienburg	2 Jan. 1940 to 9 Jan. 1943
48 312	SS-Sonderverband z.b.V. Friedenthal	10 Jan. 1943 to 26 Sept. 1943
48 312	SS-Jäger-Bataillon502	17 April 1944 to 23 Nov. 1944
48 312	SS-Jagdverband Mitte	24 Nov. 1944 to May 1945

tioned getting his men for the Mussolini rescue from the SS-Jäger-Bataillon 502. A few words should be said about SS-Sonderkommando z.b.V. Oranienburg, the embryo of the SS commando forces. SS General Hans Jüttner, the Chief of the SSFHA, was the man responsible for setting up a "Special Employment" (z.b.V.) Sonderkommando Oranienburg that would be at the disposal of the Reichsführer, the SS-FHA, and the SD.

Jüttner wanted to copy the army's Secret Service (Abwehr) commandos, the Brandenburgers. Since the start of the war in 1939, these specialized troops had spearheaded every German invasion, saving bridges and creating bridgeheads for the German army's advance. The first significant operation launched by Sonderkommando Oranienburg was a small paradrop into Persia from the Caucasus region of Russia in the fall of 1942. The operation was purely an SS affair to make contact with any groups in Persia still actively opposing the British army (after the failed anti-British coup of 1941). Once contact had been established, plans were to deliver arms and other supplies to the insurgents, creating a mass guerrilla campaign behind British lines. The Germans hoped that a protracted guerrilla war in Persia would tie up considerable British forces, making it easier for the victorious German army in Russia to cross the Caucasus and invade Persia.

The first half of 1944 kept Skorzeny busy on several projects. Not only did he locate Tito and devise a plan for his elimination (though fortunately for Tito, it was never used), but he got himself involved with the V-1 and V-2 rocket programs and even the one-man "X-boot" (X-boats). These one-man boats, armed with hundreds of pounds of dynamite, were designed to ram Allied ships. Skorzeny's first adjutant, SS-Obersturmführer Karl Radl, even took charge of the X (or ram) boat program, organizing units for deployment in the Mediterranean Sea.

The first real crisis of the second half of 1944 was an attempt to assassinate Hitler on 20 July. Skorzeny placed his commando Battalion 502 on alert at Friedenthal and eventually moved the unit south to Berlin. The commando battalion had grown considerably from its seventy-man inception in 1940. For Mussolini's rescue in September 1943, Skorzeny had chosen a platoon from a total of three rifle companies (not including the battalion Headquarters Company). By April 1944 the battalion had been expanded by the addition of a Legionnaire Company. By November, the unit contained one heavy-weapons company, three German rifle companies, and four Legionnaire companies made up of foreign personnel and ethnic Germans who had been living outside of Germany. The Headquarters Company completed the organization.

Events unfolded rather quickly during the second half of 1944. On 22 June 1944, the Russians launched their great summer offensive, virtually destroying the Army Group Center and pushing the front back to the Polish/East Prussian/Baltic area. In the south, the collapse of Romania in August, coupled with the de-

SS-Jäger-Bataillon 502 Strength Table, 30 June 1944

Officers	NCOs	Men	Total
23	108	519	615

struction of the German 8th and 6th armies, created a dangerous gap through which the Red Army, aided by the Romanians and Bulgarians, was flooding into the Balkans and eastern Hungary. If the Red Army was not stopped, it would reach the Hungarian plains before winter, threatening all of Austria.

Complicating the situation was the Hungarian regent, Admiral Horthy, who wished to copy the Romanian-Bulgarian and Italian solution—which was simply to change sides. Hitler had known for months that the regent and his son had been secretly negotiating with Stalin about a Hungarian surrender. Once again, Hitler called on Skorzeny and his commandos to perform a military miracle to stave off political disaster and imminent military defeat.

As described in Chapter 8, the plan was to topple Horthy and his clique from power and install a more pro-German government that wouldn't cooperate with the Soviets. For this operation, Skorzeny got several crack battalions, one of them the newly reorganized SS Parachute Battalion, as well as his SS-Jäger-Bataillon 502. The operation was quite hairy, but Skorzeny pulled one more miracle out of the hat, and, luckily for the Germans, the operation was a complete success.

SS-JAGDVERBAND MITTE

SS-Jäger-Bataillon 502 returned to Friedenthal in late October 1944. On 10 November, less than a month later, the unit received its final designation: SS-Jagdverband Mitte (SS Hunter Formation Center). The commando unit had grown to nine companies (including the headquarters unit). Recruits had come from all parts of Europe and Russia. The formation records state that approximately 150 Flemish volunteers served with the unit. Later, after formation of the SS-Jagdverband Nordwest, these Flemish volunteers transferred to it.

In November Skorzeny was tasked with forming a brigade of men dressed and equipped like a U.S. Army tank formation. The code name for this was Operation Greif (Griffin), a mythical animal with an eagle's head and wings and a lion's body. In Hitler's mind, this operation would allow Germany to rise from the flames and ashes of its bombed-out cities and strike the enemy a blow that would signal the resumption of the German victories of 1939 to 1942. The purpose was to deceive the enemy into thinking that Skorzeny's brigade was a U.S. formation to gain access behind the American lines and capture vital bridges that would be needed for the coming offensive (see Chapter 8 for full details). Skorzeny's brigade was to wait for the initial German attack to commence. In the confusion of those first hours of battle, his brigade would theoretically be able to infiltrate the U.S. lines.

SS-Unterscharführer (Sergeant) Andreas Meyer was a typical member of the SS-Jagdverband. Meyer had been inducted into the Waffen SS on 1 April 1942. Born on 6 March 1925, Meyer had just turned seventeen when he volunteered for service. One year later, on 20 April 1943, he finished his basic training and technical schooling, becoming an SS-Sturmmann. In the summer of 1943, he was posted as a technician to the SS-FHA in Berlin where he joined Inspectorate Number 6 (Panzer Troops). His commanding officer at this time was SS-Obersturmbannführer Willi Hardiek, who was later transferred and led the 12. SS-Panzerregiment in the Hitler Jugend SS Division.

He eventually went to the 12. SS-Panzerregiment where he was again attached to Hardiek's staff, but not before undergoing an NCO course in Sennelager-Nebenlager Staumühleühle. At this time the 12. SS-Panzerregiment was refitting around Diepholz and later to Waffen-SS Troop Training Ground Fallingbostel. Sergeant Meyer described how he eventually came to be in SS-Jägdverband Mitte.

In the Fall of 1944, Willi Hardiek was transferred to [Otto Skorzeny's] Commando Office and given the task of setting up Panzerbrigade 150 at Grafenwöhr. Hardiek's driver and I were telexed to report to Grafenwöhr. After about three weeks the Brigade was taken over by Obersturmbannführer Otto Skorzeny.

As is known, this unit was destined for commando action in the Bulge [Ardennes Offensive]. The unit was to infiltrate to [the town of] Huy during the beginning of the offensive—after the breakthrough—and, dressed as Americans, secure the bridge for the troops following to cross over the Meuse [River].

As the offensive stalled, this task could not be accomplished. My CO, Hardiek, had already fallen on the first day of the offensive as CO of Battle Group of the Brigade. His successor was Skorzeny's Ia [second-in-command], Adrian von Foelkersam. I was his battle clerk. Our battle group was made "official" as a regular [combat] unit on 21 December 1944 and was stationed near Malmedy. We were to break the road connection to the southwest [of Malmedy]. At the same time the Battle Groups X and Y of our brigade were to attack Malmedy from the northwest.

This attack failed because we were armed only with infantry weapons. Malmedy was held by the [U.S.] 30th Infantry Division supported by strong artillery. Our casualties were alarming.

At the beginning of January 1945, we were pulled out of the front. I myself was transferred to Skorzeny's headquarters in Friedenthal, by Oranienburg, in February, after completing some loose ends at Grafenwöhr. As the SS-Jagdverband Mitte had already moved to Schwedt, I received marching papers for there.

We will take up Sergeant Meyer's story further on. It should be noted here that although Meyer was a clerk, he did actively participate in the fighting. He had been promoted to SS-Rottenführer (Corporal) on November 9th,1944. For his gallantry in thsergeant on December 21st, 1944, at the height of the Ardennes offensive. He was decorated with the Iron Cross First Class, Iron Cross Second Class, and the Infantry Assault Badge in Silver.

The entire 5th Company/SS-Jagdverband Mitte had been used to help form part of the 150th Panzerbrigade. That was 175 men, not including command personnel like Skorzeny, von Foelkersam and Hardieck. The SS Parachute Battalion had contributed two more companies, while the majority of the brigade had been made up of army and air force personnel. From the very beginning, things had gone wrong. U.S. equipment had proved hard to come by. The Germans had to resort to using tin and some imagination to disguise German tanks into looking like American vehicles. When the offensive began the initial German infantry attack floundered. The small road system of the Belgian Ardennes area aided in defeating the German attack by creating the worst traffic jams the German Army had yet experienced. Skorzeny's brigade, scheduled to be "inserted" on December 17th, 1944 (a day after the attack began) could not be employed due to the tremendous traffic jams that resulted. Eventually, the brigade was committed to regular frontline fighting and spent most of its time in the Malmedy area. On 25 December, the unit was ordered withdrawn and disbanded. The command roster for the battle groups of the 150. Panzerbrigade is listed at the bottom of page 193.

When the men of Jagdverband Mitte returned to Friedenthal in January 1945, some men received leave, while others performed various garrison duties. The alarm once again sounded at the end of January, forcing Skorzeny to cancel all leave and organize his Jagdverband Mitte and Nordwest into a battlegroup for deployment on the Oder River, by Schwedt. Skorzeny also called SS-Sturmbannführer Siegfried Milius, the commander of the SS Parachute Battalion, and ordered him to follow him to Schwedt. The battle for the Schwedt bridgehead lasted from 1 February to 3 March. Thereafter, SS-Jagdverband Mitte was subordinated to a battle group under the command of SS-Sturmbannführer Milius. This battle group, made up of SS Parachute Battalion 600 and Jagdverband Mitte, was sent to the bridgehead at Zehden, just a few kilometers south of Schwedt. Mitte was still under the command of SS-Hauptsturmführer Fucker. The battle for the Zehden bridgehead lasted from 3 March to 28 March 1945. During that month, the object of the defense put up by Mitte had been the town of Niederwützen. So heavily involved was the commando unit that when the bridgehead was finally evacuated, SS-Jagdverband Mitte had to be withdrawn to Friedenthal to be re-formed.

Events moved too fast for the unit to be re-formed, however. On 31 March,

SITUATION
2 FEBRUARY 1945

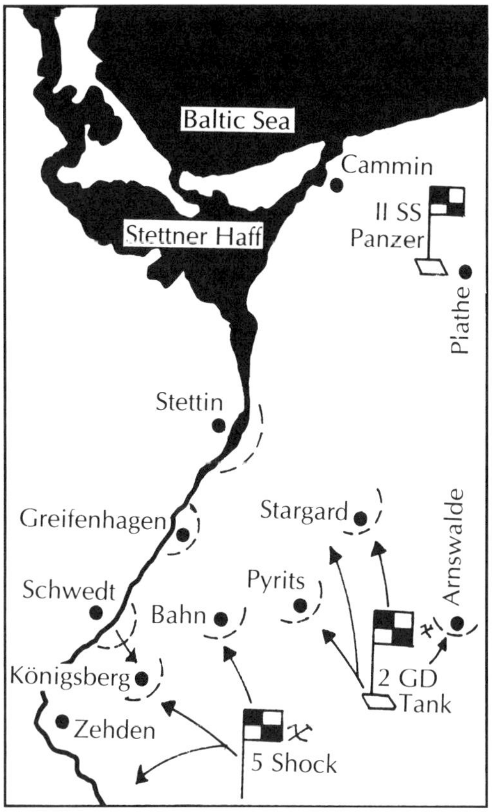

Skorzeny received orders to gather his commando forces together and move with all haste to Austria, where they would form part of the last-ditch German defense in the Alpine Redoubt. Skorzeny knew that the Oder River front needed as many men as possible, so he decided to take only his much reduced Mitte group, leaving the re-forming Ost and Nordwest groups on the Oder Front. His SS-Jagdverband Mitte had two companies with an effective strength of 250 men. SS-Hauptsturmführer Fucker was still in command. SS-Hauptsturmführer Karl

SS-Obersturmbannführer Otto Skorzeny, Commanding Officer
SS-Hauptsturmführer Adrian von Foelkersam, Brigade Adjutant
SS-Obersturmbannführer Wolf, Kampfgruppe X
SS-Hauptsturmführer Scherff, Kampfgruppe Y
SS-Obersturmbannführer Willi Hardieck, Kampfgruppe Z then
SS-Hauptsturmführer Adrian von Foelkersam (21 December)

Radl had also joined Skorzeny and his Mitte group around the middle of April, around the time the group was located at Linz, Austria. There the group had been awaiting the arrival of SS-Jagdverband Südwest and Südost. Once these other SS commando groups arrived, the plan called for them to head for the Alpine Redoubt located in western Austria. By 4 May the western Allied armies were closing in on Linz from the west and northwest, while the Red Army was advancing from the east. U.S. troops were at Passau, just seventy kilometers northwest of Linz. The Russians had taken Polten, some 100 kilometers east of Linz. It is not clear if SS-Jagdverband Mitte was committed to the fighting in Austria. It is likely that they were not employed, since they were to be used for the defense of the redoubt.

SS-JAGDVERBANDE MITTE, MARCH 1945

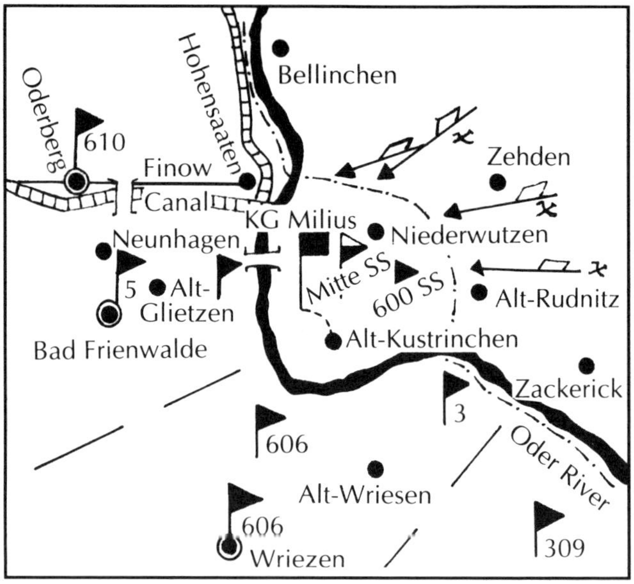

SS-JAGDVERBAND SÜDOST

The origins of Jagdverbandes Südost, Ost, Südwest, and Nordwest date back to the Brandenburg commando units. The Brandenburgers were specialized troops which had been trained by and belonged to the Abwehr (German Secret Service). After the 20 July 1944 plot to kill Adolf Hitler, it was discovered that the head of the German Abwehr (Admiral Canaris) had been involved in the plot. Thereafter Canaris was imprisoned while the Abwehr (including the Brandenburg detachments) was taken over by the SD (Sicherheitsdienst).

During the fall of 1944, the Brandenburg units were transformed into SS Jadgverbänd (hunter battalions). The nucleus for SS Jadgvervband Südost lay in the Brandenburg Streifkorps Karpaten (Raider Corps Carpathian). This commando unit had two formations roughly company-sized under its control. Einsatzgruppe Slowakei and Einsatzgruppe Rumanien had originally been formed from the Brandenburg Wildschütz units. These two groups had been formed officially on 25 August 1944 and together made up Streifkorps Karpaten. The Slowakei group contained German personnel plus one hundred ethnic Germans from Slovakia. The Romanian group also had a German cadre but only seventy Romanian nationals of German ancestry.

Initially, the two roughly company-sized formations were accommodated at Korneuberg, just northwest of Vienna, Austria. There, both groups were trained, first by the Abwehr and later by SS officers on the finer points of espionage and sabotage. They also learned how to organize resistance cells behind enemy lines. Between September 1944 and April 1945, the 70 Romanian volunteers were parachuted individually or in small groups back into Romania. One eye-witness described the attire of this small Romanian contingent at Korneuberg as being a mixture of Romanian Army,

Map, left: SS commando formations in and around Linz, Austria, 30 April 1945.

Waffen-SS, and Luftwaffe uniforms.

Eventually, at the instigation of the SS High Command, this small force was enlarged so that by December 1944 Streifkorps Karpaten had increased to five einsatzgruppen. The chart at the bottom of the page shows German Field Post Numbers for the formations.

Although the majority of the einsatzgruppen were stationed northwest of Austria, its headquarters was located at Agram (Zagreb) in Croatia, Yugoslavia. In January 1945, several additional SS companies were formed, and the term "einsatzgruppe" was dropped in favor of "jagdeinsatz." SS-Sonderkommando Südost was redesignated SS-Jagdverband Südost. The above chart on field post numbers lists the jagdverbande and their locations as they stood that month, as shown at the top of page 196.

Jagdkommando Donau (Danube) was a thirty-man platoon of mixed Waffen-SS, Vlasov Army, and Kriegsmarine personnel. The command of this specially formed subunit of SS-Jagdeinsatz Ungarn was entrusted to SS-Untersturmführer Schreiber. A Hungarian tug captain and a Russian officer of the German-sponsored Vlasov Movement completed the officer roster for the platoon. The jagdkommando had the mission of attempting to float several barges of supplies through enemy-held territory by using the Danube River. The ultimate objective of the supply mission was the trapped German and Hungarian garrison in Budapest. Leaving their training base at Dianabad, Hungary, on the night of 31 December 1944, the men traveled east by the Danube River a distance of approximately 110 kilometers.

That they left on New Year's Eve was no coincidence. The Germans were hoping that the Russians would be too busy celebrating the New Year to notice their small group, slowly moving on the Danube River, disguised as civilians. At a predetermined spot along the trip, they would change over into Red Army uniforms. The trip was begun late on the evening of Sunday 31 December and went without a hitch until the third day, 2 January 1945. Having traveled almost

Field Post No.	Unit
28 433	SS-Sonderkommando Südost (This headquarters took over for Streifkorps Karpaten.)
23 222	Einsatzgruppe Slowakei
68 081	Einsatzgruppe Serbien-Kroatien
65 637	Einsatzgruppe Rumanien (The 70-man Abwehr contingent was supplemented by an additional 175 men, who were stationed at Stockerau and Hollabrunn, northwest of Korneuburg.)
66 177	Einsatzgruppe Ungarn (also at Hollabrunn-Stockerau)
67 665	Einsatzgruppe Bulgarien (also at Hollabrunn-Stockerau)

Field Post No.	Unit	Location
41 286	SS-Jagdverband Südost	Zagreb, Croatia
23 222	SS-Jagdeinsatz Slowakei	Hollabrunn-Stockerau
68 081	SS-Jagdeinsatz Serbien-Kroatien	Zagreb, Croatia
65 637	SS-Jagdeinsatz Rumanien	Hollabrunn-Stockerau
66 177	SS Jagdeinsatz Ungarn	Hollabrunn-Stockerau
67 665	SS-Jagdeinsatz Bulgarien	Hollabrunn-Stockerau
?	SS-Jagdeinsatz Albanien	Zagreb, Croatia

ninety-three kilometers across enemy lines to their destination, their luck ran out when the barges got stuck in a sandbank. Shortly afterward, the enemy discovered them, dooming the resupply mission. The few men in the group who were lucky enough to escape managed to reach the German lines by moving through the Pilis Mountains northwest of Budapest. By 3 January, Army Group South already knew of the mission's failure.

In Yugoslavia the commando companies of SS-Jagdverband Südost continued to be employed near the Zagreb area. The officer roster for the SS Jagdverband Südost is incomplete, but officers known to have served in the commando unit are listed at the bottom of page 197.

Although most of the Jagdverband Südost units were in training during months of January and February 1945, the active commando units in Croatia were kept busy. On 8 February, an SS Commando Battle School became official, although it had been in existence since January 1945 in Zagreb.

March 1945 saw Germany lose western Hungary. The Germans had launched their last great offensive of the war during the first week in March, the objectives of which had been to recapture the Lake Balaton oil fields, destroy the mass of the Soviet 3rd Ukrainian Front in the Drava River/Danube River/Stuhlweissenburg area by using a pincer movement, and relieve the trapped German and Hungarian forces of IX. Waffen-Gebirgs-Armeekorps der-SS in Budapest. The Romanian and Bulgarian SS Commando companies had been earmarked for a series of parachute drops on the Romanian border passes as a planned follow-up breakthrough if the Lake Balaton offensive had proved successful.

In March, SS-Jagdeinsatz Serbien-Kroatien, Albanien, and the

SS-Unterscharführer Andreas Meyer shown here as an enlisted man in 1943. Photo: Andreas Meyer.

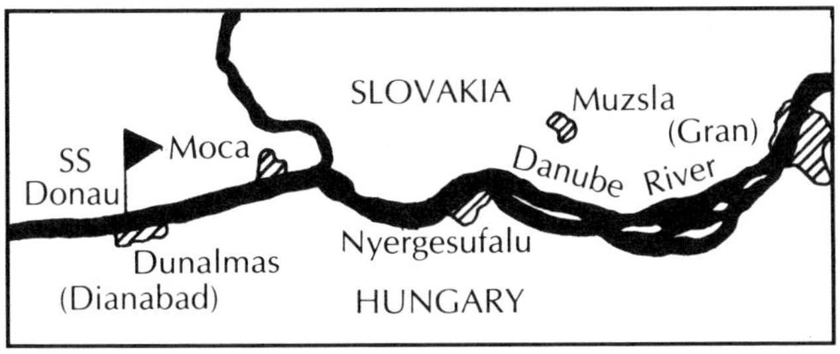

Map, left: Slovakian-Hungarian border area.

SS Kampfschule (SS Battle School) were withdrawn from Yugoslavia and sent to Stockerau, Austria, where they joined the rest of SS-Jagdverband Südost. The commando force was committed to frontline fighting at Nickelsdorf, just across the Austro-Hungarian frontier (on the Austrian side). The rest of March 1945 saw the Jagdverband making a fighting withdrawal through the Austrian towns of Berendorf, Zöebersdorf, and Singerin.

During April, SS-Jagdverband Südost fought from Singerin to St. Pölten, then St. Reith, and then through Traisen. Earlier in April, Otto Skorzeny had informed the commander of Südost of the decision to use several of the commando units (Südost included) to defend the nonexistent Alpine Redoubt in southwestern Austria. The tall commando extraordinaire arrived in the city of Linz, Austria, with two companies (about 250 men) from SS-Jagdverband Mitte. The SS Jagdverbands' Mitte, Südwest, and Südost were all supposed to meet at Linz sometime during the middle of April.

Major Auch, an ex-Brandenburger commando officer, was still leading Südost. It is not known whether he officially transferred into the ranks of the Waffen-SS or if he retained his Wehrmacht rank while serving in the Jagdverband. It is known that he had a good working relationship with his officers, especially the commander of Einsatzgruppe Rumanien, Waffen-Sturmbannführer der-SS Toba, and the German counterpart in the unit, Sturmbannführer Benesch.

Major Auch had gained notoriety in the Brandenburg commando force while leading the Legionnaire Battalion, after working his way through the ranks as company commander, staff adjutant, and, finally, commanding officer. The Legionnaire Battalion consisted of the 1st Weiss (White Russian) and the 2nd Schwarz (Eastern Russian) volunteer companies.

Another less-well-known officer in Jagdverband Südost was SS-Hauptsturmführer Priebke, who had helped Skorzeny plan Mussolini's rescue. Hauptsturmführer Priebke was killed in 1944 while fighting Tito's Partisans in Yugoslavia.

During the last three weeks of the war,

SS-Obersturmführer Schumann
Waffen-Obersturmführer der-SS Vasil Nievolov
SS-Hauptsturmführer Alexander Auch
SS-Untersturmführer Schreiber
SS-Sturmbannführer Benesch
Waffen-Sturmbannführer der-SS Toba (SS-Jagdeinsatz Rumanien)
Oberstleutnant Pfeiffer (Abwehr Korneuburg Gruppe Rumanien)
Leutnant Mueller - (Abwehr Korneuburg Gruppe Rumanien)
Waffen-Untersturmführer der-SS Virgil Popa (Abwehr Korneuburg Gruppe Rumanien)
SS-Untersturmführer Andreas Friedrich
Georgi Avksentieff Gerginoff (member of the SS-Jagdeinsatz Bulgarien; born in Sofia, Bulgaria, in 1925; possibly an officer or NCO in the Bulgarian SS Commando contingent)

any solid, organized resistance in Austria collapsed. With the Red Army rushing in from the east and the British and Americans coming from the west, the majority of the German forces began to surrender en masse. Most of the commandos, especially the eastern and Balkan volkdeutsche (ethnic Germans) disappeared into the mountainous countryside of western Austria to evade capture. Most of them knew what awaited them if they surrendered, especially to the Communists. Skorzeny himself remained in the hills until June 1945.

SUPPLY ATTEMPT OF SS-JAGDGRUPPE DONAU OF SS-JAGDVERBANDE SÜDOST JANUARY 1945

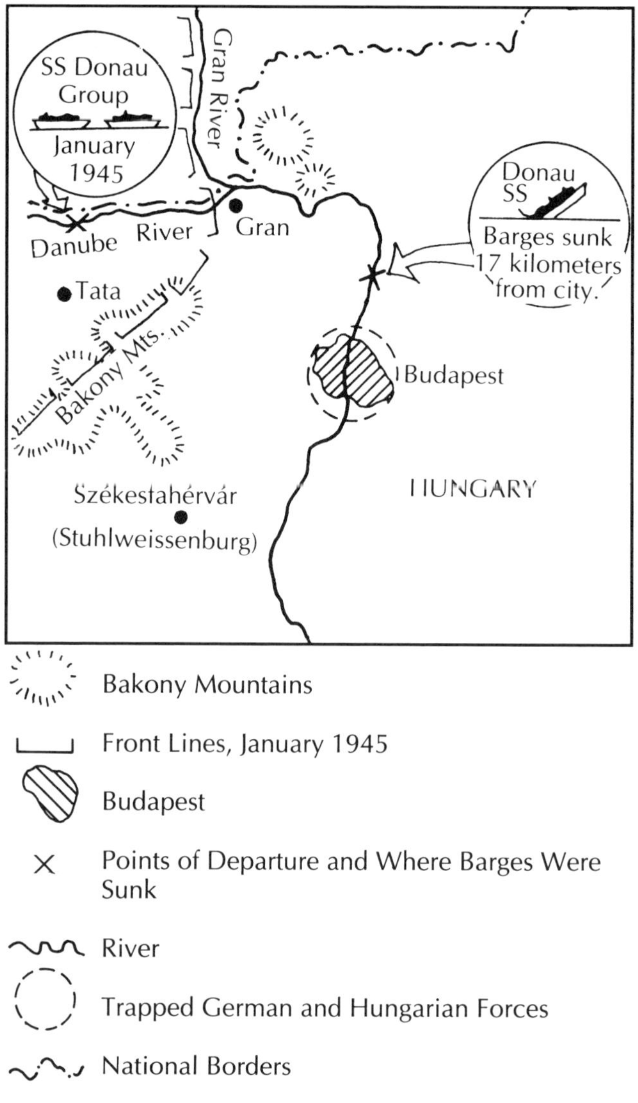

SS-JAGDVERBAND SÜDWEST

This commando group had its beginnings with the Brandenburger Streifkorps Südfrankreich. In the spring of 1944, a specialized Brandenburger unit was formed specifically for employment in southern France. The men selected came from various Brandenburger units: primarily the Legionnaire Battalion of the Brandenburg Lehr Regiment and the bulk of the 7th and 8th Companies, 2nd Battalion, 3rd Regiment Brandenburg. The Legionnaire Battalion contained two companies of volunteers from the Soviet Union. The 1st Company, nicknamed "White," contained Belorussian volunteers and a sprinkling of Ukrainians. The 2nd Company, nicknamed "Black," had Soviet citizens from the Caucasus region, including Azerbaijanis, Armenians, and Turkistanis. Seventh Company, 2nd Battalion contained Italian volunteers, while the 8th Company consisted of Spaniards.

The first contingent of Brandenburgers to form Streifkorps Südfrankreich was the Spaniards of 8th Company. From them, an Einsatzgruppe Pyrenären was formed for duty along the Spanish-French frontier. This group contained fifty Spaniards under the command of an Oberleutnant Demetrio. Einsatzgruppe Pyrenären was subordinate to Streifkorps Biscaya, which in turn reported to Streifkorps Südfrankreich. Exactly how

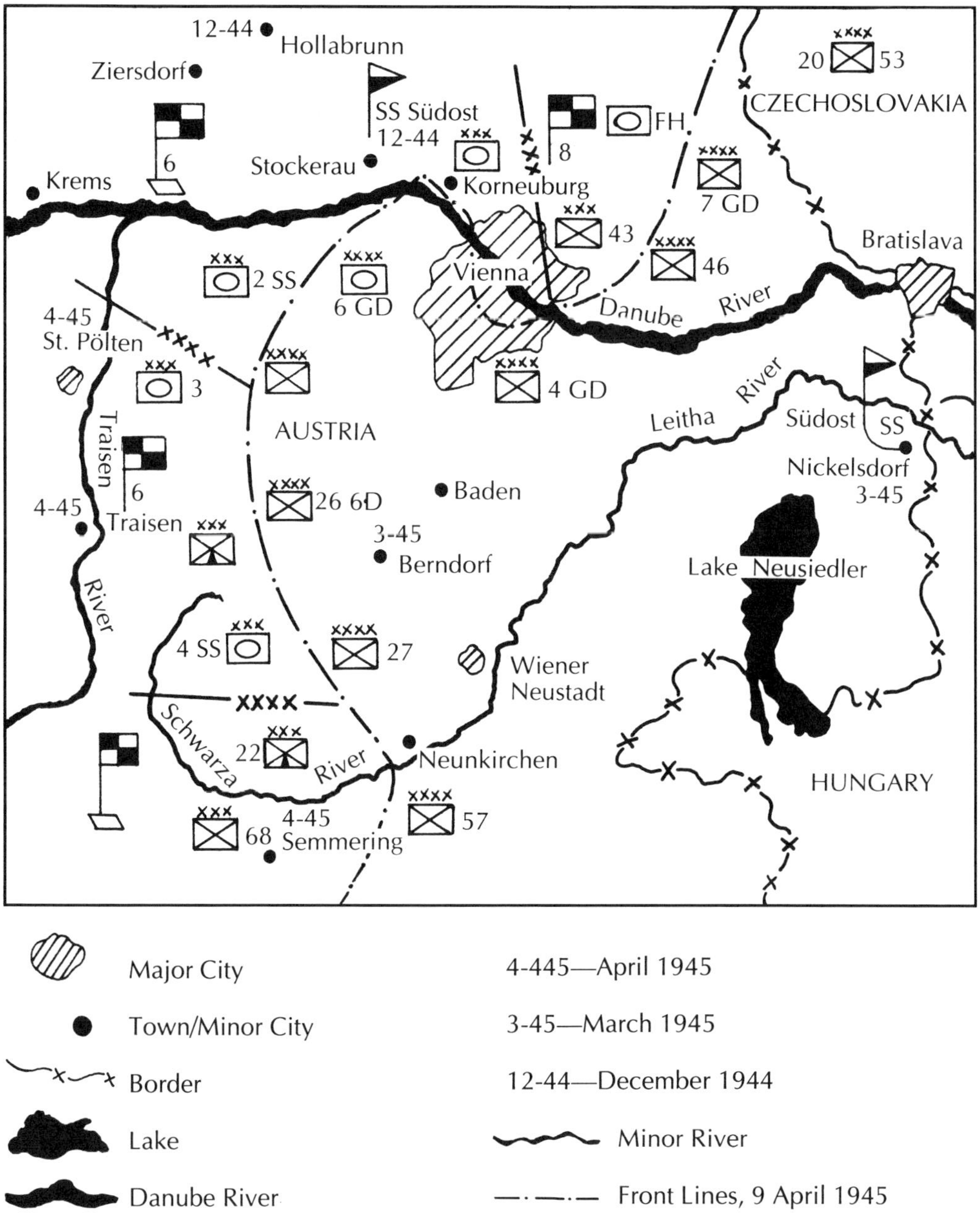

German military situation around Vienna, Austria, including the location and movements of SS-Jagdverband Südost, March-April 1945.

many Soviet citizens joined Südfrankreich is not known, although it is known that there was a significant number.

French accounts of a German assault on their FFI (French Forces of the Interior) units in the Vercors area of southern France tell of a German-Russian glider force that landed as part of the attack against them. Descriptions of these men and their actions leave almost no room for doubt that these "Russian commandos" belonged to Hauptmann Alexander Auch's Legionnaire Battalion, from which Streifkorps Südfrankreich had drawn some of its men. Additional evidence supporting this theory is the fact that the Vercors area of France bordered the French and Italian frontiers, which means that the commando operation would have fallen under the "jurisdiction" of Streifkorps Südfrankreich (South France).

The Vercors area has been described

The epitomy of the German defense in 1945: an SS Grenadier awaits the Soviet attack, grenades and Panzerfaust at the ready. Photo: Munin Verlag.

as being shaped like a giant arrowhead, thirty miles long by twenty miles wide. Its foothills are supplied with water from four rivers near the Alpine area that spreads east to Italy and Switzerland. It was in this highly defensible area that the French partisans, or maquisards, emboldened by the Allied landings in Normandy, established a "republic of Vercors." This "republic" was a self-proclaimed Allied zone in the interior of the German lines. There was no secret to it since the French FFI units expected General DeGaulle to send Free French reinforcements from North Africa. Thus the Germans were aware of the heavy, blatant FFI presence there.

It was not long before the German command turned its attention to the French presence in Vercors and attempted to destroy it. The question was: would the Free French reinforce Vercors *before* the Germans launched their assault?

They did not. On the morning of 21 July, German troops began their attack in earnest against the FFI in Vercors. The greatest defeat of the French partisans had begun.

At 9:30 A.M. the head of the Vercors FFI, Colonel Huet, received this message over the wireless radio: "Planes approaching from the south, Colonel."

"They can't be from Algiers," thought Huet, "otherwise we'd have been warned. They must be German."

The first German planes to appear over Vassieux were the FW-190 fighters and DO-217 light bombers. Approximately four hundred French personnel had been working at Vassieux that morning. The runway, designed to accommodate larger shipments of arms, was almost completed. The runway not only allowed more supplies to be flown in, but also more FFI or Allied troops.

Because the planes were coming from the direction of the Mediterranean coast to the south, the French believed that the planes were Allied. Someone yelled, "It's the Yanks! It's the Yanks!' but quickly saw the Black Balkan Cross of the German Luftwaffe and changed that to, "Run! It's the Boche!" Too late—the German fighters and light bombers were already making their descent in the attack formation. Moments later the first rounds of 20mm cannon and the explosions of the bombs rocked the stunned French in and around the town of Vassieux.

The DFS-230 gliders, carrying Streifkorps Südfrankreich, did not take long following the German support planes in, the great majority of them landing in Vassieux and, ironically, on the uncompleted runway. The runway was overrun by the fierce assault. Many French survivors described some of these paratroopers as husky men, speaking what was almost certainly Russian or one of its dialects, with the coarse and chiseled appearance of Eastern Europeans—almost certainly the men from the Legionnaire Battalion, Lehr Regiment, Brandenburg.

The objective of the landing at Vassieux was twofold. First, the Germans would secure the runway and deny its use to the Free French. Bombing from the air could not prevent its use. Second, the German glider troops were to set up a buffer to prevent FFI forces, being pushed from the north, from escaping to the southwest. The German 9th Panzer

STREIFKORPS SÜDFRANKREICH AND THE BATTLE FOR VERCORS

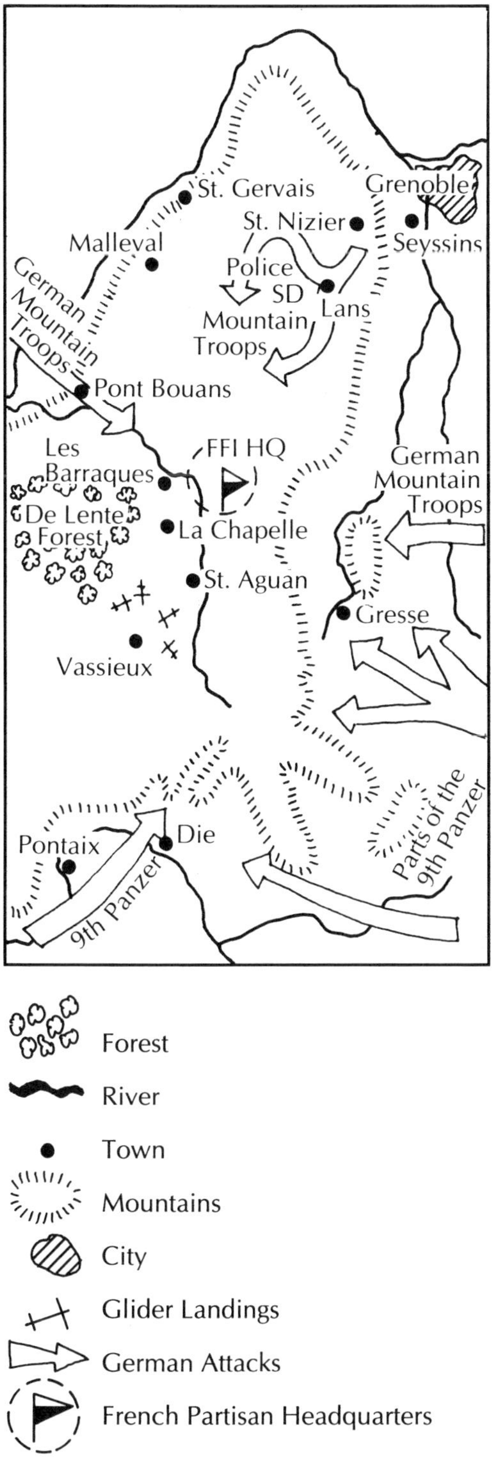

Division was driving up from the south and southeast. Soon the only escape route for the French would be Vassieux.

The precision and swiftness of the glider assault caught the French off guard. Within minutes, the FFI lost more than one hundred people at Vassieux. The FFI unit stationed at La Mure, a local hamlet just outside of Vassieux, was wiped out almost to the last man. Approximately two to three hundred glider troops landed at Vassieux. Soon the French command realized that it could not hold out against the trained German mountain troops and the tanks of the 9th Panzer Division. Indeed, those German forces, supported by the Ordnungs Polizei (Regular Police) and SD, were slowly but surely tightening the noose around the FFI troops in the Vercors.

The French quickly decided to attack the German glider force and recapture Vassieux. Eventually, the French sent out patrols to scout the German defenses at Vassieux, but they were sent scurrying by the intense fire of German machine-gun teams in the buildings at the edge of the town. The first real attack against the German defenders at Vassieux finally came during a rainy night that the French had hoped would help their attack. The assault actually managed to reach the buildings at the outskirts of the town, but, for some reason, an Office of Strategic Services (OSS) officer with the French ordered a retreat. His after-action report stated that they had been driven off by automatic-weapons fire.

On Saturday, 22 July, just one day after the glider-borne assault on Vassieux, the sky over the town darkened again with planes carrying the Balkan Cross. This time the planes were Junkers Ju-52 transport planes, fully laden with German troops. The French partisans watched with horror as the German planes landed on the very airfield that they had been building—the airfield considered too short for Allied planes to land on!

The pressure put on by the 9th Panzer Division and 157th Reserve Mountain Division eventually crushed organized resistance in the Vercors, thus preventing

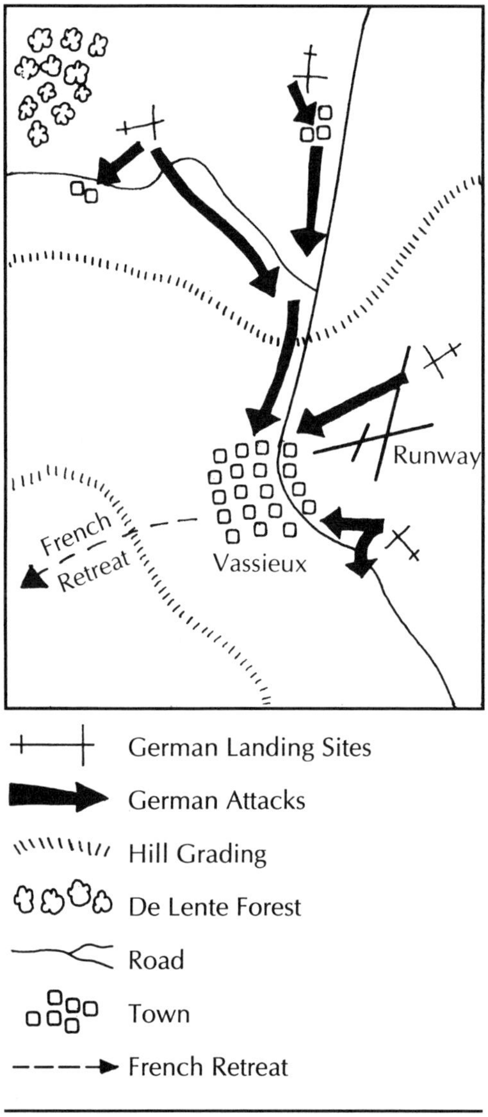

Close-up of the landing sites.

the German glider troops from being overrun at Vassieux and resulting in the death or capture of most of the FFI forces.

In September 1944, the Brandenburg commando units came under Colonel Otto Skorzeny's control. Many Brandenburgers opted to join the ranks of the Waffen-SS at that time, but others decided to stay in the army, even though their commando units were later redesignated as SS formations. One such Brandenburg officer was Major Beck, who had led the Italian contingent. In January 1945, when Streifkorps Südfrankreich was redesignated SS-Jagdverband Südwest, Beck continued to lead the Italian volunteers in that unit.

Major Beck had won fame while fighting Tito's Partisans in Yugoslavia. He, working with a Brandenburg doctor under his command and a Serbian doctor fighting for the Partisans, had been responsible for establishing a prisoner exchange at the town of Fruska-Gora. In January 1945, SS-Jagdverband Südwest contained four German and four foreign volunteer companies. These eight companies were split up into three jagdeinsatz.

The Italian officer who was considered the co-commander of Jagdeinsatz Italien, was Waffen-Hauptsturmführer der-SS "Z." Other officers in the Italian Jagdeinsatz included SS-Obersturmführer Fischer, SS-Obersturmführer Rahne, and Waffen-Obersturmführer der-SS Martiny.

Other officers in Jagdverband Südwest included Waffen-Obersturmführer der-SS Demetrio and SS-Hauptsturmführer Traege, who had previously commanded Streifkorps Südfrankreich.

After the assault on Vercors, the Streifkorps Südfrankreich withdrew in August and went to Ivrea, north of Turin, for reorganization. More recruits were picked up there. They spent the rest of 1944 operating along the French-Italian frontier, mainly against Italian partisans. By January 1945, the jagdverband was operating at Freudenstadt in the Black Forest area of Germany against the U.S. 7th Army, launching long-range reconnaissance patrols behind U.S. lines. Südwest played a role in blowing up bridges that had been left intact by retreating German forces and captured by advancing U.S. forces.

In March, the battle school that had been at Stuttgart was withdrawn to Ulm to avoid the rapidly approaching Allied troops. Meanwhile, Jagdverband Südwest retreated back along with the rest of Army Group G in a southeasterly direction toward the Austrian frontier.

On 31 March 1945, Otto Skorzeny was instructed to take as many of his

SS-JAGDERVERBANDE SÜDWEST
JANUARY 1945

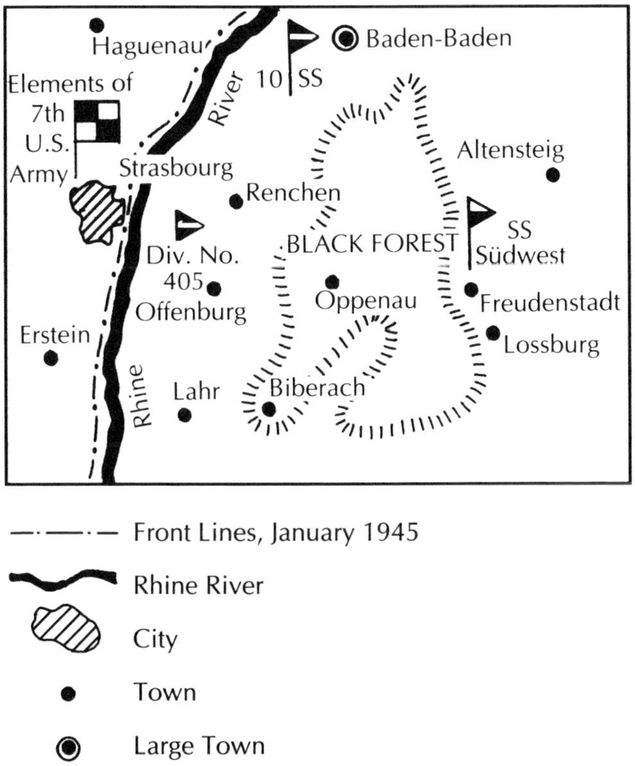

commandos as possible and head for the so-called Alpine Redoubt, where Nazism was to make its last stand. Taking 200 to 250 men of SS-Jagdverband Mitte, Skorzeny left the Oder River/Berlin area and headed south. On 11 April, Skorzeny radioed the commander of SS-Jagdverband Südwest to join his command at Linz, Austria, as soon as possible. It is not known if SS-Jagdverband Südwest reached Linz in time to link up with Skorzeny's group. What is known is that the commando group did reach the Austrian frontier and crossed it sometime during the third week in April.

Lutz Bormann was a last-minute inductee into Südwest. He had served in the SS Officer Academy at Prague from September to December 1944. From January to February 1945, Bormann served in the Waffen-SS Inspectorate Office for Female SS Members at Königsbrück, near Dresden. In March 1945, he was inducted into SS-Jagdverband Südwest and served in that formation until the end of the war. Born in Berlin on 16 May 1924, Lutz Bormann turned twenty-one just eight days after the end of the war.

SS-JAGDVERBAND OST

The origins of this commando force date back to the summer of 1944, to the Streifkorps Einsatzgruppe Baltikum, a platoon of two officers and thirty-six men of the Brandenburg commandos responsible for operations in the east. This commando unit quickly grew to company size and then to that of a battalion. One of the more famous operations carried out by this commando force was Operation Freischütz.

In the summer of 1944, the Soviets had overrun Army Group Center. Dozens of divisions had been shattered, and tens

Field Post No.	Unit
?	SS-Jagdverband Südwest
41 808	SS-Jagdeinsatz Italien
?	SS-Jagdeinsatz Nordfrankreich
?	SS-Jagdeinsatz Südfrankreich
67 010	Staff plus Staff Company, kampfschule (battle school, officially established on 8 February 1945)

of thousands of German soldiers had surrendered. Thousands more had been bypassed by the rapidly advancing Red Army, and most of these pockets of German troops surrendered. One that didn't was a 2,000-man pocket under the command of Lieutenant Colonel von Scherhorn. This battle group had survived in the woods near Minsk for nearly two months after the start of the Russian offensive (22 June 1944). Several radio messages from the beleaguered kampfgruppe had embarrassed the German army High Command, which would have preferred that Scherhorn surrender, but his pleas for aid had to be answered. A call for volunteers went out throughout the commando force.

Several of Skorzeny's men came forward. Although the Brandenburg detachment fighting in the east was not officially redesignated SS-Jäger-Bataillon Ost until 1 October 1944, already in August 1944 it had been fighting with a mixed batch of Waffen-SS and Wehrmacht personnel. It must be remembered also that it wasn't until September of that year that Skorzeny received complete control over all commando operations. The transfer of Brandenburg personnel into the Waffen-SS was slow. In many cases, those Brandenburgers chose to keep their Wehrmacht ranks even though their commando formations were transformed into SS units.

The leaders of the mission decided to airdrop several commando teams behind the Russian lines in Belorussia to locate Scherhorn's combat group. Once they found it, the SS commando team would supervise supply drops and attempt to establish a small airstrip so that the more seriously wounded could be evacuated. Eventually, they hoped to extract the battle group by guiding it all the way to Courland, Latvia, where the German Army Group North lay.

The first group sent in at the end of August was led by an SS-Oberscharführer "P." They and the second group, sent at the beginning of September and led by SS-Oberjunker (Officer Candidate) Will Linder, landed just east of Minsk. A third and fourth group parachuted in later in the area of Dserschinsk (Dzerzhinsk) and Witejka (Viteys). The third group was under the command of a Master Sergeant "M," and the fourth was under Fähnrich "R." Each group comprised two Germans and three Russians.

The first group ran into a Soviet patrol almost as soon as it had landed. Frantic radio signals reported that they were under fire. Soon no radio contact was possible, and they were never heard from again. The second group under SS-Oberjunker Linder was luckier. Fourteen days after his group parachuted in, Linder radioed Friedenthal: "Good

SS-Junker Lutz Bormann as he appeared in 1945. He served with SS-Jagdverband Südwest from March through May 1945. Photo: Munin Verlag.

landing. Group Scherhorn found." The Germans established a regular supply service using air drops, but an attempt to build an airstrip halted when the half-finished strip was discovered by Red Army troops and quickly overrun. The third group under Master Sergeant "M" was never heard from, having failed to establish contact with Scherhorn's pocket. The fourth group also failed to establish contact but was lucky enough to reach German lines. "M's" first radio call had been promising: "Excellent landing. All five men are OK." Two weeks later they had found their way to the Latvian frontier.

The Scherhorn group began its epic march toward German lines. By December 1944 the group had traveled about 250 kilometers to the Lithuanian frontier by Dünaburg (Daugavpils). Not much is known of the group after that date. Slowly, the airdrops that supplied the group began to dwindle as the needs of more important operations pre-empted. Through it all, Skorzeny kept trying to rescue the trapped men. Against his will, the supply drops eventually stopped.

Another operation conducted by the Ost commando group during the late summer of 1944 was Operation Landfried, which literally means "peaceful land." It entailed the blockade of vital passes over the Carpathian Mountains in an attempt to slow down the Red Army. The German situation in Romania had worsened in August with the collapse of the German 6th and 8th Armies, mainly because of Romania's collapse. On 31 August 1944, the Red Army entered Bucharest, and the Romanian army turned against its former German allies. If something was not done quickly, the entire German Army Group Südost would be cut off in the Balkans.

If the Red Army and its Romanian and (later) Bulgarian allies reached the vital Carpathian passes at Kronstadt, Hermannstadt, and Karlstadt, the Germans in Yugoslavia and Greece would be doomed. Hungary would also fall to the enemy and, after it, Austria. Immediately, two platoons of the 1st Company (100 men) under the overall command of SS-Untersturmführer Walter Girg were organized and dropped near the passes. They were a mixed bag of German and ethnic Germans from Romania, some dressed in Romanian army uniforms, others in Hungarian army tank uniforms, and still others in civilian clothes.

The two platoons were split into four groups. Girg's group actually ran into an entire German Luftwaffe flak regiment that had been waiting to surrender, equipment and all. Girg managed to convince a good number of these Luftwaffe troops to attempt a breakout to the west. Girg's group misdirected enemy units and even participated in a parade by Red Romanian army units through a town. His group's luck finally ran out when they were discovered by suspicious Romanian and Red Army men.

Given a quick trial, Girg and his men were lined up along a road to be executed. The men had been stripped of their Romanian army uniforms and wore only their underwear. At the instant that they were about to be shot, Girg instructed his comrades to make a run for it. He decided that if they were going to die, they would not die like sheep. The Reds reacted quickly, but, amazingly, Girg managed to reach the banks of a nearby river and to swim away. He was wounded, but he escaped. All of the other men in his group were killed. Equally amazing was Girg's epic march to the safety of his own lines.

For his bravery and actions, Girg was promoted to SS-Obersturmführer and awarded the much-coveted Knight's Cross. One month later on 9 November, Will Linder, the officer candidate who had located Gruppe Scherhorn, was also awarded the Knight's Cross. SS-Unterscharführer Fritz Winkelhake and Christof Fritsch were both awarded the German Cross in Gold on 9 October

1ST COMPANY/SS-JÄGER BATAILLON OST AND THE BATTLE FOR THE CARPATHIAN PASSES

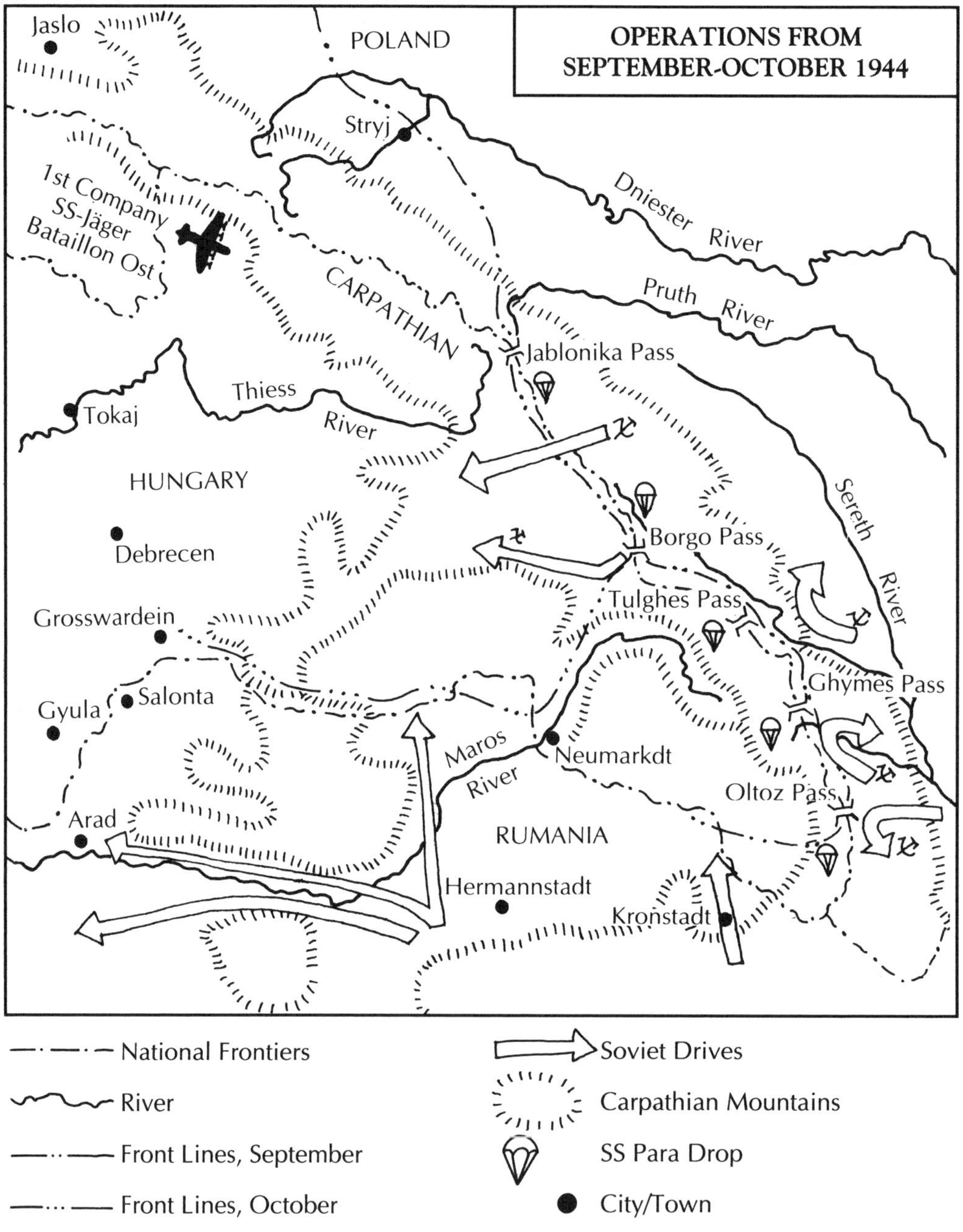

1944 for participation in Operation Landfried. What is interesting is that these two sergeants belonged to SS-Jäger Bataillon 502, implying that at least some members from this commando battalion participated in Landfried.

During the fall of 1944 the German Secret Service had been receiving reports of a Ukrainian anti-German, anti-Stalinist guerrilla movement headed by a man called Stefan Bandera. Even though Bandera's movement was strictly nationalistic, only seeking independence for the Ukraine, it appealed to the Germans since it was a real life, active group behind Red Army lines. The Germans attempt to contact the Bandera movement and to supply them with medicine and military hardware. The operation, code named "Brauner Bär" (Brown Bear), began in December 1944. It called for a dozen German NCOs and men plus about twenty Russians to infiltrate the Red Army lines in an attempt to make contact with the Bandera movement.

The command of this mission was entrusted to an SS-Hauptsturmführer Kern, who had been a captain in the Brandenburg commandos. Kern inserted his small force through eastern Slovakia in December. He moved his unit eastward and after about two weeks managed to make contact with the Ukrainian guerrilla organization in the western Ukraine.

Fourteen or fifteen days into the operation, Hauptsturmführer Kern made his first report: "B. found, no men lost." Immediately, the Germans implemented plans whereby Kampfgeschwader 200 began to supply munitions and medical supplies to the Ukrainian guerrillas. Brown Bear came to a close around the middle of March 1945, having been concluded quite successfully. Kern was surprised to have found a large number of ex-14th SS Division men among the guerrillas. It seems that many had deserted the 14th Galizien SS Division before the German withdrawal from the western Ukraine. The bloody nose received by this SS division during the Battle of Brody in July 1944 had provided the scattered stragglers who eventually found their way into the Bandera movement. Withdrawing through the Carpathian Mountains, Hauptsturmführer Kern managed to reach the German lines toward the beginning of April 1945, having lost only five of his men.

In the beginning of January 1945, Otto Skorzeny had given command of SS-Jagdverband Ost to his chief of staff, Baron Adrian von Foelkersam. Von Foelkersam had just been promoted to SS-Sturmbannführer, after having commanded Kampfgruppe 'Z' of the 150th Panzer Brigade. The assignment was a tough one. The Red Army winter offensive had begun on 12 January, and the entire German eastern front was feeling the full fury of an enemy who outnumbered them seven to one in tanks, eleven to one in infantry, and twenty to one in planes and artillery.

The choice had been von Foelkersam's. He had asked for command of the jagdverband, which now happened to be situated at Hohensalza (Inowroclaw), northeast of Posen. The front had been torn at the center on a seventy-five kilometer span.

Through this gap, the Red Army had poured an initial group of thirteen infantry and five tank corps. With no real organized front line, the Germans had to resort to garrisoning the larger cities in the way of this "Red Flood" in the hopes that these pockets of resistance would act as breakwaters. This tactic does not seem to have slowed down the Red Army advance enough to warrant the cost in German troops lost.

Sturmbannführer von Foelkersam's reasons for wanting this "lost" command were multifold. First, he wanted to return to the fight in the east, which he considered the real threat to postwar Germany. If defeat was inevitable, it was infinitely more desirable to be occupied by the Western powers than by the Soviets. Second, his brother had been taken prisoner by the Red Army, so he had a personal score to settle. Last, but certainly not least, his wife and little daughter were living in Posen, about 140 kilometers from Hohensalza. At the rate the Red Army was advancing, Posen would fall within days.

The dead of winter had arrived. Already long rows of thousands upon thousands of civilian refugees were making their way westward, through the bitter winter snows and sub-zero cold. Thousands lost their lives from the cold or from enemy action. The scene of Red

Army tankers catching up to a civilian column and crushing the hapless Germans under their tracks was repeated all too often during that terrible winter of 1945. Skorzeny had sent a Dr. Slama to Posen with orders to pick up von Foelkersam's family and take them to safety. Meanwhile, Sturmbannführer von Foelkersam managed to acquire about two dozen lorries and loaded them up with munitions and rations for his jagdverband.

He wasn't at Hohensalza a week when the Reds attacked the city with no less than two corps. The date was 18 January. By 21 January, the German situation at Hohensalza was untenable. Foelkersam radioed to Skorzeny: "Situation critical. Should I attempt a breakthrough? Foelkersam." Skorzeny quickly replied: "Breakout this very night!"

SS-JAGDVERBANDE OST, JANUARY 1945

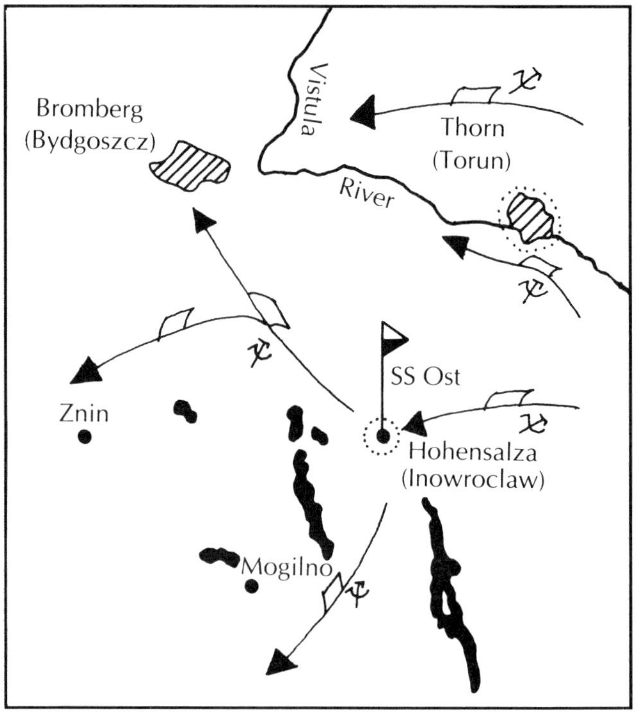

Hohensalza was taken by the Red Army on 19 and 20 January 1945. Thorn was also surrounded but held out from 22 January to 2 February 1945. Bromberg fell on 27 January.

The order was followed, and Jagdverband Ost made a breakout attempt to the west on the night of 21 and 22 January. One week later the survivors of SS-Jagdverband Ost finally arrived at Friedenthal. Of the original 800-plus men and officers at the beginning of January, only two officers and thirteen men had survived. Von Foelkersam was not among them.

He had received a shot to the head and died during the night of 22 January, in the bitterest fighting of the breakout. Skorzeny would later comment that his death had hit him hard, since he considered von Foelkersam not only his right-hand man, but the best friend he ever had. Frau Foelkersam and her little daughter managed to escape the fate of many German refugees, ending the war in western Austria, safely out of the reach of the Red Army.

SS-Obersturmführer Walter Girg, of Carpathian passes legend, was summoned to a meeting with Skorzeny at the beginning of the second week in January 1945. Skorzeny was to the point. He told Girg that the Soviets had launched their winter offensive and punctured the eastern front at numerous places. In most instances, the German High Command did not know where the main enemy columns were or where they were headed. His new mission would entail driving through the German General Government of Poland under German administration and reporting on the movement of the enemy's tank forces.

To this end, he was accompanied by twelve German personnel and an equal number of Russian volunteers. They were transported by sea to East Prussia where their soon-to-be-famous odyssey began. Girg's resourcefulness was to help the group secure adequate "transport" for their long, treacherous reconnaissance mission by dressing up as Romanian Red Army men and requisitioning Russian T-34 tanks for the long ride.

Girg's cold, calculating nerve was delicately balanced by his contempt for death and for the enemy. On one occasion, Girg and his men walked into a large Red Army unit that was bivouacking in a small village. Girg walked inside one of the houses where he met and was befriended by a "fellow" Russian officer who wined and dined Girg until the wee hours of the morning. Little did the officer realize that his uninvited guest was an SS officer. Girg's pleasure at having pulled off this deception was unbounded, but equally unlimited would have been the trouble he found himself in had his little charade been discovered.

After a 1,500-kilometer trek all the way from Danzig, Walter Girg's commando force finally reached the besieged Baltic seaport of Kolberg around the middle of March 1945. The trapped German garrison was under the command of SS-Oberführer Heinz Bertling, the former commander of the German-sponsored Indian Legion. Other top officers in Kolberg included Wehrmacht Oberst (Colonel) Fritz Fullriede and SA-Standartenführer Pfeiffer, who was in charge of the Kolberg Volkssturm.

Upon reaching Kolberg, Girg and his men "surrendered" to the German defenders, who were much surprised to see an entire Red Army tank platoon handing over their tanks and sheepishly going into captivity. The 46-year-old Bertling just couldn't believe that Girg was real. Additionally, the Red Army-sponsored Seydlitz Division, made up of captured Germans-turned-traitors, was fighting around Kolberg. This fact added to Bertling's suspicions that Girg and his men were traitors.

Bertling quickly put Girg and his men on trial, and they were found guilty of treason and sentenced to death. Luckily for Girg and his men, Skorzeny was able to radio Kolberg and tell Oberführer Bertling that indeed those were his men. It seemed that SS Obersturmführer Girg had finally used up all of his nine lives! For his tremendous exploit, Girg was promoted to Hauptsturmführer (captain) and on 31 March he received the oak leaves to his Knight's Cross. Adolf Hitler himself handed the award to Hauptsturmführer Girg, the award taking effect as of 1 April. Girg received the oak leaves exactly a week before Otto Skorzeny himself received his.

The destruction of Jagdverband Ost at Hohensalza in January 1945 was not the end of the formation. Every effort was made to re-form the commando force from the small fifteen-man cadre that remained. Realistically, the Ost Jagdverband had to be formed from scratch. In effect, the formation was a new unit, so much so that Skorzeny would refer to it as Ost II. The rest of the history of SS-Jagdverband Ost is sketchy and fragmented. In fact, the unit was broken up near the end of the war.

In April the Ost Jagdverband was sent to Neustrelitz and took part in its defense. The last report on the situation maps listed the commando force at Sachsenhausen/Nordbahn. There was one company, however, that spent the month of April fighting under Army Group Center. On 10 April, Skorzeny met the Army Group Center commander, Field Marshal Schörner, at Olmütz (Olomouc). Schörner wasted no time in telling Skorzeny that he needed his commando force for an operation in the Breslau area.

Skorzeny himself was on his way to Austria with two companies of SS-Jagdverband Mitte for the defense of the Alpine Redoubt. Although he could not spare his small 250-man force, Skorzeny managed to forward a 100-man company

SS-OBERSTURMFÜHRER WALTER GIRG'S FAMOUS DASH TO KOLBERG

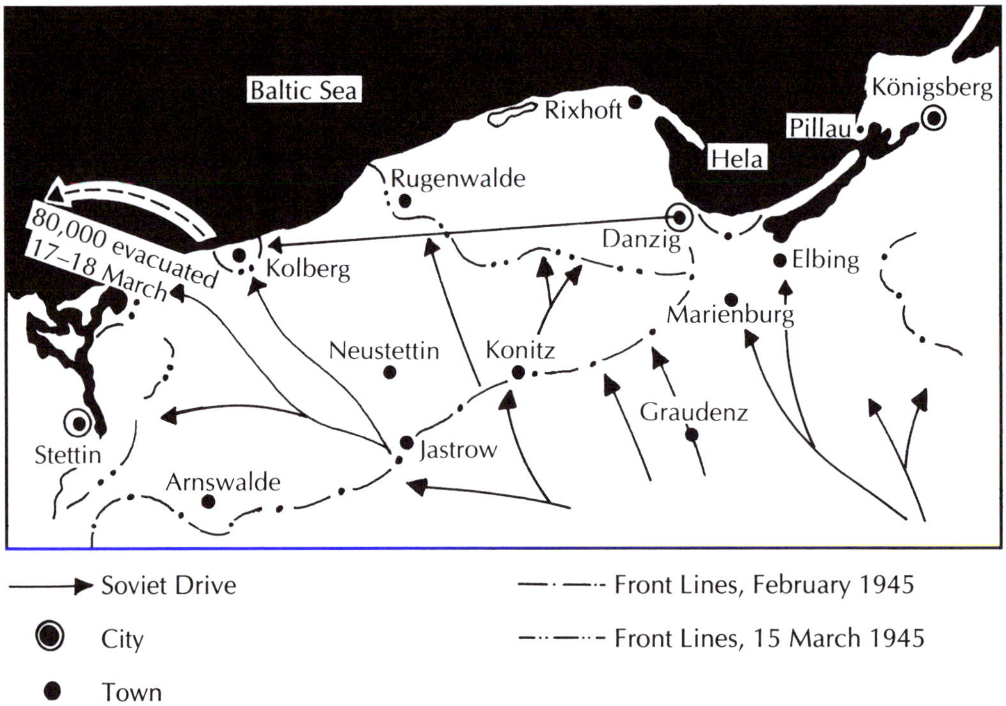

of SS-Jagdverband Ost to Schörner. It was this understrength company that was assigned to destroy an important enemy bridge near the surrounded Silesian capital. The company managed to find a path through the thick ring of Soviet troops surrounding the city. The bridge was blown up, severely limiting the flow of war materiel for a time. The operation itself was insignificant to the conduct of the campaign. The company of SS commandos had a difficult time extricating itself from behind Russian lines. It fought on under Army Group Center until the end of the war. Five days before the capitulation, it was still destroying enemy armor in Czechoslovakia. In an unfortunate turn of events, the company found itself surounded shortly before 8 May and wiped out. The Soviets took no prisoners.

SS-Jagdverband Ost was an evolutionary unit that went through several changes. Like most of the SS commando units, the formation was based on the Brandenburg Command unit, in this case the Streifkorps Einsatzgruppe Baltikum. On 1 October 1944, the SS-Jäger-Bataillon Ost absorbed the remnants of the Brandenburg Baltikum group. SS-Jäger-Bataillon Ost initially contained three rifle companies and later four. On 10 November, a little over a month later, SS-Jagdverband Ost was formed from SS-Jäger-Bataillon Ost. Jagdverband Ost had three combat groups: SS-Jagdeinsätze Ostland, Russland, and Polen. The three groups had three rifle companies and one heavy company. In addition, there were four companies made up of foreigners in all eight companies of Ost as of 16 January 1945, although the list on page 212 is incomplete.

SS-Jagdeinsatz Ostland's field post number, 34 242, had been formerly used by Streifkorps Biscaya from 4 August 1944. This leads us to speculate that perhaps this Brandenburger unit may have been used to form Ostland.

After the debacle of Hohensalza, every effort was made to reform the Ost

Jagdverband. By mid-April, the Red Army had begun its spring offensive on the Oder River front, and SS-Jagdverband Ost had been relocated to the area between Oderberg and Angermunde during the beginning of April 1945. It was there that Ost picked up some badly needed replacements from some survivors of SS-Jagdverband Mitte. A lot of men were recruited from the 7th SS-Panzergrenadierregiment that had arrived just before the Russian offensive on 16 April. SS-Unterscharführer Andreas Meyer, a member of SS-Jagdverband Mitte, found himself transferred to the newly reformed Jagdverband Ost after the battle for the Zehden bridgehead. His illuminating eyewitness account of the last days of the war under Ost describes the chaotic and hectic defense put up by the German forces on the Oder River during the spring of 1945.

> *After a few days we received new marching orders. The Russians had crossed over the Oder. I accompanied an officer from our staff in a motorcycle sidecar, and the unit withdrew in the afternoon to prepared positions by Finowfurt. Late in the evening we received our orders from Corps Headquarters,so early in the morning we again moved toward the east. In the next village there were already several houses which had been set on fire by enemy artillery fire. In the following village, we were already taking infantry fire, so we dismounted and sent out reconnaissance patrols to figure out the situation.*
>
> *In the headquarters of the divisional commander of the 25th Panzer-[grenadier?] Division, I met a Wehrmacht Major General, the commander of this rather worn-out and skeleton division. I set the reconnaissance reports and the situation maps on his table for him. Around noon all of the villages around us were occupied by the enemy. Our first ray of light came at 2 P.M. when word came that one of the villages was again reported to be free of the enemy. Through this hole, the supply columns were sent. In the middle of the night, we pulled out and by way of Beerbaum, southwest of Eberswalde, had a few skirmishes with the enemy."*
>
> *The main attack of the Russians was leading in the direction of Bernau. During the night we headed through Eberswalde in the direction of Werbellinsee [Lake Werbellin]. There we had a few days' rest and set ourselves in the defense. After a few days, I was in the headquarters (a countryhouse which belonged to the well-known Berlin gourmet Chef Kottler). Around 7 P.M. there was a call from the corps headquarters with orders for us to go to Oranienburg. I sent a reporter to the chief, SS Obersturmbannführer Schrage-Bobet, who had taken over because the commander, SS-Sturmbannführer Milius, had been wounded at Finow and was no longer with the unit.*
>
> *Two hours later I received a new call from the Corps. The earlier order was now rescinded and we were ordered to Prenzlau. The transport unit was to reach us in the morning so it wasn't until then that we moved off towards Prenzlau, which we reached by noontime. I accompanied Obersturmbannführer Schrage-Bobet into the area east of the city. There we saw the Russian tanks coming! We could make out over thirty tanks! Our unit set up a provisional defense line position. The headquarters was located in a cellar in the city on the Markt-Platz Street.*
>
> *Toward evening, the Russians reached the city limits. Against such superior forces we didn't have a chance. As the Russians advanced into the city, we slowly withdrew, skirmishing all along, way past midnight. In a small village west of Prenzlau, we*

F.P. Number	Unit
20 336	Staff plus units, SS-Jagdverband Ost
34 242	SS-Jagdeinsatz Ostland, SS-Jagdverband Ost
35 181	SS-Jagdeinsatz Russland, SS-Jagdverband Ost
?	SS-Jagdeinsatz Polen, SS-Jagdverband Ost

pulled into a farm. We immediately fell asleep and were sternly woken by the crew of our radio-car. The radiomen called to us that the Russians with their tanks had already passed us by. We immediately jumped into our kubelwagen [jeep] and drove down the main road in a westerly direction. Suddenly, left and right of the road, we saw many T-34s. One tank turned its cannon in our direction. We stopped and sprang into the ditch. Not a moment too soon! The first shot was a direct hit. Unfortunately, we had one dead. We then ran along the road ditch, and as we succeeded in getting out of clear terrain, we ran headlong into another one of our vehicles. We then drove with all haste to Burgstargard.

Around 5 P.M. the Russian infantry units arrived there so we had to pull back to Neubrandenburg. There our unit again took positions. The sturmgeschütz secured the street running from the south, from the direction of Neustrelitz. Late in the evening came the order that the sturmgeschütz must withdraw due to the overwhelming enemy forces. Besides this, the enemy had advanced with infantry forces from the east, straight into the city. SS-Obersturmbannführer Schrage-Bobet wanted the situation with the sturmgeschütz clarified and was seriously wounded by an artillery shot. He was put on an assault gun and pulled back.

At Neubrandenburg, we had more heavy losses and were practically destroyed. Corvette Captain von Behr, as senior officer, gave the order to retreat in a westerly direction and reported the unit to the division as being destroyed. We then regrouped ourselves in the Müritz Lake area, and there we counted only about 200 men left in our unit! Neubrandenburg was to be our last action. We then set off toward the west by Malchow-Parchim. The majority of our battalion became prisoners of war on 2 May 1945, near Hagenau. Eleven comrades and I went into American captivity on the evening of 2 May.

SS-JAGDVERBAND NORDWEST

Little is known about this SS commando formation. Apparently, it had its origins with the Brandenburger Streifkorps Nordfrankreich (Assault Group Northern France). Streifkorps Nordfrankreich was initially organized in 1944. SS-Jäger-Bataillon Nordwest was organized from this Brandenburg unit in October 1944. One month later, the unit's title changed to SS-Jagdverband Nordwest, which it retained through the rest of the war. The four-company SS-Jäger-Bataillon Nordwest expanded to eight companies when the Jagdverband formed in November 1944.

One reliable source reports that approximately 150 Flemish volunteers served in the Jagdverband. The unit had four German companies and four Legionnaire companies made up of foreign volunteers from Denmark, Norway, Belgium, and Holland. The formation spent a few months training and, had the German Ardennes offensive proved successful in December 1944, was scheduled to be used as an occupation unit in Holland.

SS-Jagdverband Nordwest was eventually committed to the eastern front, being sent to the Oder River front in late January. The commando unit formed part of Otto Skorzeny's defenses in front of Schwedt, holding the defense lines in and around the town of Nipperwiese. This area was the extreme left flank of Skorzeny's bridgehead and, as such, was vitally important.

The commando formation held onto that area until ordered to withdraw. The unit served near Schwedt from 1 February to 3 March 1945. (For more information about the actions of SS-Jagdverband Nordwest, see Chapter 10.) After serving in the Schwedt bridgehead, the commando formation remained in 3rd Panzer Army reserve. When the Russians launched their spring offensive in the third week of April, SS-Jagdverband Nordwest was committed once again but was pushed back by the overwhelming Red Army forces.

Officers who served in SS-Jagdverband Nordwest included SS-Obersturmführer Gutweniger, SS-Untersturmführer Karl-Heinz Gieseler, and SS-Untersturmführer DeRuyter. I have been unable to locate any others, though many more officers served.

Karl-Heinz Gieseler had an interesting history. He was born on 30 July 1925 in Schweez and joined the Waffen-SS exactly two months before his eighteenth birthday. He was attached to Ausbildung und Ersatz Bataillon 3 (Forming and Reserve Battalion No. 3), which was stationed at Warsaw. It was the replacement battalion for the SS Totenkopf Division.

One month after Gieseler enlisted, he was transferred to the Totenfopf Division. He served with this Waffen-SS unit for the rest of 1943, receiving the Iron Cross, Second Class, on 3 August 1943 for actions during the Kursk offensive. On 5 December of that same year, he received the Wound Badge in Black, for wounds sustained that November. Six days later, he received the Infantry Assault Badge, and on 14 January 1944, he was awarded the Close Combat Clasp in Bronze.

All through this period, Gieseler had been an SS-Sturmann, but his abilities began to show. In January 1944, Gieseler volunteered for service in the newly formed SS-Fallschirmjäger-Bataillon 500, receiving his Paratroop Assault Badge on 13 May 1944. Just twelve days later, he took part in Operation Knight's Move, the attempt to kill or capture Tito, the Yugoslavian Partisan leader.

For his actions during this battle Gieseler was made an SS-Junker (SS Officer Candidate) on 1 June. Two days later, he was awarded the Iron Cross, 1st Class, for actions at Drvar. Shortly after his promotion to SS-Junker, he was transferred to SS-Jagdverband Mitte, though at this time it was still called SS-Jäger-Battailon 502. His record indicates that he took part in operations in the east, probably in Skorzeny's snatching of the Hungarian president's son in October 1944 and in other operations on the Eastern Front.

On 4 December 1944, he received the German Cross in Gold, the prerequisite medal needed to attain the next highest award, the Knight's Cross. In January 1945, he was sent to the SS Officer Academy at Prague but was

SS-Hauptsturmführer Baron Adrian von Foelkersam (above center) wearing his Knight's Cross around his neck. Foelkersam was Otto Skorzeny's Chief-of-Staff. He is shown here with members of the SS Jagdverband during the Christmas celebration of 1943. Foelkersam went on to command SS-Jagdverband Ost in January 1945. He lost his life that same month leading a breakout attempt from the surrounded town of Hohensalza (Inowroclaw). Photo: Munin Verlag.

Photograph of Karl-Heinz Gieseler as an enlisted man. Gieseler was a member of SS-Jagdverband Nordwest for a time. Photo: Robert Tyler.

quickly transferred to SS-Jagdverband Nordwest, which had been attached to the defense of Schwedt. As a result of his actions in Nipperwiese, he received the Close Combat Clasp in Silver on 21 February 1945 and the Wound Badge in Silver on 29 March. On 1 April, Gieseler was promoted to SS-Untersturmführer and was posted on 17 April to 11. Freiwilligen-Panzergrenadier-Division Nordland.

Although Gieseler was still recovering from his wounds, he volunteered for frontline service, which is why he was attached to Nordland when it withdrew into Berlin during the battle for that city.

He served with Nordland from 17 April to 29 April. On 29 April, Second Lieutenant Karl-Heinz Gieseler was awarded his nation's highest military decoration, the coveted Knight's Cross.

After the war, he served as major of the reserve in the West German Bundeswehr. As recently as 1988, he also served as president of the West German Sports Bund.

Very rare photograph of Hauptsturmführer Adrian von Foelkersam as he appeared shortly before or after the start of the Battle of the Bulge in December 1944. Photo Andreas Meyer.

PART THREE

Illustration: Eric Weisman.

23.WAFFEN-GEBIRGSJÄGER-DIVISION DER-SS KAMA

21.WAFFEN-GEBIRGSJÄGER-DIVISION DER-SS SKANDERBEG

24.WAFFEN-GEBIRGS(KARST)-JÄGER-DIVISION DER-SS

30.WAFFEN-GRENADIER DIVISION DER-SS WEISSRUTHENIEN

REMINISCES OF ERIC BRORUP, A DANISH VOLUNTEER

SPANISH VOLUNTEERS OF THE WAFFEN-SS

CHAPTER SIXTEEN

Not upon a man from the colonel to the private in a regiment—both inclusive. We may pick up a marshal or two perhaps; but not worth a damn.

Arthur Wellesley
Duke of Wellington

During the summer of 1944, partisan activity in the Balkans increased in size and tempo. German military demands at the front negated the possibility of reinforcements being sent to the Balkans in adequate numbers. Hitler's solution was to order the raising of several more locally formed divisions from the native population of the Balkans. Already as early as 1942, the Germans had raised an ethnic division of Croatians of German ancestry. This division was the 7. SS-Freiwilligen-Gebirgsjäger-Division Prinz Eugen. In 1943, another Yugoslavian division was formed, the Croatian-Muslim 13. Waffen-Gebirgsjäger-Division der-SS Handschar.

Two more divisions were to be formed. One would be made up of Albanian volunteers while the other division would be formed from Croatian-Muslim recruits like the Handschar Division. (The Albanian 21. Waffen-Gebirgsjäger-Division der-SS Skanderbeg is covered in Chapter 17.) We will concern ourselves here with the ill-fated 23. Waffen-Gebirgsjäger-Division der-SS Kama. According to the OKW War Diary, Hitler gave his approval to raising this second Croatian-Muslim division on 17 June 1944. On that date, he also gave it the honorary title Kama.

The decision was also made to raise a corps command that would eventually lead the 13th and 23rd divisions of the Waffen-SS. The 9th Waffen-Gebirgs-Armeekorps der-SS was formed during the summer and fall of 1944. It went on to lead several Waffen-SS and Heer formations. In December 1944, the corps was surrounded while defending Budapest and eventually wiped out when that city fell.

Although the order to raise the Kama Division had been made on 17 June, actual transfers and recruitments for the cadre personnel had already begun on 10 June. A sizable number of German officers and NCOs were available to the division. Croatian officers and men from the Handschar Division were also transferred to Kama, including the entire 13. SS-Aufklärungs-Abteilung (Reconnaissance Battalion), who were sent en masse and became the 23. SS-Aufklärungs-Abteilung. To this core of troops was added a new batch of Croatian-Muslim recruits. At its peak strength in September 1944, the Kama Division could boast of 3,793 men (126 officers, 374 NCOs, and 3,293 men).

Because Hitler feared that the Yugoslavian Partisans would do everything in their power to disrupt and destroy the newly forming division, he decided to change the assembly site from the area bordering the Sava River-Bosna-Spreca-Drina River to the Bacska Region. This area had been annexed by Hungary from Yugoslavia, and after the war, it would again become part of Serbia.

The large regimental units of the Kama Division were located in the following areas:

1. Recruit Depot—Sambor
2. Recruit Depot—Bosujaci
3. Waffen-Gebirgsjäger-Regiment der-SS (kroatische Nr. 3) and Waffen-Gebirgsjäger-Regiment der-SS (kroatische Nr. 4) were located between Sambor and Verbas.
4. The Waffen-Gebirgs-Artillerie-Regiment der-SS (kroatische Nr. 2) was located around Kula, between Sambor and Verbas (Vrbas).
5. The old 13. SS-Aufklürungs-Abteilung, now redesignated the 23. SS-Aufklürungs-Abteilung, initially located at Centamas.

Field Post No.	Unit
19 970	Divisional Headquarters, 23. Waffen-Gebirgs-Jäger-Division der-SS Kama
33 535	Workshop Company, SS Supply Troop 23
40 349	Staff Company, Divisional Supply Troop 23
24 378	Motor Vehicle Unit, SS Supply Troop 23
25 211	Truck Repair Company, SS Supply Troop 23
29 101	Truck Column, Mixed Supply Troop 23
28 248	1st and 2nd Veterinary Company, SS Veterinary Unit 23
27 578	Field Post Office 23, SS Supply Troop 23
26 445	Staff Battery plus 1st through 3rd Batteries, SS Flak Battalion 23
43 199	Staff Company plus 1st through 3rd Companies, SS Engineer Battalion 23
39 839	Staff Company plus 1st and 2nd Companies, SS Signals Battalion 23
48 064	Staff Company plus 1st through 3rd Companies, SS Antitank Battalion 23
38 377	Staff Company, Armed Mountain Artillery Regiment of the SS No. 23
34 775	Staff Battery plus 1st through 3rd Batteries, 1st Battalion, Armed Mountain Artillery Regiment of the SS No. 23
23 497	Staff Battery plus 4th and 5th Batteries, 2nd Battalion, Armed Mountain Artillery Regiment of the SS No. 23
47 259	Staff Battery plus 6th and 7th Batteries, 3rd Battalion, Armed Mountain Artillery Regiment of the SS No. 23
32 713	Staff Battery plus 8th and 9th Batteries, 4th Battalion, Armed Mountain Artillery Regiment of the SS No. 23
45 334	Staff Company, Armed Mountain Infantry Regiment of the SS (Croatian No. 3)
36 293	Staff Company plus 1st through 4th Companies, 1st Battalion, Armed Mountain Infantry Regiment of the SS (Croatian No. 3)
20 797	Staff Company plus 5th through 8th Companies, 2nd Battalion, Armed Mountain Infantry Regiment of the SS (Croatian No. 3)
44 219	Staff Company plus 9th through 12th Companies, 3rd Battalion, Armed Mountain Infantry Regiment of the SS (Croatian No. 3)
21 356	Staff Company, 4th Battalion, Armed Mountain Infantry Regiment of the SS (Croatian No. 3)[1]
48 865	Staff Company, Armed Mountain Infantry Regiment of the SS (Croatian No. 4)
21 714	Staff Company plus 1st through 4th Companies, 1st Battalion, Armed Mountain Infantry Regiment of the SS (Croatian No. 4)
37 565	Staff Company plus 5th through 8th Companies, 2nd Battalion, Armed Mountain Infantry Regiment of the SS (Croatian No. 4)
31 597	Staff Company plus 9th through 12th Companies, 3rd Battalion, Armed Mountain Infantry Regiment of the SS (Croatian No. 4)
46 452	Staff Company, 4th Battalion, Armed Mountain Infantry Regiment of the SS (Croatian No. 4)[1]
22 397	Staff Company plus 1st through 4th Companies, SS Reconnaissance Battalion 23
41 036	Staff Company plus 1st through 5th Companies, SS Field Reserve Battalion 23
40 166	Recruit Depot Sambor, SS Kama Division
32 054	Recruit Depot Bosujaci, SS Kama Division

[1] Only the headquarters company of the 4th Battalion was formed in both mountain infantry regiments because of a lack in personnel.

This area encompassed the Danube-Tiza Canal. During its brief history, the division's area of operations would change only slightly, ending just southeast, between Vrbas and Novi Sad. The division began to take shape in earnest in July and August 1944. The Divisional Field Post Numbers give us a good description of the Kama Division on 8 September 1944.

The two mountain infantry regiments remained unnumbered until

SS-KAMA DIVISION, SEPTEMBER 1944

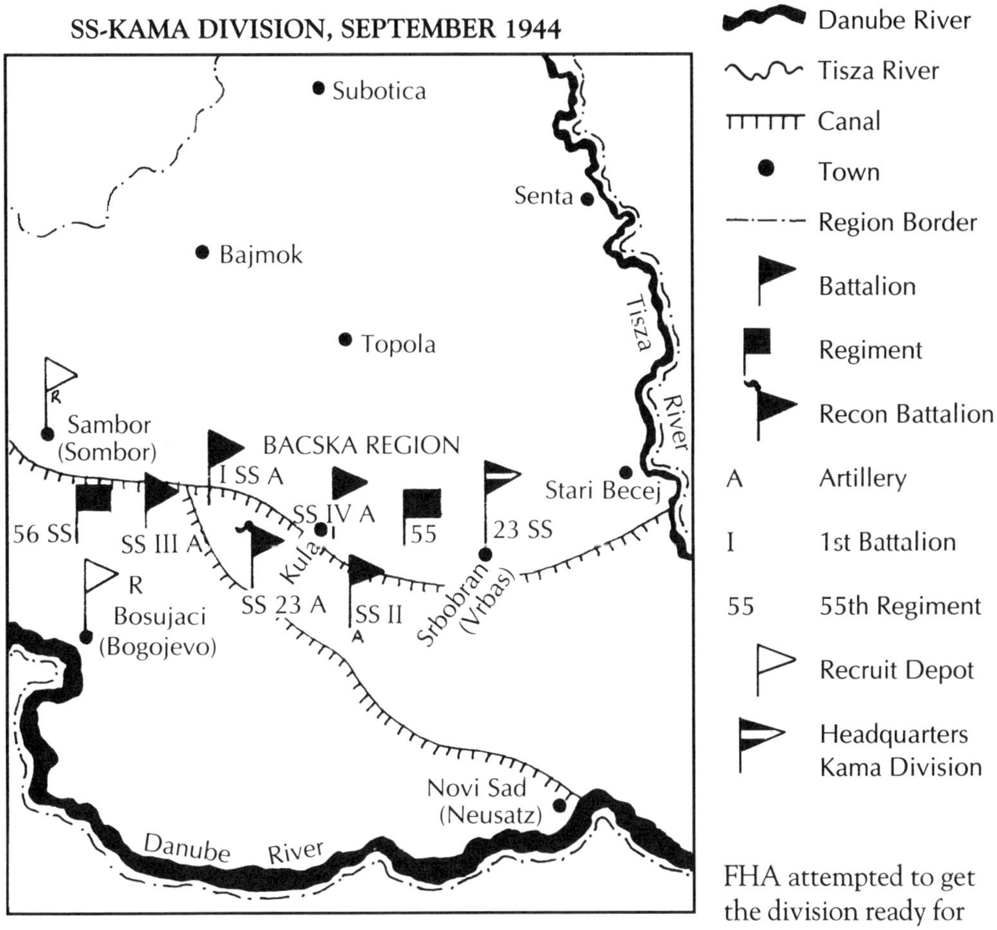

Some of the major units of the Kama Division, September 1944. The Bacska region (also pronounced "Bachka" belonged to the old Austro-Hungarian empire until 1918, when it became part of Yugoslavia. When the Axis invaded Yugoslavia in 1941, Bacska reverted to Hungarian control. In late 1944, the area was occupied by Yugoslav Communists, who expelled the 500,000 ethnic Germans by confiscating their property and denying them food. This community was made up of farmers who had been relocated from the Black Forest region of Germany during the reign of Queen Maria Theresa.

October 1944. Both regiments were then redesignated as Waffen-Gebirgsjäger-Regiment der-SS N. 55 (kroatische Nr. 3) and Waffen-Gebirgsjäger-Regiment der-SS Nr. 56 (kroatische Nr. 4).

During the months of August and September 1944, the Red Army made some dangerous advances into the Balkans and Hungary. The training bases for the Kama Division soon became precariously close to the front lines. The SS-FHA attempted to get the division ready for combat, even citing the unrealistic date of 24 September as when the unit would be ready for frontline service. The training state of the recruits was still at the basic stage, however. It would have been sheer suicide to commit the division to front-line service, especially since it had always been intended for service against the Yugoslavian Partisans and not the crack troops of a victorious Red Army.

The unrealistic date of committal passed. The SS-FHA soon realized that no time was left for the Kama Division to form. As the Red Army moved into Hungary, German military leaders decided to disband the unit and make as much use of the personnel as possible by trans-

23. WAFFEN-GEBIRGS-JÄGER-DIVISION DER-SS KAMA

Commanding Officer
SS-Standartenführer Hellmut Raithel, 37 years old[17]
Waffen-Gebirgsjäger-Regiment der-SS 55 (kroat. Nr. 3)
(?)
I.Btln./Waffen-Gebirgsjäger-Regiment der-SS 55
SS-Sturmbannführer Karl Praefcke, 43 years old[16]
II.Btln./Waffen-Gebirgsjäger-Regiment der-SS 55
SS-Sturmbannführer Heinrich Albrecht, 33 years old[15]
III.Btln./Waffen-Gebirgsjäger-Regiment der-SS 55
(?)
Waffen-Gebirgsjäger-Regiment der-SS 56 (kroat. Nr.4)
SS-Sturmbannführer der Reserve Sepp Syr, 31 years old[11]
I.Btln./Waffen-Gebirgsjäger-Regiment der-SS 56
SS-Hauptsturmführer Robert Gloning, 32 years old[14]
II.Btln./Waffen-Gebirgsjäger-Regiment der-SS 56
SS-Hauptsturmführer der Reserve Ewald Schumacher, 45 years old[13]
III.Btln./Waffen-Gebirgsjäger-Regiment der-SS 56
(?)
SS-Aufklärungs-Abteilung 23
SS-Sturmbannführer der Reserve Sepp Syr
then: SS-Hauptsturmführer Ludwig Zeitz, 28 years old[12]
Waffen-Gebirgs-Artillerie-Regiment der SS-23
SS-Sturmbannführer der Reserve Carl Dehnen, 45 years old,[2]
then SS-Obersturmbannführer Hans Zeysing, 43 years old[1]
I.Btln./Waffen-Gebirgs-Artillerie-Regiment der-SS 23
SS-Hauptsturmführer Josef Mailhammer, 29 years old[3]
II.Btln./Waffen-Gebirgs-Artillerie-Regiment der-SS 23
SS-Sturmbannführer der Reserve Carl Dehnen, 45 years old[2]
III.Btln./Waffen-Gebirgs-Artillerie-Regiment der-SS 23
(?)
IV.Btln./Waffen-Gebirgs-Artillerie-Regiment der-SS 23
(?)
SS-Panzerjäger-Abteilung 23
SS-Sturmbannführer Albert Landwehr, 37 years old[4]
SS-Nachschub-Abteilung 23
SS-Hauptsturmführer der Reserve Wilhelm Morisse, 43 years old[5]
SS-Pionier-Bataillon 23
SS-Sturmbannführer Hermann Otto, 33 years old[6]
SS-Gebirgs-Nachrichten-Abteilung 23
SS-Hauptsturmführer der Reserve Paul Barton, 32 years old[7]
then: SS-Hauptsturmführer der Reserve Albert Reimann, 45 years old[8]
SS-Feldlazarett 23
SS-Sturmbannführer Dr. Karl Matz, 35 years old[9]
SS-Sturmbannführer Dr. Karl Becker, 29 years old[10]
SS-Feldersatz-Bataillon 23
(?)

(This list is incomplete.)

NOTES ON KAMA OFFICERS

[1] SS-Obersturmbannführer Hans Zeysing (SS No. 276137) was born on 15 February 1901. He was promoted to this rank on 9 November 1942. Before being posted to Kama, Zeysing had led SS-Flak-Ersatz-Regiment, then the SS-Artillerie-Ersatz Regiment. After his tour as commander of the 23rd SS Artillery Regiment, Zeysing led the 31st SS Volunteer Artillery Regiment of the 31st SS Volunteer Grenadier Division Batchka.

[2] SS-Sturmbannführer der Reserve Carl Dehnen (SS No. 98473) was born on 10 January 1899. He was promoted to this rank on 20 April 1943. Before being posted to Kama, he commanded SS-Artillerie-Schule II (SS-Artillery-School No. 2). Dehnen was the first commander of Waffen-Gebirgs-Artillerie-Regiment der-SS 23. After SS-Obersturmbannführer Hans Zeysing took over, Dehnen was relegated to commanding the regiment's 2nd Artillery Battalion. Later, Dehnen led the 1st Artillery Battalion of the SS-Freiwilligen-Artillerie-Regiment 31 of the 31st SS Division.

[3] SS-Hauptsturmführer Josef Mailhammer (SS No. 353009) was born on 13 July 1915. After his tour of duty with Kama, he became a battalion commander in SS-Freiwilligen-Artillerie-Regiment 31.

[4] SS-Sturmbannführer Albert Landwehr (SS No. 323798) was born on 25 December 1907. He attained his present rank on 1 September 1943. Before being posted to Kama, Landwehr had been a battalion commander of SS-Flak-Ersatz-Regiment. After his stint with the SS Flak Replacement Regiment, Landwehr finished a tour of duty with the Panzerjäger elements of the Das Reich and Prinz Eugen SS divisions before seeing service with Kama. Afterwards, Landwehr led the SS Freiwilligen-Panzerjäger-Abteilung 31 of the 31st SS Division.

[5] SS-Hauptsturmführer Wilhelm Morisse (SS No. 3874) was born on 21 July 1901. Morisse was promoted to his present rank on 9 November 1942. After serving as commander of the 23rd SS Supply Battalion, Morisse went on to command the supply battalion of the 31st SS Division. Note the early SS number, indicating a long enlistment in the SS.

[6] SS-Sturmbannführer Hermann Otto (SS No. 286851) was born on 3 November 1911. He attained this rank on 30 January 1944. Before being posted to command the 23rd SS Engineer Battalion, Otto had led the SS Engineer School. After his tour of duty with Kama Division, he went on to command the 31st SS Engineer Battalion, which had been formed from the remnants of the old 23rd SS Engineer Battalion.

[7] SS-Hauptsturmführer der Reserve Paul Barton (SS No. 44774) was born on 30 April 1912. He received his present rank on 20 April 1943. He commanded the 23rd SS Signals Battalion until he was replaced by Albert Reimann.

[8] SS-Hauptsturmführer der Reserve Albert Reimann (SS No. 161815) was born on 13 December 1899. He received this rank on 30 January 1943. Reimann took over command of the 23rd SS Signals Battalion and later led the 31st SS Signals Battalion.

[9] SS-Sturmbannführer Dr. Karl Matz (SS No. 291210) was born on 21 June 1909. Promoted to this rank on 9 November 1942, Matz served in the 13th SS Handschar Division

before joining Kama. He later served in the 31st SS Division.

[10] SS-Sturmbannführer Dr. Karl Becker (SS No. 309497) was born on 5 September 1909. Like Dr. Matz, Dr. Becker received this rank on 9 November 1942. Becker also served in the Handschar Division before being posted to the Kama Division.

[11] SS-Sturmbannführer der Reserve Sepp Syr (SS No. 283029) was born on 18 October 1903. He attained this rank on 21 June 1942. Syr first led the 14th SS Fusilier Battalion of the 14th SS Division (a Ukrainian formation). He later commanded the 23rd SS Reconnaissance Battalion of the Kama Division. After a brief tenure as commanding officer to the 23rd Recon Battalion, he commanded Waffen-Gebirgsjäger-Regiment der-SS 56 of Kama Division. Later. when Kama was disbanded, Syr commanded the SS-Freiwilligen-Grenadier-Regiment 79 of the 31st SS Division.

[12] SS-Hauptsturmführer Ludwig Zeitz (SS No. 59007) was born on 22 March 1916. Commissioned on 5 January 1942, Zeitz became the commander of the 23rd SS Reconnaissance Battalion of Kama Division when Sepp Syr became the Commanding Officer of the Waffen-Gebirgsjäger Regiment der-SS 56. Zeitz later became commander of the 31st SS Fusilier Battalion, which acted as the recon unit for the 31st SS Division. Zeitz was promoted to Sturmbannführer on 9 November 1944.

[13] SS-Hauptsturmführer der Reserve Ewald Schumacher (SS No. 61328) was born on 26 October 1899. He was commissioned on 1 March 1942. He led the 2nd Battalion of Waffen-Gebirgsjäger-Regiment der-SS 56. Later, when Kama was disbanded, Schumacher became the Commanding Officer of the 2nd Battalion, SS-Freiwilligen-Grenadier-Regiment 79. Before the war, Schumacher had served as an officer in the Allgemeine (General) SS.

[14] SS-Hauptsturmführer Robert Gloning (SS No. 17238) was born on 21 August 1912. Commissioned on 1 August 1939, Gloning had various duties before joining Kama, including serving as logistics officer (lb) and then commander of the 2nd Battalion, Waffen-Gebirgsjäger-Regiment der-SS 28, of the 13th SS Handschar Division. Later, after Kama was disbanded, Gloning served as battalion commander for 1st Battalion, SS-Freiwilligen-Grenadier Regiment 79.

[15] SS-Sturmbannführer Heinrich Albrecht (SS No. 11458) was born on 2 February 1911. He received his present rank on 30 January 1944. He served at Waffen-SS Troop Training Ground Heidelager before being posted as commander of 2nd Battalion. Waffen-Gebirgsjäger Regiment der-SS 55 of Kama Division.

[16] SS-Sturmbannführer Karl Praefcke (SS No. 234960) was born on 7 July 1898. He was commissioned on 11 September 1938. Before he was posted to Kama, Praefcke worked at the SS Main Office (SS-Führungs-Hauptamt). Praefcke later served in the 31st SS Division, though his position in that division is not known.

[17] SS-Standartenführer Hellmut Raithel (SS No. 470211) was born on 9 April 1907. He received his present rank on 2 April 1944. Before taking over as commander of Kama Division, Raithel led the 28th Regiment of the 13th SS Handschar Division. He is also mentioned as having commanded the 11th Regiment in the 6th SS Division Nord, but his tenure there may or may not have been before his service in the Kama Division.

Kama Divisional Identification Shield (top).

Kama tab worn as the right collar tab (bottom). Illustrations: Antonio Munoz.

ferring them to other divisions as replacements. This decision was reached in October 1944, with most of the divisional elements going to the newly forming 31st SS Division. An order for all Muslims to return to Yugoslavia for employment in the Handschar Division was misinterpreted by the individual units within the division as orders to separate from Kama headquarters. Many recruits were lost in this way because it was easier to desert from isolated units in territories where the recruits had been raised. Still, some recruits of Kama found their way into Handschar, while the majority of the personnel went over to the 31st SS Division. The 23rd SS Recon Battalion became a corps unit of the 9th Waffen-Gebirgsjäger-Regiment der-SS Nr. 56 (kroatische Nr. 4). After the division was dropped from the Waffen-SS rolls, the number 23 went to the Dutch-SS Nederland Panzergrenadier-Brigade when it became a division in early 1945.

During its brief five-month lifespan, the Kama commander was SS-Standartenführer Hellmut Raithel. A divisional collar tab was supposed to have been prepared, depicting a round sun with its rays radiating from the center. A vehicle identification shield with a similar insignia was also reported to have been designed. During its tenure as a forming division, Kama was listed under the control of the Replacement Army in Southern Hungary. Given ample time, the Kama Division would have been used against the Yugoslavian Partisans, but the worsening German military situation required that the division's organization be aborted.

CHAPTER SEVENTEEN

An Austrian army,
awfully arrayed,
Boldly by battery besieged
Belgrade;
Cossack commanders
cannonading come,
Dealing destruction's
devastating doom.

Siege of Belgrade
The Trifler, 1817

In the spring of 1944 the Germans were beginning their third year of war against the Communist-sponsored partisans in the Balkans. While the years 1941 and 1942 had passed with relatively small skirmishes being conducted by the Communist command, 1943 saw an increased effort by Tito and his partisans against the small German and Croatian army garrisons in Yugoslavia.

Eventually, the unorthodox war spilled over into the neighboring countries of Greece and Albania. Though these two countries had started out with small bands of nationalists fighting for their country's independence, it was the British- and Russian-backed Communists from Yugoslavia who eventually found individuals in each of those two countries to represent and enlarge the war against their German and Italian occupiers.

This was especially apparent in 1943 when the Italian armistice eliminated that country's forces in the Balkans as a source of German military backing. In some cases entire Italian divisions went over to the enemy, taking with them all of their heavy military equipment. As the Italian forces disintegrated in Albania and Yugoslavia, the Communists moved in to fill up the power vacuum left by their departure.

The Communists in Albania were led by Enver Hoxha, a confirmed Stalinist Communist who had until 1943 been unsuccessful in his military campaign against the Italians. His National Liberation movement (LNC) had been a microscopic organization, at first accepting non-Communist volunteers as well as party men. Slowly, the non-Communist leaders in the LNC had been eliminated. Eventually, nationalist Albanians who were opposed to the Germans and Italians as well as Communists formed into partisan groups, independent of the LNC. Until 1943, the LNC had been largely unsuccessful in challenging the Italian Army in Albania. At most, it had numbered a few thousand men at any one time. It had also begun to take a beating from several anti-Communist Albanian tribal groups hostile to it.

That was all to change in that fateful spring of 1943. It was then that the British sent a military mission to Albania to get military arms to Albania's partisan groups. Fearing a civil war between the Communists and the Albanian Chetnik group, the British offered military assistance to Enver Hoxha's LNC, while extending to a small degree assistance to the Albanian nationalists of the National Front (BK). As a result of this policy, Hoxha's LNC became more prominent as more weapons were sent to it. Hoxha's Communists began an intensive recruiting campaign, bringing in women and children as young as ten years old.

The Italians, who up until then had garrisoned the country with only four divisions, had to send additional forces in the form of two more divisions to offset the increasing partisan presence. By the time the Italian armistice was declared, the Italian army had been confined to garrisons in the larger towns, cities and the Adriatic coastland. By now the Albanian guerrillas numbered about 20,000 (this included Communists and nationalists).

Against this background of imminent chaos, the German army was hard-pressed to supply reinforcements to its forces in the Balkans. Those in command decided to raise a volunteer mountain division of Albanians, as had been done in Yugoslavia, to help garrison that inhospitable land.

The Germans had few friends in Yugoslavia and fewer still in Albania. Of all the groups and clans in Albania, only the Catholic Albanian clan headed by Jon Marko Joni remained loyal to the Germans. In 1942 the German High Command in Yugoslavia had successfully raised an SS mountain division (the 7th) from ethnic Germans in Croatia. Two more SS mountain divisions (the 13th and 23rd) had been raised and were in various stages of development. The Germans hoped that raising an Albanian SS mountain division would not present too large a problem.

There were many obstacles, however, that eventually worked to torpedo the program, the first of which was the lack of proper arms and equipment for such a division. The SS High Command had envisioned a mountain division of about 10,000 men. There were not enough German arms to even outfit the 6,500 or so men who eventually joined the division. German uniforms and medical equipment were also in short supply. This situation was so severe that the Germans used captured British army tunics to clothe their Albanian recruits. Arms included not only German weapons, but also U.S., British, and Russian. Thirty British Vickers heavy machine guns went to the forming rifle companies.

Illustration: Eric Weisman.

Hoxa's Communists immediately attacked the recruiting effort that had begun in the Kossovo area and followed by staging two armed raids into Kossovo itself to halt the dissemination of political propaganda. In these raids, the Communists captured or destroyed hundreds of much-needed weapons and munitions, and the Germans also lost many recruits through death or, worse still, desertion to the Communists.

The Waffen-SS had requested about 10,000 men for the division. The Higher SS and Police Command in Albania, working in conjuction with the German-sponsored Albanian National Committee, came up with a list of 11,398 possible candidates, most of the Muslim faith. They hoped that the teachings of Islam, which oppose Communist ideology, would instill in the Albanian recruits enough motivation to make the division a viable antipartisan unit.

The Germans made a big mistake. First of all, the Albanian nature was to consider the interests of their local tribe before any other goals or ideals. This clannish tribal system made for unreliable soldiers when it came to serving the German cause. While blood feuds between different clans took years and sometimes decades to settle, there was a basic understanding among all Albanian tribes that it was always open season on any foreign power. Secondly, the Muslims in Albania were not as religiously fervent and faithful as their Islamic cousins in Croatia.

Many Albanian recruits who joined the division deserted from their units when the opportunity presented itself. If that wasn't bad enough, they took their arms, clothing, and equipment with them. Individual desertions were small blows to the newly forming division, but as the summer and fall of 1944 showed, in many instances whole companies were involved.

Of the 11,398 candidates suggested for the division, 9,275 were deemed suitable for drafting. Of this number, 6,491 actually appeared and were inducted. The division's recruits included a large number of sailors from the defunct Italian-sponsored Albanian Navy and an Austrian cadre of officers and noncommissioned officers. It was around this cadre of trained and combat-seasoned veterans that the division was formed. Initial command of the division went to SS-Brigadeführer und Generalmajor der Waffen-SS Josef Fitzhum, who was head of the Higher-SS and Police Command in Albania. He was to take control of initial forming and training until a suitable commander was found. Eventually, forty-three-year-old SS-Standartenführer August Schmidhuber was appointed commander. He was promoted to SS-Oberführer on 21 June 1944.

The Waffen-SS High Command had reserved two numerical listings for the two proposed mountain infantry regiments. Initially, these two regiments were listed as follows: Waffen-Gebirgsjäger-Regiment der-SS 1 (Albanische Nr. 1), and Waffen-Gebirgsjäger-Regiment der-SS 2 (Albanische Nr. 2). The 1st regiment was later redesignated Nr. 50, while the 2nd became Nr. 51 in the Waffen-SS rolls. Literally translated, their titles read "Armed Mountain Infantry Regiment of the SS," implying a non-Germanic status since the majority of its personnel were not Reich Germans. Interestingly, genealogical research has shown that the majority of the recruits came from the Ghegs of northern Albania who had apparently descended from Aryan origins. Kossovars, also found in large quantity in the division, also shared this background. This fact, however, offered little comfort to the forming division.

Originally, the artillery regiment was to have four artillery battalions of three batteries apiece, but there were too few pieces of equipment and trained men to man them. In the end, only one weak artillery battalion and a regimental staff were formed.The artillery battalion had only two weak batteries. In early October, these two batteries, plus the headquarters of the artillery "regiment," were located in the town of Gnjilane.

The reconnaissance battalion had three company-sized units, two armored-car companies, and one cavalry squadron. The armored car companies contained a few Volkswagens (the German equivalent of the U.S. Army jeep) and captured Italian armored cars and tankettes, as well as a light sprinkling of Opel trucks. Command of the division's reconnaissance battalion was entrusted to the then thirty-year-old SS-Sturmbannführer Heinz Kohler.

The division's engineer battalion had three mountain engineer companies under the command of thirty-seven-year-old SS-Hauptsturmführer Gerard Peters, who had quite an interesting career. He was born on 22 April 1907 in Karlsruhe, received his engineering degree from a university in Karlsruhe, and attended the Reserve Officer's School in Holland in 1932. After the end of his active military obligation, he entered his father's business, which took him all over Holland and even outside the country, traveling to Italy on several occasions.

In 1936, Peters was recalled to active duty as a reserve officer working in the capacity of adjutant to the district commander in Beurne. As such, he was directly subordinate to Colonel Wilson in the General Staff. Peters' work involved constructing static defenses for the Dutch army. When Holland surrendered in late May 1940, Peters was one of many who were assigned to dismantling those de-

21. WAFFEN-GEBIRGS-JÄGER-DIVISION DER-SS SKANDERBEG

Commanding Officer: SS-Brigadeführer und General-Major der Waffen-SS Josef Fitzhum[1] (April-June 1944)
SS-Brigadeführer August Schmidhuber (June-August 1944)
SS-Obersturmbannführer Alfred Graf (August 1944-May 1945)

Waffen-Gebirgs-Jäger-Regiment der-SS 50
SS-Obersturmbannführer Jakob Hauser, 44 year old
1st Btln./Waffen-Gebirgs-Jaeger-Regiment der-SS 50
SS-Sturmbannführer Richard Wolf, 39 years old
2nd Btln./Waffen-Gebirgs-Jäger-Regiment der-SS 50
Unknown
3rd Btln./Waffen-Gebirgs-Jäger-Regiment der-SS 50
Unknown
Waffen-Gebirgs-Jäger-Regiment der-SS 51
SS-Sturmbannführer Friedrich Keidel, 49 years old
1st Btln./Waffen-Gebirgs-Jäger-Regiment der-SS 51
SS-Sturmbannführer Walther Bormann, 47 years old
2nd Btln./Waffen-Gebirgs-Jäger-Regiment der-SS 51
SS-Hauptsturmführer Erich Markowz, 31 years old
3rd Btln./Waffen-Gebirgs-Jäger-Regiment der-SS 51
Unknown
Waffen-Gebirgs-Artillerie-Regiment der-SS 21
SS-Standartenführer Adolf Wunder, 48 years old
1st Btln./Waffen-Artillerie-Regiment der-SS 21
SS-Hauptsturmführer Emil Steiner, 28 years old
SS-Panzer-Aufklärungs-Abteilung 21
SS-Sturmbannführer Heinz Kohler, 30 years old
SS-Panzerjäger-Abteilung 21
SS-Sturmbannführer Fritz Bauer, 45 years old
SS-Pionier-Bataillon 21
SS-Hauptsturmführer Gerard Peters, 37 years old
SS-Versorgüngs-Abteilung 21
SS-Hauptsturmführer Hermann Talbot von Pistor, 38 years old
SS-Gebirgs-Nachrichten-Abteilung 21
SS-Hauptsturmführer der Reserve Wilhelm Wiersch, 34 years old
SS-Sanitäts-Abteilung 21
SS-Obersturmbannführer Dr. Robert Schrader, 40 years old
SS-Obersturmbannführer Dr. Adolf Murthum, 37 years old
SS-Sturmbannführer Dr. Karl Matz, 33 years old
SS-Sturmbannführer Dr. Paul Rosner, 38 years old

[1] *(Fitzhum is known to have taken personal command of the Skanderbeg Battle Group in November 1944, but I have been unable to ascertain whether it was a temporary or permanent posting. Graf did remain with the battle group until May 1945).*

fenses. In July 1940, he decided to join the then-minuscule Waffen-SS and was immediately sent for engineer training to SS Engineer Training and Replacement Battalion 1 at Dresden. Between 1940 and 1941, he worked in the company of Max Schaefer, a soldier who later figured prominently in the ranks of the Waffen-SS. Peters was later transferred to other engineer companies within the battalion as a first lieutenant.

In April 1942, he was transferred to the newly formed 7th SS Engineer Battalion. This battalion belonged to the Croatian-German 7th SS-Freiwilligen-Gebirgsjäger-Division Prinz Eugen. The commander of the 7th SS Engineer Battalion was SS-Sturmbannführer Zimmerman. Peters served as an officer of that battalion's ski-lift commando, and was later promoted to SS-Hauptsturmführer on the battalion's staff. Toward the end of 1943 he left for the Waffen-SS Commanders' School in Antwerp.

In May 1944 he completed his courses and was assigned to command the newly forming 21st SS Engineer Battalion. He held this post until November 1944, when he was transferred.

The beginning of 1945 found Peters attached to the staff of the Higher SS and Police Command Adriatic Coastland. He was first charged with instructing the engineer elements of the Circassian and Caucasian SS cavalry formations then in Northern Italy. He was later transferred to the SS Karstjäger Brigade (see history of the 24th SS Division, Chapter 18), where he took charge of the engineer company. This brigade was one of numerous formations charged with delaying the British and U.S. forces in Italy while the bulk of the German forces withdrew to Austria. Later it would keep open the passes over the Karawanken Alps, separating Yugoslavia from Austria. By doing this, they allowed the retreating German divisions to escape capture by Tito's Partisan army.

Other officers assigned to the Albanian division are listed on page 230.

Besides the Italian armored cars and tankettes found in the reconnaissance battalion, the only other armored support found in the division was an assault-gun battery that belonged to Sturmbannführer Fritz Bauer's antitank battery. The rest of Bauer's unit was made up of the German 75mm and 50mm antitank guns (the towed versions). The 13. Waffen-Gebirgs-Division der-SS Handschar gave up a reinforced battalion of riflemen of Albanian extraction. These men were incorporated as a whole unit and became the 3rd Battalion, Waffen-Gebirgs-Jäger-Regiment der-SS Number 1 (later 50). This particular battalion was especially troublesome for the division and is discussed later on.

In keeping with Waffen-SS practice, a distinctive arm patch, designed by the SS Main Office, was issued to the division. Photographs show it being worn by at least some members of the division. The arm shield had a black, double-head-

SS-Brigadeführer August Schmidhuber, Skanderbeg commander from June – August 1944. Photo: Munin Verlag.

A typical member of SS-Skanderbeg Division: SS-Sturmann Karl-Heinz Damerow. Born in Neustettin in 1925, Damerow was reported MIA near Belgrade in 1944. Photo: Munin Verlag.

ed Albanian national eagle on a red background. Above the eagle, the word "Skanderbeg" was embroidered in white. This arm shield was worn on the left sleeve, approximately six inches below the shoulder.

It was also at this time that the division received the official title Skanderbeg. This was the name given by the Turks to Albania's greatest national hero, George Kastriotis. In 1443 he rallied the Albanians and proclaimed a holy war against the Turks. In 1449 Sultan Murad II sent an army under his best general to crush Skanderbeg but was defeated at the frontier. Afterwards, invading Turkish armies appeared almost every year and were regularly destroyed.

Standard practice in the Waffen-SS was for the left collar patch to show rank, while the right collar patch indicated whether the soldier who wore it belonged to a Germanic or non-Germanic formation. The Germanic formations wore the double thunderbolt "SS" runes, while the non-Germanic units wore emblems all their own. The Albanian SS division had a right collar tab with the helmet of Kastriotis. The helmet (with a goat's head on top) was done with white threading. The background of the collar tab was of standard black cloth.

These emblems were supposed to give the Albanian SS men a sense of belonging and to raise their morale. It proved to be a wasted effort. By June 1944, the Skanderbeg SS Mountain Division had been partly formed and large-scale field maneuvers scheduled. This first exercise would be attended by the entire 14. SS-Freiwilligen-Gebirgsjäger-Regiment from the 7th SS Mountain Division Prinz Eugen. Other units that would participate were the 2,000-man Albanian militia unit under the command of Hauptmann der Schützpolizei Spruny, a panzergrenadier battalion, and an artillery battery from the Brandenburger Lehr Regiment.

The site of these maneuvers was to be the area between the towns of Berane and Andrijevica in Montenegro, just north of Albania, . The units that Skanderbeg employed were the 1st and 2nd Battalions/Waffen-Gebirgsjäger-Regiment der-SS Nr. 1 (aka Nr. 50) and the 1st Battalion/Waffen-Gebirgsjäger-Regiment der-SS Nr. 2 (aka Nr. 51).

The first real test for the division came in August 1944 when the nationalist guerrillas began mounting attacks against units of Skanderbeg in the Kossovo area. This initial combat action yielded the following casualties:

	Missing	Wounded	Dead	Total
German	14	20	8	42
Albanian	152	72	48	272
Total	166	92	56	314

In August, events in the Balkans and on the Russian front conspired against the Germans. Romania changed sides and soon began offensive operations against its former ally. The Bulgarians also did so on 29 August, making the German position all the more precarious. The Albanian and Yugoslav guerrillas sensed that the moment had come for an all-out offensive against the German army in the Balkans. Beginning in October, they gave the withdrawing German and Croatian divisions no respite. The Albanian Communist LNC began to swell in size. During the withdrawal of the German forces from Albania, the guerrillas even encircled (for a time) the 21st Mountain Corps in the central part of that country.

September was also catastrophic for the efforts to form the Skanderbeg Division. The desertion rate increased drastically to epidemic proportions. The division moved into Yugoslavia in early September to help garrison Macedonia. The Bulgarians had kept troops in

FIELD MANEUVERS IN MONTENEGRO
14-30 JULY 1944

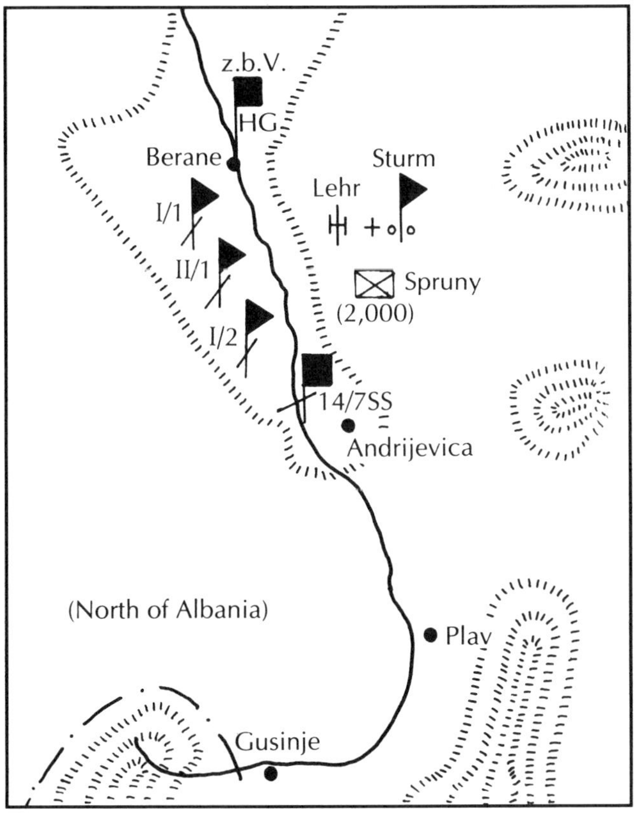

Skanderbeg Regimental HQ for special employment

Lehr Battery and Assault Battalion (motorized)

14th SS-Mountain Infantry Regiment/7th SS Mountain Division

2,000-man volunteer militia under Hauptman der Schutzpolizei Spruny

I/1 1st Battalion/Gebirgsjäger der-SS Regiment 1
II/1 2nd Battalion/Gebirgsjäger der-SS Regiment 1
I/2 1st Battalion/Gebirgsjäger der-SS Regiment 2

Macedonia since 1941, and those troops threatened the retreat route of the German forces in Greece. The Skanderbeg Division was ordered into the areas surrounding the towns Skopje-Kumanovo-Presevo-Bujanovce. While crossing the Albanian-Macedonian frontier in the Tetovo-Gostivar area, the division received a serious blow; 1,000 men from the Waffen-Gebirgsjäger Regiment der-SS Nr. 50 deserted with their guns and equipment. The Albanian battalion, which had been transferred from the 13. Waffen-Gebirgsjäger-Division der SS Handschar, accounted for 697 of these desertions. Since that battalion had been used to form the 3rd Battalion/50th Regiment, it seems that most of the desertions were confined to one unit in the regiment.

The Partisans operating in Macedonia were not going to give the Skanderbeg Division any respite. Approximately 1,500 of them attacked the division's reconnaissance battalion at Gjakowica (also spelled Djakowica). They struck on 11 September, surprising the men of the Skanderbeg.

Heavy fighting ensued as the guerrillas tightened their noose around Djakowica. Trapped inside the town were the recon battalion under Sturmbannführer Heinz Kohler and one signals company of the division.

Pressing their attack, the Partisans managed to enter the outskirts of town, destroying a fuel depot and capturing several guns that they quickly turned against their former owners. Kohler turned the Skanderbeg radiomen into infantrymen and launched a counterattack on 13 September. He had coordinated the date of the counterattack with the expected arrival of the German 181st Infantry Division, which radio communication had indicated would be

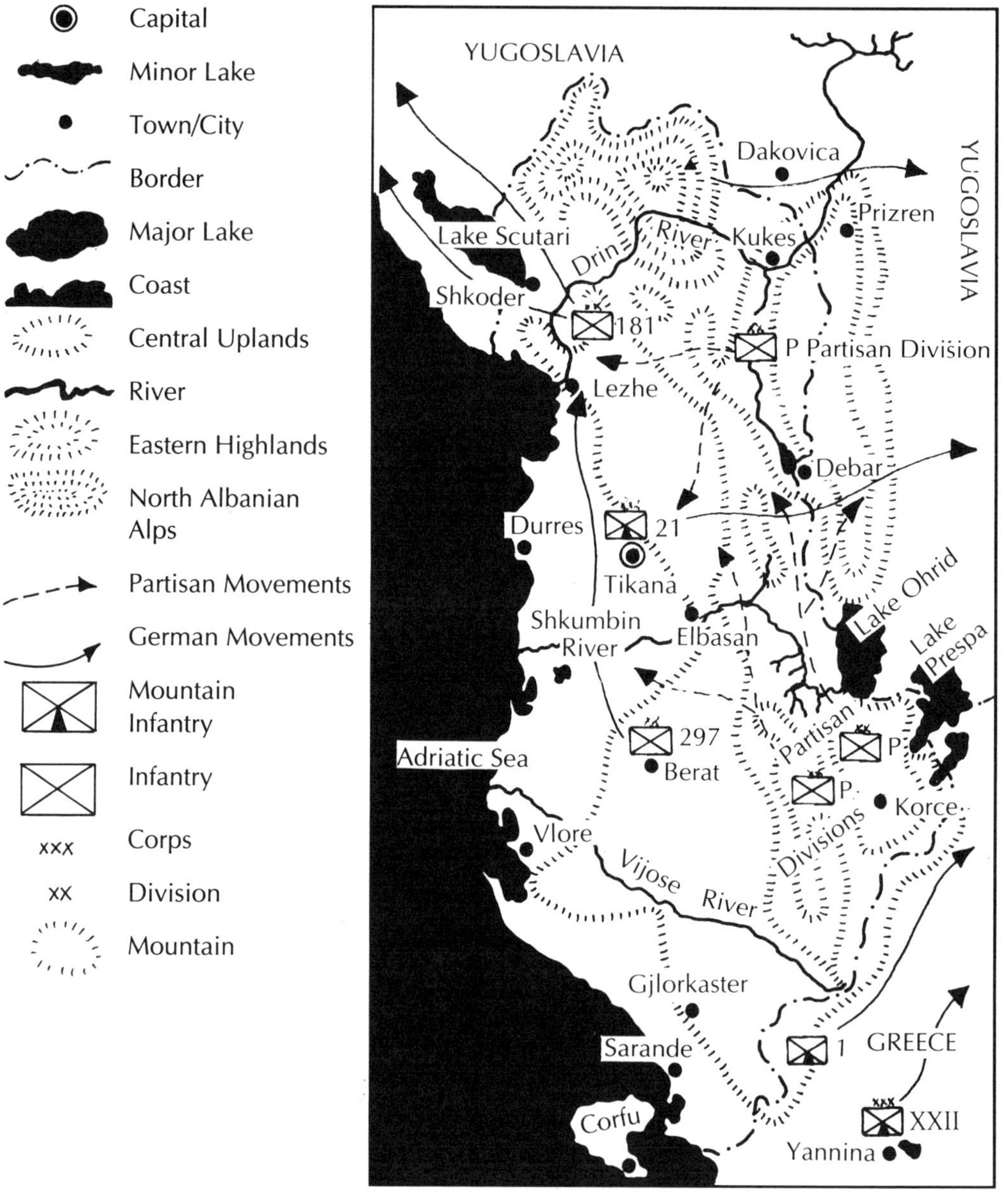

This map depicts the German withdrawal from Albania, which began in August 1944. The German 21st Mountain Corps was cut off several times by Enver Hoxa's Communist Partisans before it could extricate itself and reach Macedonia, Yugoslavia. The Skanderbeg Division withdrew safely, but it, too, was under pressure.

about that time. In a coordinated effort, the reconnaissance battalion tied the guerrillas down while two battalions of the 181st Infantry Division maneuvered to encircle the enemy.

The trap was too late, however, and most of the Partisans withdrew safely. A report from Kohler to Skanderbeg headquarters stated that during the three-day engagement, his total losses had been a microscopic 10 men dead versus 104 enemy guerrillas. This seems unlikely, but apparently the report was not questioned by the divisional commander.

The decision to disband the forming Skanderbeg Division came in late September. The Germans had had enough. Plans were now put into effect

Skanderbeg Strength, 1 Oct 1944	Officers	NCOs	Men (R)	Deserters
Divisional staff	17	41	69 (51)	1
1st Btln./Waf.Geb.Jag.Rgt.Nr.51	8	66	68 (295)	404
2nd Btln./Waf.Geb.Jag.Rgt.Nr.51	6	42	167 (284)	123
1st Btln./Waf.Geb.Art.Rgt.Nr.21	6	18	8 (539)	450
SS-Nachrichten-Abteilung 21	5	27	67 (299)	14
SS-Panzerjäger-Abteilung 21	6	23	14 (414)	65
Cavalry Squadron/SS Recon Btln.	5	32	43 (232)	380
SS-Feld-Ersatz-Bataillon 21	15	126	253 (781)	854
SS-Versorgungs-Abteilung 21	6	37	53 (168)	–
Veterinärs-Kompanie	1	6	13 (166)	34
1. Sanitäts-Kompanie	5	15	11 (128)	42
Kranken-Kraftwagen-Zug	–	3	3 (31)	3
Kampfgruppe Mitrovica[1]	1	9	50 (0)	–
Kampfgruppe Novi Pazar[1]	1	10	50 (20)	–
Gruppe Hauser[2]	2	5	5 (0)	1,055
Gruppe Klingenberg	2	7	13 (2)	–
Penal Platoon[3]	–	–	(112)	–
TOTAL:	86	467	887 (3,504)	3,425

[1] These two battle groups represented a much-decimated 3rd Battalion/Waffen-Gebirgs-Jäger-Regiment der-SS Nr. 1 (aka Nr. 51).
[2] This was Waffen-Gebirgsjäger-Regiment der-SS Nr. 1 (aka Nr. 50) under the command of SS-Obersturmbannführer Jakob Hauser.
[3] This platoon had recruits who were pending disciplinary cases.
R = Recruits in unit

that would use the reliable elements as an attached battle group to the 7th SS Mountain Division.

As stated earlier, the months of August and September were disastrous ones for the German army in the Balkans. On 28 September, the Russian 57th Army's advance units began crossing the Romanian-Yugoslav frontier. The bulk of the 57th Army and the rest of the 3rd Ukrainian Front followed close behind. Just six days earlier, Marshal Tito had gone to Romania to meet with top-ranking Soviet officers to coordinate their invasion of Yugoslavia with his Partisan forces.

Shortly before leaving, Tito's forces had overrun Draza Mihailoviĉ's main body of Chetnik troops (royalist forces loyal to exiled King Peter). In this engagement, the Partisans almost wiped out Mihailoviĉ's troops, caused the remainder to scatter, and almost captured Mihailoviĉ himself. By the end of September, the Partisans had broken through into Serbia. The Chetniks, routed in their last bastion (Serbia), began to disintegrate as a military force, and a number of the rank and file deserted to the Communists. What few units remained loyal to the king stayed close to units of Nediĉ's collaborationist Serbian State Guard Troops, avoiding for the most part any military operations.

Against this chaotic backdrop, the remnants of the Skanderbeg Division began military preparations for action against Tito's Partisans and the advancing Russian, Romanian, and Bulgarian forces. Skanderbeg's strength on 1 October 1944 showed that of the remaining 4,944 men, only 1,440 were combat veterans and 3,504 were still recruits. The Germans attempted to increase the unit's strength by adding about 3,500 German sailors who had withdrawn from the Aegean and had made their way north through Communist-infested northern Greece. By the middle of October, the unit had fil-

Photo of an NCO from the Skanderbeg Division. He is wearing the camouflage smock and grey-green M1943 style mountain cap with the Eidelweiss cloth insignia on the side. This insignia depicts an Austrian flower found in the Alps. From 1943 onwards, it was worn by all German mountain troops (right).

A very rare photograph of the members of the 21. Waffen-Gebirgs-Division der-SS Skanderbeg. The men in the photo were German NCOs attached to the Division (bottom right). Photos: Munin Verlag.

tered out its unreliable Albanian recruits and sent them on their way. The few Albanians who remained continued to serve in the Skanderbeg formation till the end of the war.

Plans to incorporate the Skanderbeg formation into the 14. SS-Freiwilligen-Gebirgs-Regiment of the 7th SS Mountain Division were cancelled. Instead, the battle group served as an independent unit alongside the 7th SS Division. For a time the SS-Feld-Ersatz-Bataillon 21 fought alongside one battalion each of the German 181st and 22nd Infantry divisions as an independent battle group. On 19 October, this kampfgruppe was in the process of attacking Bulgarian army units just south of the town of Kursumlija. It seems that SS-Brigadeführer und Generalmajor der Waffen-SS Josef Fitzhum took command of the Skanderbeg battle group in October, attaching what few police battalions he had withdrawn from Albania.

It is not known whether Fitzhum received only temporary control of the Skanderbeg Battle Group from Kurt Graaf, but all information points to the fact that command of the unit eventually returned to Graaf. The end of October saw a shifting of the frontlines for the German forces in Macedonia and Serbia. The withdrawal from southern Yugoslavia was under way. Skanderbeg followed that German withdrawal, making a fighting retreat north through Peć and Mitrovica and then turning west at Ćaćak and withdrawing through Visegrad, some sixty-five kilometers east of Sarajevo. By mid-December the Skanderbeg Battle Group was located around Zvornik on the west bank of the Drina River. Throughout its withdrawal, it had to encounter Partisan divisions blocking its advance, while fighting rear-guard actions with elements of the 1st and 2nd Bulgarian Armies.

Between June and September 1944, the Skanderbeg formation had operated as part of the German 21st Mountain Corps under 2nd Panzer Army, Army Group "E." By 15 October, it was still fighting under the 21st Mountain Corps, but by then that corps was directly subordinated to it, which in turn was subordinate to Army Group "F." By the beginning of December, Skanderbeg had been transferred to the V. SS-Freiwilligen-

21. WAFFEN-GEBIRGS-DIVISION DER-SS SKANDERBEG, 1 OCTOBER 1944

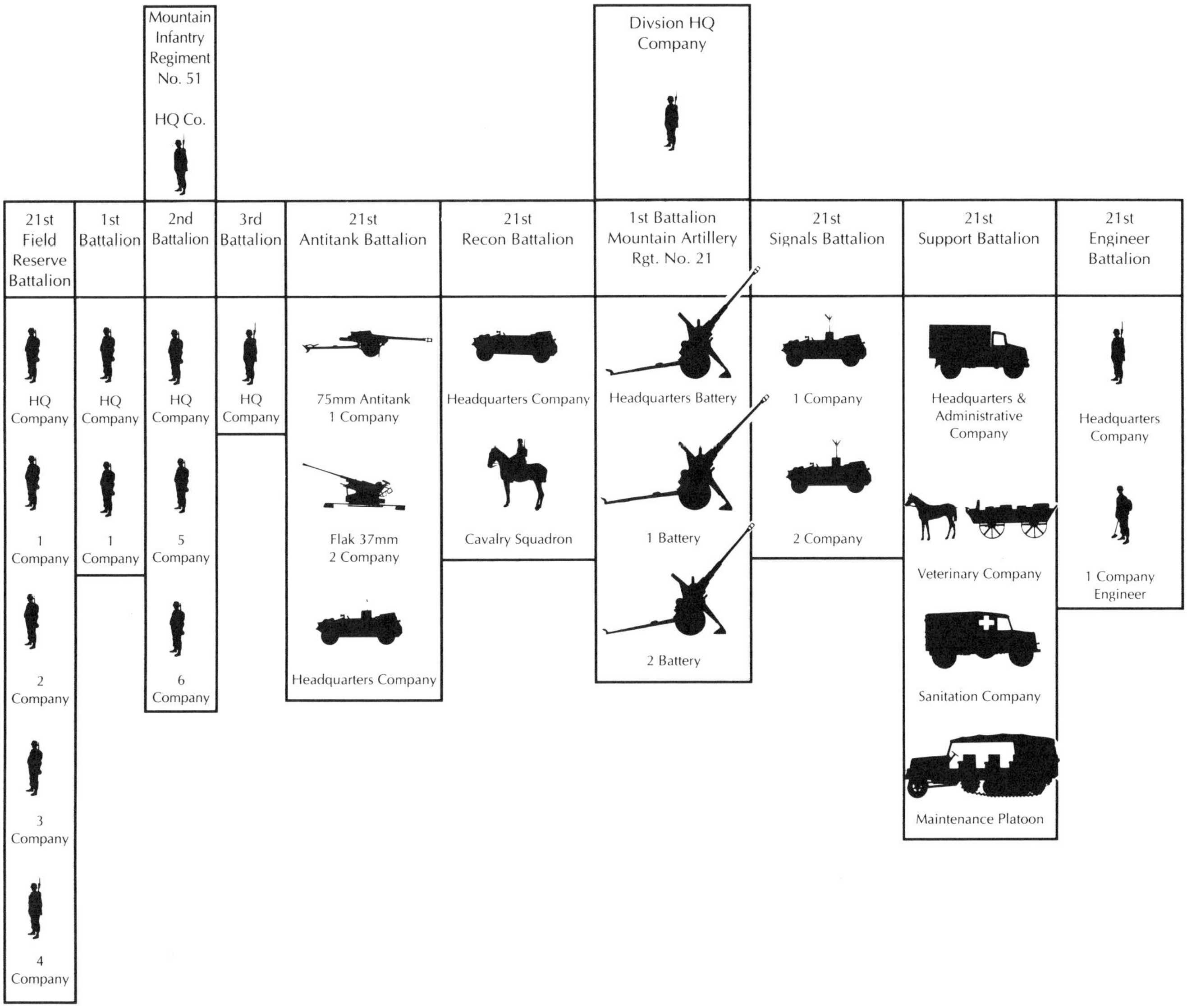

Gebirgs-Korps and was fighting alongside the 7. SS-Freiwilligen-Gebirgs-Division Prinz Eugen. The only Skanderbeg unit that remained separate from the battle group was the 21. SS-Panzer-Jäger-Abteilung, which was northwest of the Skanderbeg positions, between the towns of Brod and Vinkovci. The battalion was in reserve behind the German 34th Army Corps lines. This battalion was soon committed near Vinkovci during the third week of December to support the German 118th Jäger Division. The Russian 52nd and 223rd Infantry divisions were pressing hard their attacks on Vukovar and Vinkovci. Eventually the 118th Jäger Division had to withdraw because of Russian pressure and the presence of a Partisan division moving north into the rear of 34th Army Corps near Samac, where the Drava meets the Bosna River.

Eventually the 7. SS Prinz Eugen Division and the Skanderbeg Battle Group withdrew in a northwesterly direction, finding themselves just east of Brćko by the Drava River in mid-January,

SKANDERBEG DISPOSITIONS, 1 OCTOBER 1944

I/2 = 1st Battalion/Gebirgsjäger-Regiment der-SS 2
II/2 = 2nd Battalion/Gebirgsjäger-Regiment der-SS 2
III/2 = 3rd Battalion/Gebirgsjäger-Regiment der-SS 2

A.A. = Aufklärungs-Abteilung (Reconnaissance Battalion)
PzJgr. = Panzerjäger-Abteilung (Antitank Battalion)

1945. At this time, both units were under the command of the 34th Army Corps and faced the bulk of the Bulgarian 3rd, 8th, and 9th Infantry divisions.

The final blow came in January when the Waffen-SS High Command requested that Skanderbeg give up all of its remaining heavy equipment, after which

WITHDRAWAL OF SKANDERBEG OCTOBER-DECEMBER 1944

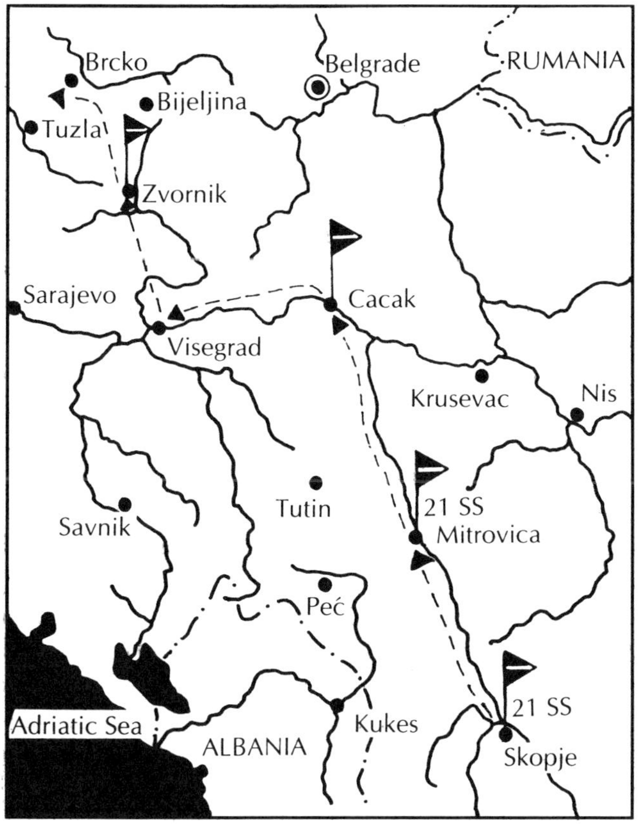

SKANDERBEG LOCATION, 18 DECEMBER 1944

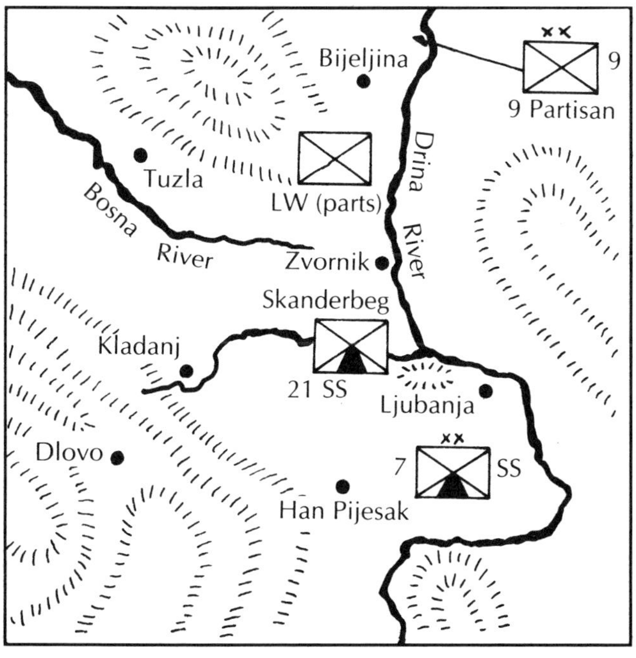

the remaining Skanderbeg personnel were to be incorporated into the 14th SS Volunteer Mountain Infantry Regiment of the Prinz Eugen Division. A list of that equipment, as well as the Waffen-SS units who were to receive it, was reported in one of the last orders of the formation:

- One mountain infantry howitzer (type-36) to go to the 23. Waffen-Gebirgsjäger Division der-SS Kama (Croat No. 2)
- Eighteen light trucks to go to 18. SS-Freiwilligen-Panzer-grenadier-Division Horst Wessel
- 100mm artillery pieces (type-18) to go to the Waffen-SS Artillery School Beneschau
- Four light vehicles to go to 18. SS-Freiwilligen-Panzer-grenadier Division Horst Wessel
- Two trucks full of artillery ammunition to go to the Waffen-SS Artillery School Beneschau
- Two cooking kettles (large) to the Waffen-SS garrison units at Prague
- One hundred machine guns (various types) to go to the 30. Waffen-Grenadier Division der-SS and 22. SS-Freiwilligen-Kavallerie-Division Maria Theresa
- Two 75mm PAK antitank guns (type L-48) to go to the 22.SS-Freiwilligen-Kavallerie-Division Maria Theresa
- One truck full of 75mm antitank ammunition to go to the 22. SS-Freiwilligen-Kavallerie-Division Maria Theresa
- Five large veterinary instru-

34TH ARMY CORPS SITUATION, 20 DECEMBER 1944

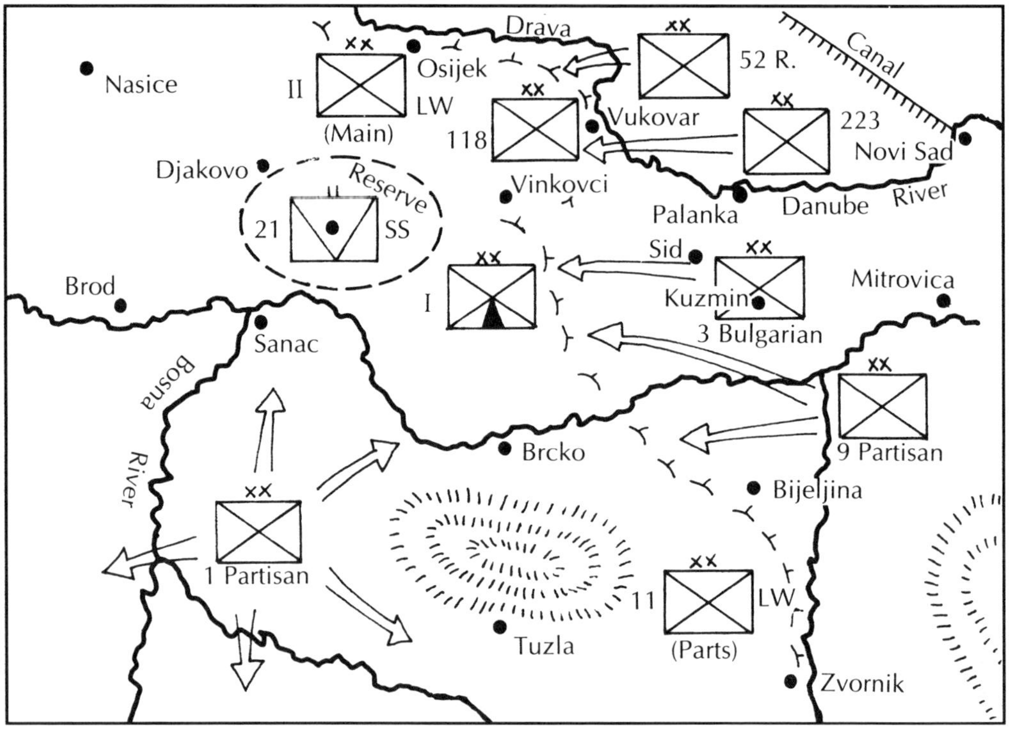

34TH INFANTRY COPRS DISPOSITIONS 15 JANUARY 1945

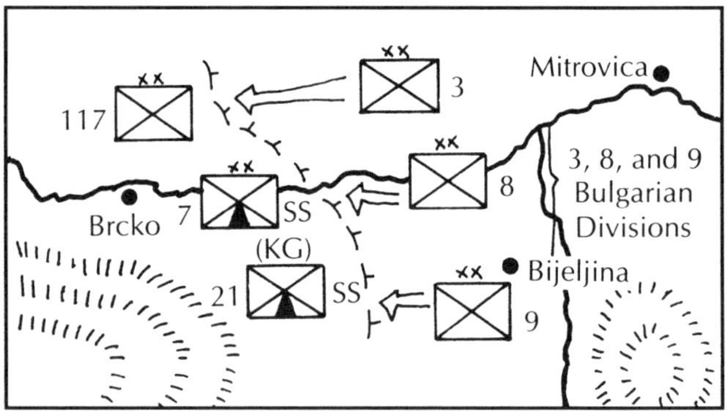

ments to go to the Waffen-SS Veterinary Station in Vienna.

Since the 22. SS-Freiwilligen-Kavallerie-Division Maria Theresa had already been trapped in Budapest in late December 1944, it is doubtful that they received the equipment assigned to them. Apparently the list had been drawn up in early December, before the 22nd was surrounded. Likewise, it is highly improbable that the 30. Waffen-Grenadier-Division ever received any armaments from Skanderbeg, since the decision to reduce the 30th from a division to a brigade occurred in January 1945. It is probable that the equipment that could not be delivered to the 22nd and 30th divisions went to other formations, quite possibly the 18th or 7th SS divisions.

In late January, the V. SS-Freiwilligen-Gebirgs-Korps was sent from Yugoslavia to the newly forming Army Group Vistula just east of Berlin. Some accounts have mentioned that former Skanderbeg personnel were sent to this army group along with the V. SS Volunteer Mountain Corps. While this may have been true in individual cases, most Skanderbeg personnel served with the 14. SS-Freiwilligen-Gebirgsjäger-Regiment of the Prinz Eugen right up to the end of the war. The Prinz Eugen withdrew into Austria in late April or early May 1945, hoping to surrender to

the Allies rather than get caught by Tito's Partisans. This they did, but the British immediately returned all remaining troops to the Partisans, who promptly put all of them to the sword.

Top: Albanian militia, who became the initial cadre of recruits for the Skanderbeg Division. Note the mixture of Albanian and Waffen-SS attire. Photo: Munin Verlag.

Bottom: Skanderbeg Division cuff band worn on the lower left sleeve. Photo: Antonio Munoz

21. WAFFEN-GEBIRGSJÄGER-DIVISION DER-SS SKANDERBEG
ORDER OF BATTLE

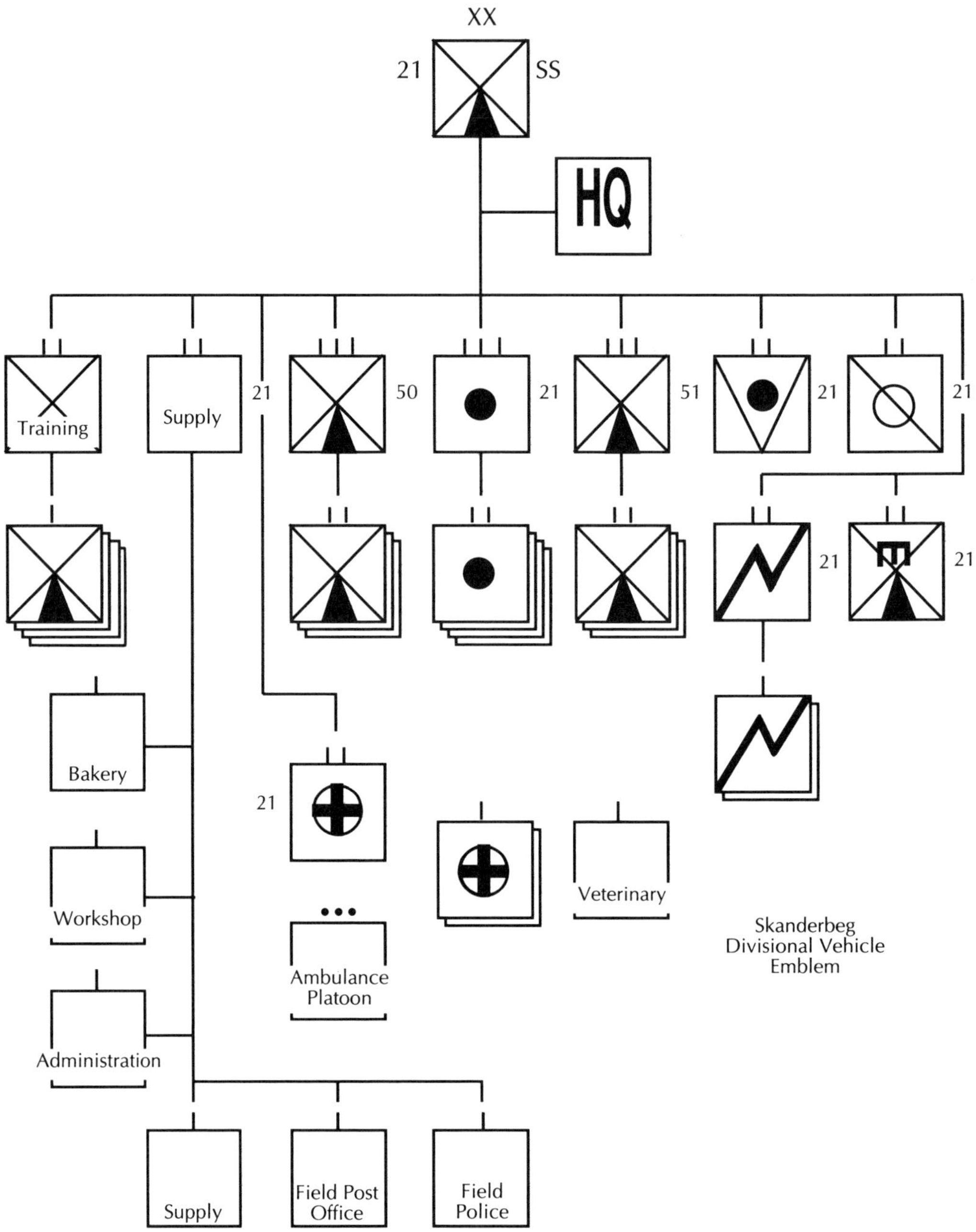

This was the table of organization developed for the Skanderbeg Division, but the full organization was never attained, as the strength table for the division on 1 October clearly shows. In spite of this, the Germans had already assigned field post numbers to every unit in this order of battle. At one time, Skanderbeg had the strength of a brigade.

SKANDERBEG FIELD POST NUMBERS

12 939	Skanderbeg Headquarters
16 034	Waffen-Gebirgsjäger-Regiment der-SS Nr. 1 (aka Nr. 50)
25 306	I. Battalion/Waffen-Gebirgsjäger-Regiment der-SS Nr. 1
21 739	II. Battalion/Waffen-Gebirgsjäger-Regiment der-SS Nr. 1
18 230	III. Battalion/Waffen-Gebirgsjäger-Regiment der-SS Nr. 1
22 892	Waffen-Gebirgsjäger-Regiment der-SS Nr. 2 (aka Nr. 51)
27 508	I. Battalion/Waffen-Gebirgsjäger-Regiment der-SS Nr. 2
24 736	II. Battalion/Waffen-Gebirgsjäger-Regiment der-SS Nr. 2
22 466	III. Battalion/Waffen-Gebirgsjäger-Regiment der-SS Nr. 2
25 618	SS-Aufklärungs-Abteilung 21
23 943	SS-Panzerjäger-Abteilung 21
17 493	Waffen-Gebirgs-Artillerie-Regiment der-SS Nr. 1 (aka Nr. 21)
28 006	I. Battalion/Waffen-Gebirgs-Artillerie-Regiment der-SS Nr. 1
30 292	II. Battalion/Waffen-Gebirgs-Artillerie-Regiment der-SS Nr. 1
23 065	III. Battalion/Waffen-Gebirgs-Artillerie-Regiment der-SS Nr. 1
29 593	IV. Battalion/Waffen-Gebirgs-Artillerie-Regiment der-SS Nr. 1
16 808	SS-Pionier-Bataillon 21
30 055	1 and 2. Companies/SS-Nachrichten-Abteilung 21
20 114	SS-Feldersatz-Bataillon 21
20 837	1. Gebirgs-Veterinärs-Kompanie - SS-Sanitäts-Abteilung 21
21 384	l. Kranken-Kraftwagon Zug - SS-Sanitäts-Abteilung 21
20 584	l. Gebirgs-Sanitäts-Kompanie - SS-Sanitäts-Abteilung 21
28 486	2. Gebirgs-Sanitäts-Kompanie - SS-Sanitäts-Abteilung 21
	SS-Versorgungs-Abteilung 21
26 629	Bakerei-Kompanie - SS-Versorgungs-Abteilung 21
28 954	Werkstatt-Kompanie - SS-Divisions-Nachschub-Trupp 21
31 182	Verwaltungs-Kompanie - SS-Verwaltungs-Trupp 21
23 611	Versorgungs-Kompanie - Divisions-Nachschub-Trupp 21
23 357	SS-Feldpostamt 21
24 409	SS-Feldgendarme Trupp 21

CHAPTER EIGHTEEN

. . . the devil taketh him up into an exceeding high mountain . . .

Matthew 4:8

During the summer of 1942, the Reichsführer-SS Heinrich Himmler ordered the creation of a counterinsurgency unit that could operate against the increasing partisan activity in the alpine and cisalpine arc, from the South Tirol to Istria. The Tirol area encompasses that part of Austria that touches the Italian, Swiss, and German frontiers. It extends across the Austrian border into Italy as far south as Lake Garda. In 1939, the German-speaking Italian subjects on the Italian side of the Tirol numbered 267,000. These ethnic Germans were spread out in the Italian provinces of Bolzano, Trentino, Belluno, and Udine. In 1939, these ethnic Germans were given the choice of retaining Italian citizenship or becoming German citizens. Almost 70,000, or 25 percent, chose to join the Reich, leaving for Germany that summer.

The Tirolean region is noted for its high alpine peaks and rugged limestone rock formations. Mountain passes allow movement from one valley to another, but deep crevices and sinkholes abound. Overall, it is difficult terrain to patrol and move around in. For this reason and because of increasing incidents of partisan attacks, the Waffen-SS organized the SS-Karstwehr-Kompanie. The company was initially formed and equipped at the SS training grounds near Dachau. In the beginning, it had two platoons.

Command of this specialized mountain company was entrusted to SS-Standartenführer Dr. Hans Brand. Brand was a professor of geology and, as such, an authority on this Karst region.

The company received its title on 10 August 1942, and training began in earnest. Recruits came from various parts of Europe. They included ethnic Germans from the Austrian and Italian sides of the Tirol; Romanian Volkdeutsche from the Banat and Transylvania; and ethnic Germans from Hungary, Ukraine, and Yugoslavia (Slovenia, Croatia, and Serbia). The German NCO and officer cadre was composed of Austrians as well as men from the Posen Treskau Military Academy and Breslau in Silesia. In 1944, even a company of Spaniards from the Brandenburg commandos was added to the formation.

The decision to raise the unit from company to battalion strength occurred sometime in late summer. At that time, the Karstwehr unit was placed under the jurisdiction of the newly created Waffen-SS Panzer Command Staff that was forming in Pottenstein in Oberfranken. This formation would be the first pure Waffen-SS corps command, and it would be led by the best Waffen SS field commander—SS General Paul Hausser.

Attaching the forming Karstwehr battalion to this command was purely an administrative affair, since the intention was to employ the battalion in the Tirol, while the panzer command was destined for the Russian Front. By November 1942, the battalion had three mountain infantry companies and a small staff unit. Its top officers are listed on the following page.

By the beginning of the summer of 1943, the battalion had been fully formed and was posted to Arnoldstein, in the Carinthian region of Austria, for field training. This area encompassed the region where the Austrian, Yugoslavian, and Italian frontiers meet. The Thörl-Maglern pass is located just south of Arnoldstein, which, incidentally, was the site of an important Waffen-SS sport school. The commander of this school in 1943 was SS-Standartenführer Hinterstöser.

The battalion was billeted in the following villages: lst Company, Selciak; 2nd

SS-KARSTWEHR-BATAILLON STAB, NOVEMBER 1942

Stab

SS-Standartenführer Dr. Hans Brand, Ia
SS-Sturmbannführer Erich Wieland, Ib

1st Company

SS-Hauptsturmführer Mehrwald

2nd Company

SS-Sturmbannführer Kühpantner

3rd Company

SS-Sturmbannführer Josef Berschneider

Total strength: 480 men[1]

[1] Quoted from *Lo Sterminio Mancato* by P.A. Carnier, Mursia Publishing, 1982, pp. 76-82.

Company, Pöckau; and 3rd Company, Hohenthurn.

At the end of June 1943, the SS-Karstwehr-Bataillon went on alert because of the worsening military situation in Italy. On 25 July, Benito Mussolini resigned as head of the Italian government after a vote of no confidence by his own Fascist Grand Council. Mussolini's own son-in-law, Count Ciano, who was a member of that council, had voted against him. Marshal Pietro Badoglio took over as head of the new Italian government, at the urging of the king of Italy, whose power had disappeared after his abdication but who still wielded much influence.

(See Chapter 1 for more information on the situation in Italy.)

Hitler slowly began moving divisions into Italy on the pretext that they eventually would be committed to the fighting in Sicily and southern Italy. Numerous ad-hoc training and school forma-

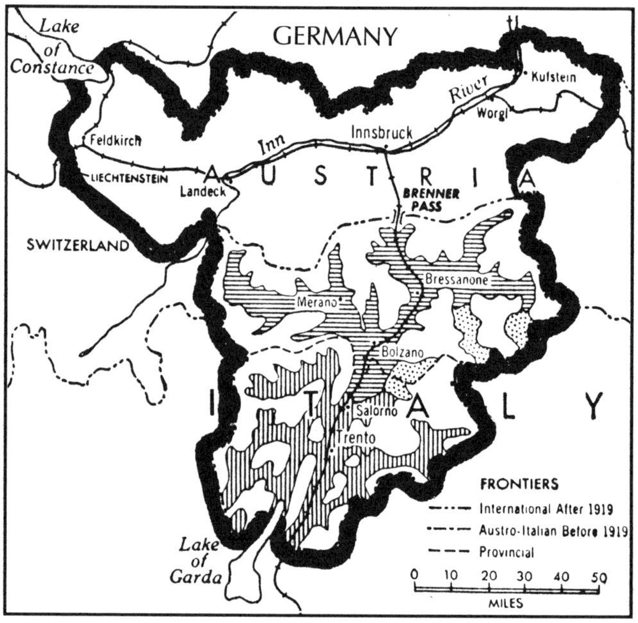

Distribution of principal languages around Tirol at the Italo-Austrian frontier. (Reproduced with permission from "Tirol" in *Encyclopedia Britannica,* 14th edition, 1963, by Encyclopaedia Britannica, Inc.)

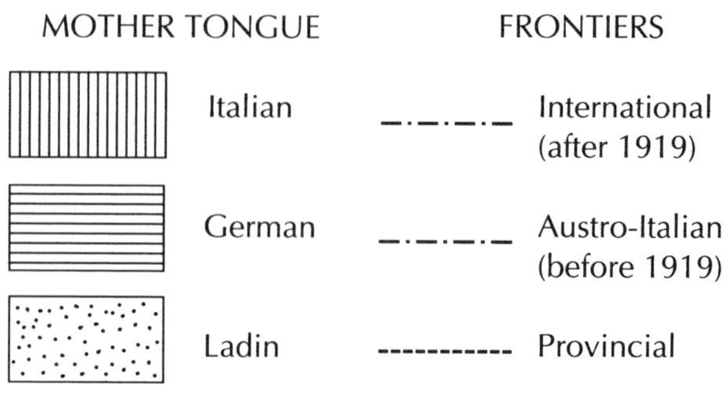

GENERAL AREA OF OPERATIONS OF THE KARSTWEHR FORMATION

General area of operations of the Karstwehr formation.

Road — River

Frontier — Town/City

Pass — Coast

tions had been forewarned about their possible employment in an occupation of Italy. The Karstwehr Battalion was one of a few regular units scheduled to take part in this occupation should the need arise.

That need came on 8 September 1943 when Italy suddenly surrendered unconditionally to the Allies. The following day, during the early hours of 9 September, the Karstwehr Battalion moved across the Austrian-Italian frontier. Crossing through the Thörl-Maglern Pass, the battalion quickly reached the town of Tarvisio, which was located on the Italian side of the pass. That area was heavily populated by ethnic Germans, so throngs of people greeted the men of the Karstwehr with cheers and flowers. The small Italian army contingent at Tarvisio simply surrendered. The towns of Camporosso and Boscoverde also welcomed them.

The battalion was not so lucky when they advanced on the Italian garrison that defended Ugovizza. After a stiff but brief battle, that town also fell to the men of the Karstwehr, but not before ten of them had lost their lives capturing it. One of those killed was SS-Oberscharführer Gedicke. After this, the Karstwehr Battalion was ordered to the Canale (Kanal) Valley, where Italian Communist bandits had begun to step up their attacks on the local populace loyal to Mussolini.

The Karstwehr Battalion was transferred to the Istrian Peninsula, where it successfully recruited many volunteers, picking up foreigners and ethnic Germans from Istria, Slovenia, Croatia, and even Serbia. These enlistments allowed the Karstwehr Battalion to attain full strength, including headquarters and supply companies. A supply depot for the Karstwehr unit was established on the Italian side of the Thörl-Maglern Pass for weapons, clothing, ammunition, and other materiel. The headquarters for the battalion was established in Gradisca d'Isonzo. The structure of the Karstwehr unit and the commanding officers were as shown on page 248.

The battalion stationed units at Gradisca and Sagrado, with detachments at Ugovizza, Malborghetto, Pontebba, Tolmezzo, and other towns and villages as far as the northwestern zone of the littoral. The battalion received the task of operating against the Communists in the Isonzo Valley and in the eastern part of the Julian Alps. These battles were hard and gruesome. The men of the Karstwehr faced an enemy that did not fight by the rules of the Geneva Convention, as the atrocities later committed by the guerrillas showed.

In November 1943, while temporarily attached to the V. S-Freiwilligen-Gebirgsarmee-Korps, the battalion made

SS-KARSTWEHR-BATAILLON, DECEMBER 1943

Stab Company
SS-Standartenführer Dr. Hans Brand, Ia
SS-Sturmbannführer Erich Wieland, Ib
SS-Sturmbannführer Werner Hahn, Stab Kp. Chef

1st Company
SS-Hauptsturmführer Mehrwald

2nd Company
SS-Sturmbannführer Kühpantner

3rd Company
SS-Sturmbannführer Josef Berschneider

Nachschub Company
SS-Untersturmführer Ubald Flake[1]

[1] Flake continued to lead the supply company throughout its existence. He was promoted to SS-Obersturmführer in April 1945 and married a Tirolese woman on 25 April in the Kanal Valley. Quoted from *Lo Sterminio Mancato* by P.A. Carnier, Mursia Publishing, 1982, pp. 76-82.

a sweep of the town of Piedicolle but was ordered out on November 23. The battalion could not withdraw, however, until 3 December because some parts of the battalion baggage train had not arrived. If the combat units of the battalion had left on 23 November, the supporting units might have been attacked by the Communist partisan groups still in the surrounding countryside.

No mercy could be expected from these bandits. Many petty criminals now assumed the title of "partisans" to disguise their underworld crimes. General Frido von Senger und Etterlin, the famous German defender at Cassino, aptly described these bandit groups fully by showing the situation that faced the Germans in the city of Bologna in 1945.

> *The most pressing problem was to cope with the partisans. They were the underworld and the gangsters who ruled the town. Before I took over the sector, they had made an armed raid on the leading hotel, firing indiscriminately on the guests in the hall, most of whom were German officers or Italian Republican Army adherents. Our own security service came to the conclusion that it would not be possible to arrest any large number of partisans in the town, since they could always go underground again . . .*
>
> *Murders even among the civil population were a daily occurrence. The partisans not only struck at their political opponents, but were often actuated by purely personal motives of hate or revenge. They sought to cover their traces by pinning to their dead victims a label with the inscription, "Spia Tedesca" (German spy). Although this accusation might deceive a few credulous Italians, it was not so with us who had means of checking up on the perpetrators.*[1]

In early 1944, the Communist parti-

[1] Quoted from the 1989 Presidio Press edition of *Neither Fear nor Hope: The Wartime Memoirs of the German Defender of Cassino*, by General Frido von Senger und Etterlin, with permission of the publisher.

THE ISTRIA PENINSULA

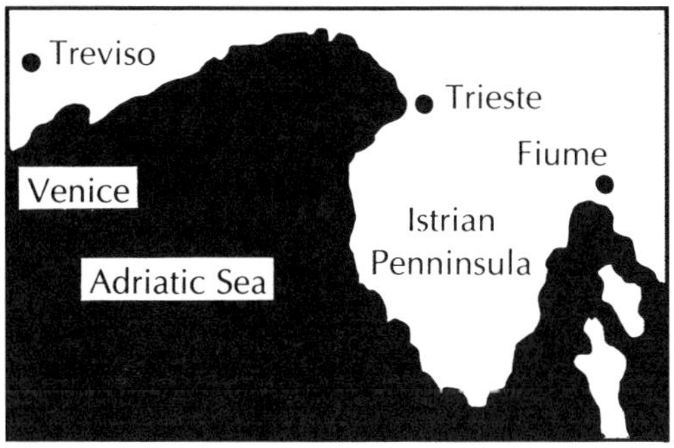

sans committed an atrocity against members of the Karstwehr Battalion that became infamous. The Karstwehr unit had been transferred to the Udino-Frioli area, placing its headquarters unit in the town of Moggio. In February, after what was described as a mopping-up operation in Selva di Tarnova, a squad of fifteen Karstwehr soldiers, most of them recent inductees between the ages of eighteen and nineteen, failed to return to base after a patrol; the squad had been ordered to reconnoiter north of the town of Cividale. Soon, an anonymous letter arrived at Moggio, which said that the fifteen soldiers had been tried and executed. A detachment of the Karstwehr was sent out immediately in response. The fact that the squad had not returned added credibility to the terrorist letter.

The Karstwehr detachment found the mutilated bodies of the soldiers. The Communists had first tortured the young recruits in the most hideous ways, finishing their gruesome work by decapitating them and sticking the heads on the bayonets of their victims. This particularly gruesome incident typified the character and nature of the partisan war that the Karstwehr troops faced. Is it any wonder, then, that the Karstwehr troops often failed to give quarter in return to captured partisans? The partisans had long ago demonstrated only contempt for the Geneva Convention agreements; they meant nothing whatsoever to these Communist freedom fighters.

These acts of atrocities failed to cow the men of the Karstwehr unit; instead, they induced fear and terror into the defenseless civilian population still loyal to the duce. Preservation of life now became more important to the peasants than any political or national ideal. The battalion recovered the bodies of the dead soldiers and wrapped each in his camouflage poncho. Then, in a solemn military procession, the men of Kartswehr wound their way through Gradisca in the middle of a heavy snowstorm, carrying their dead comrades. A train delivered the bodies to the old Austro-Hungarian war cemetery at Fogliano. There they were buried together in a common grave. As a final farewell, an SS officer ended his eulogy with this statement: "... we aren't among men any longer, we are among beasts!"[2]

Recently, the names of the fifteen soldiers were learned and are shown by company at the top of page 250.

(The only member of the Karstjäger formation whom I could interview while writing this manuscript was Mr. Günther Witt. He related the story of the massacre to me and mentioned an Albert[o] Horeth. Perhaps the passage of time dimmed his memory somewhat—although he came very close to remembering the exact name. Mr. Witt has since passed away.)

The Karstwehr unit remained in the Udino-Frioli area during the spring and summer of 1944. For reasons unknown, the battalion commander, SS-Standartenführer Brand, had a long standing feud with Higher-SS and Police Leader Globocnik. The sixty-three-year-old Brand eventually lost and was replaced as battalion commander by the former commander of the 3rd Company, SS-Sturmbannführer Josef Berschneider. Berschneider had a thirty-five-year-old Muslim aid-de-camp by the name of Alia,

Divisional insignia of the 24. Waffen-Gerbirgs-(karst)jäger-Division der-SS. Illustration: Antonio Munoz.

[2] Quoted from *Lo Sterminio Mancato* by P.A. Carnier, Mursia Publishing, 1982, pp. 76-82.

2nd Company
- Franz Meierhofer
- Hans Krellner

3rd Company
- SS-Unterscharführer Albert (Alberto) Heinrich (not Horeth)
- Johan Ramsauer
- Ernst Bormann
- Mathias Kirmeier
- Gerhard Hirsch
- Julius Benz
- Willi Tellmann
- Hans Schoger
- Mustafa Moratowitsch

4th Company
- Werner Messinger
- Herbert Schmidt
- Andreas Doerfler
- Stefan Gubansk

who had been recruited from the Yugoslavian province of Bosnia-Herzegovina. The men of the Karstwehr Headquarters Company soon came to know him as "Mohammedan."[3]

At the beginning of the summer, Himmler began to receive increasing reports about the stepped-up partisan activity along the Italian-Yugoslavian frontier. He responded by ordering the immediate expansion of the SS-Karstwehr Battalion. The Karstwehr had a numerical strength of about 1,000 men; Himmler hoped that through a vigorous recruiting campaign, it could be formed into a full-strength division, with a TOE strength of 6,600 men. Himmler signed the order on 18 July 1944, exactly five months to the day after the fifteen Karstwehr soldiers had been buried at Fogliano.

Overall supervision for raising the Karstwehr Division went to SS-Gruppenführer und Generalleutnant der Polizei Odilo Globocnik, who headed the Higher-SS and Police Command Adriatic Coastland. SS-Sturmbannführer Werner Hahn, the commander of the Staff Company, took over for SS-Sturmbannführer Berschneider as commanding officer. Heavy recruiting drives were conducted all over the Tirol area—Udine-Friuli, Istria, and even the western parts of Slovenia.

The Karstwehr formation was a hodgepodge of different European nationalities, all mixed together: Romanians like volkdeutscher Rotpeher, Gross, and Gustav Klosius (the first two were Unterscharführers, while the latter was a Rottenführer); South Tiroleans like SS-Männer Giovanni and Max; Austrian Oberscharführer Ilg; and Croatian Oberscharführer Alia. The Karstwehr formation has been described by postwar historians as third-rate. While it is true that it was never a top-of-the-line armored or armored infantry formation, its description as a mediocre mountain unit is undeserved.

Discipline was absurdly strict in the formation. Insubordination and other offenses were punishable by hard labor and prison time, and the German military prison at Pontebba had its share of Karstjäger men. Civilians would often see them—guarded by other Karstjäger men—on long treks around the area, carrying rucksacks filled with heavy stones. Gerhard Prelop, a twenty-three-year-old German, was executed by a firing squad made up of men from his own platoon after he was convicted of stealing a small amount of money. Only one rifle in the platoon carried a live round so no one would know who actually carried out the execution.[4]

The men of the Karstjäger tried to get along with the locals. German bread, intended for the Karstjäger unit, would often find its way to the needy families or street urchins of Tolmezzo. The SS High

[3,4] Quoted from *Lo Sterminio Mancato* by P.A. Carnier, Mursia Publishing, 1982, pp. 76-82.

The rugged Tirol area, noted for its very high peaks and rough limestone rock formations. Note the Karstjäger soldier in the center of the photograph leaping from one rock to another (far left).

A member of the Karstjäger Division during an antipartisan drive, rifle at the ready and rope slung over his shoulders. Note also the camouflaged uniform (left). Photos: Munin Verlag.

Command Adriatic Coastland eventually found out about this practice and forbade any more donations. Part of the Karstjäger baggage train, stationed north of Udine in the village of Collerumiz, halfway between Collalto and Artegna, often loaned the local farmers mules and horses to plow the fields. Maria Viviana Paladino, a former barmaid at a pub in Gorizia frequented by men of the Karstjäger, had this to say about them:

> *. . . they were discreet; they cared for the perfect uniform, the crease. Sometimes, after the actions, they kept silent. Obviously someone had been missing. However, a word on the actions carried [out] or on those to be carried [out] never came to be known from their conversations at the club.*
>
> *The secret was total: in that respect, they were like a wall. They often left all of a sudden by truck, but nobody ever knew where they were going to. Sudden departures occurred by night, too: one heard the noise of the steps, made heavy by the mountain hobnailed boots, the roar of the trucks. They had the alpine star on the headgear. There were many Rumanians from Transylvania, Southern Tiroleans, Istrians . . .*[5]

Because of the mainly ethnic German composition of the formation, it received a non-Germanic classification. First titled Waffen Gebirgs(karst)jäger-Division der-SS, it received the ancillary number "24" before its title on 9 September 1944. Divisional support units formed after that date had this number.

Two rock-climbing mountain infantry regiments were earmarked for the division. They received the numbers 59 and 60 from the Waffen-SS regimental list. Because of a lack of recruits, however, the 60th Regiment was never able to form. The worsening military situation, which began to take a heavier toll on the Karstjäger formation, and terrorist intimi-

[5] Quoted from *Lo Sterminio Mancato* by P.A. Carnier, Mursia Publishing, 1982.

Men of the 24. Waffen-Gebirgs(karst)jäger-Division der-SS, during an anti-Partisan drive near the Italo-Yugoslavian frontier. Note the particularly rugged terrain of this area and the atypical machine gun slung over the shoulder of the soldier being assisted up the cliff. Photo: Munin Verlag.

dation inhibiting recruitment, made it impossible to expand the Karstjäger unit into a division. During late summer and early fall, the unit was referred to as "SS-Gebirgs-Karstjäger-Regiment 59," but this was incorrect since it signifies a unit. Regardless of titles, the formation was never employed in the regimental sense of the word. Instead, the Karstjäger unit was employed into two self-supporting jagdkommandos (hunter commandos). These jagdkommandos would be attached to whichever higher army or Waffen-SS/Police command took charge of the anti-Partisan operation. While under these commands, the jagdkommandos remained only nominally attached to their own headquarters.

The units of the Karstjäger operated over a wide area. In practice, they operated between Selva di Tarnova and the Cansiglio plateau or further. In one instance, they prevented British intelligence from contacting Tito's Partisans on the islands of Quarnaro, leading to the complete elimination of the British mission.

From the beginning, the Karstwehr unit had been issued specially trained Alsatian dogs skilled in the art of tracking. One such dog, nicknamed Attus, had become so famous with the Communists that a reward of 50,000 lire had been offered for its head.[6]

Late in the summer of 1944, a new divisional commander, SS-Obersturmbannführer Carl Marx, was appointed for the Karstjäger formation. As had every previous commander, Marx led the unit in a positive manner, thus making the Karstjäger formation one of the most effective anti-Partisan units in the whole Adriatic littoral area. Carl Marx left the Karstjäger unit before the end of 1944. He went on to command the massive training program for the reorganizing Ukrainian 14th SS Division.

The esprit d'corps these men possessed had been built by men like SS-Unterscharführer Gustel Müller, an NCO of the Waffen-SS High Mountain Training School. Between the drills and climbs, the recruits sang a motivational song specially composed for the Karstjäger, which spoke of soldiers who feared nothing, laid ambushes, and faced death: "We Karst riflemen go in the woods and don't know what apprehends us. . ." These songs were inspired by the type of fighting they encountered in the high-alpine region.

[6] Quoted from *Lo Sterminio Mancato* by P.A. Carnier, Mursia Publishing, 1982, pp. 76-82.

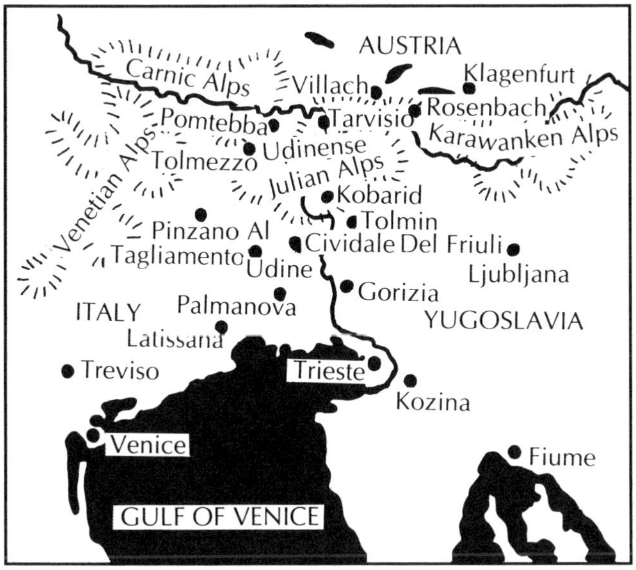

This map shows the general area of operations of the Karstwehr unit.

THE DEATH OF GOTTFRIED MÜLLER

Gottfried Müller was born on 24 March 1924 in Schässburg/Siebenburgen. He was drafted in 1943 and eventually posted to the Panzer Company (Field Post No. 64 584) of 24. Waffen-Gerbirgs(karst)jäger Division der-SS. He died at the age of twenty-one on 1 or 2 May 1945 in Tricesimo (by Udine) in an Allied air attack. This death, though one of many suffered by the Karstjäger Division, leads to speculation that at least the Panzer Company of the Division/Brigade was still on Italian soil in May 1945.

During the last three months of 1944, the strength of the Karstjäger varied between 2,000 and 3,000 men. One source states that the strength was more between 3,000 and 4,000, but this is unsubstantiated. The new commanding officer of the Karstjäger was SS-Obersturmbannführer Wagner, who led the formation until the end of the war.

The decision to downgrade the formation from division to brigade was made in November 1944, and Himmler signed the order on 5 December. The order officially retitled the Karstjäger "division" as the "Waffen Gebirgs(karst)jäger-Brigade der-SS." However, the German Field Post would continue to refer to it as a division until 27 February 1945, and its vehicles continued to use the divisional emblem, formed by the runic "TYR" entwined with the imperial Tirolean hunter insignia.

The history of the Karstjäger Brigade at the beginning of 1945 is scarce to say the least. Little is known except that it continued to engage the partisans around the Italian, Austrian, and Yugoslavian frontiers. One source states that it was eventually committed to frontline fighting against the advancing British army in Italy, delaying some of its units for a time. By the end of March 1945, the German withdrawal from Italy had begun, and the brigade was moved back to Istria.

By late April, the Karstjäger Brigade was already in Austria in the area between Klagenfurt and Villach. The brigade was then turned into an ad-hoc battle group consisting of replacement troops from the 7. SS-Freiwilligen-Gebirgs-Division Prinz Eugen, plus the SS-Officer Academy cadets from SS-Junkerschule Klagenfurt. The only unit that did not join was the Training Company, which was sent to Vienna in early April to help in that city's defense against the advancing Red Army.

At this time, the engineer company in the Karstwehr Brigade received a new commander, SS-Hauptsturmführer Gerard Peters, who had seen action in the Balkans as commander of the Albanian 21st SS Engineer Battalion. He had recently acted as engineer instructor to the Circasian and Caucasian SS troops then in Northern Italy. Another newcomer to the forming kampfgruppen to which the

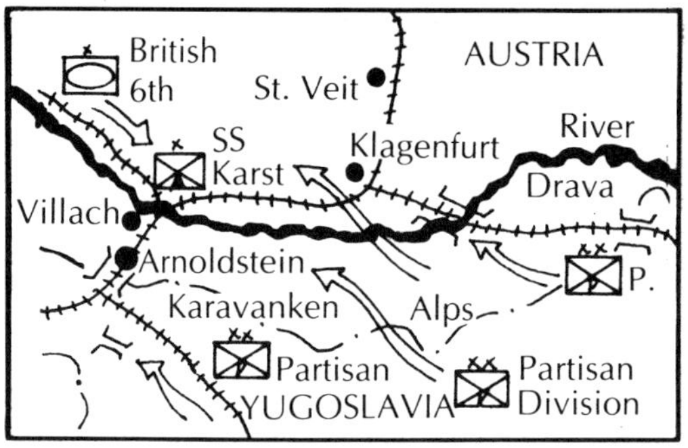

The last known area of operations of the SS Karstjäger Brigade, April 1945.

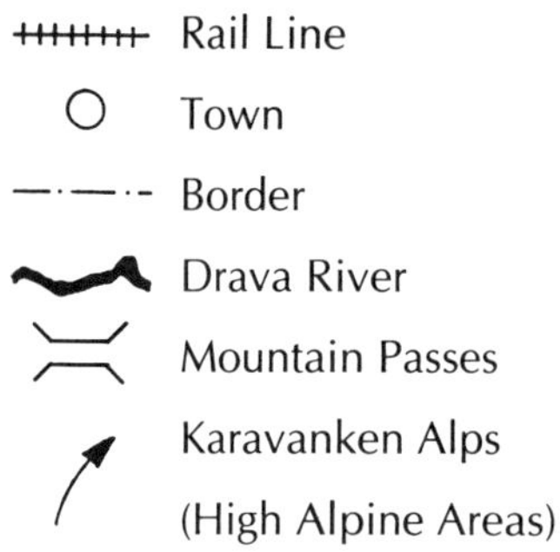

Karstjäger Brigade now belonged was SS-Brigadeführer Heinz Harmel, who had previously led the 10. SS-Panzer-Division Fründsberg. This kampfgruppe came under the operational control of the German 22nd Mountain Corps.

The entire German army in the Balkans was attempting to extricate itself from the jaws of the enemy by making a fighting withdrawal to the Yugoslavian-Austrian frontier. Blocking their retreat into southern Austria and safety were the Karawanken Alps. SS-Brigadeführer Heinz Harmel, in charge of the battle group, moved his force south to secure the river and mountain passes for the withdrawing German army. The Germans were making a frantic run for the border. An uncertain future awaited any German who surrendered to the Yugoslav Partisans.

The last days of April and first days of May found the Karstjäger Brigade fighting not only the Partisans but the advancing British army as well. The brigade finally surrendered to the British 6th Armored Division in the Rosen Valley (near Villach) on 9 May 1945. Thus ended the brief history of the 24. Waffen-Gebirgs(karst)jäger Division/Brigade der-SS.

Officers in the unit included SS-Sturmbannführer Dr. Siegfried Liebow, SS-Hauptsturmführer der Reserve Dr. Heinz Habisreutinger, and SS-Obersturmbannführer Dr. Harald Waegner. The veterinarian in the unit was SS-Obersturmbannführer Dr. Rudolf Sechser. Günther Witt, already mentioned, served as an SS-Unterscharführer in the 3rd Battalion, Waffen-Gebirgs-Karstjäger-Regiment der-SS Nr. 59.

WAFFEN-GEBIRGS(KARST)JÄGER-DIVISION DER-SS
ORDER OF BATTLE, 31 AUGUST 1944

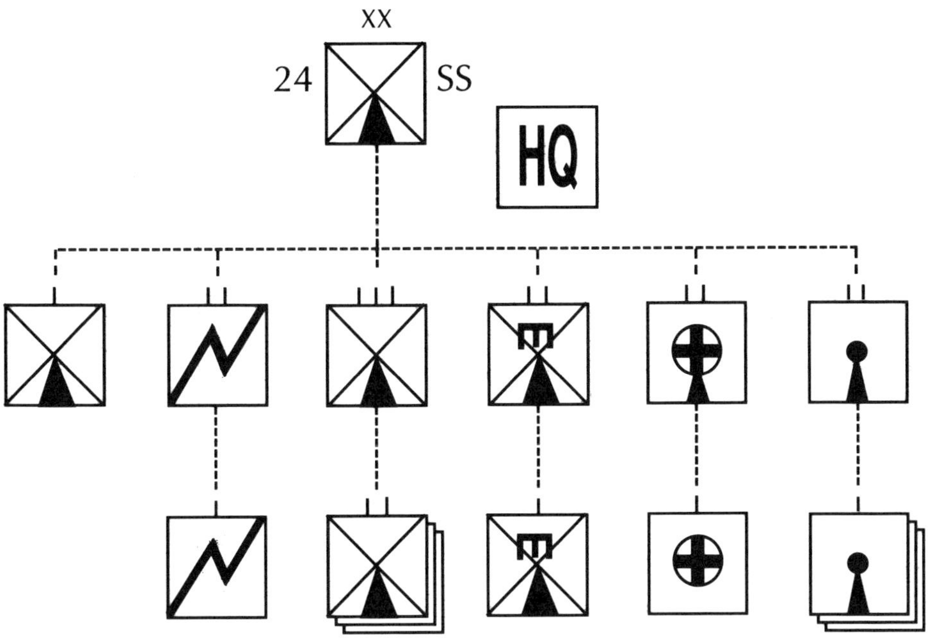

Field Post No.	Formation
40 731	Staff HQ
05 219	Staff plus 1st through 4th Companies, I. Battalion
28 657	Staff plus 5th through 8th Companies, II. Battalion
33 627	Staff plus 9th Company, III. Battalion
28 007	1st Company, Gebirgs-Pionier-Bataillon 24
32 137	1st Company, Gebirgs-Sanitäts-Abteilung 24
34 445	1st Company, Gebirgs-Nachrichten-Abteilung 24
64 384	Jäger Company
22 224	Staff plus 1st through 3rd Batteries, I. Battalion, Gebirgs-Artillerie-Regiment der-SS

A second mountain regiment had been planned for the 24th Division, but a lack of recruits made that impossible. Slovenes, Croats, and Serbs were inducted into the unit. Gone were the days back in 1942 when potential recruits were turned down because of physical or racial requirements. (Their recruits had been overage SS men from the SS office at Ahnenerbe, who were deemed inadequate for the rough mountain training of the SS Karstwehr Company.) The artillery regiment never got beyond battalion strength.

24. WAFFEN-GEBIRGS(KARST)JÄGER-DIVISION DER SS

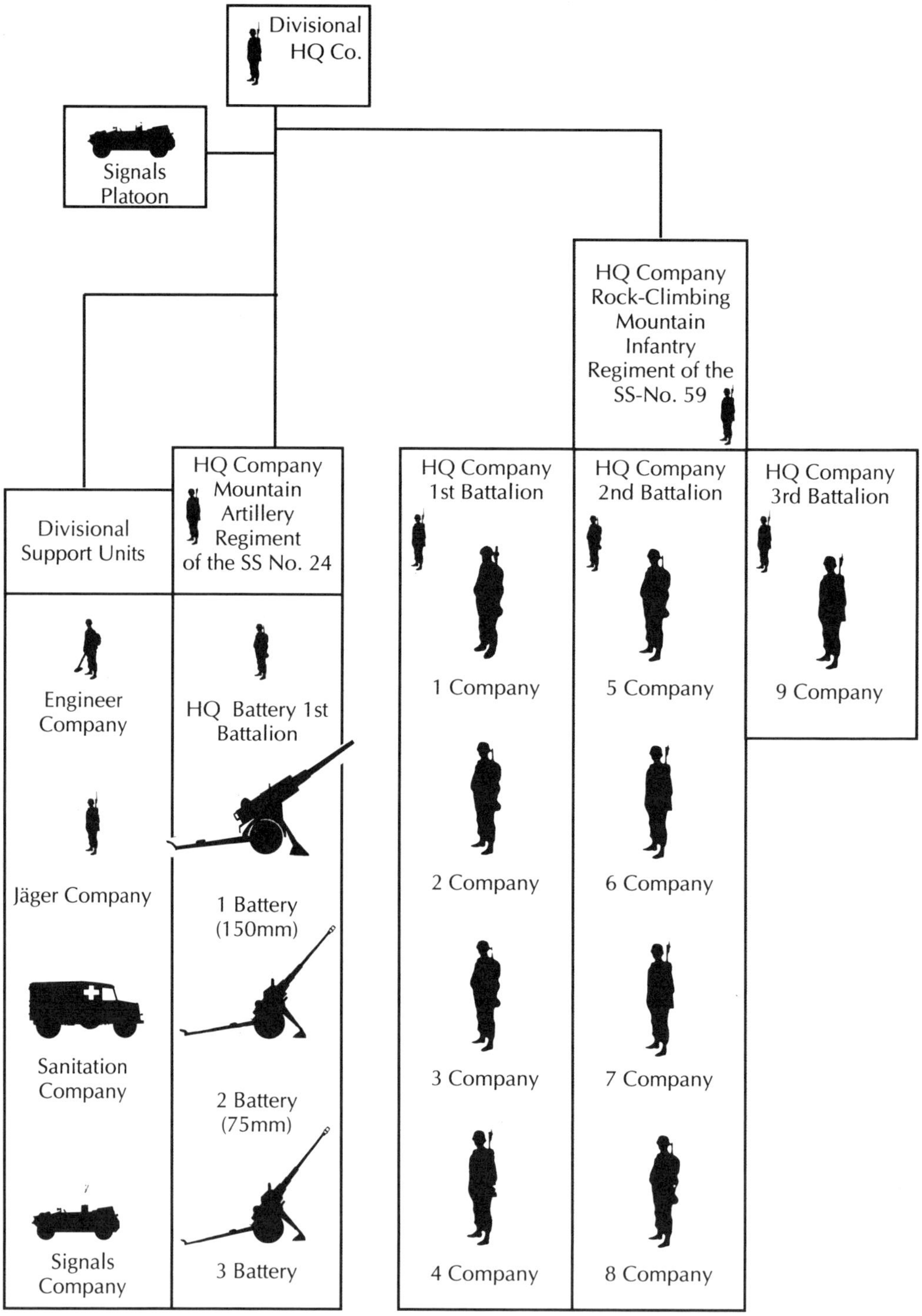

Illustrated diagram, September 1944

WAFFEN-GEBIRGS(KARST)JÄGER BRIGADE DER SS

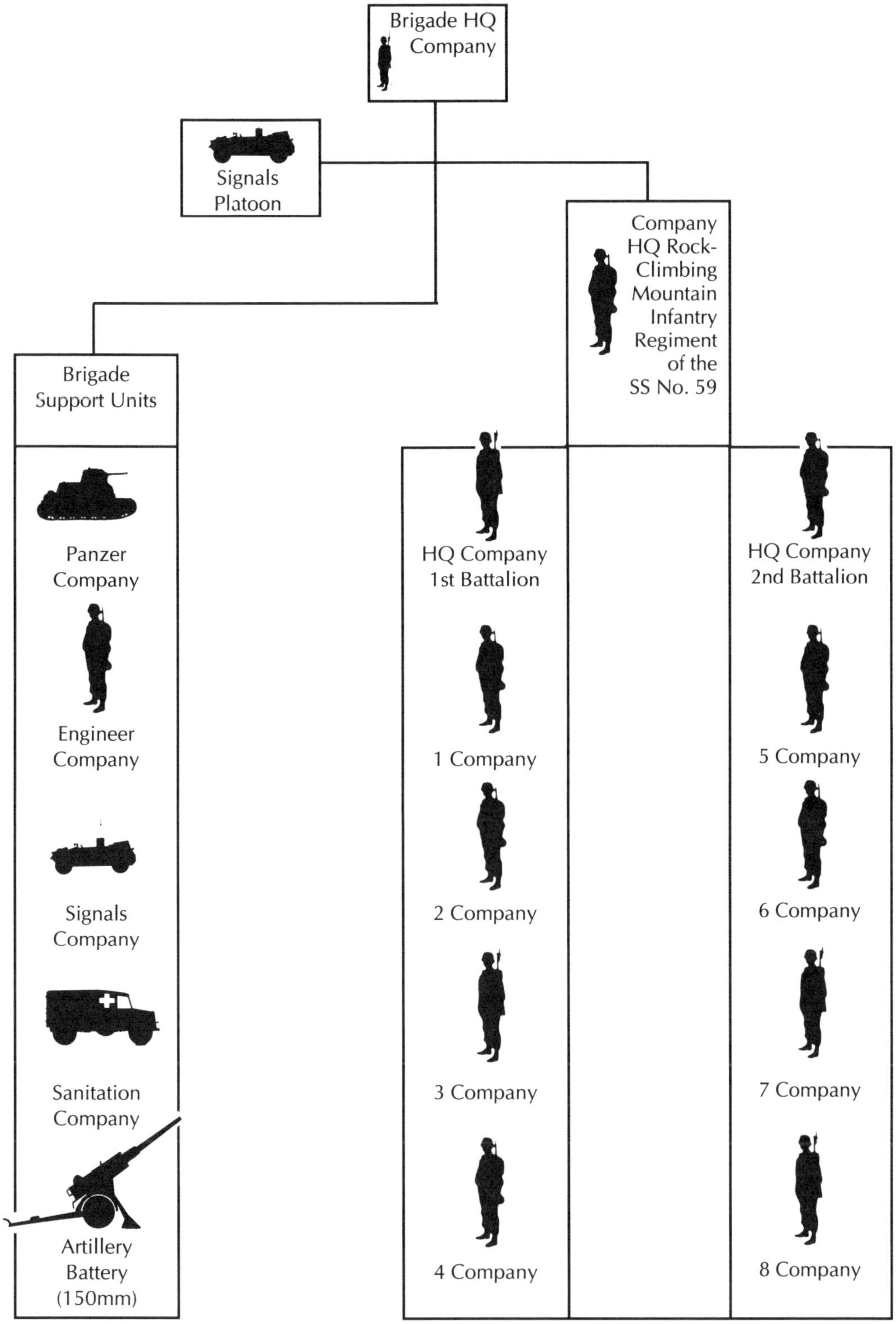

Illustrated diagram, November 1944

Karstwehr/Karstjäger Strength	Officer	NCOs	Men	Total
SS-Karstwehr Btln. 31 December 1943	8	117	821	946
SS-Karstwehr Btln. 30 June 1944	12	132	813	957
Waffen-Gebirgs-(Karstjäger) Division der-SS 18 July 1944	12	132	730	874
24. Waffen-Gebirgs-(Karstjäger) Division der-SS 20 September 1944	27	163	1,799	1,989
Waffen-Gebirgs-(Karstjäger) Brigade der-SS December 1944	?	?	?	about 3,000

WAFFEN-GERBIRGS(KARST)JÄGER BRIGADE DER-SS ORDER OF BATTLE, 5 DECEMBER 1944

Field Post No.	Formation
40 731	Brigade Staff HQ
57 542	Staff Company
05 219	Staff plus 1st through 4th Companies, I. Battalion
28 657	Staff plus 5th through 8th Companies, II. Battalion
64 584	Panzer-Kompanie[1]
22 224	Gebirgsbatterie
32 137	Sanitäts-Kompanie
34 445	Nachrichten-Kompanie[2]
28 007	Gebirgs-Pionier-Kompanie[3]

[1] This was the former Jäger Company in the proposed division. It was armed with an assortment of captured Italian tanks and armored cars.
[2] By the end of February 1945, the signals company had been disbanded. Officially, it was struck from the records on 1 March 1945.
[3] This engineer company was finally disbanded during the middle of February 1945. Officially, it was struck from the records on 27 February 1945.

NOTE: The Third Battalion was never fully formed, containing only the 9th Company, which was disbanded by late November 1944.

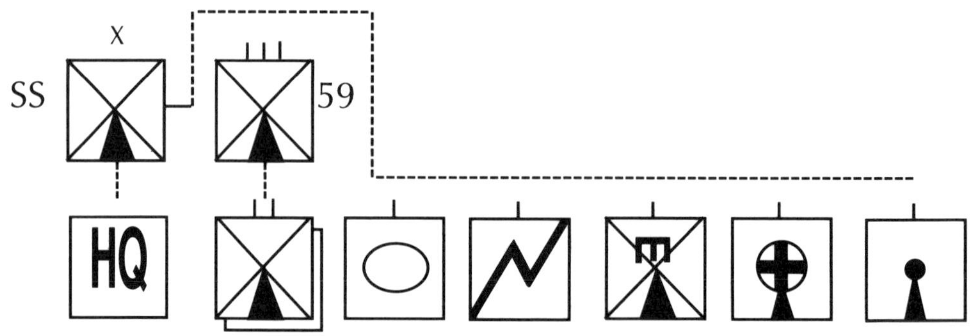

CHAPTER NINETEEN

What men dare, I dare;
Approach thou like
the rugged Russian bear,
The arm'd rhinoceros of
the Hyrcan tiger,
Take any shape but that,
and my firm nerves
Shall never tremble.

William Shakespeare
Macbeth

On 22 June 1944, the Russians launched their summer offensive, mocking the starting date of the German invasion of Russia exactly three years before. The objective of the offensive was nothing less than the complete encirclement of the German Army Group Center and its destruction. Events occurred quickly, and soon the German front was shattered at numerous points.

The German retreat across the expanse of Belorussia (White Russia) was nothing more than a semiorganized rout. The badly depleted German divisions had been too few and weak to halt the Soviet war machine. The withdrawal of the front-line divisions was precipitated by the retreat of the German rear-area formations in White Russia. This included not only garrison, antiaircraft, signal, repair, and other support troops, but also special antipartisan forces that had been engaged behind the German lines fighting the guerrillas.

These antipartisan forces were not only German police and SS-police units, but auxiliary Russian formations and even the better-organized Schützmannschaft (Schuma) Bataillonen, or "self-defense force." These Schuma battalions were made up of Russian volunteers and were organized along regular military lines for combat against partisans and on the front lines, and the officers were usually German nationals—although, in some cases, a Soviet NCO or, rarely, officer would be present.

The fighting capabilities of these Soviet volunteer formations depended on the motivation of the men in them. In cases where opportunists were in the majority, the units performed well, so long as the Germans suffered no major setback. If the battalion contained a majority of men with a true hatred of Stalinism, there was nothing their German officers could ask that they wouldn't willingly do. This particular aspect of the volunteers' nature made them unreliable when it came to fighting in a situation where the Germans were taking a beating. An attempt was made to sort out the troublemakers and to distribute the reliable locals to all of the units, but it could never guarantee the complete collusion of the unit in question. The volunteers often hid their true feelings behind rhetoric.

As a result, the volunteer formations were almost always kept behind the front lines, fighting partisans. The employment of single volunteer battalions fighting alongside German units at the front was common, but those battalions had gone through numerous tests before they were employed as combat formations. Time and experience had shown that unreliable units, when confronted with the mass of seasoned Soviet troops at the front, coupled with a German reversal, often led to disaster and collapse.

For this reason, in July 1944, these Russian volunteer forces were grouped together under one command, with the hope that—given ample training and a sufficient German officer and NCO cadre—they would perform better in a frontline situation. The massive losses which the Wehrmacht was taking in France and Russia in the summer of 1944 demanded the immediate call-up of all available reserves. The Germans needed their auxiliary forces to remain a viable combat force more than ever, and they believed that their immediate withdrawal to Poland for re-forming was essential. The stage was now set for the formation of a unique and obscure division of the Waffen-SS: the 30. Waffen-Grenadier-Division der-SS Weissruthenien.

Thus, numerous Russian volunteer units began making their way to Poland and gathering in the district area of Warsaw during July. Initially, the overall control of these formations was entrusted to the Higher SS and Police Command White Russia, whose headquarters had also been withdrawn westward to Lomza, Poland. The commanding officer chosen to lead the forming Russian volunteer brigade was SS Obersturmbannführer und Oberstleutnant der Schützpolizei Hans Siegling, who had gained the attention of his superiors while a captain. He had commanded Schützmannschaft Bataillon 57, and on 11 April 1944, he had received the much-coveted German Cross in Gold for continuous gallantry in combat.

During the first two weeks of July, the Russian battalions continued to gather together, arriving continually. The units were at varying levels of combat worthiness. For example, the Ukrainian Schützmannschaft Bataillon 61, which had belonged to SS-Police-Regiment 31, arrived on 19 July. This "battalion" only had two officers, two NCOs, and ninety-eight men—a far cry from being battalion strength. In contrast to the 61st, the 65th Schuma Battalion had four companies, one of which had ninety-eight Germans and 167 Russian volunteers.

Typical member of the 30. Waffen-Grenadier-Division der-SS. Notice the right collar tab showing the divisional emblem, a white double cross on a black background. Illustration: Eric Weisman.

Four rifle regiments, numbered 1st through 4th, began to take form. They were not redesignated until October 1944. In addition, numerous support battalions were also established. Waffen-Grenadier-Regiment der-SS Number 1 came from Schuma battalions 57, 60, 61, and Police Command Minsk. Waffen-Grenadier-Regiment der-SS Number 2 was formed from Schuma Battalion 62, 63, 64, and Police Commands Glebokie and Lida. Waffen-Grenadier-Regiment der-SS Number 3 used numerous Ordnung (Official) Police formations, plus Police District Commands Slusk, Baranowitschi, and Wileika. Waffen-Grenadier-Regiment der-SS Number 4 was formed from Schuma Battalion 101, Police District Command Slonim, and Police Leader Pripjet Sumpf (Pripet Marsh).

The nucleus for a future artillery regiment came in the form of Schützmannschaft-Artillerie-Abteilung 56. A reconnaissance battalion derived from the Schützmannschaft-Reiter (cavalry) Abteilung 68. Schuma Cavalry Battalion 68 had its origins in SS Sonderkommando Fieguth, an auxiliary-SS cavalry formation that operated against the partisans in the Pripet Marshes. In July, the 68th Battalion became the Kavallerie-Abteilung (30. SS-Division). This title did not change until November, when it was officially called "Aufklärungs-Abteilung" (Reconnaissance Battalion). The ancillary number 30 may have been given to the battalion-level support units of the formation. One reference states that indeed this ancillary number was au-

RUSSIAN INTERPRETERS IN THE 3RD BATTALION, 4TH REGIMENT, 12 AUGUST 1944

Ethnic Group	Languages
White Russians	
Waffen-Untersturmführer der-SS Leo Koserew[1]	Russian, Polish, German
Waffen-Oberscharführer der-SS Sergei Radiejewski[2]	Russian, Polish, German
Armenians	
Waffen-Scharführer der-SS Nikolai Orduchanov[1]	Russian, Armenian, German

[1] Belonged to 12th Company/3rd Battalion
[2] Belonged to 10th Company/3rd Battalion

thorized, but I have not found any evidence of this in the few remaining records of the Weissruthenien Division.

On 20 July, the day on which an attempt to kill Hitler failed, the forming brigade received its initial title, Schützmannschaft Brigade Siegling. Organization of Brigade Siegling continued through the rest of July. A Feldersatz-Bataillon (field-training battalion) was established for the brigade by merging Schuma Training Battalion Vretje and November together. Certain Schuma battalions were large enough to be redesignated as battalions in the Grenadier regiments of the brigade. Thus, Schuma Bataillon 65 became the 3rd Battalion/ Waffen-Grenadier-Regiment der-SS Nr. 3 (Russian Nr. 6). Schuma Bataillon 101 became 2nd Battalion/Waffen-Grenadier-Regiment der-SS Nr. 4 (Russian Nr. 7).

Initially, each grenadier regiment contained only two battalions. The only exception was Waffen-Grenadier-Regiment der-SS Nr. 4, which was able to form three rifle battalions. The ethnic makeup of the Siegling Brigade was quite interesting. Not only were there Belorussians, Russians, and Ukrainians in the brigade, but Armenians, Tatars, and even Poles. Documents show that there was a high concentration of Poles in the 1st and 2nd battalions Waffen-Grenadier-Regiment der-SS Number 2. Additionally, that regiment also had quite a number of Polish nationals in its PAK (Antitank) Company, and its Headquarter Company. It is quite probable that some of these Polish nationals may have been of German extraction, but documents from the unit clearly lists many of these volunteers as purely Polish. In at least one instance, a Czechoslovakian national served in the unit.

As can be imagined, this multilingual conglomeration of European peoples created a communication problem among the volunteers and for the German cadre personnel in the brigade. Therefore, translators were needed within the brigade. Ideally, each battalion and regiment would have at least six to ten translators, but the usual number was actually three to five. The 3rd Battalion/Regiment No. 4 for example, only had three.

The condition of the volunteer battalions varied greatly, not only as a result of motivation (or lack of it), but also because of the state and quantity of the equipment available to them. The Schuma battalions had been armed with a motley assortment of captured equipment, a practice continued in the Siegling Brigade that later figured prominently in the unit's poor performance. The battalions were equipped with French, Czech, Russian, and German weapons, the supply and maintenance of which must have presented a nightmare to the German officer in charge of maintaining the unit's equipment. Taken from the records of the 4th Regiment, the table

HEAVY WEAPONS IN THE 4TH REGIMENT 13 AUGUST 1944

	Russian-Made Machine Guns
1st Battalion	60
2nd Battalion	31
3rd Battalion	53
	French-Made Mortars
1st Battalion	1[1]
2nd Battalion	none
3rd Battalion	3[2]

[1] 6,000 rounds available
[2] 4,500 rounds available

SPECIALISTS IN THE 3RD BATTALION/4TH REGIMENT 1 AUGUST 1944

Antitank:	3	Engineers:	3[1]
Mortars:	18	Horse Riders:	8
Drivers:	0	Radiomen:	1

[1] By 15 August 1944, this number had risen to 16 trained engineers.

above lists the mortars and machine guns in the 4th's three battalions.

The brigade suffered from a lack of sufficiently trained specialists as well as a lack of proper equipment. Using the 3rd Battalion/4th Regiment as an example, we can see certain deficiencies, such as the lack of radiomen and drivers.

The decision to transfer the unit to France had already been made in July. Controversy still exists, among the few historians who have studied this unit, over the reason for its transfer. After having studied what few documents remain, I have concluded that the Germans believed that committing the unit on the Eastern Front would prove its undoing. They hoped that by committing the formation to the Western Front—far from any possibility of deserting and blending back into the Russian hinterland—the Russian volunteers would feel compelled to fight.

The Germans may have believed the transfer would cut down desertions, but the results proved that volunteers who had desertion on their minds would carry that thought out whether they were near their country or as far away as France. In addition, the volunteers who joined the Germans to fight Stalinism (but not necessarily democracy) found themselves thousands of miles from what they considered the real enemy, fighting French partisans, who, in their eyes, were merely men fighting to gain their independence. Though this thinking was rather simplistic and naive, it nevertheless proved to lower the morale of the more reliable elements in the formation.

In late July the brigade was transferred to Susz (Rosenberg) in East Prussia. The brigade had not completed training, but the approach of the Red Army forced its withdrawal. In fact, the formation actually spent most of its existence fighting the French maquis, while at the same time going through numerous stages of development and training.

Motorization in the brigade was almost nonexistent. The rifle companies had no motor transport at all. The battalion commanders were better off, with perhaps a light LKW truck and a staff car. The strength of a typical rifle company is shown above.

The nomenclature of rank in the Waffen-SS was devised to indicate quickly the racial status of an individual or formation. Purely German formations were prefixed with the SS initials, while Germanic or ethnic German units included the term "Freiwilligen" (volunteer) following the SS. Units made up almost entirely of non-Germans were prefixed by the title "Waffen" (armed). Thus, Waffen-Grenadier-Regiment der-SS (Armed Grenadier Regiment of the SS) signified that this unit was a non-Germanic infantry regiment in the service of the SS.

Individual rank titles were similarly

STRENGTH OF 11TH COMPANY/3RD BATTALION/4TH REGIMENT 2 AUGUST 1944				
Officers	NCOs	Men	Panjewagons (Russian wagons)	Horses
1	18	125	40	45

phrased. For example, the rank of SS-Hauptsturmführer was used for German officers holding the rank of captain; non-Germanic volunteers with the rank of captain had the title of Waffen-Hauptsturmführer der-SS. This nomenclature was not strictly for the purpose of differentiating between Aryans and non-Aryans; it was instituted mainly to instill in German commanders an appreciation of the fighting quality of the SS formation under their command. Non-Germanic SS units were believed to fight with less vigor than purely German units; as a matter of fact, the number of Germans in a division was often used as a measuring stick by which German commanders assessed the fighting quality of a formation. The table at the bottom of page 263 shows the rank breakdown of the 3rd Battalion/4th Regiment.

By 3 August, SS-Hauptsturmführer Deutschbein, the IA of the divisional staff, issued an order that converted Brigade Siegling into a full-fledged Waffen-SS division. The order also dealt with the proper nomenclature of Germanic and non-Germanic volunteers in the division.

As was common with Waffen-SS volunteer regiments, each regiment raised was numbered sequentially, beginning with "1," according to nationality. This designation was always carried in parenthesis after the regiment's title. For example, Waffen-Grenadier-Regiment der-SS Nr. 1 of the Siegling Brigade/Division carried in parenthesis its nationality and sequential number: (Russ. Nr. 4). Thus, the full title of this regiment was Waffen-Grenadier-Regiment der-SS Number 1 (Russ. Nr. 4).

You will note that three other Russian volunteer grenadier regiments had been formed, although these were not attached to the Siegling Division (soon to be the 30th Division on the Waffen-SS rolls). Those initial three regiments had been formed from the 3,300 man Kaminski Brigade. This brigade, made up of Russian volunteers from the central sector of Russia, had been formed from collaborationist forces in 1942. Also called the RONA Brigade (the initials of

OFFICERS, NCOS AND ENLISTED MEN IN THE 3RD BATTALION/4TH REGIMENT (BY RANK) 31 JULY 1944

Officers		**NCOs**	
SS-Hauptsturmführer	3	Waffen-Sturmscharführer der-SS	9
Waffen-Haupsturmführer der-SS	1	Waffen-Hauptscharführer der-SS	15
SS-Obersturmführer	5	Waffen-Oberscharführer der-SS	23
Waffen-Obersturmführer der-SS	1	Waffen-Scharführer der-SS 42	
SS-Untersturmführer	9	Waffen-Unterscharführer der-SS	4
Waffen-Untersturmführer der-SS	3	Waffen-Rottenführer der-SS	45
Enlisted Men			
Waffen-Sturmann der-SS	114	Waffen-Grenadier der-SS	319

the political party sponsoring it), this antipartisan brigade successfully fought Soviet partisans behind German lines and was supported by the people in the areas in which it operated. Once the German withdrawals from Russia began, however, it was forced to retreat to areas where the population was not supportive. At that point, the efficiency of the brigade began to drop, and some mutinies occurred. To stop these mutinies, Kaminski, the brigade's leader, flew to every regimental headquarters and personally shot the would-be mutineers.

This demoralized unit withdrew to Poland during the summer of 1944 to fight the Polish Home Army (PHA) in Warsaw. Plans were drawn up to turn this unit into the 29th division of the Waffen-SS, with the hope that by reforming the brigade and giving it German officers, better use could be made of it. These plans never reached actual employment, and the unit was a paper division within the Waffen-SS for only a very short time.

Its short-lived status resulted from its premature withdrawal from action against the PHA. The cause was not a lack of motivation but the reprehensible behavior of the men against the Poles. Individual and group lootings and rapings prevailed in the brigade. The German commander of the Warsaw operation lost control of the unit, mainly because Kaminski himself was guilty of the same outrageous behavior as the enlisted men.

After numerous complaints about the unit's behavior by German officers near the brigade's area of operation, the Waffen-SS High Command ordered the brigade to withdraw. Kaminski was tricked into leaving his men so German authorities could deal with him. He was shot, but his death was made to appear as the result of a car accident so his men would not turn against the Germans. The more reliable men in the unit were then sent to the dubiously faithful Russian Liberation Army (RLA), headed by the Russian turncoat, General Vlassov. The uncontrollable elements of the brigade were sent to German POW camps. Shortly after this, the Waffen-SS dropped the Kaminski Brigade from its rolls, and its divisional number, 29, went to the Italian SS Assault Brigade, which was in the process of forming into a full division.

The fact that the Waffen-SS carried the Kaminski Brigade as a division on its rolls, albeit on paper and for a mere month or so, later caused it major problems during the Nuremberg trials. All of the brigade's misdeeds and inhuman behavior were cited as evidence of the Waffen-SS' war crimes.

The fate of the second Russian SS division, the 30th (Siegling), was not as disastrous as the more unruly Kaminski Brigade, but it was less than successful. The problem of deserters within its ranks was endemic and plagued the division throughout its existence. A case in point is that of SS-Untersturmführer Wüst's 12th Company, from the 3rd Battalion/4th Regiment, from which five men had been reported missing on 4 August. Acting company commander Wüst received a phone call during the afternoon of 5 August from the commanding officer of the local Feldgendarmerie (Field Police), reporting that his men had detained five Russian volunteers who apparently belonged to the 30th.

At Field Police headquarters, Wüst learned that none of the men had any orders to be where they were found, still in the uniform of the Scützmannschaft. They had told the police that they belonged to the 30th Division, hoping to be let go. It was apparent to Wüst that they had attempted to escape. Six days later, on the night of 11 August, four other men deserted from the same company, and at the end of August, a much larger desertion was attempted—one that would lead to the mutiny of some of the division's battalions.

3RD BATTALION/4TH REGIMENT STRENGTH, 12 AUGUST 1944

	Officers	NCOs and Men	Panjewagons (Russian wagons) and Horses
German	8	109	270 wagons and 338 horses[1]
Russian	4	541	

[1] The 3rd Battalion had one Pkw, StKw.4 light vehicle.

The month of August was spent training and organizing. Early on, the Higher SS and Police Command gave up control of the Siegling Brigade/Division to the SS-FHA (Waffen-SS High Command). News of the division's impending transfer to France soon reached the ears of the Russian volunteers. Slowly, the resentment began to simmer just beneath the surface of most of the volunteers.

Individual desertions continued to haunt the division, though the divisional staff received more volunteers. The 10th Company of 3rd Battalion/4th Regiment had a strength of 141 officers, NCOs, and men, plus thirty panjewagons and thirty-five horses on 11 August. Nine days later, the Germans had increased the number to 182 men.

At the beginning of August, commanders decided to disband the troubled 4th Regiment. This was in keeping with the Waffen-SS (and German Heer) practice of having three infantry regiments in the rifle divisions. Further, although volunteers were still forthcoming, the number of deserters continued to increase.

On 12 August, the 3rd Battalion had moved to Viereck so the Headquarter's Company could have better control of the battalion. It was apparent that its officers had lost control of the situation. The 3rd Battalion was still a full-strength unit even though the regiment was scheduled to be split up among the rest of the division. German command felt that tighter control of the battalions would decrease or stop the desertion rate altogether. On the night of 12 August, Waffen-Unterkorporal Paul Dubak of 9th Company, 3rd Battalion/4th Regiment deserted from Viereck.

Certainly the most troubled rifle company in the 4th Regiment's 3rd Battalion was the 12th. On 14 August, it had a strength of 92 German and 167 Russian men. Six days later, that number had decreased to 117. This fifty-man difference was not the result of transfers to other regiments in the division, since the battalions of the 4th Regiment were assigned in early September to the other regiments.

On 18 August, the SS-FHA redesignated the Siegling Brigade/Division the 30. Waffen-Grenadier-Division der-SS (Russ. Number 2), backdating the order to 1 August. Two days later the division's staff headquarters issued a divisional order indicating the imminent transfer of 4th Regiment's battalions to the rest of the division, which did not take place until 9 September.

On 27 August, two mutinies erupted within the division, taking the divisional staff by surprise. The divisional staff had expected something to happen, but they expected it to occur in the 3rd or 4th Regiment. Instead, the Russian and Polish personnel of 2nd Battalion/1st Regiment and 1st Battalion/2nd Regiment rebelled, killing as many of their German cadre as they could find.

The division ruthlessly put down the revolt, killing all of the ringleaders, and disbanded 2nd Battalion/1st Regiment. The division was to begin the immediate transfer of individual battalions from regiment to regiment in an attempt to maintain the regiments up to strength. The personnel who mutinied were removed

3RD BATTALION/4TH REGIMENT STRENGTH 4 SEPT 1944					
German Officers	4	2	0	2	2
Russian Officers	1	0	0	0	0
German NCOs	13	–	–	–	–
German and Russian NCOs and Men	12 R.	101	150	132	112
Horses	44	29	31	20	65
Wagons	24	24	22	21	16
Trucks	1	0	0	0	0
	Btln. Staff	9th Co.	10th Co.	11th Co.	12th Co.

from the division, including other men from the rest of the division who had proved troublesome. In this way, 2,300 of the more unreliable elements were sent to Transport Command Karlsruhe. They were formed into two construction regiments and employed building up the defenses of the Siegfried Line.

The transfer of the 30th Division to the Western Front had begun in the last week of August, the division's units moving individually by rail. The first advance units, which were to organize billeting and other support duties, arrived in France on 17 August. Most of the division would be moved by rail following the route: Susz (Rosenberg)/Ilawa (Deutsch Eylau)/Kowalevo (Schonsee)/Torun (Thorñ/Inowroclaw/Frankfurt-am-Oder/Frankfurt-am-Main/Mannheim/ Strasbourg/Colmar/Mulhouse/Altkirch/Montbéiard/Besançon and St. Hippolyte. The divisional headquarters was located in Mulhouse.

At least one battalion in the division, the 3rd Battalion/4th Regiment, made a roundabout transfer to France. Leaving by train on 4 September, it found itself near Dachau (north of Munich) by the second week of September. A train roster for this battalion on 4 September listed its strength.

On 12 September, just sixteen days after the mutiny, the division received a battalion of rehabilitated Russian personnel from the 2,300 men who had been sent to Karlsruhe as punishment. It seems unlikely that all of those men sent back under the command of Waffen-Sturmbannführer der-SS Murawjew were rehabilitated. Likely, they had a choice between hard labor in the construction battalions versus duty in the division. This battalion was used to form the 2nd Battalion/1st Regiment. You will recall that the lst Regiment's 2nd Battalion had been disbanded. Thus, to keep that regiment at a strength of two battalions, the

3RD BATTALION/4TH REGIMENT STRENGTH

	20 August	9 September	20 August	20 August
3rd Battalion Headquarters[1]	15	17	27	16
9th Company	148	101	25	41
10th Company	182	150	21	33
11th Company	145	132	25	33
12th Company	117	112	46	74
	Officers, NCOs & Men	Officers, NCOs & Men	Wagons	Horses

[1] The Headquarters Section contained two light PKW vehicles.

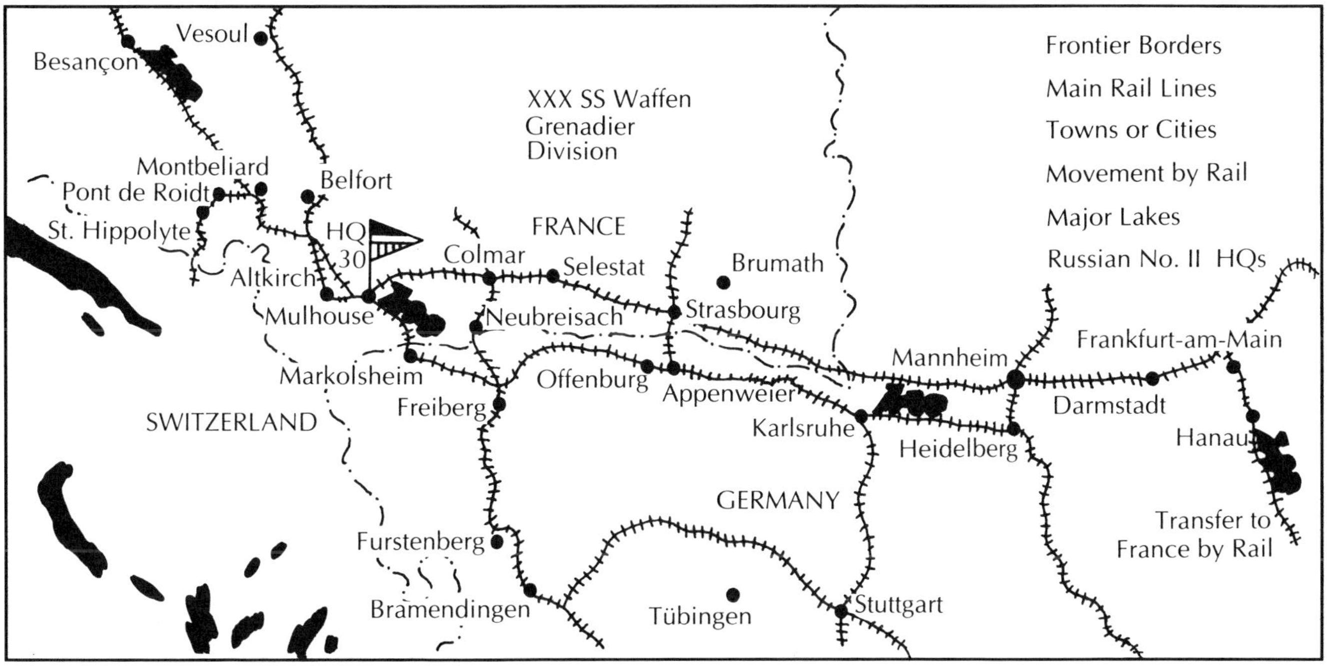

Transfer to France where forming, training, and security duty continued.

Frontier Borders

Main Rail Lines

Towns or Cities

Movement by Rail

Major Lakes

30. Waffen-Grenadier-Division der-SS
Russian No. II HQs

returning Russians were ordered grouped in whole as its new 2nd Battalion.

Throughout its existence, this battalion would still be referred to as Battalion Murawjew. Eventually, this battalion would be detached from the lst Regiment and would become the 1st Battalion of Waffen-Grenadier-Regiment der-SS Nr. 77. By mid-September, the 1st Regiment once again could count on two battalions. A third battalion was added to the regiment by removing the 2nd Battalion/2nd Regiment and turning it into the 3rd Battalion/Regiment Nr. 1.

Since the 2nd Regiment had only the unreliable 1st Battalion to work with, an effort was made to bring the regiment back up to strength by turning the disbanded 4th Regiment's 3rd Battalion into a new 2nd Battalion/2nd Regiment. In addition, the independent Russian Infantry Battalion Nr. 654 was now added to the 2nd Regiment, becoming its 3rd Battalion. As in the case of Battalion Murawjew, this battalion would still be referred in the divisional documents as Battalion 654, instead of 3rd Battalion/2nd Regiment. Eventually, the 654th Battalion would again be transferred and used to form the 2nd Battalion/Waffen-Grenadier-Regiment der-SS Nr. 77.

By mid-September the 4th Regiment had already been disbanded. The 3rd Regiment was in the process of disbanding. Three Waffen-SS regimental numbers had been set aside for the 30th Division—75, 76, and 77. These numerical listings would not be given to the divi-

Divisional insignia of the 30th Waffen-Granadier-Division der-SS. Illustration: Antonio Munoz.

30. WAFFEN-GRENADIER-DIVISION DER-SS
DIVISIONAL ORDER OF BATTLE
SEPTEMBER 1944

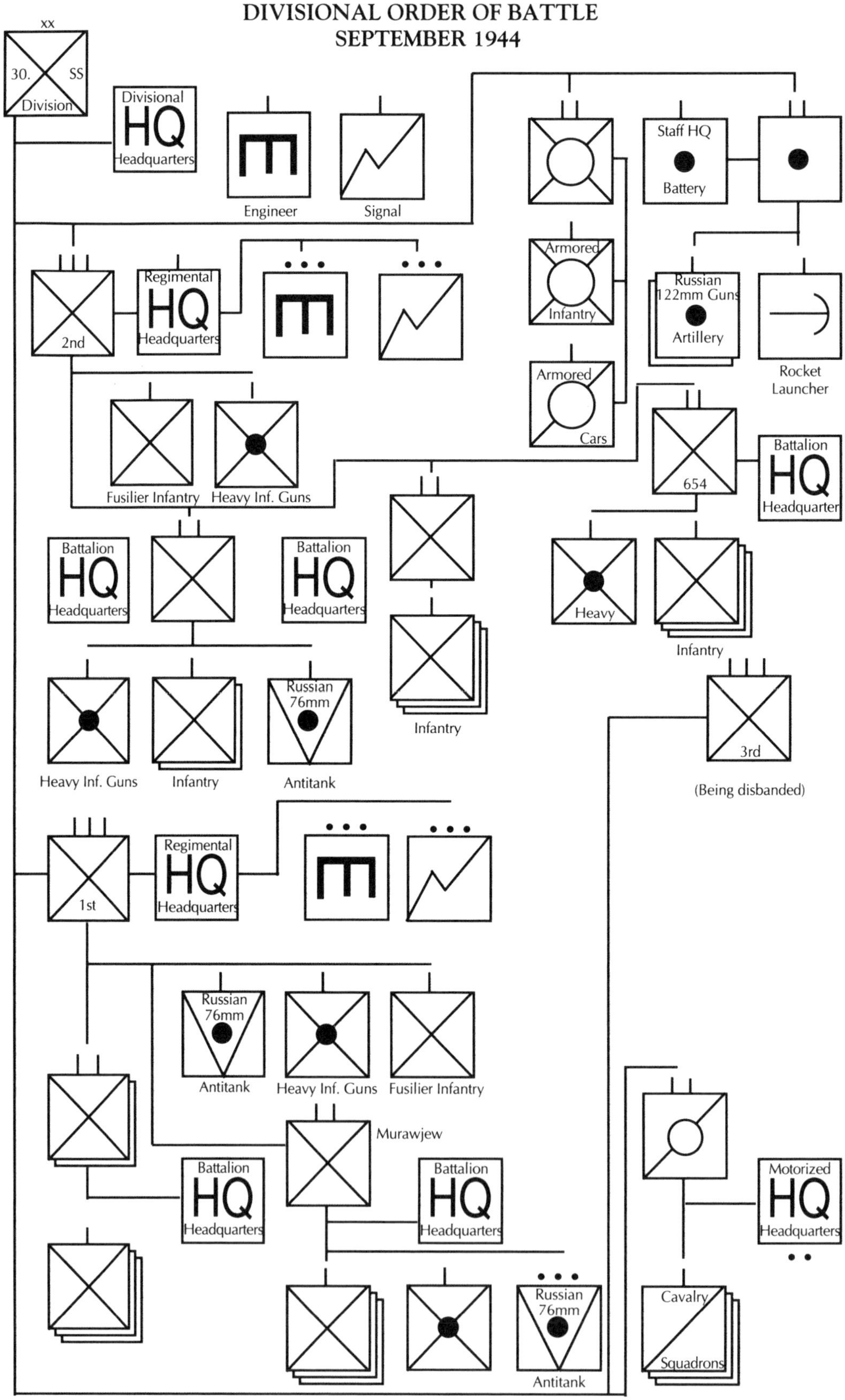

1.	Divisional Staff	Markolsheim
2.	Versorgungs Regiment (Supply)	Markolsheim
2a.	Gruppe Kraftwagon Kolone (Truck Column)	Markolsheim
2b.	Verwaltungs Kompanie (Administrative Company)	Markolsheim
2c.	Krankenkraftwagon Zug (Truck Repair Platoon)	Markolsheim
2d.	Nachrichten Kompanie (Signals Company)	Markolsheim
2e.	Feldgendarme Trupp (Field Police Company)	Markolsheim
3.	Waffen-Grenadier-Regiment der-SS Nr. 1 Staff	Mussig
4.	Waffen-Grenadier-Regiment der-SS Nr. 2 Staff	Saasenheim
5.	Panzergrenadier-Bataillon (Armored Infantry)	Elsenheim
6.	Aufklärungs-Abteilung (Reconnaissance)	Mackenheim
7.	Artillerie-Abteilung	Hessenheim
8.	Russische Bataillon Nr. 654	Bootzheim
9.	Feldersatz-Bataillon und Deutsche-Ausbildungs-Abteilung (Russian und German Training Battalions)	Pionier-Übungsplatz Neubreissach
10.	Bataillon Murawjew	Wasserübungsplatz Streitkopf
11.	Pionier-Kompanie (Engineer Company)	Heidelsheim

sion's regiments until 19 October. With two regiments up to three battalions in strength (the 1st and 2nd), the division now had the problem of raising a third infantry regiment. This problem was solved by mid-October, when Waffen-Grenadier-Regiment der-SS Nr. 77 was formed from Battalion Murawjew and Battalion 654. In the meantime, however, the division would have to carry out its antipartisan duties in the Mulhouse-Besancon-Dijon area with only two operational regiments.

On 16 September, the 3rd Battalion of the disbanded 4th Regiment left Dachau by train for France. At this time, it had 7 German and Russian officers, and 100 German and 280 Russian NCOs and men. The battalion had lost a considerable number of personnel since it had left for the west twelve days earlier. Most of them had been transferred to Russian Construction Regiment's 1st and 2nd battalions. They boarded the train about 5 P.M. without incident, but the train did not leave until 2:30 A.M. on the following day.

Their first destination was Augeburg, then Ulm, Esslingen, and Stuttgart. By 1:05 P.M. on the 17 September, they had arrived at Stuttgart, where they were allowed to detrain briefly. They were treated to coffee and donuts by the local N.S.V. organization on the train platform. By 1:40 P.M. they were on their way again, passing Eruchsal, Germershein, and Landau. On 18 September at 2:00 A.M., they passed through Strassbourg on their way to Neubreissach. At 6:30 A.M. they finally arrived at Neubreissach and immediately unloaded and

Illustration: Eric Weisman.

dispersed, fearing an Allied air attack.

By 28 September, the 30th Division had already established itself. The divisional order of battle for that day (shown at the top of page 269) listed the units and the towns where they were located.

CHAPTER TWENTY

The fatal futility of Fact.

Henry James
The Spoils of Pyonton

The activities of the 30th Division in the months of August and September had been completely taken up with the technical and logistical problems of its transfer from Poland and East Prussia to the west. Beginning from the end of August, the elements of the division that had already arrived in Alsace, France, were hit and constantly harassed by the French maquis.

Small clashes abounded: reconnaissance platoons approaching farmhouses were showered with submachine-gun fire, supply trucks were ambushed, and Russian guards were found dead at their posts in the morning. These clashes continued throughout the month of September with ever-increasing scale and boldness. The Germans had hoped that by pitting the Russian troops against the resistance fighters, the division could safely pick up some battle experience. In addition, the fact that the division needed time to train meant it couldn't be placed in the front lines, where it was sure to be destroyed before it had a chance to organize.

The French resistance command had not failed to notice the low morale and poor state of equipment of the division. Armed with this information from their sources in the garrison towns, the French began to attack them with much more vigor and presence. During the month of September, they hit 30th Division with everything they had. The relatively "quiet occupation duty" behind the lines they had been led to expect was not to be. The Russian volunteers and their German officers received no rest. The French attacks were not completely disastrous, but they did sap enough strength to further demoralize the unit. Adding to the division's problems was the fact that, because of low morale and the number of desertions, the division had not been properly trained before arriving in France, and the training schedule continued to be interrupted, further delaying the division's eventual deployment as a combat-ready formation.

Not all of the Russians fought half-heartedly. While some companies staggered under resistance assaults and fell apart, others fought the French irregulars to a standstill. Hard defensive battles were fought in and around towns whose French names were unpronounceable to the Russian volunteers. Venarey, Armançon, and Auxonne (east of Dijon) became part of the division's history, which, until those battles in September, had been rather lamentable. The German officers and the Russians who still had faith in the Germans welcomed the few successes. They came at a high cost in lives, however. High casualties and desertions in the midst of the fighting and the training took their toll.

By the beginning of October, the artillery complement in the division was still not past the battalion level. Its single battalion had only two batteries of captured Soviet 122mm guns and one battery of German, towed, six-barreled, rocket-launcher guns. The division's reconnaissance battalion had three cavalry squadrons (equivalent to a company each), plus a headquarters company. The headquarters company had bicycles. The division's "secret weapon" was its Panzergrenadier battalion, which was in the process of organizing. Formed in the beginning of September, it contained mostly Germans, with some Russian volunteers who were highly motivated and unswervingly faithful to the Germans. The unit only had a motorized infantry company outfitted with an assortment of French, Soviet, and German vehicles. The other company-sized formation in this battalion was an armored car company made up of captured Soviet cars armed with either a 76mm cannon or 7.62mm Kalashnikov machine gun.

LOCATION OF 30TH SS-DIVISION SUBUNITS, EARLY OCTOBER 1944

4 October 1944	**Location**
Waffen-Grenadier-Regiment der-SS Nr. 1	Mussig
Waffen-Grenadier-Regiment der-SS Nr. 2 Staff	Saasenheim
1st Battalion/Waffen-Grenadier-Regiment der-SS Nr. 2	Saasenheim
5 October 1944	
2nd Battalion/Waffen-Grenadier-Regiment der-SS Nr. 2	Sundhausen
3rd Battalion/Waffen-Grenadier-Regiment der-SS Nr. 2	Wittisheim
6 October 1944	
Artillerie-Abteilung	Hessenheim
Pionier-Kompanie	Heidolsheim
30. Waffen-Grenadier-Division der-SS Staff	Markolsheim
7 October 1944	
Panzergrenadier-Bataillon	Enselheim
Russische-Bataillon 654	Fackenheim
Aufklärungs-Abteilung	Fackenheim
8 October 1944	
Bataillon Murawjew	Pionier-Übungsplatz Neubreissach
Ausbildungs-Bataillon 30	Pionier-Übungsplatz Neubreissach
Deutsche-Ausbildungs-Bataillon	Pionier-Übungsplatz Neubreissach

On 2 October, the Feldersatz-Bataillon, responsible for supplying replacements for the division, was renamed once again. Containing a staff company and four rifle companies, it became known as the Ausbildungs-Bataillon 30. The division's location changed little between 4 through 8 October, but the shuffling of some units from one town to another continued well into the middle of October. A divisional roster for the same period (above) showed the disposition of the division's units.

On 15 August, the Allies landed in southern France. After a short but stiff defense, the German 1st and 19th armies began their withdrawal from the Mediterranean coast, the Bay of Biscay, and the French-Spanish border. The Battle of the Falaise Pocket reached its zenith, wiping out the flower of the German Panzer army in France. The long retreat to the German frontier began. The U.S. 6th Army Group, including the 1st French Army, was fast approaching Alsace. The Allied plan was to link the advance of the 1st French Army from the south with General George S. Patton's 3rd U.S. Army from the west.

By mid-September, the French 1st Army was threatening Dijon, the southernmost city of 30th Division's area of operations. As a result, the division was ordered to withdraw to the northeast. During the retreat, they fought stiff, rearguard actions against the French resistance between Belfort and Mulhouse. By the beginning of the third week in September, the French 2nd Armored Division of General Le Clerc had captured the town of Epinal. By late September, the division relocated to the north of Mulhouse, in the Colmar area between the Rhine River and the Vosges Mountains.

October was hectic for the divisional staff. The rapid advance of the Allied armies meant that they did not have much time left to prepare the division for eventual employment in the front lines.

Located in the Colmar area, division commanders thought that, at last, they were free to train without fear of partisan attack, but another problem arose to in-

30. WAFFEN-GRENADIER-DIVISION-DER-SS AREA OF OPERATIONS

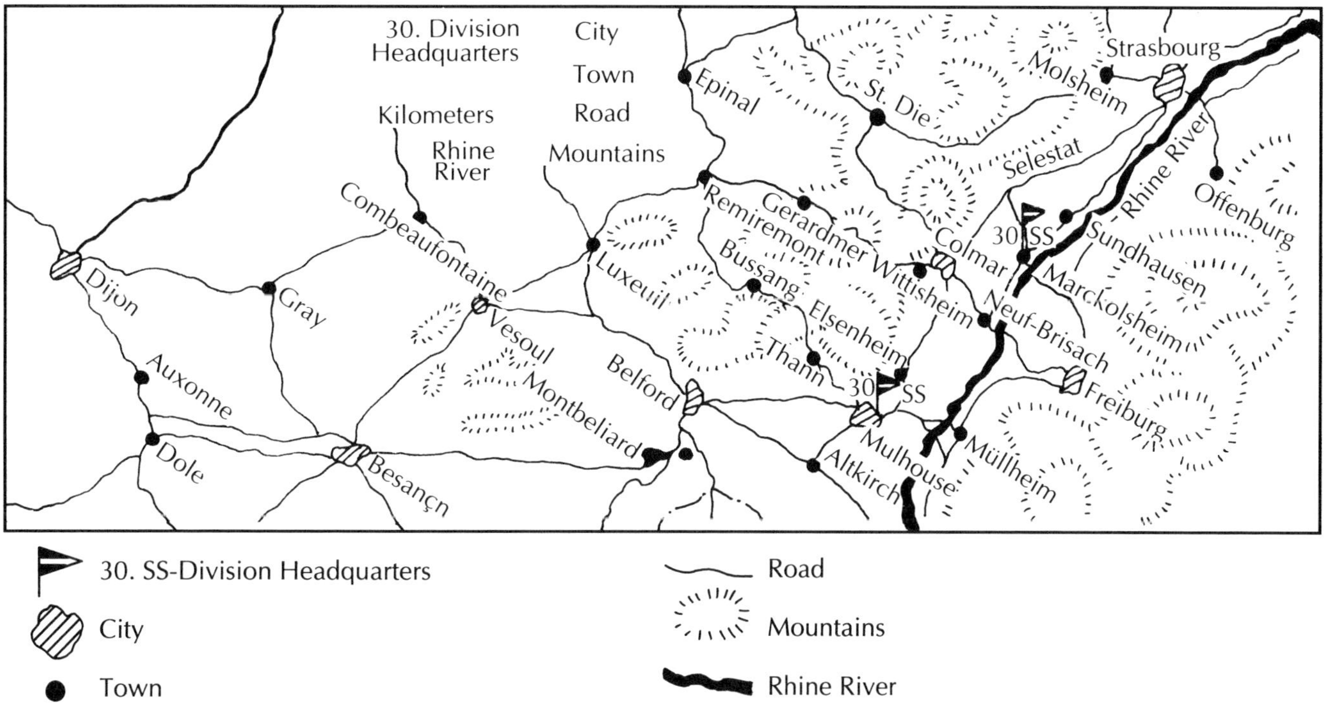

terrupt the training schedule. The German command was desperate for reinforcements for the depleted German armies withdrawing from France. The commander of the 30th Division, SS-Obersturmbannführer Hans Siegling, received word that he was to implement standing orders for the possible employment of his division to the front lines—this despite the fact that his men were not ready for such a mission.

Thus, the corps headquarters, which had control of Siegling's division, continuously conducted alerts designed to prepare the battle-worthy units in his division for quick deployment, but they only disrupted what little time was left for the division to prepare itself. Time was running out.

The training schedule on page 274 for the 3rd Battalion/Waffen-Grenadier-Regiment der-SS Number 2 gives us a good picture of those hectic days in October 1944.

As can be seen from the training schedule, the division had more and more alarm calls as the month of October progressed. Eventually, the division assumed bridge-security duty on the Rhine in anticipation of withdrawal by the German forces still on the western half of the river. During October, the division began to quickly take form, though the divisional staff knew that more time was needed to make the unit combat ready. A divisional roster for the period 9 through 14 October (page 275) shows the units' locations.

An order dated 18 October stated that the number of battalions in each regiment would be increased from two to three; however, this was easier said than done. The divisional staff had already decided to use the 654th Russian Battalion and Battalion Murawjew to form Waffen-Grenadier Regiment der-SS Nr. 77. This meant that those battalions had to be withdrawn from the other two regiments and replaced. Thus, Russian Battalion 654, which had become the 3rd Battalion/Waffen-Grenadier Regiment

TYPICAL SCHEDULE OF THE 3RD BATTALION WAFFEN-GRENADIER-REGIMENT DER-SS NR. 2

5 October 1944 Forming.

6 October 1944 Forming. Officers and NCOs meeting with foreign personnel officers and NCOs of the 1st and 2nd battalions in the regiment.

7 October 1944 Forming. Meeting with battalion and company leaders of the lst Regiment. Movie.

8 October 1944 Forming. 9th and 12th companies detached to help look for an Allied pilot shot down near Hilsenheim by the Flakinstandsetzungs-Kompanie. Wounded pilot found and given to the Wehrmacht.

9 October 1944 Forming. Promotion meeting of battalion and regimental officers. Sports exercises among the companies.

10 October 1944 Forming.

11 October 1944 Forming. March exercise by the 11th Company to Wittisheim, where it was quartered. The commanding officer of 9th Company, SS-Obersturmführer Völkel was transferred to the staff company of the regiment. Ninth Company taken over by SS-Untersturmführer Randuhn. SS-Untersturmführer Hesse gave a list of his new platoon leaders in 10th Company.

12 October 1944 Forming.

13 October 1944 Leaders Class in Saasenheim given by the regimental commander. Part of the battalion given the day off.

14 October 1944 No forming. Day off for the men.

15 October 1944 Forming. Divisional orchestra gave a performance in Markplatz and Wittisheim in the evening. The best group to sing was the 12th Company.

16 October 1944 Forming. Speech by regimental commander at Saasenheim and at Sundhausen.

17 October 1944 Regimental order requiring the battalion to force march, bivouac. Reconnaissance done by the division's cavalry squadron. Assault practice against a bunker. Visit by the divisional commander.

18 October 1944 Alarm practice, forming suspended for today. Leaders School in Saasenheim. German nationals given classes on how to deal with foreign personnel.

19 October 1944 Forming. Film at night, "The White Tram." Twenty-four hour notice given warning of impending alarm call.

20 October 1944 Forming cancelled. Battalion prepared. Alarm sounded at 20:00 hours. Battalion prepared to move in the direction of Wittisheim.

21 October 1944 Forming.

22 October 1944 Alarm practice, training canceled today.

25 October 1944 Divisional order changes the regiment's title to Waffen-Grenadier-Regiment der-SS Number 76 (russ. Nr. 5).

26 October 1944 Forming. Twenty-four-hour notice giving warning of impending alarm call.

27 October 1944 Forming canceled. Ninth Company marched through Sundhausen in the direction of Strausberg. Battalion moved to Sassbach on bridge-security duty. Enemy breakthrough thought possible. Regiment marches at 8:45 A.M. Moved through Schwobsheim-Hessenheim-over the canal-Artolsheim highway-Macrkolsheim and the Sasbach highway. The march was done by 1st and 2nd battalions of the regiment, plus the regimental staff. Our battalion [3rd Battalion] went separately through Bootzheim and the Rhine dam. Battalion performed bridge security duty on the Rhine River. Our battalion marched fifty-two kilometers today.

LOCATION OF 30TH SS-DIVISION SUBUNITS, MID-OCTOBER 1944

Unit	Location
9 October 1944	
Ausbildungs-Bataillon 30	Pionier-Ubungsplatz
Neubreissach	
Bataillon Murawjew	Kuenheim
10 October 1944	
Aufklärungs-Abteilung	Mackenheim
Russische-Bataillon 654	Mackenheim
Waffen-Grenadier-Regiment der-SS Nr. 2 Staff	Saasenheim
I/Waffen-Grenadier-Regiment der-SS Nr. 2	Schönau
11 October 1944	
II/Waffen-Grenadier-Regiment der-SS Nr. 2	Sundhausen
III/Waffen-Grenadier-Regiment der-SS Nr. 2	Wittisheim
12 October 1944	
Waffen-Grenadier-Regiment der-SS Nr. 1 Staff	Musig
I/Waffen-Grenadier-Regiment der-SS Nr. 1	Musig
13 October 1944	
II/Waffen-Grenadier-Regiment der-SS Nr. 1	Baldenheim
14 October 1944	
Artillerie-Abteilung Staff	Schwobsheim
1 Batterie/Artillerie-Abteilung	Schwobsheim
2 Batterie/Artillerie-Abteilung	Hessenheim
Pionier-Kompanie	Heidolsheim
Panzergrenadier-Bataillon with one Panzer platoon and the motor-school class	Elsenheim

der-SS Number 2, was detached and replaced with the old 1st Battalion/Waffen-Grenadier-Regiment der-SS Nr. 3.

Bataillon Murawjew was also detached from Waffen-Grenadier-Regiment der-SS Nr. 1 (it had been serving as its 2nd Battalion), leaving the 1st Regiment with only two battalions, the 1st and 3rd. A third battalion was ordered formed to become the regiment's new 2nd Battalion, but the regiment could not form until 1945. Thus for the rest of 1944, the 1st Regiment (soon redesignated Regiment Nr. 75) operated with only two battalions, and on 18 October, the 3rd Regiment finally disbanded. The divisional support units merged into a regimental supply section, SS-Versorgungs-Regiment Number 30.

As stated earlier, Bataillon Murawjew and the 654th Battalion were used to form the 77th Regiment. An authorized third battalion was never formed. A regimental order of battle for 26 October at the top of page 276 illustrates the situation.

An order from the SS-FHA (dated 19 October but received by the regiments on 25 October) renumbered the regiments from 1, 2, and 3 to 75, 76, and 77. On the 27 October, the division was activated for

Waffen-Grenadier-Regiment der-SS Nr. 75
1st Battalion/Waffen-Grenadier-Regiment
3rd Battalion/Waffen-Grenadier-Regiment

Waffen-Grenadier-Regiment der-SS Nr. 76
1st Battalion/Waffen-Grenadier-Regiment
2nd Battalion/Waffen-Grenadier-Regiment
3rd Battalion/Waffen-Grenadier-Regiment[1]

Waffen-Grenadier-Regiment der-SS Nr. 77
1st Battalion/Waffen-Grenadier-Regiment[2]
2nd Battalion/Waffen-Grenadier-Regiment[3]

[1] The old 1st Battalion/Waffen-Grenadier-Regiment der-SS Number 3
[2] Bataillon Murawjew
[3] Bataillon 654

security duty on the Rhine River. The 75th Regiment moved through Heidelsheim-Elsenheim to Grussenheim, where it was deployed for combat.

The regiment's 2nd Battalion was the first to reach the area and was ordered to set up a defensive position covering the bridge near the town. The 75th Regiment was then ordered to cover the defense areas from Kuehnheim to Marckolsheim. The division's Reconnaissance Battalion made a sortie west of both towns in the direction of the enemy positions. The division faced elements of the French 1st Army. The entire German 19th Army was trapped west of the Rhine River in the Colmar pocket, and the mission of the 30th Division was to allow as many German units as possible to cross the Rhine, while preventing the French and U.S. forces from capturing any intact bridges.

The 76th Regiment, minus its 3rd Battalion, moved to the area around Artolsheim-Bootzheim-Marckolsheim-Artenheim and prepared for an attack. They fought near the towns of Grussenheim (near the area of operations of the 75th Regiment) and Jabsheim-Muntsenheim. The division thought that it was facing regular ground combat troops but soon learned that it was opposing elements of an Allied glider unit. This unit had apparently landed near the Rhine Canal on the night of 26 October. Its mission was to capture or destroy the four bridges over the Rhine Canal between Marckolsheim and Kuehnheim. If not for the rapid deployment of the 30th Division, it might have succeeded.

The 75th and 76th regiments took so many losses that the then-forming 77th Regiment was disbanded (just a week and a half after it was ordered formed) and used to reinforce them. The 1st and 2nd battalions of the 76th Regiment had suffered the most casualties and were disbanded on 2 November, five days after being committed. The 2nd Battalion/Waffen-Grenadier-Regiment der-SS Nr. 77 (the old 654th Battalion) became the new 1st Battalion of the 76th Regiment. The 1st Battalion, 77th Regiment (the old Murawjew Battalion) became the 2nd Battalion of the 76th Regiment. Thus ended the very brief history of the 77th Regiment. It had been formed and disbanded in less than two weeks, without having fired a shot.

On 10 November, the Ausbildungs-Bataillon 30 was redesignated as the SS-Feldersatz-Bataillon 30. Previously, it had contained a staff company, plus four rifle companies. This reorganization added a 5th Rifle Company. That same day, the Kavallerie (cavalry) Abteilung was officially redesignated Aufklärungs-Abteilung 30.

Events in November sealed the fate of the 30th Division. Despite its demoralized state and lack of sufficient armaments and time to form, the division was finally committed to battle against the forward elements of the 1st French Army. The division fought the vanguard of the French troops, the Vosges and the Rhine units, making a desperate fighting withdrawal through Altkirch, Thann, Gebweiler, and Sélestat. In a particularly nasty encounter in the middle of November, it suffered greatly at the hands of lavishly equipped troops of the 1st French Armored Division. Lacking proper antitank equipment, the division attempted to halt the mechanized columns, and for two days, it held up the 1st French Armored Division. But, despite the individual acts of bravery as told by the survivors, the French equipment was too much for the division. One of the most incredible stories of bravery involved the Russian volunteers who threw themselves in the path of the French tanks and leaped onto the chassis, armed only with hand grenades.

After these November battles, the division withdrew behind the Rhine River. It had been badly mauled and was in no condition to offer any type of defense; so it was ordered to the German-Swiss border for frontier watch until January, when it was sent to Oberpfalz for re-forming. There was no longer any question of keeping the unit at division level. The division was to be converted to a brigade, as evidenced by an order dated 1 January, although the remnants of the division did not get the order until 11 January. Once again, the unit was combed for unreliable elements, and about 5,000 Russians were transferred to the Vlassov Army. They were later used as part of the Russian Volunteer 600th Panzergrenadier Division.

On 27 February 1945, the unit was redesignated Waffen-Grenadier-Brigade der-SS Weissruthenien, with only one active regiment, the 75th. The recon battalion was reduced to one cavalry squadron, Aufklärungs-Reiter-Schwadron/Waffen-Grenadier-Brigade der-SS Weissruthenien SS-Feldersatz-Bataillon 30 became Feldersatz Bataillon/Waffen-Grenadier-Brigade der-SS Weissruthenien. The rest of the brigade was made up of the artillery and antitank battalions, the latter having been formed by confiscating all weapons in the unit capable of stopping enemy armor. This included not only the few bazookas in the division but also some Russian-made 76mm antitank guns.

In late February, the brigade

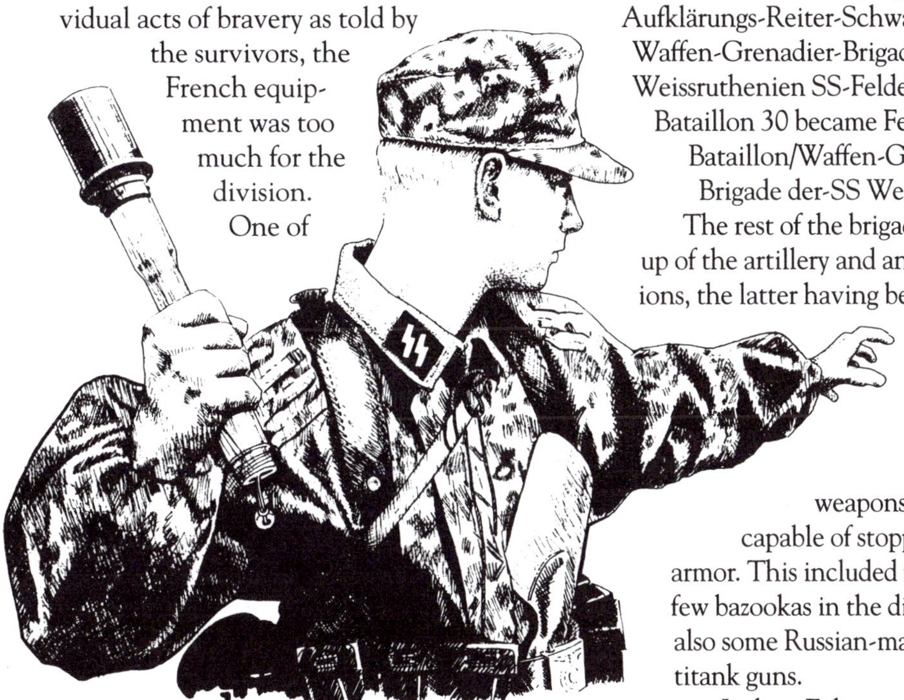

Illustration: Eric Weisman.

Date	Unit	Area of Operations
1 August 1944	Higher SS and Police Leader Beylorussia	Susz (East Prussia)
17 August 1944	64 Armeekorps, H.G. "G"	Mulhouse-Dijon-Besancon
6 September 1944	Armeekorps, H.G. "G"	Southeast of Dijon area
16 September 1944	58 Panzerkorps, H.G. "G"	Venarey-Armançon
28 September 1944	Special Employment, "G"	Auxonne
13 October 1944	58 Panzerkorps, H.G. "G"	Ognon-Belfort
21 November 1944	53 Armeekorps, 19 Armee	Guémar-Ostheim
26 November 1944	63 Armeekorps, 19 Armee	Neu Breissach-Markolsheim
4 December 1944	63 Armeekorps, 19 Armee	Northwest of Mulhausen
31 December 1944	Armee Gruppe Oberrhein	German-Swiss frontier
1 January 1945	Replacement Army	Oberpfalz
? February 1945	Replacement Army	Grafenwöhr
1 March 1945	Replacement Army	Grafenwöhr
9 March 1945	Replacement Army	Grafenwöhr
28 April 1945	Replacement Army	Grafenwöhr

Date	Notes
1 August 1944	Forming into a brigade. Some units still in Poland.
17 August 1944	Forming, training, security duty.
6 September 1944	Training, forming, antipartisan operations.
16 September 1944	Forming, training, antipartisan operations.
28 September 1944	Withdrawing, rear-area antipartisan operation.
13 October 1944	Forming. Division reported in a mutinous mood by Army.
21 November 1944	Forming, reorganizing.
26 November 1944	Fighting withdrawal.
4 December 1944	Fighting withdrawal.
31 December 1944	Frontier guard duty on the German-Swiss border.
1 January 1945	Disbanding. Forming into a brigade.
? February 1945	Forming into a brigade.
1 March 1945	Forming into brigade.
9 March 1945	Redesignation to a division.
? April 1945	Surrender to Allied troops at Grafenwöhr.

was moved to Grafenwöhr Training Camp with orders to disband. As a result of a critical shortage of equipment, the German Command had decided to do away with its least-active foreign units and to use the equipment to form other units. The German cadre in the brigade were used to form the 25. Waffen-Grenadier-Division der-SS Hunyadi and the 38th SS

WAFFEN-GRENADIER-BRIGADE DER-SS WEISSRUTHENIEN ORDER OF BATTLE 1 MARCH 1945

Field Post Number	Formation
14 471	Brigade Staff
01 069	Waffen-Grenadier-Regiment der-SS Nr. 75 Staff Company
64 868	1st Battalion/Waffen-Grenadier-Regiment der-SS Nr. 75
67 283	2nd Battalion/Waffen-Grenadier-Regiment der-SS Nr. 75
66 415	3rd Battalion/ Waffen-Grenadier-Regiment der-SS Nr.75
05 607	Artillerie-Abteilung (artillery battalion)
67 002	Panzerjäger-Abteilung (antitank battalion)
31 251	Reiter-Schwadron (cavalry squadron)

Panzergrenadier-Division Nibelungen. The Nibelungen was the last SS division formed before the war ended.

On 9 March the brigade was redesignated a division, though it isn't clear the intent of the German command. Apparently, nothing came of the redesignation. Remnants of the brigade were still at Grafenwöhr when the Allies entered the area. The end of the Weissruthenien had finally come.

The active combat life of the 30th Division had been just a little over a month, not including its actions against the French partisans. Suffering from poor morale, as well as a lack of arms and proper training, it was eventually sent to the front lines. The outcome of such a decision had already been foreseen by its German staff.

In frontline action, the unit quickly fell apart. This move was prompted by the rapid Allied advance and the dwindling German reserves. Theorizing that if adequate supplies and equipment had been allotted and enough time had been given to the division to organize and train, we can speculate that 30th Division might have stood a chance of surviving for several months. Crippled with the deficiencies that it had, it is surprising that the division held its own for an entire month against superior armored forces. It was a surprising feat from a division that was always listed as a third-class formation. The chart on page 278 shows the higher-echelon commands under which the 30th Division served.

As can be seen from the schematic diagrams on pages 280-282, the infantry regiments underwent several changes throughout their history. The Artillery Battalion remained in the unit all along, while the Reconnaissance (cavalry) Battalion was reduced to a mere squadron (company) when the division was reduced to a brigade. The Signal Company, Supply Regiment, and Panzergrenadier Battalion were all disbanded and handed over to other Waffen-SS formations once the division was disbanded.

One last part of the history of the 30th SS Division has not been covered, the officer roster. It was impossible to make a complete list of every officer in the unit; however, I was able to compile a list of quite a number of the officers, including their assigned units within the division. In some cases it was impossible to place an officer in any specific company or battalion. Still, the list on page 283 offers the most complete roster to date.

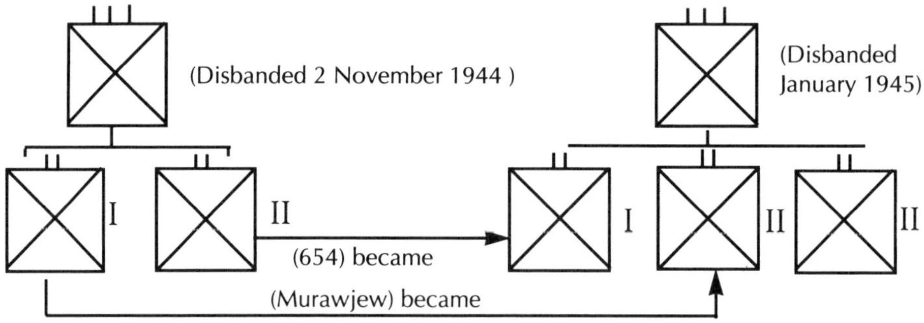

The regiments and battalions in the 30th Division had field post numbers, but they are very deceptive since not all of the units were active at the same time. It is better to follow the development of the division through the following schematic diagrams to get a truer picture of the condition of the unit at any given time. The diagrams illustrate the great difficulty that the Germans encountered in raising the 30th Russian Division.

WAFFEN-GRENADIER-REGIMENT DER-SS NR. 3 (RUSS. NR. 6)

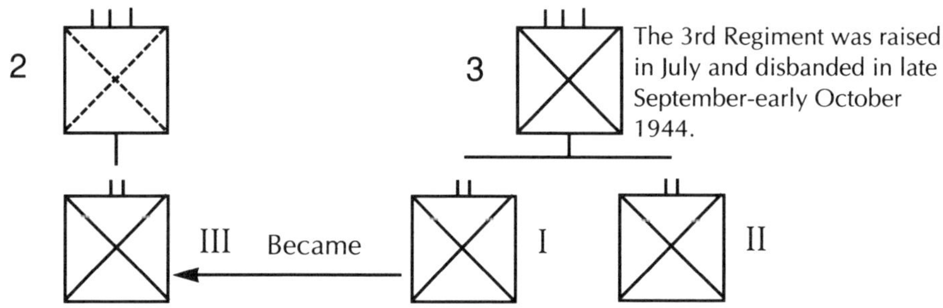

WAFFEN-GRENADIER-REGIMENT DER-SS NR. 4 (RUSS. NR. 7)

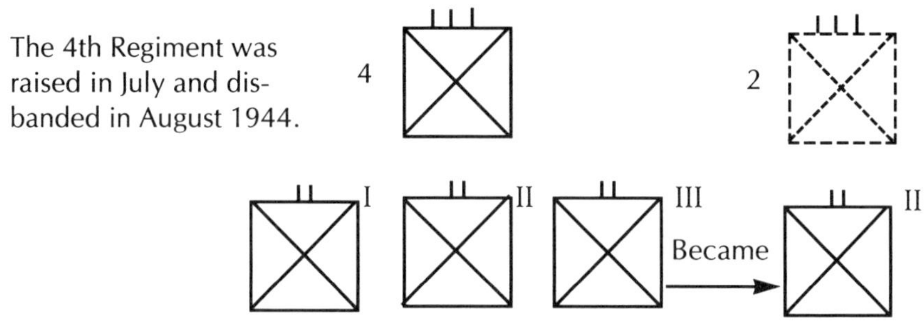

WAFFEN-GRENADIER-REGIMENT DER-SS NR. 1 (THEN NR. 75) (RUSS. NR .4)

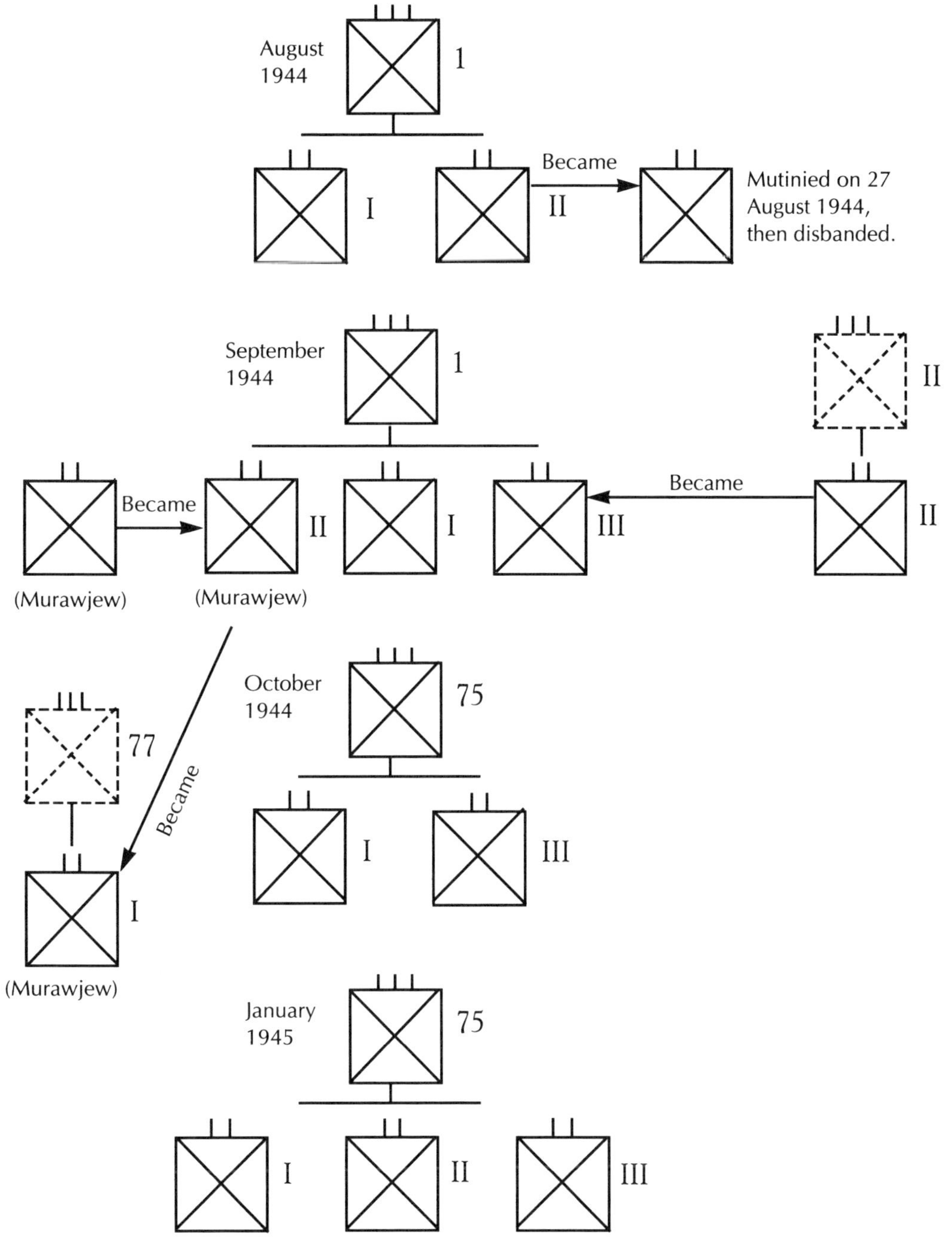

WAFFEN-GRENADIER-REGIMENT DER-SS NR. 2 (THEN NR. 76) RUSS NR. 5

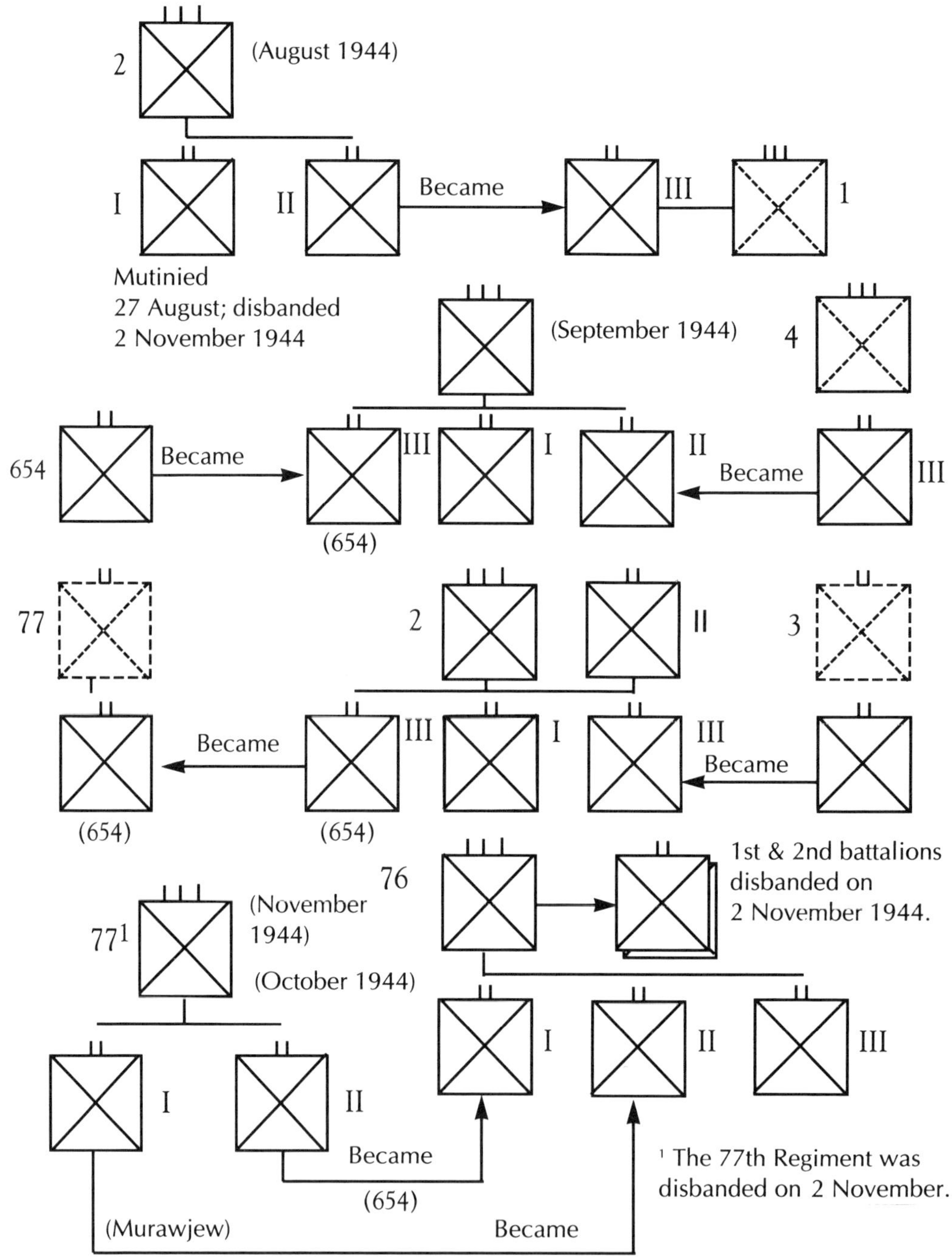

OFFICER ROSTER (INCOMPLETE) FOR THE 30. WAFFEN-GRENADIER-DIVISION DER-SS (RUSS. NR.2)

SS Grenadier Blasius Zauner. An ethnic German who was born in Russia in 1915. Zauner served in the 30th SS Division and was later transferred to the Vlasov Army. Photo: Munin Verlag.

1. Divisional Headquarters Staff
SS-Obersturmbannführer Hans Siegling, commanding officer
SS-Hauptsturmführer Schneider, division adjutant
SS-Obersturmbannführer Noffz, section VI chief
SS-Haupsturmführer Löffler, section Ia chief
SS-Hauptsturmführer Platz, unknown
SS-Sturmbannführer Rosenstock, division supply chief
SS-Obersturmführer und Oberleutnant der Veterinar Dr. Heuer, division veterinary officer
SS-Obersturmführer Simoneit, section IVz chief
SS-Hauptsturmführer Deutschbein, chief staff officer
SS-Sturmbannführer Hennigfeld, section IV chief[1]
SS-Obersturmführer Meschkat, section V chief
SS-Hauptsturmführer Streubel, staff officer
SS-Obersturmführer Karsch, division administrative officer

[1] SS-Sturmbannführer Hennigfeld was transferred to the divisional staff after serving in Feldersatz-Bataillon 30.

2. Waffen-Grenadier-Regiment der-SS Nr. 76 Headquarters Staff
SS-Hauptsturmführer Schmidt, commanding officer[2]
SS-Hauptsturmführer Ernst Sieloff, commanding officer[3]
SS-Hauptsturmführer Schäfer, regimental adjutant
SS-Obersturmführer Degenberger, regiment staff
SS-Untersturmführer Heinrich Hardt, regiment staff
Waffen-Untersturmführer der-SS Nor, regiment staff
SS-Untersturmführer Karl Tartsch, regiment staff[4]

[2] Commanding officer as of 3 August 1944.
[3] Commanding officer as of 26 November 1944.
[4] SS-Untersturmführer Tartsch headed the division's Engineer Company until it was disbanded.

3.Waffen-Grenadier-Regiment der-SS Nr. 76 (3rd Battalion) 9th Company
SS-Untersturmführer Ernst Branduhn, commanding officer[5]
SS-Obersturmführer Cancy, commanding officer[6]
SS-Untersturmführer Albert Schlipf, platoon leader
SS-Untersturmführer Gerdey

[5] Commanding officer as of 12 August 1944.
[6] Commanding officer as of November 1944.

10th Company
SS-Obersturmführer Johannes Merten, commanding officer
SS-Untersturmführer Herbert Hesse, platoon leader

SS-Untersturmführer Kein, platoon leader
Waffen-Untersturmführer der-SS Vladimir Miasnik, platoon leader Waffen-Untersturmführer der-SS Wasura, platoon leader[7]

[7] Waffen-Untersturmführer der-SS Wasura later became regimental adjutant in October 1944.

11th Company
SS-Obersturmführer Alfred Käseweiter, commanding officer
SS-Untersturmführer Süss, platoon leader[8]
SS-Untersturmführer Meier, platoon leader[9]
Waffen-Untersturmführer Johann Gralewski, platoon leader[10]

[8] As of 23 August 1944.
[9] As of 16 August 1944
[10] As of 8 August 1944

12th Company
SS-Obersturmführer Franz Erdl, commanding officer[11]
SS-Untersturmführer Wilhelm Brandt, commanding officer[12]
SS-Untersturmführer Ludwig Wüst, commanding officer[13]
SS-Untersturmführer Willi Stünkel, platoon leader
Waffen-Untersturmführer der-SS Dimitri Hnätenko, platoon leader
Waffen-Untersturmführer der-SS Leo Koserew, platoon leader

[11] As of October 1944.
[12] As of 14 August 1944.
[13] Commanding officer until 13 August 1944 ,then became platoon leader.
SS-Obersturmführer Johann Berberich, platoon leader. Berberich was transferred in late August to the 2nd Battalion, 2nd SS-Waffen-Grenadier-Regiment.

4. Schützmannschaft-Bataillon 654
SS-Obersturmführer der Reserve und Oberleutnant der Schützpolizei Reserve Schirrmacher, commanding officer

5. Waffen-Grenadier-Regiment der-SS Nr. 4 (2nd Battalion)
Waffen-Untersturmführer Awrean Katschan, platoon leader[14]

[14] Served in Schützmannschaft-Bataillon 101, which later became the 2nd Battalion, Waffen-Grenadier-Regiment der-SS Nr. 4.

6. Waffen-Grenadier-Regiment der-SS Nr. 4 (3rd Battalion)
SS-Untersturmführer und Leutnant der Schötzpolizei Walter Kramer, platoon leader

7. Waffen-Grenadier-Regiment der-SS Nr. 4 (battalion unknown)
SS-Untersturmführer Leo Tessmann, platoon leader[15]

[15] SS-Untersturmführer Leo Tessmann became battalion adjutant of the 2nd Battalion,

Waffen-Grenadier-Regiment der-SS Nr. 76.

8. 30.Waffen-Grenadier-Division der-SS Divisional Staff
Hauptsturmführer der-SD Regitschnig[16]

[16] It was common to find an SD (Sicherheitsdienst) officer in each Waffen-SS division. They were usually assigned to some administrative task on the divisional staff but still held a title similar to a Soviet political commissar, although unlike that commissar, he had no weight when it came to military matters or decisions.

9. Feldersatz-Bataillon 30
SS-Sturmbannführer Franz Hennigfeld, commanding officer[17]
SS-Untersturmführer Willi Wilbrenner, platoon leader
SS-Hauptsturmführer Anton Bindl, company commander
SS-Hauptsturmführer Otto Fiebig, company commander
SS-Untersturmführer Werner Plorian, platoon leader
SS-Obersturmführer Wilhelm Grelschak, platoon leader[18]
SS-Untersturmführer Max Schroder, platoon leader[18]

[17] Hennigfeld later joined the divisional staff.

[18] Grelschak and Schroder served in 1st Battalion, 75th SS Waffen-Grenadier-Regiment at one time.

10. SS-Versorgungs-Regiment 30
SS-Obersturmführer Robert Schallert, unknown
SS-Untersturmführer Josef Ryschka, unknown

11. Waffen-Grenadier-Regiment der-SS Nr. 75 (3rd Battalion Staff)
Waffen-Sturmbannführer der-SS Murawjew, commanding officer[19]

[19] Waffen-Sturmbannführer der-SS Murawjew was the highest-ranking non-German national in the 30th Division. He was commanding officer of Battalion Murawjew, which later became 3rd Battalion, 75th SS Waffen-Grenadier-Regiment.

12. Waffen-Grenadier-Regiment der-SS Nr. 75 (1st Battalion) Waffen-Untersturmführer der-SS Erich Biebl, platoon leader[20]

[20] Erich Biebl was a Czech national who had served in Schützmannschaft-Bataillon 57 (a Ukrainian battalion).

13. Waffen-Grenadier-Regiment der-SS Nr. 75 (battalion unknown)
SS-Untersturmführer Bruno Bürger, platoon leader
SS-Obersturmführer Kurt Ansorg, company commander
SS-Hauptsturmführer Walter Tuttas, company commander
SS-Hauptsturmführer Karl Zöphel, company commander
SS-Hauptsturmführer Kurt Mischlewitz, company commander

SS-Untersturmführer Karl Franzl, platoon leader
SS-Obersturmführer Haiden, company commander
SS-Obersturmführer Kurt Angers, company commander
SS-Untersturmführer Max Schrocher, platoon leader

14. Waffen-Grenadier-Regiment der-SS Nr. 76 (battalion unknown)
SS-Untersturmführer Matthias Pire, platoon leader
SS-Untersturmführer Erich Skowronek, platoon leader
SS-Obersturmführer Lenz, company commander
SS-Hauptsturmführer Ernst Niepel[21]

[21] Ernst Niepel went on to command the 3rd Battalion, 76th SS Waffen-Grenadier-Regiment.

15. Aufklärungs-Abteilung
SS-Obersturmführer Dr. Merkulow, commanding officer

16. 30.Waffen-Grenadier-Division der-SS (location unknown)
SS-Sturmbannführer der Reserve Kurt Haag (SS No. 105572) is known to have been attached to the 30th Division. The forty-three-year-old Haag had previously served in the Higher-SS and Police Command Ukraine. In December 1944, he was still carried on the rolls of the division.

17. Waffen-Grenadier-Regiment der-SS Nr. 76 (regimental staff)
Waffen-Hauptsturmführer der-SS Mikula, regimental welfare officer

18. SS-Artillerie-Abteilung 30
SS-Hauptsturmführer Fritz Glaser. commanding officer. Glaser was thirty-one years old in 1944.

ILLUSTRATIVE SCHEMATIC DIAGRAM OF THE 30. WAFFEN-GRENADIER-DIVISION DER-SS, SEPTEMBER 1944

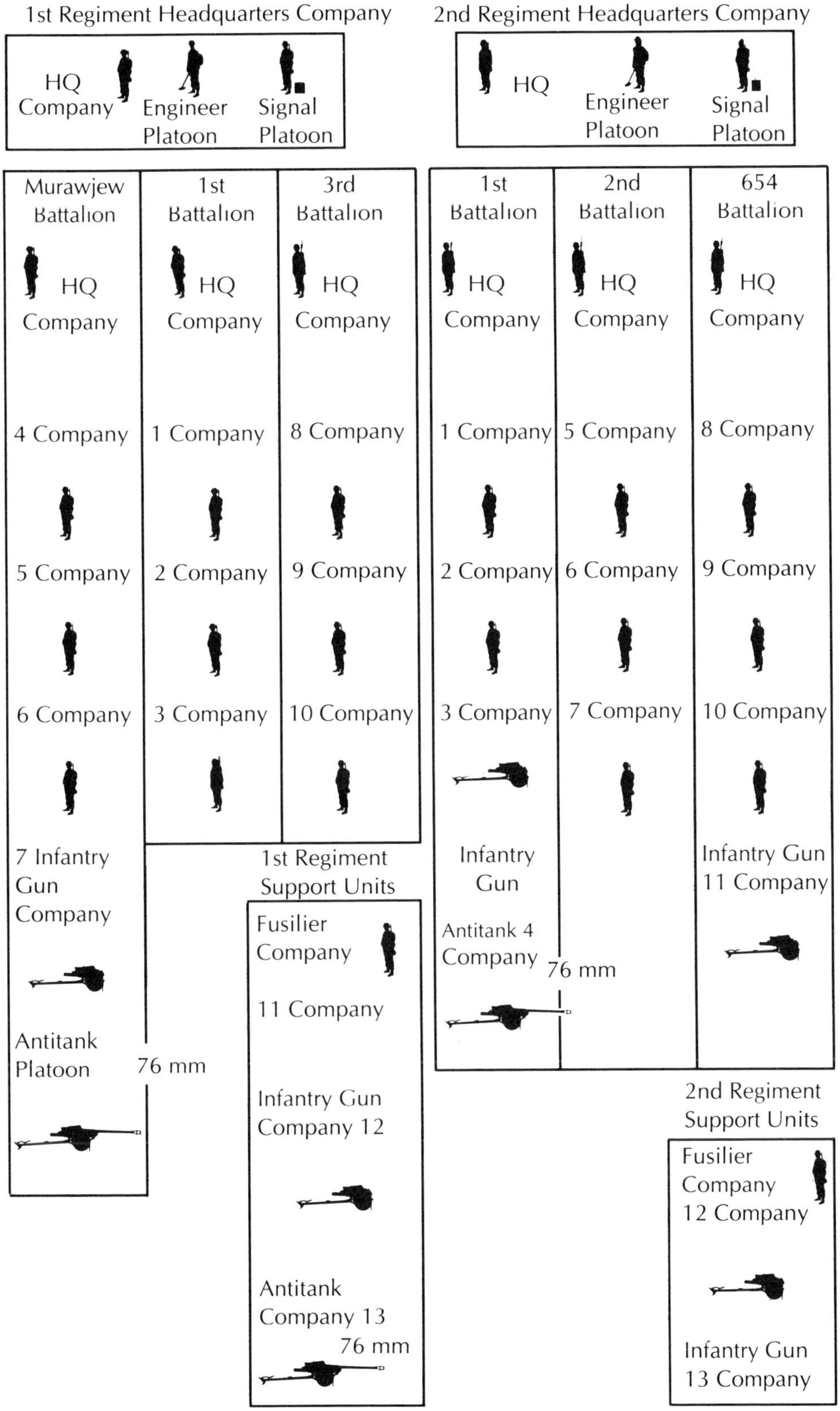

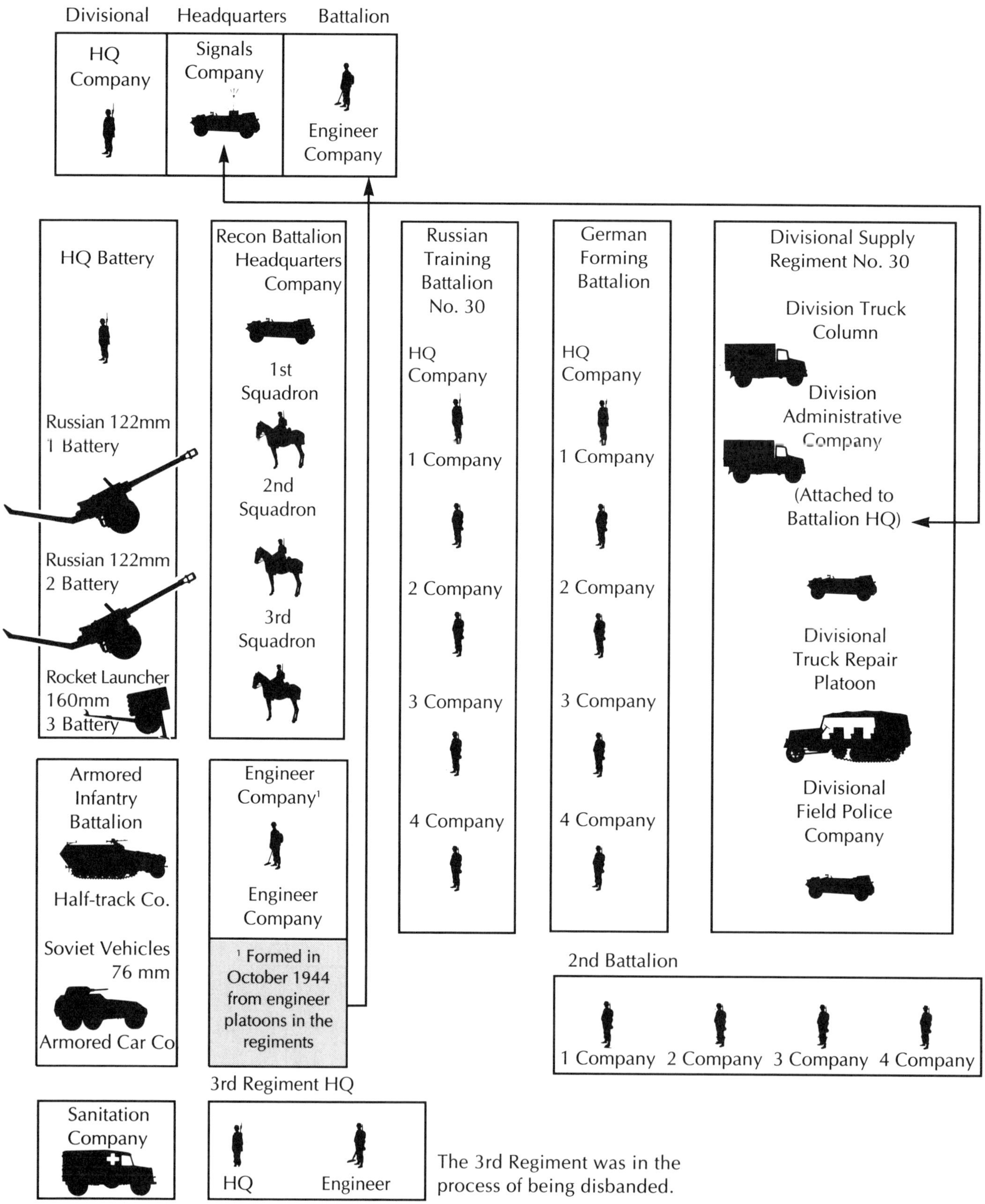

Divisional Headquarters Battalion
HQ Company
Signals Company
Engineer Company
HQ Battery
Russian 122mm 1 Battery
Russian 122mm 2 Battery
Rocket Launcher 160mm 3 Battery
Recon Battalion Headquarters Company
1st Squadron
2nd Squadron
3rd Squadron
Russian Training Battalion No. 30
HQ Company
1 Company
2 Company
3 Company
4 Company
German Forming Battalion
HQ Company
1 Company
2 Company
3 Company
4 Company
Divisional Supply Regiment No. 30
Division Truck Column
Division Administrative Company
(Attached to Battalion HQ)
Divisional Truck Repair Platoon
Divisional Field Police Company
Armored Infantry Battalion
Half-track Co.
Soviet Vehicles 76 mm
Armored Car Co
Engineer Company[1]
Engineer Company
[1] Formed in October 1944 from engineer platoons in the regiments
2nd Battalion
1 Company 2 Company 3 Company 4 Company
3rd Regiment HQ
HQ
Engineer
Sanitation Company
The 3rd Regiment was in the process of being disbanded.

CHAPTER TWENTY-ONE

Eric Brorup's story is that of a Danish volunteer in the Waffen-SS, including service in the Freikorps Danmark as well as the SS parachute battalion in Papa, Hungary.

Brorup was born in Copenhagen, Denmark, on 14 December 1917. He served in the Danish army, enlisting in the Danish Mounted Cavalry in October 1937. Eleven months later, he was made an ensign, or sublieutenant. He had planned to attend a military academy, but the German occupation had cut short his plan. He was laid off from the Danish army, something which bothered him since he liked soldiering.

Before being discharged, he had attended a couple of lectures on the role of the Waffen-SS in the Western campaign in 1940, by his former tactics instructor from the Danish cavalry. This had apparently sparked an interest in that German branch of service. In those lectures, the instructor had mentioned the actions of a German officer who in the future would become Brorup's commanding officer (Knight's Cross winner, Hauptsturmführer Fritz Vogt).

Brorup sought permission to join the Waffen-SS for a two-year term. He got permission from his former unit, the Danish Jydske Dragoon Regiment, and formal approval from the Danish government, which had legally allowed the Germans to recruit its citizens (a legal right rescinded by the government after the war). Because of his qualifications, Brorup was given the opportunity to attend the SS Officer Academy at Bad Tölz on 15 July 1941. He had signed his two-year contract on 25 April.

In the meantime, the Waffen-SS had created a single unit consisting of a large batch of Danish volunteers. That formation, named Freikorps Danmark, was destined for action on the Eastern Front. The Danish unit was short on qualified cadre with military experience (it had been formed from large numbers of politically motivated volunteers who lacked military expertise). For this reason, Eric Brorup's orders to the SS Officer Academy were rescinded on 19 July, and, instead, he found himself in the Freikorps Danmark.

Brorup had joined the Waffen-SS because he liked Germany: the country, people, and culture. He also liked soldiering and was not politically motivated whatsoever. As a result, he resented his tour of duty with his fellow countrymen in the Freikorps. Several other Danish volunteers were in the same boat, but, luckily for Brorup, his stint with the Freikorps lasted only from July 1941 to January 1942, when he was finally sent to Bad Tölz. He spent the first half of 1942 undergoing training and graduated on 8 May 1942. He was then sent to the Training and Replacement Battalion for the Freikorps Danmark and was attached to that unit from 9 May to 15 July at Posen-Treskau. On 16 July, he received the transfer to the cavalry that he had sought and was ordered to report to the 4th Squadron, SS Cavalry Regiment Nr. 3 of theFlorian Geyer SS Cavalry Division.

He was commissioned an SS-Untersturmführer at Waffen-SS Troop Training Grounds Debica on 1 September 1942. On 25 November 1942, he joined the 2nd Squadron of the Florian Geyer Reconnaissance Battalion. In due time, he received both the Iron Cross 1st and 2nd Class. He served with that reconnaissance unit (in particular, with a ski platoon from that battalion) until 19 April 1943, when he was once again transferred to another unit, this time the newly forming SS-Freiwilligen-Panzergrenadier-Regiment Nr. 24 Danmark (which was mostly being formed from Danish volunteers).

After seeing action with the 11. SS-Freiwilligen-Panzergrenadier-Division Nordland in Croatia (where he was wounded) and Estonia (on the Narwa front), he was posted to the replacement battalion for the Nordland Division, which on 16 April 1943 was located in Graz, Austria. There he trained new recruits while recovering from his wounds. Brorup had always been interested in aviation and had some flying experience in civilian life. When a call had gone out for more volunteers for the SS Parachute Battalion, he immediately volunteered. He was attached to the SS-Parachute-Training-Company between 21 July and 8 September 1944. Though his tenure with the unit was brief, it nevertheless sheds some interesting light into the history of the SS parachute battalion.

My tenure with the SS-Fallschirmjäger-Ausbildungs-Kompanie in Papa, Hungary, was very short—between late July and early September 1944. Because of my aviation background and also because I did not feel comfortable in the infantry (I had been in the cavalry/reconnaissance outfits), I volunteered for the paratroopers. Furthermore, SS-Panzergrenadier-Regiment Nr. 24 Danmark, to which I belonged before being transferred to the SS parachute force, still had too many "political residues" from the defunct Freikorps Danmark for my taste, so I requested the transfer, which was granted.

When I reported to the SS Parachute Training Company, none of the personnel in the company were "B-Schütz," meaning that they were transfers from a penal outfit. The majority were volunteers. Other units stationed by the airfield at Papa included:

1. A transport squadron from the Luftwaffe (HE-111 and Ju-52 planes)

2. Cadre personnel from Parachute School Number Three (Luftwaffe)

3. Airport administration

4. A Hungarian (Honved) parachute battalion, which was commanded by a Hungarian major.

Because of fuel shortages, we never got started in jump training (SS-Obersturmführer Leifheit, however, was already qualified). We were issued parachute equipment, i.e., jump pants, jump boots, camouflage smocks, jump helmet and the popular "Kappmesser" (cutting knife), which we, of course, wore with pride and a little bit of "show-offishness."

The duty was not hard; weather was mostly fine. I looked after tactical and weapons training, and time off was spent at the swimming pool. The days (and nights) were also spent at local hotels, trying out the good wine, eating "Gulyas" and listening to Gypsy music.

A couple of events come to my mind. I cannot pinpoint the accurate time of occurrence, but it must have been late in August or early September 1944. Rumania had pulled out of the war, and the Hungarians were gung-ho to go in and snatch Siebenbürgen (Transylvania). [This region is still today disputed by both countries.]

Our local Honved parachute battalion laid on a jump exercise. The commanding officer, a Hungarian major, invited me along. He said that I could use his car to reach the drop zone. I told him I would prefer to fly along, so I came aboard a Hungarian Ju-52 [transport] as jumpmaster. When SS-Hauptscharführer Blum de Lorne saw that, he asked me if I could get him along too.

He was also made a jumpmaster in another Hungarian Ju-52. We had great fun that day! Later on that night at a party, the German pilots told me that I had been plumb crazy to fly with those "weird" Hungarians who believed that Ju-52s were made for aerobatics! Legally, I should not have been allowed to be a crewmember on a foreign aircraft, even an Axis one, but nothing came out of it.

The other occurrence was slightly more dramatic. One morning a green flare was fired from the airport control tower. This was the same as an air-raid warning. There were no shelters, so standard operating procedure was for each leader to take his squad out in the terrain and hide somewhere.

This particular morning, I was standing at the airport together with Fritz Leifheit, Hajo Dräger, a Luftwaffe captain, and others. A radio van (from the Luftwaffe) was monitoring the enemy radio traffic nearby.

We looked up and observed hundreds of contrails [streams of vapors from plane's engine exhaust] from a USAAF bomber formation (B-24 Liberators). They were flying north, apparently heading for Wiener-Neustadt and the Messerschmitt factories there. I had just mentioned to the other officers that it was a dirty shame that all those bombers could fly unhindered in our airspace, when we saw several, maybe thirty to forty, small, shining points above the bomber streams—our own fighters!

In no time flat, they attacked, right above as you could even hear reports of the crewmen [remember that the van was monitoring the U.S, bomber communication] and see the spent brass coming down. The end result was five U.S. bombers shot down and three German fighters lost.

Suddenly an ME-109 [fighter] came over the airfield (he had just nailed one B-24), made a pass, wiggled his wings (a sign of victory), and landed. He taxied up to where we stood, opened the hood, and climbed out. We had expected one of the famous aces with his Knight's Cross, etc., but he was a young fellow of nineteen years, a Gefreiter (Private First Class).

With a big grin he came over to us, came to attention, saluted and reported: "Gefreiter _______ von Feindflug zurück ein vier-mot abgeschossen." Then he added that it was his fourth victory in the last three weeks and said that would make his old man proud! His father happened to be a test pilot. PFC or not, we invited him to the Officer's Mess and treated him in style.

He had landed because his gas tank was almost empty and he was out of ammo. The airport had fuel for the ME-109, but a few minutes later, a "Feldwebel" landed his FW-190, also low on gas. He could not be helped because he had to find another "service station" with the proper octane gas.

Some of the crews of the B-24s had gone down with the aircraft. One had a whole wing shot off and spun around. Nobody got out. All aircraft blew up on impact, all had full bomb loads. Several crew members bailed out and later were collected by the Hungarian (Honved) army.

Later that day I was ordered to come to the Honved barracks where the Americans were being held by the Hungarians. I was told it was rumored that I could speak English and would I please question the prisoners. I said OK, but the interrogation was the Luftwaffe's job, but we could always have a shot.

Some of our officers were present in the Honved barracks, among them an SS-Obersturmführer who was the intelligence officer. I think SS-Obersturmführer Fritz Leifheit [and] SS-Untersturmführer Draeger were there too and, of course, the Honved soldiers.

When the American crew members came in, I noted that they were shoeless and wore summer khaki dress. I asked them where their shoewear was and was told that they were blown off during the bail-out. This happens all of the time; for instance, my own flight boots from the Canadian army air force service have leather straps with buckles at the ankle to prevent their loss . . . but I digress.

The Americans were apparently a little nervous when they saw our Totenkopf [Death's Head] badges at our headgear; otherwise we were in shirtsleeves under, with shoulder boards indicating our ranks (army style), jump pants and boots, and, of course, our sidearms—which for officers was usually the 7.65 mm (32 caliber) semi-automatic, like the Walther PPK.

Our Yankees thought that they were in for tough times. I had to put down their fears and informed them that we were not policemen or Gestapo, but ordinary fighting soldiers. I asked each individually for name, date of birth, and place of birth, which they willingly gave, together with rank.

I then inquired about their unit, their aircraft, and target. No reply. Then, a 2nd Lieutenant spoke up and said: "Sir, if you were in our place, would you answer those questions?"

"Certainly not!" I replied, "and I do not expect you to do it."

They asked me what would happen to them now; would their next-of-kin be informed as well as the International Red Cross? I said that they probably would be turned over to the German Luftwaffe, and it was the Luftwaffe's responsibility to look after that end.

I also informed them that we knew their target, anyway, as well as their aircraft type (B-24). We had witnessed this particular plane's defect at the hands of the Gefreiter. I don't know if they were embarrassed by being shot down by an enlisted man! I also told them that if they had to be put out of action, they were lucky to be captured by regular soldiers. If they had bailed out in the target area, the civilians whose homes were destroyed by an air attack usually took a dim view of that kind of activity.

Sometimes they might hang people who came down in a parachute. For instance, the abovementioned German fighter pilot of the ME-109 wore an armband (brassard) with the words, "DEUTSCHE LUFTWAFFE." This was to prevent him from being torn apart by irate citizens, if he should have the tough luck to bail out and land among his compatriots.

A few days after this incident, I was ordered to report to the SSFHA in Berlin (Main SS Command Headquarters). A Brigadeführer there gave me a hard time about "fraternizing," as he said, with prisoners of war. I replied that I did not do such a thing, but treated the POW the same way I would expect to be treated if taken prisoner.

I was then ordered to report to the 1st Battalion, SS-Panzergrenadier-Regiment Nr. 24 Danmark at Hammerstein and went with that battalion to the Eastern Front. Shortly after that, I was nevertheless promoted to SS-Obersturmführer and came to my favorite outfit, SS-Panzer-Aufklärungs-Abteilung Nr. 5 Wiking, in which I ended the war.

The real reason for my transfer from the SS paratroopers, however, was that Leifheit's company was slated for the Western Front and I was a Scandinavian volunteer (with family ties in the USA) and was not supposed to be there.

I later met Otto Skorzeny, who told me that I had been on the SHAEF "wanted" list.[1] I had been on that wanted list, but Skorzeny had told the

[1] Supreme Headquarters Allied Expeditionary Force, this list included the unit rosters listing men who belonged to the disguised German brigade under the command of Otto Skorzeny, acting as Americans.

Americans that I was not in Operation Greif. Someone must have been remiss in removing my name from the list.[2]

Eric Brorup ended the war with the SS reconnaissance battalion of the Wiking SS Panzer Division. His unit surrendered to the Americans on 8 May 1945. There followed a period of two and one-half years in a U.S. then British POW camp. On 9 November 1947, he was handed over to the Danish government, who quickly sentenced him to another two and one-half years in prison for having volunteered in the German Armed Forces—even though it had been legal to do so at the time. His sentenced was reduced, however, due to the years he had already spent in Allied POW camps. He was released from jail on 2 June 1948, after spending eight months in a Danish cell.

He harbored an understandable bitterness toward the Danish government. Shortly after his release from jail, he took up flying once again. By 1955, he was employed as a commercial pilot in Denmark and West Germany. Leaving the homeland that fathered then rejected him, he emigrated to Canada in November 1956. He became a bush pilot in northern Canada and the Arctic regions.

He served with the Canadian Army Reserve in 1959 and 1960 (with the 19th Alberta Dragoons, Royal Canadian Armoured Corps). He received the Army Flying Badge from that unit. In 1969 and in 1970, he served as pilot/instructor with the rank of lieutenant in the Royal Canadian Air Cadets and was on the directing staff, CFB Trenton (1969) and CFB Penhold (1970). He holds a teacher's certificate and B.A. and B.S. degrees. He married in 1949 and fathered four daughters, who have given him fourteen grandchildren. His hobbies include cross-country skiing, hiking, bicycling, photography, and military history.

Brorup is now retired but is still active in his favorite sport, cross-country skiing. In a recent letter, he stated proudly that his experiences with the skiing unit of the reconnaissance battalion, Florian Geyer Cavalry Division, were his most exciting and demanding. He stated further that most of these actions were done behind the Russian lines, and he considers them to be, and I quote, "the ultimate biathlon!"

[2] Brorup is referring to the fact that he was still being carried on in the rolls of the SS-Fallschirmjäger-Ausbildungs-Kompanie as late as 1 January 1945, even though he had long left the unit. He states that Leifheit's training company was used in the Ardennes offensive as a whole unit. More interesting is that he also states that the SS-Fallschirmjäger-Ausbildungs-Kompanie (the training company) was redesignated as the new 1st Kompanie in the SS Parachute Battalion 600. This explains the disappearance of the training company from the picture in late 1944.

SS-Obersturmführer Eric Brorup taken in May 1944 (top).

Below. A group picture of the officers and NCOs of the SS-Fallschirmjäger-Ausbildungs-Kompanie by the base pool, summer 1944. From left to right are: SS-Untersturmführer Werner Droste, SS-Obersturmführer Eric Brorup, five men (unknown), and extreme right, SS-Oberscharführer Aussenthal. Kneeling, right: SS-Untersturmführer Walter Scheu; kneeling left, unknown. This picure was taken at Papa, Hungary. Photos: Eric Brorup.

CHAPTER TWENTY-TWO

Anyone who has read about the great Russo-German War in any detail has heard of the famous Spanish volunteer division that fought alongside the German Wehrmacht, predominantly in the northern sector of the Russian Front. In the German divisional roster, it was the 250th Infantry Division, raised in the summer of 1941 and sent to the Russian Front later that year.

When the war turned against Germany in the summer of 1943, Generalissimo Francisco Franco, fearing that his involvement with the Third Reich would cause the Allies to invade Spain and thus threaten his dominion of that country, began to remove himself and his country from the conflagration raging all over Europe and North America. He distanced himself from Hitler by also withdrawing his contribution to the war effort against Stalin and the Soviet Union—the tough, formidable 250th (Blue) Division.

The Germans not only saw what Franco was up to (the sly old fox had evaded Hitler's attempts at getting Spain into the war in 1940) but also resented the withdrawal of one of their most seasoned and battleworthy divisions. After many palavers between the German embassy in Madrid and the Spanish Ministry of War, the two countries reached an agreement wherein the Spanish 250th Infantry Division would be withdrawn, but a volunteer "legion" of a few thousand men would remain as a token force.

Attracting volunteers for the "Russian adventure" had never been a problem for the Blue Division. The Spanish Falange Party saw to that. Additionally, Franco had never been a fascist in the true sense of the word. He was and had always been a conservative. The Spanish Falangist Party, which before the Spanish Civil War had been headed by Jose-Antonio Primo de Rivera, had been as much of a threat to the conservative Spanish Army and wealthy Spaniards, as the dreaded left (the Communists, Socialists, and Anarchists).

As a matter of fact, a stroke of luck at the beginning of the Spanish Civil War helped to eliminate the Falangist leader from the Spanish picture. It seems that Jose-Antonio Primo de Rivera had been jailed by the Republican government just before the outbreak of the military insurrection. When the fighting broke out, there were many prisoner exchanges between the Republic and the insurrectionists.

General Franco and General Mola, who at first headed the military junta against the Republic, were in no hurry to release the Falangist leader from almost certain liquidation at the hands of the leftists. They saw the potential threat that his party represented to the army. Jose-Antonio was never exchanged; he was executed by firing squad that fall of 1936. With his death, the Falangist Party leadership was left in chaos.

As is the case in most one-leader parties or governments, cutting off the head causes the body to die also. After the mysterious death of General Mola in an airplane crash, General Franco took over as head of the junta. In effect, he became the caudillo, or führer, of the Spanish rightists. His first move was to consolidate his power by merging all of the distinctly different factions within the conservative block. In turn, the Carlists, the CEDA adherents, and the Falangists were all pushed into the Spanish army fold. In many instances, the caudillo had to nudge or prod some elements into line.

Franco rendered the Falangists Party powerless by removing from command all the leaders who opposed him and replacing them with "yes men" who could be counted to follow like faithful dogs. The Russian "cruzada" (crusade) also gave Franco the chance to get rid of the more active and reactionary members within the rank and file of the

Falangist Party by sending them to the Russian Front (as all of them blindly volunteered to do). At best, they would be killed fighting Russians; at any rate, they were not in Spain, fomenting trouble. Franco was a cunning and intelligent opponent, and he understood and manipulated what motivated the Spanish people. For instance, he had pictures of the dead Jose-Antonio plastered on the white walls of Spanish towns and villages, right alongside posters of his own, showing him to be very much alive and vital. Franco's strategies worked—the Spanish dictatorship remained in power until his death in 1975.

With the changing tide of the war, Franco's Falangist problem became secondary to the threat of being pulled into the German Götterdämerung. Besides, by 1943, the Falangist Party was, in essence, no more.

Thus, at the close of 1943, the Spanish Blue Legion was formed from 3,000 hardcore volunteers. The returning Blue Division veterans had mixed feelings. Most were happy to see their homes and loved ones again, though they sensed that they had left something undone in the East.

The tenure of the volunteer Spanish Blue Legion was very brief, however. It was withdrawn from the Russian Front in February 1944 and officially dissolved in March. By that time, however, the Wehrmacht had begun the secret recruitment of Spanish volunteers, against the wishes of el caudillo. Word of the secret recruiting spread from volunteer to volunteer, in the gatherings of ex-Blue Division members and later on to the Falangist Party groups in all of Spain.

Crossing the closely guarded Franco-Spanish frontier became hazardous for the would-be volunteers. The Spanish border guards had been ordered to shoot first and ask questions later, and, although many of these border guards were sympathetic to the volunteers, they had to worry about their own lives and careers. Many a Falangist volunteer was shot as he tried to make the treacherous journey.

The Wehrmacht's Sonderstab "F" (Special Staff "F") began processing the Spanish volunteers by forming a fifty-man recruiting unit at the end of January 1944, headed by Dr. Edwin Haxel, the old liaison official for the Germans in the Spanish Blue Division. Staff headquarters were at Lourdes, France, but offices were also located at Hendaye, in the tiny nation of Andorra, on the Franco-Spanish frontier.

Other officers in Sonderstab "F" included Oberst von Heyde, and Hauptmanns Täggert and Keller. The staff also kept up contacts with the SD (Sicherheitsdienst, or State Secret Police) through SS-Brigadeführer Hansen of the SD.

The recruits were eventually sent to various training centers, including Stablack and the Tyrolian town of Hall. At first these recruits were attached to the Heer. Groups of platoon- and company-strength replacements went to the 357th German Infantry Division, with the 3rd German Mountain Division among the authenticated units. Two companies formed at Stablack Training Camp received antipartisan duty in Yugoslavia in mid-August 1944. They traveled to Calje in the triangle formed by the towns of Maribor, Zagreb, and Ljubljana and established headquarters at Zalec, just east of Celje, to begin operations. These two Spanish companies were part of the German Brandenburg unit, which by 1944 boasted more than three regiments and several battalions.

At the beginning of September, one of these companies was sent to the Carpathian Mountains area and used as a source of replacements for the 3rd German Mountain Division. The other Spanish company was later sent to Klagenfurt Training Camp in Austria, the home of several foreign-raised Waffen-SS formations. It was this

DIVISION BRANDENBURG, 14 JULY 1944

Subject: Formation of Raiding Corps France
Liaison Officer—Army High Command West
Major Hollman

A Raiding Corps France has been formed for the disposal of Army High Command West with the following sub-units:
Raiding Corps South France, Raiding Corps Biscayne, and Raiding Corps North France. They have ben formed from the following units:
8th Company, 2nd Battalion, 3rd Regiment "Brandenburg" has released units for the formation of Raiding Corps South France.

Composition: One headquarters column with a signals platoon and two assault platoons.

Commander: Raiding Corps South France Hauptmann Traege

8th Company has also released one mixed platoon of German and Spanish legionnaires for the formation of:
Raiding Platoon Pyrenees, attached to Raiding Corps Biscayne.
Commander: Raiding Platoon Pyrenees, Leutnant Demetrio[1]

Leutnant Pawel is charged with the formation of Assault Platoon Britany and Assault Platoon Flanders, both of Raiding Corps North France.

For the Division Brandenburg
1st General Staff Officer
Frankfurth.

[1] Lieutenant Demetrio was a Spanish officer.

Spanish company, raised for the Brandenburg Division, that eventually became the Spanische-Freiwilligen-Kompanie der-SS 101. By the end of 1944, it was transferred to another Waffen-SS training camp (Stockerau, near Vienna), where it completed its training.

The company had four rifle platoons and a staff platoon. Eventually a second Spanish company (Nr. 102) was formed in the spring of 1945. The Spanish 101st Company was first sent to the Belgian-manned 28. SS-Freiwilligen-Panzergrenadier-Division Wallonie in Pomerania in February 1945. Apparently, they were incorporated in the 1st Battalion of the 70th SS-Freiwilligen-Panzergrenadier-Regiment of that division. The total number of Spaniards in that battalion has been estimated at only 140 men.

Spanish officers in the company seemed to include Waffen-Untersturmführer der-SS Ocaña, Botet, and LaFuente, with Waffen-Obersturmführer der-SS Valdamos as acting company commander.

Let us now take a look at another

Spanish volunteer group that was first assigned to another part of the Brandenburg Division before being incorporated into the Waffen-SS commando force. A document found in the German Bundesarchiv and dated 14 July 1944 clearly shows the presence of Spanish volunteers in the Brandenburg Division and in a sub-unit of that division in France.

This document tells us that a sizable Spanish contingent existed in the ranks of the Brandenburg Division, more specifically in the 8th Company/2nd Battalion/3rd Regiment Brandenburg—the regiment that had been engaged in antipartisan operations in northern Italy in May 1944. Its movements had taken the 2nd/3rd Brandenburg through the towns of Avezzano, Rome, Arsoli, Carsoli, Terni, Bevagna, Perugia. By 19 June 1944, the battalion was located around Perugia. On 18 July, it was situated by Arezzo and Cita di Castello. It is almost certain that the Spanish contingent fought in these operations before being withdrawn to France. I speculate that a portion of them also remained with this regiment, since there is enough documentary evidence of a Spanish company attaching itself temporarily to the 24. Waffen-Gebirgs(karst)jäger-Brigade der-SS during the latter half of 1944. That same company is noted as having been commanded by a Leutnant José Ortiz. In April 1945, it was operating in northern Italy and was scheduled to defend the phantom Alpine Redoubt, the supposed last-ditch area for the Nazi elite. The end of war probably found this Spanish company on Austrian soil, though their final fate remains a mystery.

Let us return our attention to the Spanish contingent in Raiding Corps South France, and more specifically, Raiding Platoon Pyrenees. That platoon operated on the Franco-Spanish frontier during the summer of 1944. Quite a number of former Spanish Republican men were actively engaged in the French resistance, perhaps several thousand. They operated all over southern France, particularly on the Franco-Spanish border. Raiding Platoon Pyrenees did their best to neutralize the Spanish Maquisard efforts in that area.

The fate of the fifty Spaniards in Einsatzgruppe Pyrenären (Raiding Platoon Pyrenees) is linked with its parent unit, Streifkorps Südfrankreich (Raiding Corps South France). In September 1944, all of the Brandenburg commando units were transferred to SS-Oberststurmbannführer Otto Skorzeny's Waffen-SS commando force. It was in January 1945 that Streikorps Südfrankreich was redesignated SS-Jagdverband Südwest. Einsatzgruppe Pyrenären became part of a sub-unit of Südwest SS-Jagdverband, SS-Jagdeinsatz Südfrankreich. Leutnant Demetrio became Waffen-Obersturmführer der-SS Demetrio.

The history of SS-Jagdverband Südwest is shrouded by uncertainties from the many years that have passed since the end of the war. It is known, however, that it operated in the Schwarzwald (Black Forest) area of southern Germany against the U.S. 7th Army. It had its headquarters for a time in the town of Freudenstadt. Its actions included blowing up bridges and conducting long-range reconnaissance patrols behind Allied lines.

At the beginning of April 1945, Skorzeny ordered SS-Jagdverband Südwest to head into Austria, where it would link up with the remnants of SS-Jagdverband Mitte and Südost. The plan was to use the commando forces for the defense of the phantom Alpine Redoubt. It is not known if Südwest reached the staging area in Linz, Austria. In any case, the end of the war, Hitler's death, and the nonexistent redoubt ended the commando plan for a last-ditch defense. Those who did not surrender at the end of the hostilities took off for the Austrian hills—literally.

We return now to the Spanish volunteers in the 28. SS-Freiwilligen-Panzergrenadier-Division Wallonie. After the battles of Pomerania ended, the much-depleted Wallonie formation withdrew west, past the Oder River, into a hastily formed defense line that became part of the last German defense before Berlin. The Wallonie unit was to the north of Berlin, much closer to the port of Stettin.

The story of the Waffen-SS Spanish contingent now gets hazy. There is less corroborating evidence as to the sequence of events. The Spanish 101st Company (its remnants) ended up in Berlin for that city's defense under the 11. SS-Freiwilligen-Panzergrenadier-Division Nordland. It fought especially hard in the defense of Möritz Platz, the large plaza surrounding the German Air Force Ministry and the Ministry of Propaganda. The Latvian 15th Fusilier Battalion of the Waffen-SS fought alongside the Spaniards toward the end of the battle for Berlin. What is definitely known is that Waffen-Obersturmführer der-SS Miguel Ezquerra took command of these die-hard Spaniards during Berlin's final agony.

As to the fate of Spanische-Freiwilligen-Kompanie der-SS 102, a lot less information is available. Several sources mention a Spanish company that was ordered to Austria during late March/early April 1945 for the last-ditch Nazi defense, but the story has not been proven.

I should also mention here the presence of a Spanish contingent within the ranks of the German 357th Infantry Division in Czechoslovakia in March and April 1945. This fact has been substantiated and just adds more confusion to the story. We cannot rule out the possibility that this second Spanish-SS company could have possibly made its way to Berlin to join their other comrades in a Teutonic Götterdämerung.

In any case, sufficient evidence exists to prove that such Spaniards as there were (perhaps 500 to 1,000) served in the ranks of the German army, and later in the Waffen-SS between 1944 and 1945—long after Spain's involvement with Hitler had supposedly ended.

As Professor Carlos Caballero-Gurado's monograph on these volunteers attested, to this day the Spanish volunteers of the Wehrmacht and Waffen-SS were men who did not exist officially. Their unit was the phantom battalion that still remains obscure and esoteric in nature.

EPILOGUE

The first half of this book dealt mainly with obscure Waffen-SS formations that, although microscopic in size, were considered elite units. The latter part of the book concentrated on obscure units that, for the most part, were made up of foreign elements. There is no need to explain why the Germans fought so bitterly to the end. Faced with the thought of a divided Germany and the fear of what awaited their families at the hands of a vengeful Red Army, German soldiers resisted bitterly in the east, while their counterparts facing the Allies began to surrender in the west.

But what about the thousands of foreign volunteers who had served on the German side? For most Germans, the end of the war meant simply removing their uniforms and returning to the *fliessband* at the local factory, but for the Eastern European or Russian volunteer, the end of the war meant repatriation and death—at the very least, the Russian volunteer could expect to spend twenty years of hard labor in the Soviet gulag.

General Koestring, a German officer who during the war had been heavily involved with the foreign volunteer movement, had the following to say regarding the German and Allied attitudes toward these tragic men and women at the end of the war.

> *We Germans, owing to our lack of reason [and] our limitless appetite, inability, and ignorance, have lost the greatest capital that existed and can exist in the fight against Bolshevism. In the imagination of countless Russians, we have thrown the picture of European culture into the mud, and, yet, we have left certain capital which in [the] future could grow.*
>
> *You will not understand me today when I tell you that during the last few weeks, you have destroyed this capital for the second time, not only in the material sense, but also in the souls of all those who had counted on your help and understanding, after the Germans let them down. It may easily happen that in the near future, you will be calling for what is now perishing.*

Lieutenant General Wladyslaw Anders was the Allied commanding officer of Polish forces in exile during World War II. Anders fought the Axis from the beginning of the war, seeing the country he loved prostrate and defeated. He worked hard to destroy the Third Reich and liberate Poland. Yet in a monograph he wrote after the war dealing with Russian volunteers in Hitler's army, he was quite objective and frank in describing the reasons and causes that had sparked so many foreigners to join. In retrospect, he clearly described the attitude that prevailed during and shortly after the war.

> *In those days, all Soviet citizens who took up arms against the USSR were, in the eyes of the West, traitors to their country, who did not deserve leniency. This was, of course, a much too simplistic way of looking at the whole question.*
>
> *From the moral point of view, the Ukrainians, Belorussians, Cossacks, the Georgians, Armenians, and Turkomans, and the members of all the other non-Russian nations, were not traitors. No matter under which government they were born and in which part of the world, they fought against a government and against a country which was not their country, but which had enslaved them.*
>
> *If, in the eyes of many people, Germans who fought against Hitler were not traitors, why should the Russians who fought against the Soviet system be traitors?*

How very little the West understood the hatred that many Soviet citizens felt against Stalin, and his brand of communism was shown when Allied intelligence prepared the surrender leaflets. They urged all Russians serving under the Germans to surrender, promising swift repatriation to the Soviet Union as a reward! The effect was completely the opposite than what the Allies had sought. After the promise of repatriation, the Russian volunteers fought on even more stubbornly, many till death, rather than surrendering and being returned to Russia. A perfect example of this was an Armenian battalion that was virtually destroyed at Normandy. General Anders described what happened at the end.

> *Deprived of their fatherland, scorned by their protectors, regarded generally as traitors, although in their consciences they were not traitors, they fought often for an alien and hateful cause; the only reward which they eventually received for their pains and toil was death, mostly in a foreign land, or "repatriation" to the hell from which they had tried to escape.*

Perhaps the best description of those repatriations came from an English sailor, present when the Cossacks were handed over to Stalin.

> *I took part in the evacuation of Dunkirk. Our soldiers felt very badly. I helped to fish out Germans from the sunken Bismarck, which received the greatest number of torpedoes in history. I saw the population of Malta being bombed incessantly and deafened by explosions of bombs and shells. They were exhausted from the constant explosions and alarms. I lived through the sinking of my own ship. I know about jumping into water at night, dark without bottom, and terrifying shouts of the drowning, and then the boat, and looking for the rescue ship. It was a nightmare.*
>
> *I drove German prisoners captured during the invasion of Normandy. They were almost dying from fear.*
>
> *But all that is nothing. The real, terrible, unspeakable fear I saw during the convoying and repatriation of people to Soviet Russia. They were becoming white, green, and gray with fear that took hold of them.*
>
> *When we arrived at the port and were handing them over to the Russians, the repatriates were fainting and losing their senses. And only now I know what a man's fear is who lived through hell, and that it is nothing compared with the fear of a man who is returning to the Soviet hell.*

The final tragedy of the Waffen-SS is that, in the name of the anti-Communist crusade, hundreds of thousands of Germans and foreign volunteers, most of them decent men, unwittingly helped to sustain a regime that killed six million Jews and millions of other peoples. They paid for the crimes of the criminals by forever being linked to the concentration camp spectre, even though most of them fought on the front lines and never committed any war crimes. Guilt by association was the verdict handed down at Nuremberg. It is not up to this author to judge the fairness or unfairness of that. That job is left to God.

APPENDIX A

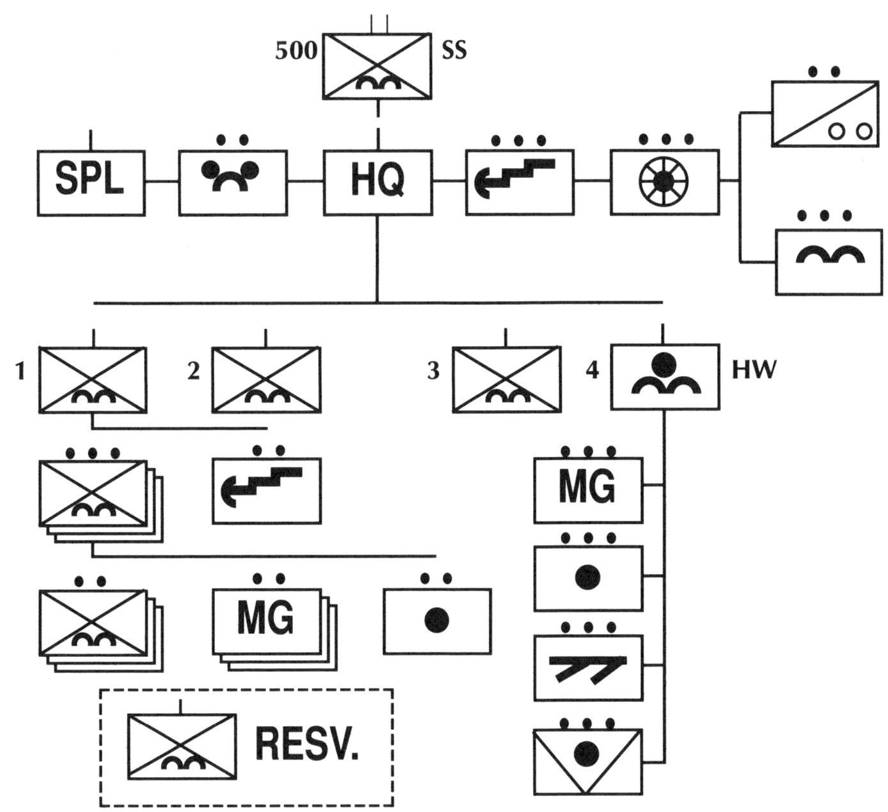

This schematic diagram shows the full structure of the SS-Fallschirmjäger-Bataillon 500 on 25 May 1944. Total strength amounted to around 1,000.

	Parcahute Infantry
HQ	Headquarters
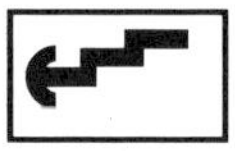	Signals
	Maintenance
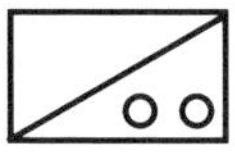	Motorcycle Dispatch
	Parachute Maintenance
	Section III (Legal)
SPL	Supply
MG	Machine Gun
	Mortar
	Flamethrower
	Antitank
RESV.	Reserve
II	Battalion
I	Company
• • •	Platoon
• •	Section/Squad
HW	Heavy Weapons

APPENDIX B

ORDER OF BATTLE FOR OPERATION MARGARETHE
19 MARCH 1944

Reserve Units
100th Jäger Division (in Belgrade, Yugoslavia)
21st Panzer Division (in France)
507th Schwere Panzer Abteilung (Tiger)
301th Panzer Abteilung (in Nettuno bridgehead, Italy)
679th Pionier Regiment Stab (motorized; in Yugoslavia)

These reserve units were allocated to Operation Margarethe in the event that they were needed.

Main Combat Units
XXII. Gebirgs-Armee-Korps

Kampfgruppe Hildebrandt

- Grenadier-Regiment 92 (motorized)
- III./4.Regiment Brandenburg
- Pionier-Bataillon 45 (motorized)
- Panzer-Abteilung 202
- Sturmgeschütz-Brigade 201
- Brüko B 658 (motorized)
- SS-Polizei-Regiment 5
- II./Flak-Regiment 25 (motorized)

Kampfgruppe Brauner

- Elements of 42.Jaeger-Division
- I./Flak-Regiment 6 (motorized)
- 1.Ko./Pionier-Bataillon 45 (motorized)
- I./Jäger-Regiment 54
- Panzerzug 64
- Brüko B 2./402 (motorized)

Kampfgruppe Streckenbach

- Motorized elements of 8.SS-Kavallerie-Division Florian Geyer

LXIX.Armee-Korps z.b.V.

Kampfgruppe Stettner

- 1.Gebirgs-Division
- Brüko B 42 (motorized)
- Panzerzug 23
- Panzerzug 65

Kampfgruppe Zwade
367.Infanterie-Division
Parts of 18.SS-Freiwilligen-Panzergrenadier-Division Horst Wessel
Brüko B 815 (motorized)

LVIII.(Reserve)-Panzer-Korps
Kampfgruppe Bayerlein
130.Panzer-Lehr-Division
Schwere-Artillerie-Abteilung 997
Bau-Pioneer-Bataillon 789
Landschützen-Bataillon 503

Kampfgruppe Simon
16.SS-Panzer-Grenadier-Division Reichsführer-SS
I./5. SS-Panzer-Regiment Wiking (Panther tanks)
Brüko B 853 (motorized)

LXXVIII.Armee-Korps z.b.V.
Grenadier-Regiment (motorized) 1029 Grossdeutschland
Grenadier-Regiment (motorized) 1030 Feldhernhalle
Alarm-Regiment/Division Brandenburg
Bau-Pionier-Bataillon 799
Landschützen-Bataillon 837 Brüko B2/412 (motorized)

Operation Trojan Horse
SS-Fallschirmjäger-Bataillon 500
One battalion from the Brandenburg Division
SD units (battalion strength)
Operation Trojan Horse was the occupation and pacification of the Hungarian capital by commando units dropped over and landed in Budapest.

Objectives

XXII.Gebirgs-Armee-Korps
Capture of the Tisza River bridges. Occupation of Szeged, Kecskemét, Cegléd, and Szolnok. Capture of the airfields near these towns and the neutralization of Hungarian forces in these areas. Road and railway bridges over the Drava River in Neusatz (Novi Sad). Capture of the rail and industrial areas of Maria Theresiopel. Capture of Kiskunhalas and Baja; capture of Danube Flotilla. Capture of the Danube bridges at Baja and Dunaföldvär.

LXIX.Armee-Korps z.b.V.
Capture of Pécs, Kaposvár, and Stuhlweissenburg (Székesfehérvár). Capture of the oil fields at Lispe and the large industrial areas near Aika and Urkut on the north bank of Lake Balaton.

LVIII.(Reserve)-Panzer-Korps

Capture of Raba, Komárno, Papa, Sopron, Sarvar, and Veszprem. Capture of industrial areas in and around Budapest, Bakony, Tata, and Bánhida. Capture of Budapest itself and relief of the Trojan Horse units.

LXVIII.Armee-Korps

Capture of Rožňva (Rosenau) and Kosice (Kaschau), with their important industrial areas. Capture of Tisza bridges between the Czechoslovak border and Szolnok area. Occupation of eastern half of Budapest.

Operation Margarethe went peacefully, negating the employment of the SS-Fallschirmjäger-Bataillon. When Admiral Horthy returned to Budapest the day following his meeting with Hitler, a German honor guard met him at the railway station of his own capital!

APPENDIX C

PANZERBRIGADE 150

Kampfabteilung 2150/I

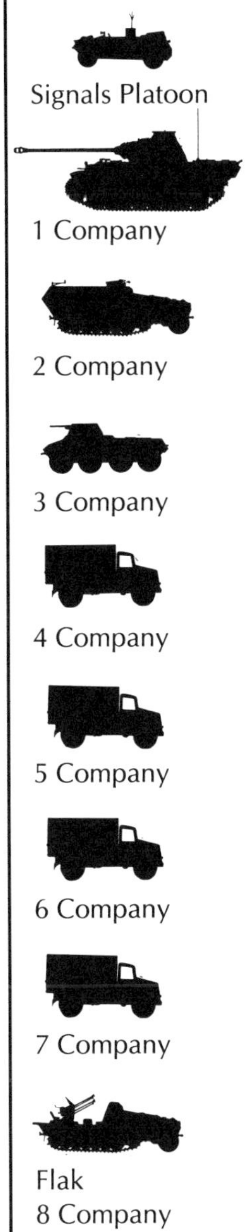

Kampfabteilung 2150/II

Support Elements

Brigade Staff

NOTES: The 1st Company was supposed to contain 22 Panther tanks.

The 2nd Company (half-tracks) was supposed to have two 81mm mortars, two 75mm antitank guns, and two 20mm guns.

The 3rd Company was to have eighteen armored cars carrying the 20mm gun.

Each motorized rifle company was to have two 81mm mortars for support.

The 8th Company (antiaircraft) was to have six half-tracks carrying the 88mm flak gun and three half-tracks carrying the quadruple 20mm flak gun.

The Combat Engineer Company had two 81mm mortars in addition to the light bridging equipment.

The Artillery Battery had six of the heavy 150mm artillery pieces.

Bridging Column 2150 had bridging equipment for a sixty-ton capacity.

Otto Skorzeny formulated this order of battle based on the projected needs of the brigade, which, in turn, were based on the assigned mission. The document was dated 25 November 1944. Lack of equipment eventually forced Skorzeny to reduce this powerful structure to far less than he hoped for (see Organization of Panzerbrigade 150 on 16-25 December 1944, next page). SS-Fallschirmjäger had 380 men who served in this Panzer brigade, while 175 men of SS-Jagdverband Mitte served.

ORGANIZATION OF PANZER-BRIGADE 150, ARDENNES OFFENSIVE 16–25 DECEMBER 1944

Kampfgruppen "X" and "Y" had the following units each:

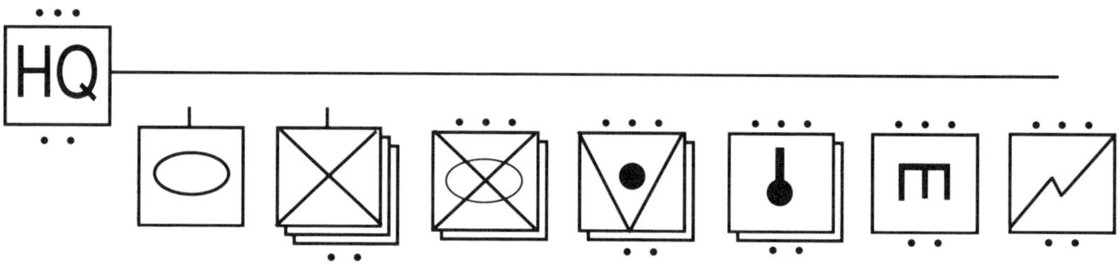

Kampfgruppe "Z" had the same units, minus the tank company.

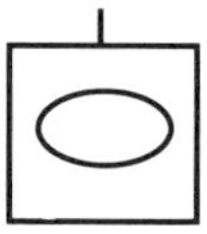

Tank Company
10 Panther and Sturmgeschütz armored vehicles

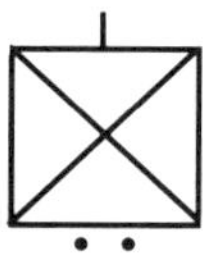

Motorized Infantry Company
120 to 150 men each in an assortment of U.S. and German vehicles

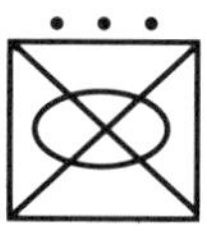

Armored Infantry Platoon
30 to 40 men in captured U.S. armored cars and German half-tracks

Antitank Platoon Motorized
Six 120mm mortars with accompanying mortarmen

Heavy Mortar Platoon Motorized
Six 120mm mortars with accompanying mortarmen

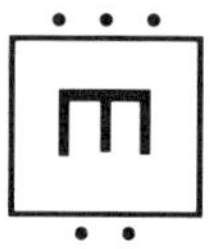

Engineer Platoon Motorized
50 combat engineers, no bridging equipment

Signals Platoon Motorized
About 25 to 30 men with radios and signals equipment

APPENDIX D

A recruiting poster for the Flemish-SS Infantry Division (top left).

Bottom left is a recruiting poster for the Volksturm (Home Guard).

A recruiting poster for the Italian-SS Division (left).

A recruiting poster for the Hitler Youth (below). Photos: Antonio Munoz.

APPENDIX E

ORDER OF BATTLE: OPERATION PANZERFAUST
OCTOBER 1944

Commanding Officer
SS-Sturmbannführer Otto Skorzeny

Chief of Staff
SS-Hauptsturmführer Adrian von Foelkersam

Main Combat Units
22.SS-Freiwilligen-Kavallerie-Division Maria Theresa
Mission: the combat group from the Maria Theresa Division is to occupy the railway station and all other important buildings and installations in Budapest. In addition, units are to be positioned around the Burgberg (Castle Hill) and are to provide backup for Sturmbannführer Skorzeny's combat group.

Junkerschule Wiener Neustadt
Mission: the battalion of officer cadets from Wiener Neustadt is to blow open the iron gates leading to the gardens of the Palais on Castle Hill, occupy the gardens, and pin down any Hungarian forces in the area.

SS-Fallschirmjäger-Bataillon 600
Mission: attack and capture of the underground passage leading from the bank of the Danube River to the War Ministry on Castle Hill.

SS-Jagdverbande Mitte
Mission: the units of SS-Jagdverbande Mitte, supported by Tiger tanks from SS-Schwer-Panzer-Abteilung 503, are to drive up the two parallel roads leading to Castle Hill, under the direct command of Sturmbannführer Skorzeny. Indiscriminate rifle fire at the motorized column is to be ignored. The troops should excercise discipline by not showing any aggressive action until the signal from Sturmbannführer Skorzeny.

Sturm-Pioneer-Bataillon 500
Mission: a platoon from the battalion is to assist the main column under Sturmbannführer Skorzeny. Goliath remote-controlled demolition vehicles are to move up Castle Hill with the main combat group.

Commanding Officers
Combat group of the Maria Therese Division: SS-Standartenführer August Zehender
SS-Fallschirmjäger-Bataillon 600: SS-Hauptsturmführer Siegfried Milius
SS-Jagverbande Mitte: SS-Hauptsturmführer Fucker
Junkerschule Wiener Neustadt: Unknown

Date of Operation
16 October 1944, 6:00 A.M.

APPENDIX F

The clothing and equipment issued to the SS paratroopers was the same as that issued to the German air force parachute formations.

The helmet was the standard German army mold, minus the rim. Wearing the standard rimmed helmet while jumping created far more injuries to the neck and shoulders, so the SS paratroopers used the rimless model. (The superiority of this design was recently affirmed by the U.S. Army's adoption of a similar helmet. U.S. soldiers credit this rimless design to the German army, affectionately calling the helmet "the Fritz.")

The parachutists wore the helmet with or without the geometric rope netting (used to support foliage). The forage cap (style M-43 or M-44), worn by the SS paratroopers when they were not in combat, sported the standard Death's Head emblem on front. Pif covers, with the same emblem, were also used.

The paratroopers did not wear standard waterproof overalls; instead, they wore the standard gray-green blouse and trousers. For camouflage, they chose one of the many Luftwaffe camouflage patterns. Most notably, their camouflage took the form of the smock shown at right in the pattern illustrated. The smock went over the uniform. This later version smock had front-fastening zippers. The specifications for the smock were: length of sleeve, 25 cm; length of smock, 10 cm; depth of breast pocket, 7 cm. The smock had five front-opening concealed bottoms, a metal neck fastener, and twelve metal fasteners on the bottom (six on each side) to draw it together. It had two bottoms on the back to help support the cloth holster. All front pockets had zippers under the pocket flaps. Photographs suggest that some SS troops wore this smock like a shirt—that is, they tucked the lower half into the trousers.

In April 1945, the Americans found hundreds of these SS camouflage smocks in a clothing factory in Dachau. Other U.S. soldiers found more smocks in a bombed-out train; apparently these were the first of the SS parachute clothing and had been destined for SS troops. No photographs show the SS camouflage smock being worn, so it is likely that they never reached the SS parachute battalion. Most of the confiscated smocks ended up in the United States, Britain, or Canada as war souvenirs; their rarity makes them one of the most sought-after and valuable war memorabulia from the war. I was fortunate enough to locate and photograph one of the smocks, which is owned by an Irish collector.

The standard Waffen-SS belt buckle and Y-angled leather harness were worn over the smock. Thick rubber pads for the knees and elbows were worn when available. The Luftwaffe breast eagle was apparently not worn on the camouflage smock and was likely removed by the paratoppers themselves.

The parachute used by the Fallschirmjäger in their only jump at Drvar, Yugoslavia, was the RZ-20, an improved model first used during the Crete campaign. A static line replaced the rip cord because it was safer at lower heights (300 feet for static lines versus 1,000 feet for the rip cord). The static line's only disadvantage was that jumpers had less control during descent.

Personal equipment included the standard army canteen and entrenching tool, though the latter was not worn during jumps. The paratroopers received various small arms, including the Gewehr-98 bolt-action rifle, the MG-42 submachine gun, the MP-40 machine pistol, and later the MP-43 machine pistol. The "kamppistole" flare gun was standard issue as well.

Close-up of camouflage pattern. Photo and Illustrations: Antonio Munoz.

This sketch shows an SS paratrooper wearing a parachute cammo smock.

Front

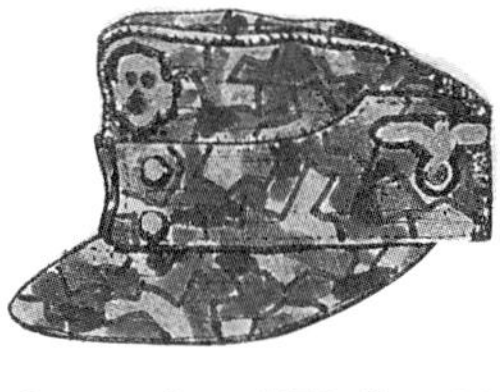

Camouflaged Waffen-SS soft cover.

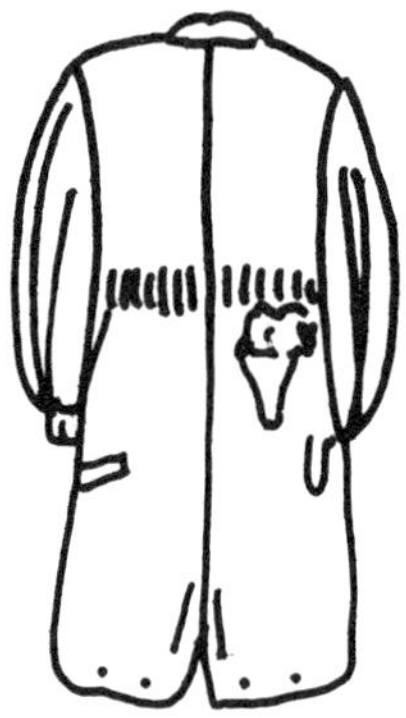

Back

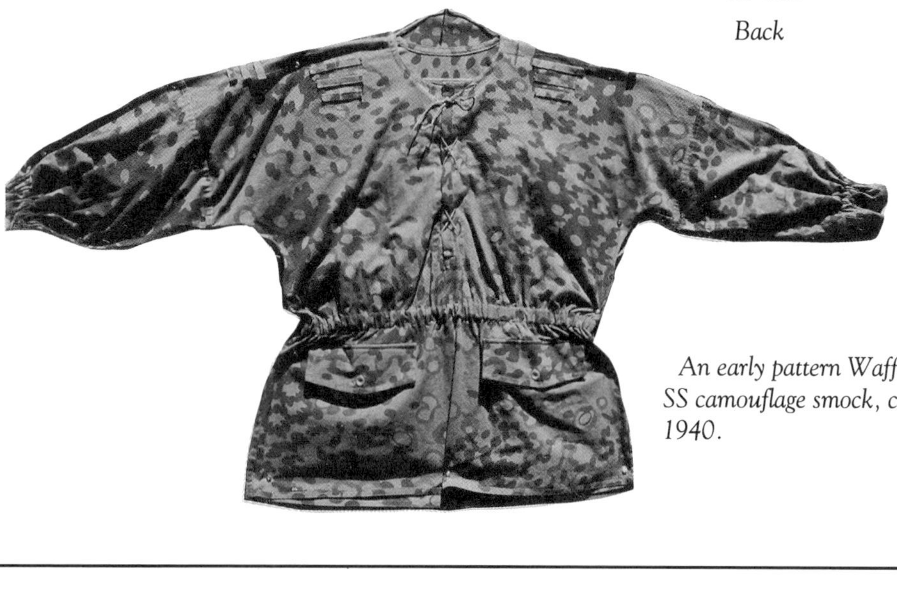

An early pattern Waffen-SS camouflage smock, circa 1940.

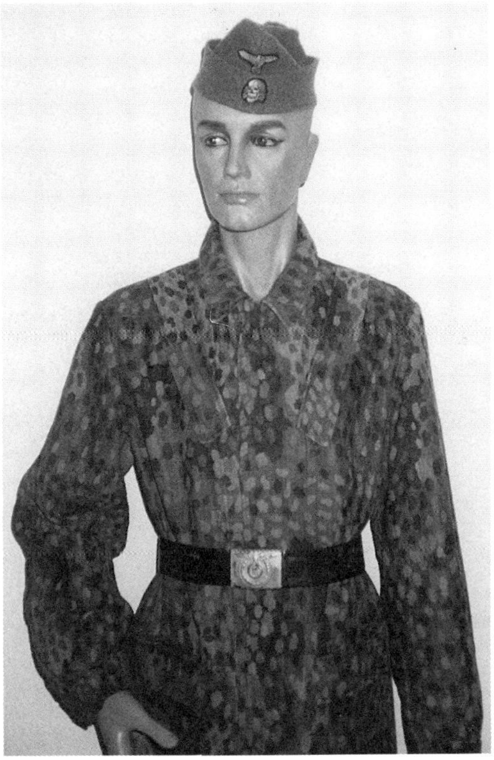

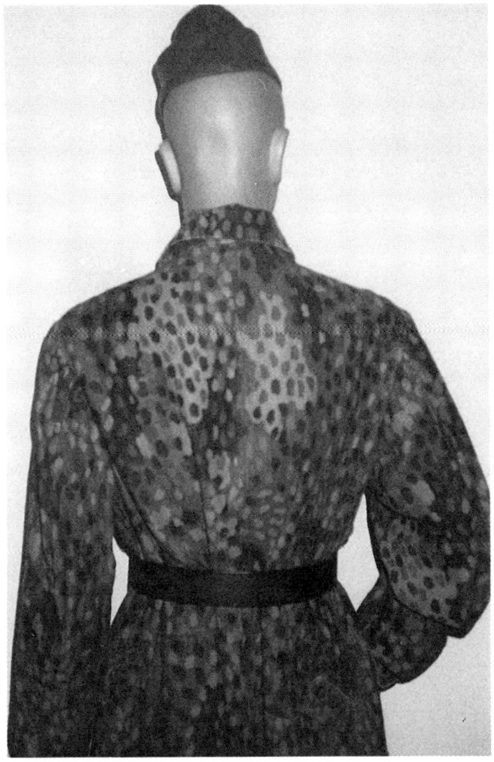

Left, the very rare and never used Waffen-SS paratrooper smock. The first batch of smocks were sent out by train in April 1945, but the train was bombed. Photos: Justin Horgan.

Waffen-SS paratrooper cuffband was worn on the lower left sleeve of the field blouse (below). Photo: Dr. Einar Himma.

APPENDIX G

TABLE OF COMPARATIVE WAFFEN-SS, GERMAN ARMY, AND U.S. ARMY RANKS

Waffen-SS	German Army	U.S. Army
Commissioned Ranks	**Commissioned Ranks**	**Commissioned Ranks**
Reichsführer-SS	Generalfeldmarschall	General of the Army
SS-Oberstgruppenführer	Generaloberst	General
SS-Obergruppenführer	General	Lieutenant General
SS-Gruppenführer	Generalleutnant	Major General
SS-Brigadeführer	Generalmajor	Brigadier General
SS-Oberführer	None	None
SS-Standartenführer	Oberst	Colonel
SS-Obersturmbannführer	Oberstleutnant	Lieutenant Colonel
SS-Sturmbannführer	Major	Major
SS-Hauptsturmführer	Hauptmann	Captain
SS-Obersturmführer	Oberleutnant	1st Lieutenant
SS-Untersturmführer	Leutnant	2nd Lieutenant
Noncommissioned Ranks	**Noncommissioned Ranks**	**Noncommissioned Ranks**
SS-Sturmacharführer	Stabsfeldwebel	Sergeant Major
SS-Standarten-Oberjunker	Oberfähnrich	None
SS-Hauptscharführer	Oberfeldwebel	Master Sergeant
SS-Oberscharführer	Feldwebel	Technical Sergeant
SS-Standartenjunker	Fähnrich	None
SS-Scharführer	Unterfeldwebel	Staff Sergeant
SS-Unterscharführer	Unteroffizier	Sergeant
Enlisted Ranks	**Enlisted Ranks**	**Enlisted Ranks**
SS-Rottenführer	Stabsgefreiter	None
SS-Oberschütz	Obergefreiter	Corporal
SS-Schütz	Gefreiter	Private First Class
	Oberschütz	Private
	Schütz	

Note: Comparisons below commissioned officer grades are all estimates. German office candidate grades have been included; there are no U.S. Army equivalents for these ranks.

APPENDIX H

WAFFEN-SS SUPPLY AND DEPOT CENTERS

City/Town	Military District/Area	Years	Center
Wehlan	I	1944	TSD
Schröttersburg	I	1944	CF/CD
Bütow (West Prussia)	II	1944	CD
Ravensbrück/Fürstenburg	II	1940-1945	SC/CD
Stettin-Stolzhagen	II	1942-1945	CD
Berlin	III	1934-1945	MSD/TSD
Brandenburg-an-der-Havel	III	1944-1945	SH/CD
Oranienburg	III	1944-1945	CD/CF
Sprember/Lausitz	III	1944-1945	CD
Lieberose-über-Cottbus	III	1944-1945	CD
Frankfurt-an-der Oder	III	1943-1944	MSD
Prettin-an-der Elbe	IV	1943-1945	CD
Baden Baden	V	1944-1945	MSD
Rastatt	V	1944-1945	MSD
Kircheim-unter-Teck (Stuttgart)	V	1943-1945	TSD
Iffezheim/Baden	V	1943-1945	MSD
Opladen	VI	1943-1945	TSD
Straubing	VII	1940-1945	CF
Dachau	VII	1935-1945	MSD/CD
München (Munich)	VII	1938-1945	MSD
Breslau	VIII	1941-1945	MSD
Auschwitz	VIII	1940-1944	MSD
Weimar-Buchenwald	IX	1940-1944	SC
Arolsen-Waldeck	IX	1944-1945	CD
Gotha	IX	1942-1944	TSD
Kassel/Breitenbach	IX	1944-1945	MSD
Lüneburg	X	1942-1945	TSD
Danzig	X	1943-1945	CD
Langfuhr	X	1943-1945	TSD
Braunschweig	XI	1940-1945	TSD/CD
Bergen/Belsen	XI	1940-1945	CD
Mürchingen	XII	1943-1945	TSD
Mainz	XII	1943-1945	TSD
Schlackenwerth	XIII	1941-1945	CD
Bayreuth	XIII	1943-1945	CF
Wien (Vienna)	XVII	1938-1945	TSD/MSD
Thorn	XX	1943-1945	TSD
Sandweier	XXI	1943-1944	MSD
Prague/Reuth	Bohemia-Moravia (Czech)	1943-1945	TSD
Beneschau	Bohemia-Moravia (Czech)	1941-1945	TSD
Warsaw	Poland	1940-1944	CD
Poniatowa	Poland	1941-1944	CF

Rowno	Poland	1940-1944	CD
Lublin	Poland	1944-1945	CD
Krakau (Cracow)	Poland	1940-1944	CF
Radom-Blizyn	Poland	1940-1944	CF
Lemberg	Poland	1942-1945	TSD
Trawniki	Poland	1942-1944	TSD/CF
Lomscha (Lomza)	Poland	1940-1944	CD
Litzmannstadt	Poland	1943-1944	CD
Bobruisk	Central Russia	1942-1944	TSD/SH/CD
Kiev	Southern Russia	1942-1943	CF
Dnepropetrovsk	Southern Russia	1942-1944	SH/CD
Reval	Estonia	1943-1944	TSD
Riga	Latvia	1942-1944	TSD/SH/CD
Merano	Italy	1943-1945	CD

KEY: CF-Clothing Factory; CD-Clothing Depot; TSD-Troop Supply Depot; MSD-Main Supply Depot; SH-Supply Headquarters; SC-Supply Camp.

NOTE: Depots and centers in France, Holland, Belgium, Hungary, Finland, Norway, and Yugoslavia not shown.

APPENDIX I

INSIGNIA OF THE OSTTURKISCHER-WAFFENVERBANDE DER-SS

The arm patch of the SS-Waffengruppe Krim had a red inverted blunt trident on a blue background. There is no proof that it was ever worn, though it may have been produced. It was supposed to have been worn by Crimean Tatar volunteers.

The arm patch on the SS-Waffengruppe Idel Ural featured crossed sabres with a perpendicular arrow pointed up, on a dark blue (upper) and green (lower) background. They were worn by Volga Tatar volunteers during their service in the Wehrmacht and Waffen-SS.

The arm patch of the SS-Waffengruppe Turkistan had a mosque with a large blue and white dome, flanked by two white towers topped off by gold domes, on a blue-green background. It was worn by Turkestani volunteers. The top of the arm patch had the following inscription: "BIZ ALLA BILEN." This patch was worn during the volunteers' service in the Wehrmacht and was carried over into the Waffen-SS. A new patch for the Turkestani volunteers was supposed to have been made, containing the national colors of Turkestan, though I have never seen any evidence that it was worn.

This is the right collar tab attributed to the Tatar volunteers of the Waffen-Gebirgsjäger-Regiment der-SS Tatar. This regiment was disbanded in December 1944 and became the basis for the SS-Waffengruppe Krim. The patch may never have been issued. It had a white wolf's head on a black background.

This shows a white sword and shield on a black background. It has been attributed to both the Osttürkischen and Kaukasicher SS formations. It could have been worn by the SS-Aserbaijani Regiment, which was later transferred to the Kaukasicher-Waffen-Verband der-SS.

Illustrations: Antonio Munoz.

This is the arm patch of the 30. Waffen-Grenadier-Division der-SS Weissruthenien. Issued in 1945 when the division became a brigade, the patch had a perpendicular St. George's Cross in red on a black background, flanked by two white areas.

The arm patch of the 21. Waffen-Gebirgsjäger-Division der-SS Skanderbeg had a double-headed Albanian (national) eagle in black on a red background. Photographs exist that show this patch was issued and worn by members of the division.

This is the vehicle identification shield of the 24. Waffen-(Karst)-Jäger-Division der-SS.

Photographs still exist that document the right collar tab of the 30. Waffen-Grenadier-Division der-SS Weissruthenien.

This right collar tab is also attributed to the Weissruthenien Division, although some photographs suggest that it was worn by members of the 29. Waffen-Grenadier-Division der-SS (Russian No. 1).

This right collar tab was of the Albanian 21. Waffen-Gebirgsjäger-Division der-SS Skanderbeg. It featured a medieval helmet with a goat's head, supposedly worn by Prince Skanderbeg, a legendary Albanian hero.

No arm badge is thought to have existed for the 23. Waffen-Gebirgsjäger-Division der-SS Kama (Croat No. 2), but this right collar tab is believed to have been produced—although it is not certain whether it was actually issued or worn.

NOTE: When non-Germanic units of the Waffen-SS were not issued special collar tabs, they wore blank collar patches of black. More often than not, the double thunderbolt SS runes were worn as the right collar tab, even though non-Germans were prohibited from wearing them.

APPENDIX J

WAFFEN-SS COMMANDING OFFICERS
DECEMBER 1944

Formation	Commander
1 SS LSSAH	SS-Standertenführer Wilhelm Mohnke
2 SS DR	SS-Brigadeführer und Generalmajor der Waffen-SS Heinz Lammerding
3 SS "T"	SS-Brigadeführer und Generalmajor der Waffen-SS Hellmuth Becker
4 SS Polizei	SS-Standertenführer Walter Harzer
5 SS Wiking	SS-Oberführer Karl Ullrich
6 SS Nord	SS-Gruppenführer und Generalleutnant der Waffen-SS und Polizei Karl Heinrich Brenner
7 SS PE	SS-Brigadeführer und Generalmajor der Waffen-SS Otto Kumm
8 SS FG	SS-Oberführer Joachim Rumohr
9 SS "H"	SS-Brigadeführer und Generalmajor der Waffen-SS Sylvester Stadler
10 SS "F"	SS-Brigadeführer und Generalmajor der Waffen-SS Heinz Harmel
11 SS "N"	SS-Brigadeführer und Generalmajor der Waffen-SS Joachim Ziegler
12 SS HJ	SS-Standartenführer Hugo Kraas
13 SS Handsch.	SS-Oberführer Desiderius Hampel
14 SS Gal.	SS-Brigadeführer und Generalmajor der Waffen-SS und Polizei Fritz Freitag
15 SS Lett #1	SS-Oberführer Herbert von Obwurser
16 SS RF-SS	SS-Oberführer Otto Baum
17 SS GB	SS-Standartenführer Hans Lingner
18 SS HW	SS-Oberführer Wilhelm Trabandt
19 SS Lett #2	SS-Brigadeführer und Generalmajor der Waffen-SS Bruno Streckenbach
20 SS EST	SS-Brigadeführer und Generalmajor der Waffen-SS Franz Augsberger
21 SS Skanderbeg	SS-Obersturmbannführer Alfred Graaf
22 SS MT	SS-Brigadeführer August Zehender
23 SS Kama	Division already disbanded
24 SS Karst	SS-Obersturmbannführer Wagner
25 SS Hunyadi	SS-Brigadeführer Josef von Grassy
26 SS Hungaria	SS-Oberführer Zoltan von Pisky
27 SS Lngnrk	SS-Oberführer Thomas Müller
28 SS Walne	SS-Standartenführer Léon Dégrelle
29 SS Ital	SS-Brigadeführer Peter Hansen
30 SS "W"	SS-Obersturmbannführer und Oberstleutnant der Schütz-Polizei Hans Siegling
LN SS Brigade	SS-Oberführer Martin Kohlroser
Dirlewanger SS	SS-Gruppenführer und Generalleutnant der Waffen-SS Dr. Oskar Dirlewanger
Nodrland SS Brigade	SS-Brigadeführer Juergen Wagner
33 SS Charl.	SS-Brigadeführer Dr. Gustav Krukenberg
Indian-SS	SS-Oberführer Heinz Bertling
Serbian-SS	General Kosta M. Musicki
XV. SS-Cossack	1st Cossack Cavalry DivisIon: Oberst Wagner 2nd Cossack Cavalry Division: Oberst von Schultz Plastun Brigade: Colonel Kononov
Tatar SS Brigade	SS-Standartenführer Fortenbacher
Osttwvd-SS	SS-Standartenführer Harun-el-Raschid Bey
SS-Falls Battalion	SS-Sturmbannführer Siegfried Milius
150 Panzer Brigade	SS-Obersturmbannführer Otto Skorzeny
Rum-SS No. 1	SS-Sturmbannführer Gustav Wegner

Kaukasischer-SS	Kaukasischer-Waffen-Verbande der-SS: SS-Standartenführer Arbed Theuermann
SS-Panzer Abteilung 502	SS-Sturmbannführer Hartrampf
SS-Panzer Abteilung 503	SS-Sturmbannführer Fritz Herzig
Spanische SS Kp. 101	Waffen Obersturmführer der-SS Valdamos
SS-WrfrAbt.505	SS-Sturmbannführer des Courdes
SS-Panzer Grenadier Schill	SS-Obersturmführer Klotz
Bulgarian-SS No. 1	Waffen-Grenadier-Regiment der SS Nr. 1 (bulgarische Nr. 1): SS-Standartenführer Gunther Anhalt
SS-Flak Abteilung B	SS-Flak-Abteilung B Obersalzberg: SS-Sturmbannführer Hallmann
SS-Flak Abteilung P	SS-Flak-Abteilung Prague: SS-Hauptsturmführer Anton Heider

APPENDIX K

WAFFEN-SS ORDER OF BATTLE
DECEMBER 1944

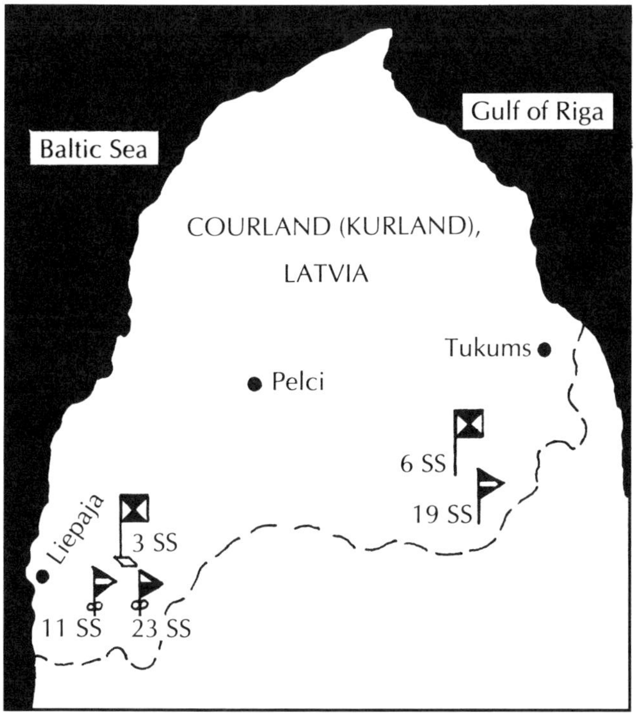

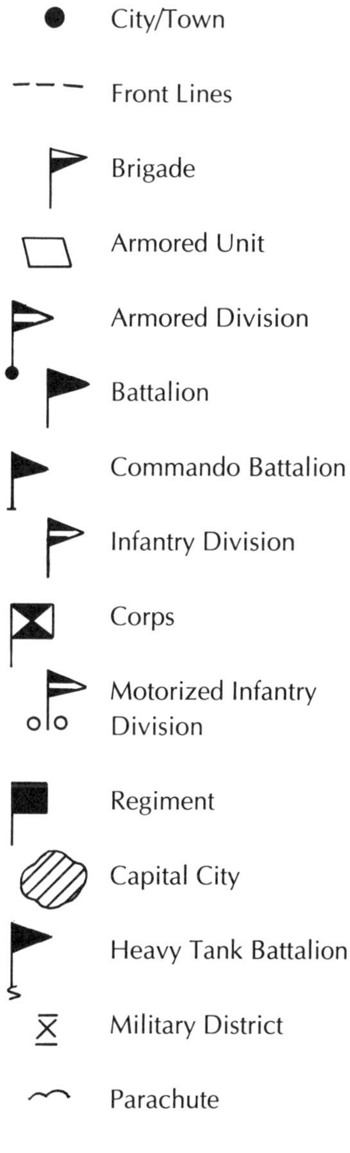

During the last quarter of 1944, the Waffen-SS topped out at about 950,000 men. Well over half of this number was made up of foreigners, including ethnic Germans who had lived for some time outside Germany and people who had no ethnic or national tie to Germany. This appendix lists the main Waffen-SS formations outside of Germany and inside the borders of the Third Reich in December 1944. The maps of Germany also show the German military districts. Cities that are capitalized represent the headquarters of the military district and/or the capital of a country. Towns and cities noted on the maps represent lower military headquarters or training camps. Cities and towns outside the German borders are shown for points of reference.

LOCATION OF WAFFEN-SS FORMATIONS IN YUGOSLAVIA, DECEMBER 1944

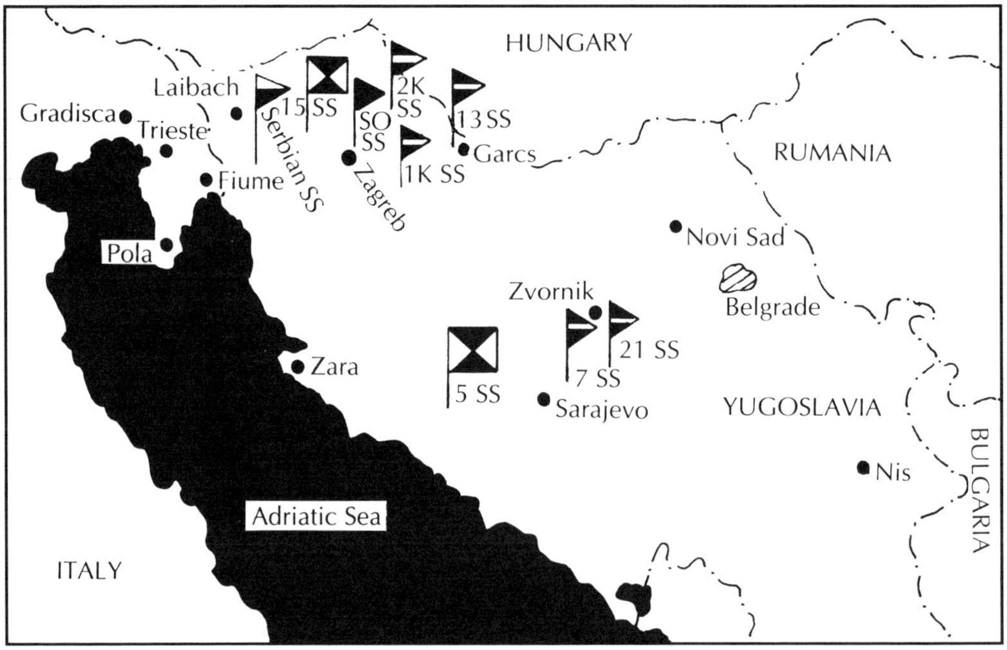

COURLAND, LATVIA

Under the 18th Army, III. (Germanic) SS Panzerkorps:
23. SS-Freiwilligen-Panzergrenadier-Division Nederland
11. SS-Freiwilligen-Panzergrenadier-Division Nordland[1]
121. Infantrie-Division
SS-Korps-Flak-Abteilung 503 (two batteries)
SS-Nachrichten-Abeilung 103
SS-Werfer-Batterie (after November, used as infantrymen because of lack of munitions)

Under the 16th Army, VI.Waffen-Armeekorps der-SS (lett.):
19. Waffen-Grenadier-Division der-SS (lettische Nr. 2)
4. Panzerdivision
12. Panzerdivision
12. Luftwaffen-Feld-Division
93. Infanterie-Division
227. Infanterie-Division
SS-Korps-Flak-Abteilung 506 (five batteries)
Waffen-Grenadier-Regiment der-SS (lettische Nr. 7)
SS-Lehr-Bataillon 506

Under the commander, Army Group North:
Latvian Construction Battalions of the Waffen-SS, Nrs. I-VII, Klavins and Zvaigzne

[1]The 11. SS Division had the SS-Werfer-Battery 521 attached to its 23rd Regiment. Ammunition shortages forced it to be turned into an infantry unit.

YUGOSLAVIA
Under the Higher-SS and Police Leader Trieste
XCVII. Armeekorps z.b.V.
188. Reserve-Gebirgs-Division
237. Infanterie-Division
Serbische-SS-Freiwilligen-Korps[2]
Regiment 1, I-III Batallions (twelve companies): Postojna
Regiment 2, I-IIIrd Batallions (twelve companies): Pivka
Regiment 3, I-III Batallions (twelve companies): Mućićih
Regiment 4, I-III Batallions (twelve companies): Ilirska Bistrica
Regiment 5, I Batallion(four companies): Prem

Under the Replacement Army
XV. Kosaken-Kavallerie-Korps der-SS (Zagreb)—under Army Group "F"
1. Kosaken-Kavallerie-Division[3] (Croatia)

2. Kosaken-Kavallerie-Division[4] (Croatia)
SS-Korps-Nachrichten-Abteilung 11 (three companies)[5]
Kosaken-Pioneer-Bataillon 1 (three companies)
Kosaken-Aufklärungs-Abteilung 55 (three cavalry squadrons)
Kosaken-Artillerie-Regiment 1 (1st Kuban and 2nd Don battalions)

Under the Higher-SS and Police Leader Zagreb
SS-Jagverbande Sudost[5] (parts only, rest of the group in Stockerau, Austria)

Under the 2nd Panzer Army
LXVIII. Armeekorps
Regimentsgruppe 13. Waffen-Gebirgs-Division der-SS Handschar
Waffen-Gebirgsjäger-Regiment der-SS 27 (Kroat Nr. 1) Waffen-Gebirgsjäger-Regiment der-SS 28 (Kroat Nr. 2)
SS-Korps-Flak-Abteilung 509 (two batteries)
SS-Nachrichten-Abteilung 109 (one company)
Panzergrenadier-Division Brandenburg
Grenadier-Brigade-(motorized) 92
44. Infanterie-Division Hoch und Deutschmeister
31. SS-Freiwilligen-Grenadier-Division
1. Gebirgs-Division
117. Jäger-Division

[2] On the map, this unit is represented by a brigade symbol with the words "Serbian SS."
[3] On the map on page 330, this unit is represented by a division symbol with the initials "1K SS."
[4] On the map on page 330, this unit is represented by a division symbol with the initials "2K SS."
[5] On the map on page 330, this unit is represented by a battalion symbol with the initials "SO SS."

NOTE: The XV. Kosaken-Kavallerie-Korps der-SS was in the process of forming a Cossack plastun (infantry) brigade of two regiments.

Under Army Group "E"

V. SS-Freiwilligen-Gebirgs-Korps: no divisional units at this time.

SS-Aufklärungs-Abteilung 505 (three companies)
SS-Korps-Flak-Abteilung 505 (six batteries)
Schwer SS-Artillerie-Abteilung 505 (three batteries)
SS-Sturmgeschütz-Abteilung 105 (two batteries)
SS-Nachrichten-Abteilung 105 (two companies)

XXXIV. Armeekorps

7. SS-Freiwilligen-Gebirgs-Division Prinz Eugen
Regimentsgruppe 21. Waffen-Gebirgs-Division der-SS Skanderbeg SS-Polizei-Regiment 5
SS-Polizei-Regiment 18
11. Luftwaffe-Feld-Division
117. Jäger-Division
264. Infanterie-Division (bodenstandig) [static]: only parts of this unit. 1. Gebirgs-Division (only one regimental group)
Gruppe Koruc z.b.V.
Gruppe Stephan z.b.V.

HOLLAND

Forming directly under 25. Armee control.
34. SS-Freiwilligen-Grenadier-Division Landstorm Nederland

ITALY

Under Army Command Ligurien

29. Waffen-Grenadier-Division der-SS (italienische Nr. l)

Under l0. Armee

LXXIII. Panzerkorps-

16. SS-Panzergrenadier-Division Reichsführer-SS
356. Infanterie-Division
162.(Turkistaner)-Infanterie-Division (only one regiment)
114. Jäger-Division
710. Infanterie-Division (motorized; parts only)

Under Direct Army Group "C" control

Waffen-Gebirgs-(Karstjäger)-Brigade der-SS[6]

WESTERN FRONT

6. Panzerarmee

II. SS-Panzerkorps (had the SS-Werfer-Abteilung 502 plus)
2. SS-Panzer-Division Das Reich
9. SS-Panzer-Division Hohenstaufen
560. Volksgrenadier-Division

[6] On the map on page 339, this unit is represented by a brigade symbol with the words "SS Karst."

5. Panzerarmee (directly under 5.Panzerarmee control)
12.SS-Panzer-Division Hitler Jugend
340. Volksgrenadier-Division

XXXIX.Panzerkorps
1. SS-Panzer-Division Leibstandarte SS Adolf Hitler
3. Panzergrenadier-Division
26. Volksgrenadier-Division

6. Panzerarmee (no divisional units at this time)
Schwer SS-Artillerie-Abteilung 501 (three batteries)
SS-Werfer-Abteilung 501 (three batteries)
SS-Nachrichten-Abteilung (five companies)
SS-Korps-Flak-Abteilung 501 (three batteries)
SS-Korps-Sicherungs-Kompanie (one Field Police company)

15.Armee
XII. SS-Armeekorps
59. Infanterie-Division (bodenstandig)
Division Nr. 176
183. Volksgrenadier-Division
SS-Korps-Nachrichten-Abteilung 112 (two companies)
SS-Feldgendarme-Trupp 112 (Field Police, one platoon)

Directly under Army Group "H" control as a reserve:
10. SS-Panzer-Division Frundsberg

Army Group "G"
1. Armee (directly under 1. Armee control)
17. SS-Panzergrenadier-Division Götz von Berlichingen
21. Panzer-Division
25. Panzergrenadier-Division (battle group)
36. Volksgrenadier-Division
Division Nr. 526
559. Volksgrenadier-Division

XIII. SS-Armeekorps
19. Volksgrenadier-Division
347. Infanterie-Division (bodenstandig)
410. Volksgrenadier-Artillerie-Korps (five battalions)
?. Volksgrenadier-Artillerie-Korps (five battalions)
7. Volksgrenadier-Werfer [rocket]-Brigade
    Werfer-Regiment 83
    Werfer-Regiment 84
18. Volksgrenadier-Werfer-Brigade
Werfer-Regiment 21
Werfer-Regiment 22
SS-Korps-Nachrichten-Abteilung 513 (two companies)

Panzerjäger-Abteilung 566[7]
SS-Korps-Artillerie-Abteilung 513
9. Flak-Division

Oberkommando Oberrhein
XIV. SS-Armeekorps
716. Infanterie-Division (bodenstandig)
Division Nr. 405
SS-Polizei-Nachrichten-Kompanie 41

XVIII. SS-Armeekorps
Grenadier-Bataillon I/Wehrkreis V
Grenadier-Bataillon IV/Wehrkreis V
Grenadier-Bataillon V/Wehrkreis V
Grenadier-Bataillon VI/Wehrkreis V
Grenadier-Regiment "G"/Wehrkreis V (with two battalions)
Grenadier-Regiment "K"/Wehrkreis V (with two battalions) Regimental Staff Battle Groups
  Kampfgruppe "L"/Wehrkreis V
  Kampfgruppe "M"/Wehrkreis V
  Kampfgruppe "N"/Wehrkreis V
  Kampfgruppe "O"/Wehrkreis V (staff + schw.kp.)
  Kampfgruppe "P"/Wehrkreis V (staff + schw.kp.)
Frontier Battalion V[8]
Volkssturm-Bataillon XIII[8] (Alarm)
Volkssturm-Bataillon Wartburg[8]
Pz.Jg.Kp.[8]
II./Luftwaffe-Brigade Oberrhein[9]
Polizei-Bataillon Deggingen[9]
Volkssturm-Bataillon XIII[9] (Baden)
Pz.Jg.Kp.[9]

Leichte (light) SS Panzerjäger Kompanie 95[10]
Leichte Artillerie Abteilung 905[10]
Pioneer Kompanie 905[10]

Nachr. Kp. 805[11]
Freiwilligen-Stamm-Regiment 3[11]
Regiment 5 Oberrhein (with Baden Volkssturm-Bataillon XI, XII, and XX)[11]
Regiment-Stab (staff)/Grenadier-Regiment 1212[11]

[7] This tank-destroyer battalion was armed with the superheavy Jagdtiger tank destroyer carrying the 128mm antitank gun.
[8] All these units became part of the 6th Regiment Oberrhein, raised in Wehrkreis V, forming part of newly raised Brigade 1005 in January 1945.
[9] All these units became part of the 7th Regiment Oberrhein, raised in Wehrkreis V, forming part of newly raised Brigade 1005 in January 1945.
[10] All these units formed Division Nr. 905, raised in Wehrkreis V in January 1945.
[11] All these units formed Division Nr. 905, raised in Wehrkreis V in January 1945.

Regiment Boineburg
19. Armee

LXIII. Armeekorps
30. Waffen-Grenadier-Division der-SS (russische Nr. 2)
159. Infanterie-Division
269. Infanterie-Division
338. Infanterie-Division

EASTERN FRONT
Army Group "A"
SS-Panzergrenadier-Regiment Schill (directly under army group command)[12]
17. Armee
XI. SS-Armeekorps
78. Volksgrenadier-Division
320. Volksgrenadier-Division
545. Volksgrenadier-Division

Army Group South
IV. SS-Panzerkorp (directly under the control of Army Group South, recently arrived from Warsaw, Poland)
3. SS-Panzer-Division Totenkopf
5. SS-Panzer-Division Wiking
711. Infanterie-Division (bodenstandig)
118. Jäger-Division
96. Infanterie-Division
211. Volksgrenadier-Division

8. Armee
IV. Panzerkorps
4. SS-Polizei-Panzergrenadier-Division
18. SS-Freiwilligen-Panzergrenadier-Division Horst Wessel
24. Panzer-Division
46. Infanterie-Division
SS-Sturmbrigade Dirlewanger[13]

Forming under the Replacement Army
Waffen-Gebirgsjäger-Brigade der-SS (Tatar Nr. 1, in Hungary)[14]
Osttürkischen Waffen-Verbande der-SS (in Myjava, Slovakia)[15]

[12]This regiment is not shown on the situation map.
[13] This unit was later designated the 36th Division of the Waffen-SS. On the situation map, it is represented by a divisional symbol with the inscription "36 SS."
[14] On the situation map, this formation is represented by a brigade symbol and the words "Tatar SS."
[15]On the situation map, this formation is represented by a brigade symbol and the inscription "Osttürk W. Vbd. SS."

Army Group Balck (also known as 6. Armee)
IX. Waffen-Gebirgs-Armeekorps der-SS (directly under Balck control and trapped in Budapest).
SS-Polizei-Regiment 1 (also known as Kampfgruppe Doerner)
8. SS-Kavallerie-Division Florian Geyer
22. SS-Freiwilligen-Kavallerie-Division Maria Theresa
Panzergrenadier-Division Feldherrnhalle (also known as 60. Panzergrenadier-Division).
13. Panzer-Division
10. Ungarische-Infanterie-Division
12. Ungarische-Ersatz-Infanterie-Division
Parts of the following:
- 1. Ungarische-Panzer-Division
- 271. Volksgrenadier-Division[16]
- 1. Ungarische-Kavallerie-Division[17]

SS-Gebirgs-Pioneer-Bataillon 509 (one company)
SS-Gebirgs-Artillerie-Regiment 509 (1st through 4th battalions with a total of nine artillery batteries)
SS-Aufklärungs-Abteilung 509 (three companies and one cavalry squadron)
SS-Panzerjäger-Abteilung 509 (three companies)

The following units were forming and undergoing training inside the Reich:

SS-Schwer-Panzer-Abteilung 502 at Camp Senne near Paderborn
15. Waffen-Grenadier-Division der-SS (lettische Nr. l) at Torun (Thorn) in Poland, as an Army Group North reserve. Re-forming.
26. Waffen-Grenadier-Division der-SS Hungaria (ungarische Nr. 2), southwest of Lodz, Poland, committed in January 1945 before fully formed.
25. Waffen-Grenadier-Division der-SS Hunyadi (ungarische Nr. 1), at Neuhammer Training Camp, Silesia
26. Waffen-Ski-Jäger-Bataillon der-SS (ungarische Nr. 2), at Neuhammer Training Camp, Silesia
25. Waffen-Ski-Jäger-Bataillon der-SS (ungarische Nr. 1), at Neuhammer Training Camp, Silesia
2nd Battalion, SS-Panzer-Abteilung Herman von Salza, at Grafenwöhr Training Camp, East Prussia
2nd Battalion, SS-Panzer-Regiment 12, at Paderborn Training Camp 20. Waffen-Grenadier-Division der-SS (estnische Nr. 1), at Neuhammer Training Camp, Silesia
14. Waffen-Grenadier-Division der-SS (galizien Nr. l), at Sillein, Slovakia
28. SS-Freiwilligen-Panzergrenadier-Division Wallonie, at Süd/Hann in Braunschweig
33. Waffen-Grenadier-Division der-SS Charlemagne, at Wildflecken Training Camp, about twenty-five miles northwest of Schweinfurt. Indische-Freiwilligen-Legion der-SS at Klagenfürt Training Camp in Austria.

[16]As Kampfgruppe Schmidthuber
[17]As Brigade Kesseoe. This is the unit that many historians have mistakenly identified as the mythical 33. SS-Hungarian-Cavalry Division.

Kaukasicher-Waffen-Verbande der-SS at Klagenfürt Training Camp
SS-Waffengruppe Armenien at Klagenfürt Training Camp
SS-Waffengruppe Aserbeidschan at Klagenfürt Training Camp
SS-Waffengruppe Georgien at Klagenfürt Training Camp
SS-Waffengruppe Nordkaukasus at Klagenfürt Training Camp

Waffen-Grenadier-Regiment der-SS (bulgarische Nr. l), at Dollersheim Training Camp, Austria
Waffen-Grenadier-Regiment der-SS (rumänische Nr. l), at Dollersheim Training Camp, Austria
Waffen-Grenadier-Regiment der-SS (rumänische Nr. 2), at Dollersheim Training Camp, Austria
SS-Panzer-Jäger-Abteilung 16, Böhmen-Mähren Troop Training Ground in the Czech Protectorate
SS-Artillerie-Abteilung 503 (three batteries), in Altenkirchen/Westerwald
SS-Werfer-Abteilung 505, at Troop Training Ground Kurmark
SS-Schwer-Panzer-Abteilung 503, at Paderborn Training Camp, Germany
Spanischen-Freiwilligen-Kompanie der-SS 101, in Stockerau, near Vienna, Austria

WAFFEN-SS UNITS ON SPECIAL ASSIGNMENT

Two companies of the SS-Fallschirmjäger-Bataillon 600, attached to the 150. Panzer-Brigade
1st Battalion, SS-Panzergrenadier-Regiment 23 Norge, attached to the 5. SS-Panzer-Division Wiking
1st Battalion, SS-Panzergrenadier-Regiment 24 Danmark, attached to the 5. SS-Panzer-Division Wiking
2nd Battalion, Waffen-Artillerie-Regiment der-SS 19, its three batteries attached to the 21. Luftwaffen-Feld-Division in Courland, Latvia

WAFFEN-SS SPECIAL GUARD BATTALIONS

SS-Wach-Bataillon 1 Berlin, stationed in Berlin
SS-Wach-Bataillon 2 Prague, stationed in Prague, Czechoslovakia
SS-Wach-Bataillon 4, stationed at the Waffen-SS Troop Training Ground Kurmark. Like the two battalions above, this guard unit had a staff company and four rifle companies. In late 1944 it was disbanded, and its personnel were used as replacements for frontline Waffen-SS combat units.
SS-Wach-Bataillon 5, stationed at the Waffen-SS Troop Training Ground Westpreussen
SS-Wacht-Bataillon Böhmen-Mähren, stationed in the Czech area of Bohemia and Moravia. It only had a staff company (stationed in Prague) and two rifle companies stationed at Brunn and Thresienstadt.
SS-Wacht-Bataillon Debica, stationed in Troop Training Ground Debica (or Heidlager). Had five companies. It was disbanded when the camp was evacuated, and its personnel were sent to other Waffen-SS units.
SS-Wacht-Bataillon Obersalzberg/3.SS-Gebirgsjäger-Bataillon Obersalzberg. This battalion guarded Hitler's residence in Obersalzberg. In November 1944, it was redesignated 3. SS-Gebirgsjäger-Bataillon Obersalzberg and eventually committed in April 1945.

SS-Wach-Kompanie 7, stationed in Munich
SS-Wach-Kompanie 8, stationed in Vienna
SS-Wach-Kompanie 10, stationed in Budapest, Hungary. It was trapped in Budapest and destroyed along with the rest of the German and Hungarian garrison in early February 1945.
SS-Wach-Kompanie Bohmen, stationed at Troop Training Ground Böhmen-Mähren

OTHER WAFFEN-SS FORMATIONS

Spanische-Freiwilligen-Kompanie der-SS 101 underwent training in Stockerau, near Vienna. It contained four rifle platoons, plus a staff platoon made up of Spaniards. Another Spanish SS company was later formed at the beginning of 1945. It eventually served under the 11. SS-Freiwilligen-Panzergrenadier-Division Nordland in Berlin. The SS-Ausbildungs und Ersatz-Regiment 14 had three Ausbildungs and one ersatz battalion. The Ausbildungs battalions had four rifle companies, while the Erstaz Bataillon had two rifle companies. This unit served as the replacement unit for the 14. Waffen-Grenadier-Division der-SS. It was stationed at Ordca, Slovakia.
SS-Flak-Abteilung Obersalzberg, stationed as air defense for Hitler's retreat in Austria. It had the following units:
1st Battalion (Waffen-SS) with six batteries
2nd Battalion (RAD) with five batteries
3rd Battalion (Waffen-SS) with three batteries
SS-Flak-Abteilung Prag, stationed in Prague, Czechoslovakia. It had three batteries.
SS-Flak-Ausbildungs-und-Erstaz-Regiment Munchen, stationed in Munich. It provided replacements for Waffen-SS flak units.

***

A final note should be made here. This Waffen-SS order of battle is incomplete, but it is fairly comprehensive. There were other Waffen-SS training formations that were not included. Among them is the Latvian SS Field Reserve Brigade that had two regiments with a total of forty companies. It was located in West Prussia, possibly at the Troop Training Ground Könitz. There was also a two-battalion formation in Bohemia-Moravia entitled SS-Grenadier-Regiment Klotz. Regiment Klotz was a combat formation.

LOCATION OF WAFFEN-SS UNITS IN EUROPE, DECEMBER 1944

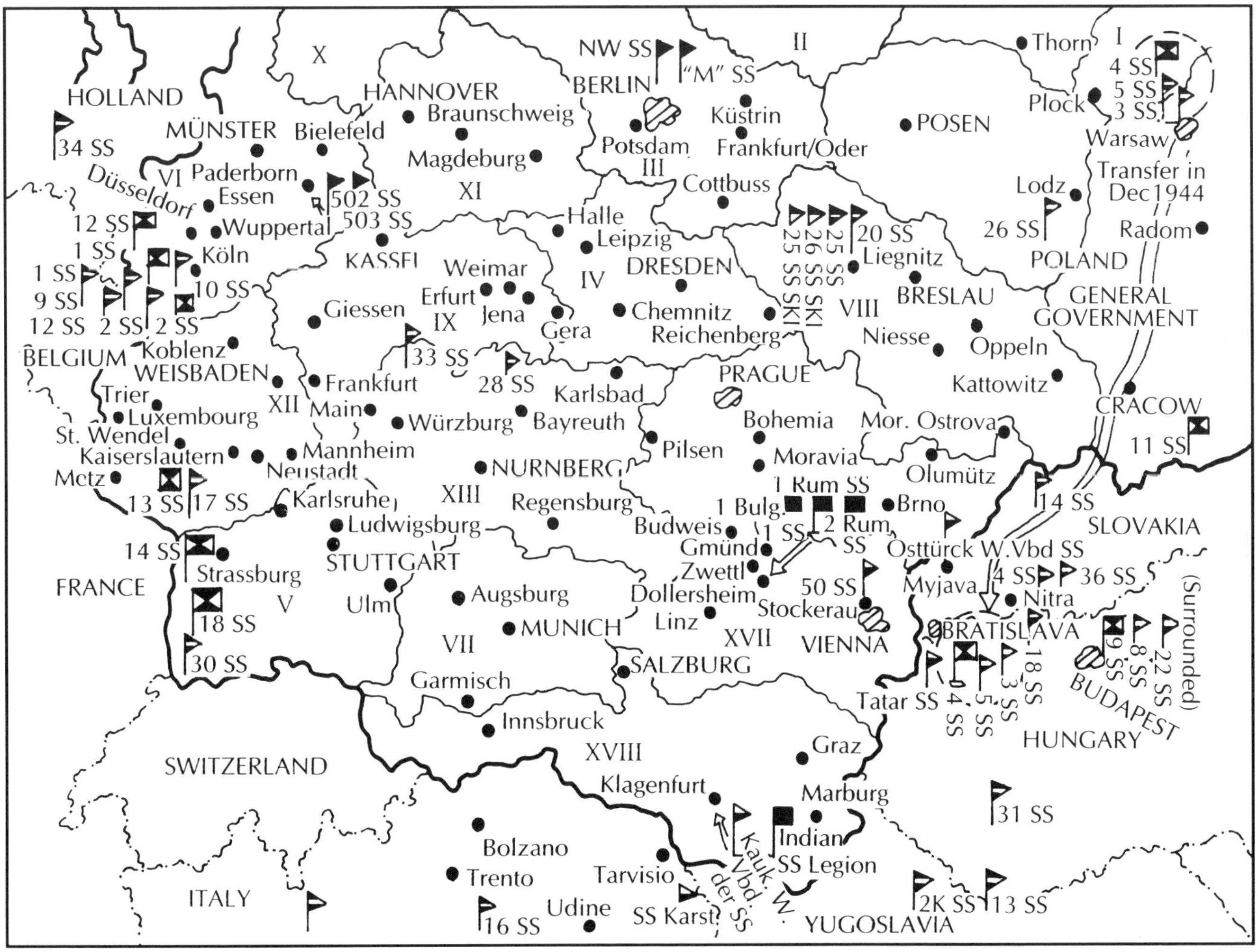

LOCATION OF WAFFEN-SS UNITS IN NORTHERN EUROPE, DECEMBER 1944

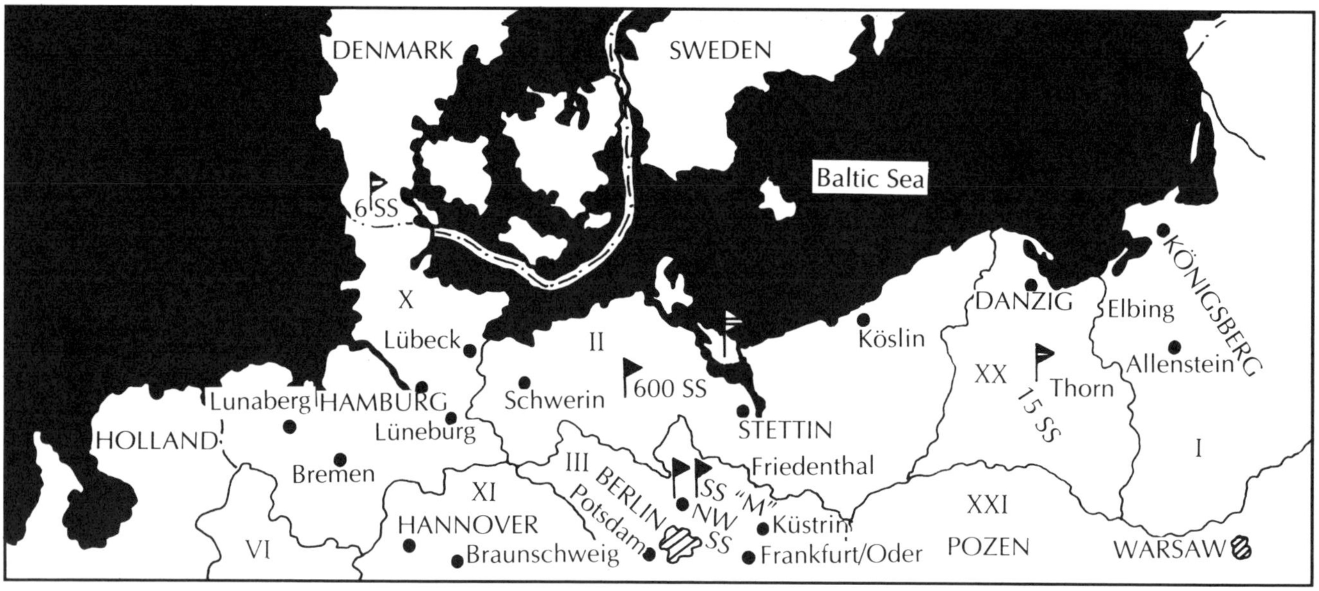

Under the German Replacement Army:
15. Waffen-Grenadier-Division der-SS (lettische Nr. 1): in West Prussia
27. SS-Freiwilligen-Grenadier-Division Langemarck (Flamische Nr. 1): in Luneburgheide
Danish-German Border: 6. SS-Gebirgs-Division Nord (in transit)

Under the command of Kommando-Stab Reichsführer-SS:
SS-Fallschirmjäger-Bataillon 600 (Neustrelitz)
SS-Jagdverbande Nordwest (Friedenthal)
SS-Jagdverbande Mitte (Friedenthal)

APPENDIX L

MAIN WAFFEN-SS TROOP TRAINING GROUNDS, FALL 1944

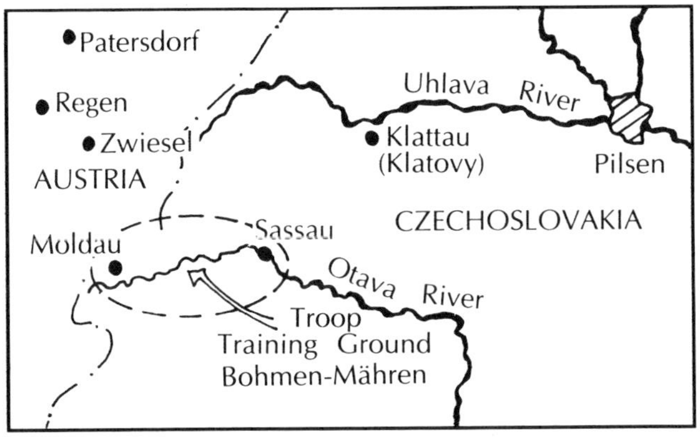

1. TRUPPEN-ÜBUNGSPLATZ BÖHMEN-MÄHREN

Established on 1 November 1941, this camp was originally called Troop Training Ground Beneschau. The name was changed to Böhmen-Mähren in September 1943. The camp contained the following schools: SS-Panzergrenadier-Schule, SS-Pionier-Schule, SS-Artillerie-Schule, and SS-Sturmgeschütz-Schule.

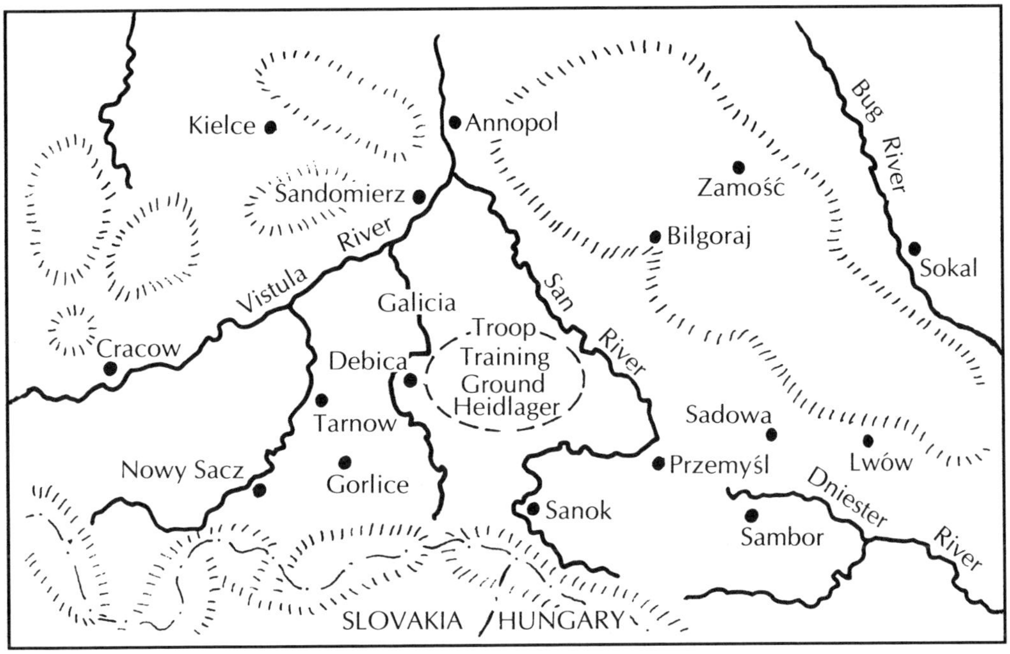

2. TRUPPEN-ÜBUNGSPLATZ HEIDLAGER

Ordered in December 1939 and established on 26 June 1940, this camp was not fully operational until the spring of 1941. Originally called Truppen-Übungsplatz Ostpolen, its name was first changed to Debica and then Heidlager. The camp was authorized to be manned by two reinforced infantry regiments. When the Red Army moved into the area in 1944, the camp staff was disbanded, and the camp itself was burned to the ground.

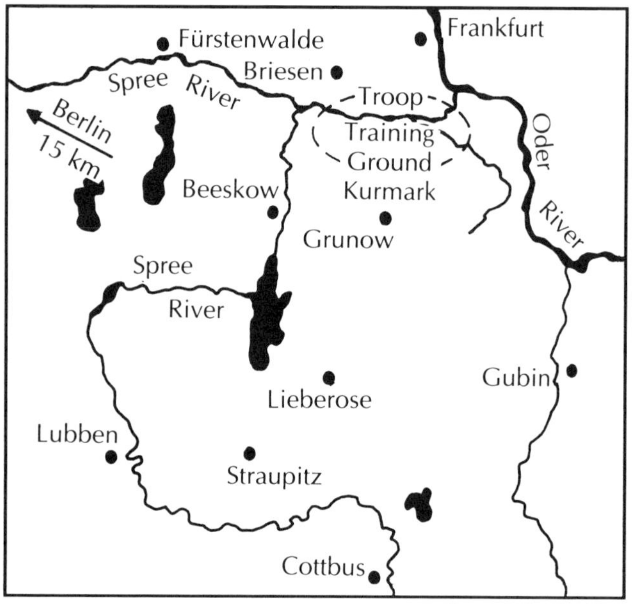

3. TRUPPEN-ÜBUNGSPLATZ KURMARK

Established during the summer of 1943, this camp was located just east and southeast of Berlin, between the towns of Briesen and Grabow. In the beginning of 1945, its schools were used to help form the 32. SS-Panzergrenadier-Division 30-Januar. Among other training units, the SS-Werfer-Ausbildungs-und-Ersatz-Abteilung with eight training batteries was located in this Waffen-SS maneuver area.

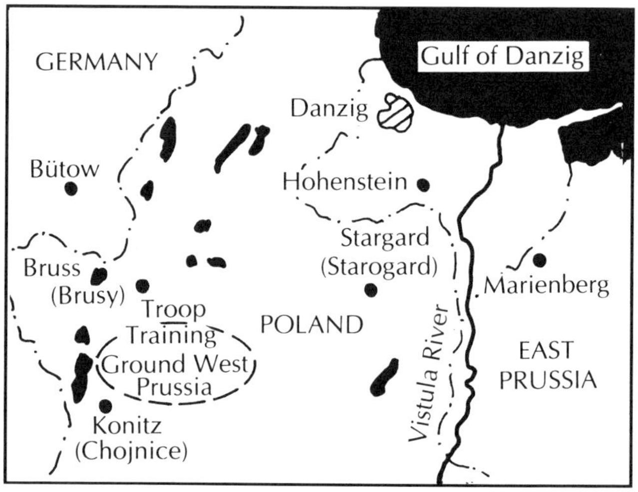

4. TRUPPEN-ÜBUNGSPLATZ WESTPREUSSEN

Formed in the fall of 1943 in West Prussia (what was then known as the Polish Corridor—to the Baltic Sea), this camp had its command structure in the Kelm-Gut district at Bruss (Brusy). The training camp itself, usually referred to as "Konitz," lay between Konitz (Chojnice) and Bruss. SS-Sturmbannführer der Reserve Gustav Fricke led the Guard Battalion at this camp.

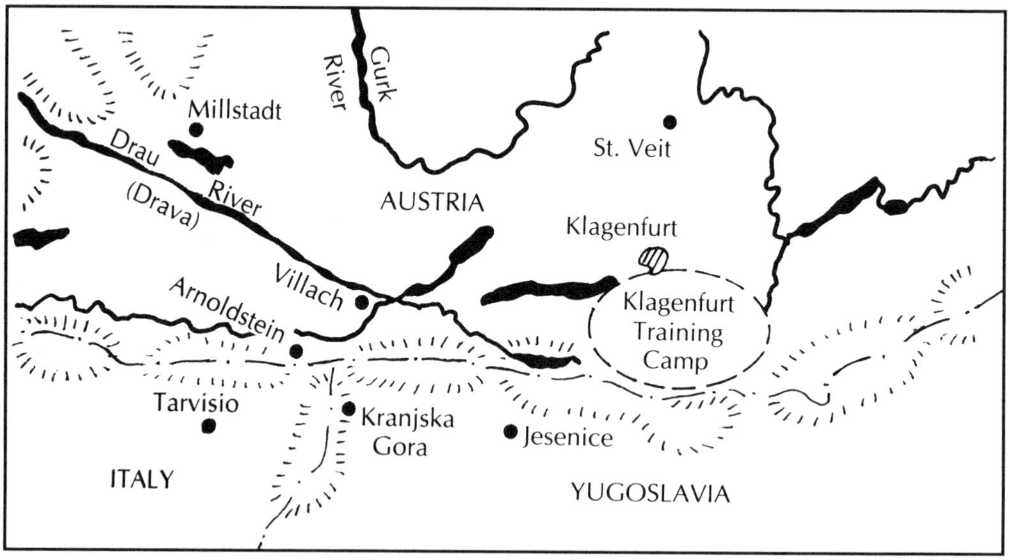

5. TRUPPEN-ÜBUNGSLAGER KLAGENFURT

This camp was not a maneuver area like the first four camps, but rather a training camp for newly forming units. Many foreign volunteer units of the Waffen-SS formed at this site. It remained active until the end of the war. The camp was located in Austria, near the Yugoslavian-Italian border.

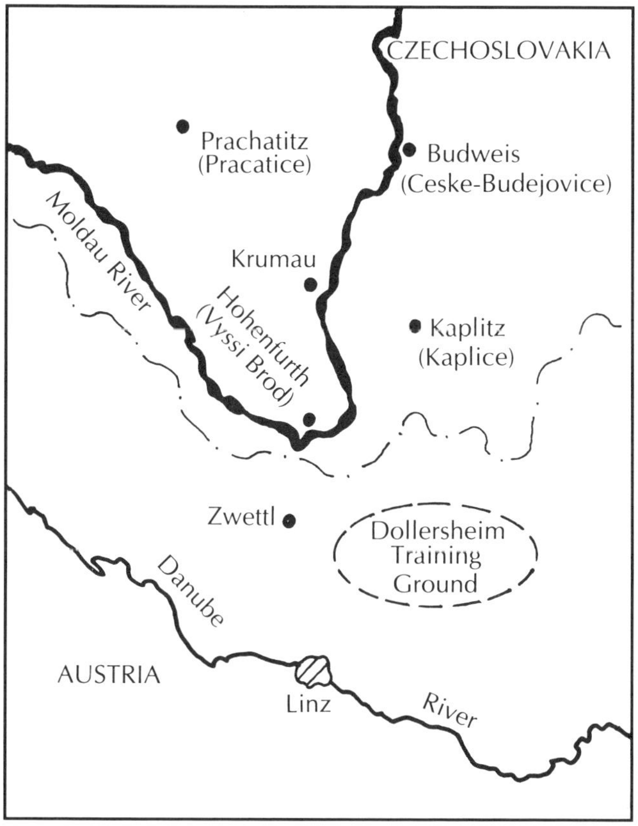

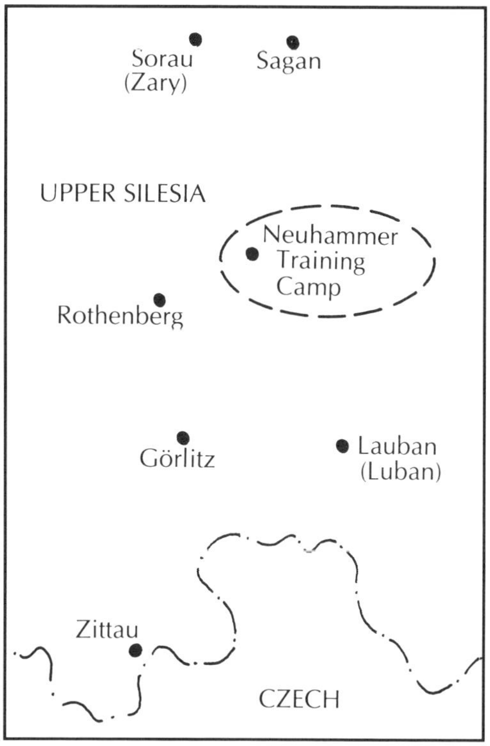

6. TRUPPEN-ÜBUNGSLAGER DOLLERSHEIM

Like Klagenfurt, this camp was also not a maneuver area; it was a training camp. Located twelve kilometers east of the Austrian town of Zwettl, near the Czech border, this camp also saw heavy use by forming Waffen-SS units.

7. TRUPPEN-ÜBUNGSLAGER NEUHAMMER

This training camp was active in preparing forming and re-forming combat units of the Waffen-SS. It was located in Upper Silesia, about twenty-five kilometers northeast of Görlitz.

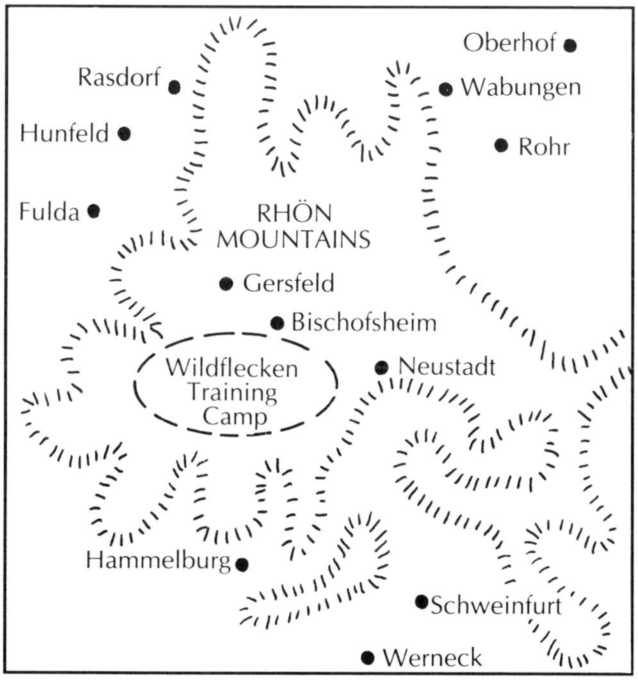

8. TRUPPEN-ÜBUNGSLAGER WILDFLECKEN

This camp, located about twenty-five kilometers northwest of Schweinfurt in the German province of Hessen, was the site of the newly forming 33. Waffen-Grenadier-Division der-SS Charlemagne in December 1944.

APPENDIX M

"In June 1944 the Spanish volunteers who refused to go back to Spain formed the Volunteer Unit Stablack (Freiwilligen-Einheit Stablack) in the strength of the company. (By the way, Stablack is the name of a German training center in East-Prussia.)

"In March 1945, out of all Spanish volunteers, two companies were formed: SS-Volunteer Company 101 (Spanish) and SS-Volunteer Company 102 (Spanish).

"In the lists of the field units of the Waffen-SS only these two companies exist, not a battalion. Also the name of the commanding officer, SS-Haupsturmführer Miguel E. Sanchez, is not known.

"The members of both the companies became split up and served in several units of the Waffen-SS.

"It is known that the Spanish volunteers of the Waffen-SS wore the army-type arm shields and the standard collar patches of the Waffen-SS. No cuff title was given out."[1]

The above extract clearly indicates that the Spaniards at Training Camp Stablack were eventually formed into Spanish SS rifle companies Nos. 101 and 102. The officer in question, SS-Haupsturmführer Miguel E. Sanchez (should be properly referred to as Waffen-Haupsturmführer der-SS Miguel E. Sanchez), is quite possibly Miguel E. (Ezquerra) Sanchez. It is common in Spain to list an individual by both the father's and mother's last names. In any case, Ezquerra/Sanchez did not take command of these volunteers until 1945.

[1] Taken from *Der Freiwillige* magazine, September, 1980, page 31.

DOCUMENTS RELATING TO THE FORMATION OF A SPANISH VOLUNTEER BATTALION AT TRAINING CAMP STABLACK

Anlage 7:

a) OKH/Chef H.Rüst u BdE/AHA/Stab Ia(3)
Nr. 51 925/44 geh. v. 14.11.44 = Seite 1
b) OKH/Chef H Rüst u BdE/AHA/Stab Ia(3)
Nr. 56561/44 geh. v. 14.11.44 = Seite 1
c) OKH/Chef H Rüst u BdE/AHA/Stab III
Nr. 3545/44 g.Kdos. v. 7.6.44 = Seite 1-5

Anlage 7

S p a n i e r

Abschriit der Verfügungen

a) OKH/ Der Chef der Heeresrüstung und Befehlshaber des Ersatzheeres/AHA/Stab la(3) Nr. 51925/44 geh. v. 14.11.44.
Bezug: 1) OKH/Ch H R u BdE/AHA/Stab /Chef Abt.Nr. 45199/44 geh. v. 22.9.44
2) OKH/Ch H R u BdE/AHA/#tab Ia(3) Nr. 56561/44 geh. v. 14.11.44.
Betr.: Kürzungen bei Freiw. Ers.Btl.(span.) und Freiw.Ausb. Btl.(span.)

b) Der Chef der Heeresrüstung und Befehlshaber des Ersatzheeres AHA/Stab la(3) Nr. 56561/44 geh. v. 14.11.44
Bezug: OKH/Ch H R u BdE/AHA/Stab /Chef Abt.Nr.45199/44 geh. v. 22.9.44.
Betr.: Verlegungen und Umbenennung des Freiw.Ers.Btl.und Frei.Ausb.Btl.

1.) Die Einheiten werden umbenannt in
Freiw.Ausb.Btl.(span.)
Freiw.Ers.Btl. (span.)
2.) Es werden verleft:
a) Freiw.Ausb.Btl.(span.) nach Stockerau b) Freiw.Ers.Btl. (span.) nach Hellabrunn
3.) Dienststempel und Dienstsiegel bleiben unverändert.
Die Bezugsverfügung ust zu vernichten.

c) OKH/Chef H Rüst u BdE/AHA/Stab III Nr. 3545/44 g.Kdos. v. 7.6.44
Betr.: Aufstellung der Freiwilligen-Einheit Stablack. An

W.Kdo. I

A.A l l g e m e i n e s .

1.) Durch W.Kdo.I ist unter Auflüsung des Ersatz-Kommandos der Span.Legion die

Freiwilligen-Einheit S t a n b 1 a c k

mit Stab Ers-u-Ausb.Btl. gem. K.St.N. 6011a v. 1.5.43 und 1 Schützenkp.gem.K.St.N 131n v. 1.12.43 (deutsches Rahmenpersonal ist gem.Anl.l)das restl.Personal aus Freiwilligen zu stellen) aufzustellen. Die Aufstellung weiterer Kompanien wird von Fall zu Fall befohlen.

2.) In der Freiwilligen-Einheit Stablack sind all Freiwilligen spanischer Nationalität zu erfassen.

Der Btl.Stab übernommt die Aufgaben eines Ausb.-Stabes.

3.) Feldverwendungsbereitschaft der Kp. ist bis 31.7.44 herzustellen.

B. Durchführungsbestimmungen.

1.) Personelle Bestimmungen
Das Personal des Ersatz- Kommandos der Span.Legion steht zur Verfügung. Geeignetes Personal ist zu übernehmen. Nicht geeignetes milit.personal steht W.Kdo. I zur Verfügung. Nicht geeignetes Fachpersonal ist den zuständigen Dienststellen des OKH zu melden.

Jahrgangmässice Zusammensetzung:

Stab Ers.u.Ausb.8tl.:	Nach den Bestimmungen der Verfügung OKH/Chef H Rüst u BdE/AHA/Stab III Nr. 12 000/44 geh. v. 18.5.44 An1.2 Nr. 3.
Schützenkomp:	Nach den für das Feldheer gültigen Bestimmunge.

a) Offz.Stellenbesetzung erfolgt durch H.P.A. auf Vorschlag W.Kdo.l.
In den Stab Ers.u.Ausb.Btl. der Freiw.-Einheit Stablack sind zu übernehmen.
Hptm. G r a e f e, Führer des Ers.Kdo.d.SPan.Le8. Kt. P a b t h e r Ers.Kdo.D.Span.Leg.
OKW/W.Pr. wird geheten, den Sd.Fhr.(Z) F i n k h mit Wirkung vom !4.6.44 zur Freiw.-Einheit Stablack zu kommandieren

b) Beamte werden durch OKH/Chef H Rüst u BdE/VA zugewiesen.

c) Uffz. und Mannschaften (deutsches Rahmenpersonal) stellt W.Kdo.l unter Rückgriff auf Genesende.

d) Dolmetscher stellt W.Kdo. 1 Aufteilung:
Stab Ers.u.Ausb.Btl.: 2 Dolmetscher
Schützenkomp.13 Dolmetscher

2.) Materielle Bestimmungen:
a) Waffen, Munition und Gerät usw- werden durch die zuständigen Stellen des OKH ohne Anforderung zugewiesen.
b) Wirtschaftsgerät: stellt Fz.Kdo.l
d) Uberzählige Kfz. sind an der nächsten H.K.P. abzugeben.

3.) Feldpostnummer für die Schützenkp. ist bei Gen.St.d.H./Gen.Qu./ Qu 2 anzufordern.

4.) Dienststempel sind bei H.Za. Spandau anzufordern.

C Sonstige Bestimmungen:

1) Vereidigung.
Die Freiwilligen sind in der l.Woche nach ihrem Eintreffen in Stablack nach der Eidesformel gem.An1.2) zu vereidigen.

2) Ausbildung:
Die Ausbildung dauert 10 Wochen und erfolgt auf der Grundlage der gültigen Vorschriften mit dem Ziel: Aufstellung einer frontverwendungsfähigen Schützenkp.

3) Beförderung:
Jeder Freiwilliger wird ohne Rücksicht auf seinen ehemaligen Dienstgrad als Schütze in die Freiw.-Einheit Stablack eingestellt.
Beförderungen bis einschl. zum Offizier erfolgen grundsätzlich nach den Beförderungsbestimmungen der deutschen Wehrmacht. Nach Bewährung können Beförderungen in Anrechnung der Dienstzeit auch ausserhalb der deutschen Wehrmacht und ohne Einhaltung der in den Beförderungs bestimmungen geforderten Fristen ausgesprochen werden.

4) Erfassung:
Hinsichtlich der Ausgabe von Erkennungsmarken, Soldbüchern usw. gelten dieselben Bestimmungen wie för deutsche Soldaten.

5) Debührnisse:
a) Die Abfindung regelt sich den Bestimmungen des Erlasses A.H.N.41. Ziff. 787.
Kriegsbesoldung ist auch den Freiwilligen zu gewähren, die den Wehrsoldgruppen 15 und 16 angehören,
b) Auszahlung der Gebührnisse erfolgt durch die Standortgebührnisstelle Stablack.
c) Das Verfügungsrecht über die Kriegsbezüge im In land und die Uberweisung von Ersparnissen in die Heimat regelt sich nach dem Erlass HV.B1.43. Teil B, Ziff 179.

6.) Versorgung
Die Versorgung regelt sich nach den Erlass A.H.M.41 Ziff 892 und seinen Ergänzungen.

I.A.
gez.Unterschrift.

Eidesformel

Ich schwöre bei Gott diesen heiligen Eid. dass ich im Kampf gegen die bolschewisfischen Fiende meiner Heimat dem Obersten befehlshaber der Deutschen Wehrmacht, Adolf H i t l e r undedingten Gehorsam leisten und bereit sein will, jederzeit för diesen Eid mein Leben einzusetzen.

–·–

Abweichungen der S p a n i e r — Verfülgungen von denjenigen der Ostvölker.

Eidesformal: Anderer Wortlaut als bei Ostvölkern.
Beförderung: Ubernahme in die Deutsche Wehrmacht grundsätzlich als Schütze.
Kriegsbesoldung:
Wird auch gewährt för Angehörige der Whersoldgruppen 15 und 16.

Top right is Armin Blesinger's SS Paratrooper Soldbuch (note his name on the cover).

Blesinger's photograph on the inside front cover. Page one shows the rank of SS-Unterscharführer (sergeant) as of 2 July 1944 (below right). Photos: Michael J. Belles Collection.

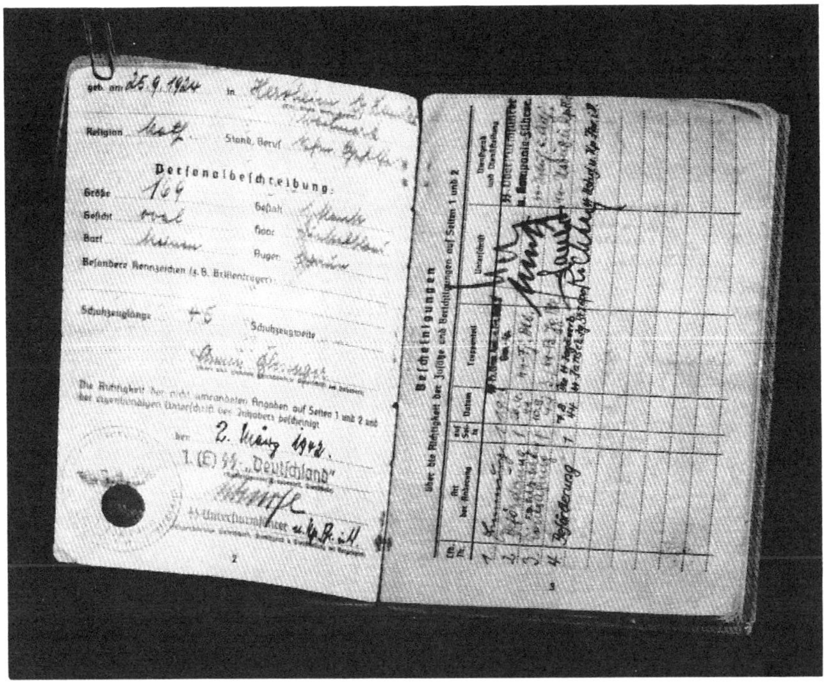

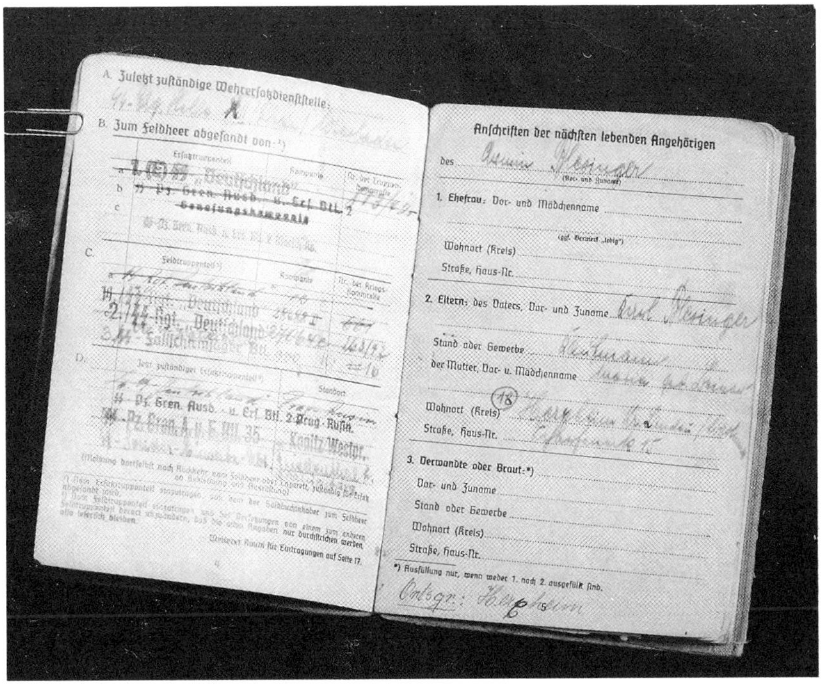

Note the top photograph on page two. The top left page gives a description of the individual who holds this soldbuch plus his date of birth (25 September 1924). At the bottom of page two is the date that the SS soldbuch was issued (2 March 1942) and the unit issuing it. In this case, it was the First Company, SS-Panzer-grenadier-Ausbildungs und Ersatz Abteilung 1 Deutschland.

Page three on the top right lists the units under which Armin Blesinger served in sequential order. From his training battalion, he was posted to the SS-Fallschirm-jäger-Bataillon 500 on 20 April 1944. He was later attached to that Battalion's 3rd Company on 10 September 1944 when he was posted to the SS-Jagdverband/SS-Fallschirm-jäger-Bataillon 600. This date is probably an error and should probably read 1945 not 1944. The parachute battalion came under the jurisdiction of the SS-Jagdverband only in October 1944 so the real date is probably 1945.

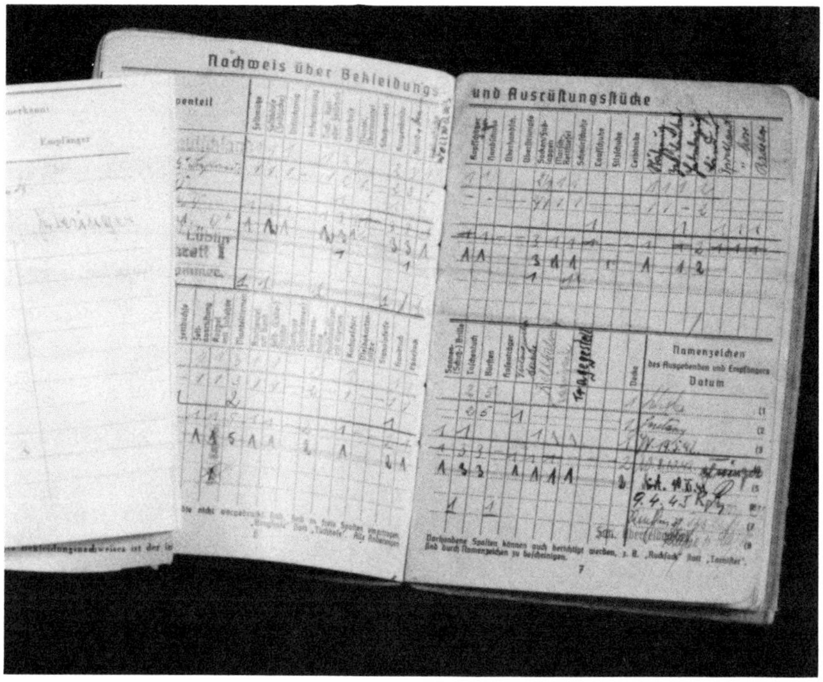
Nachweis über Bekleidungs
und Ausrüstungsstücke
Namenzeichen
Datum

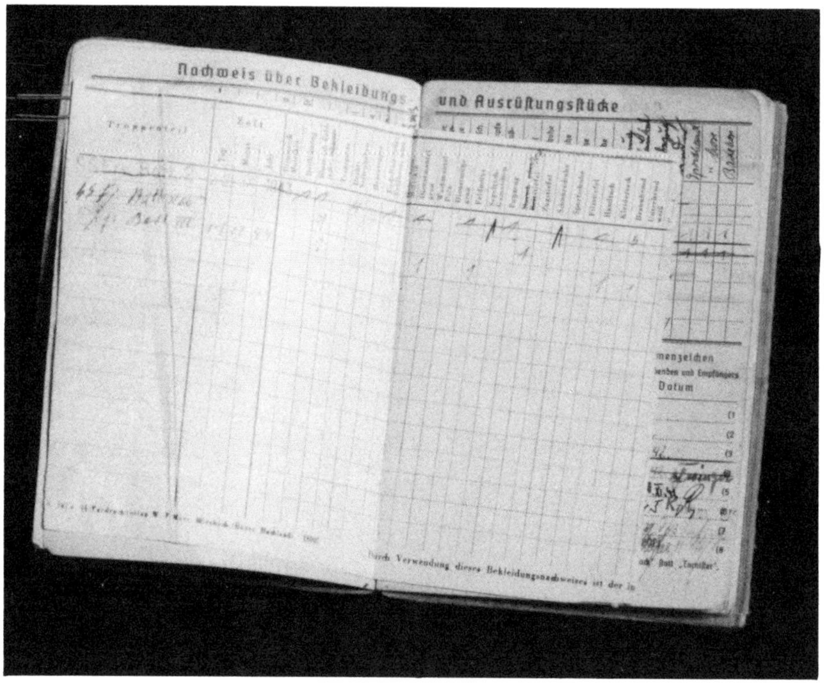
Nachweis über Bekleidungs
und Ausrüstungsstücke
Datum

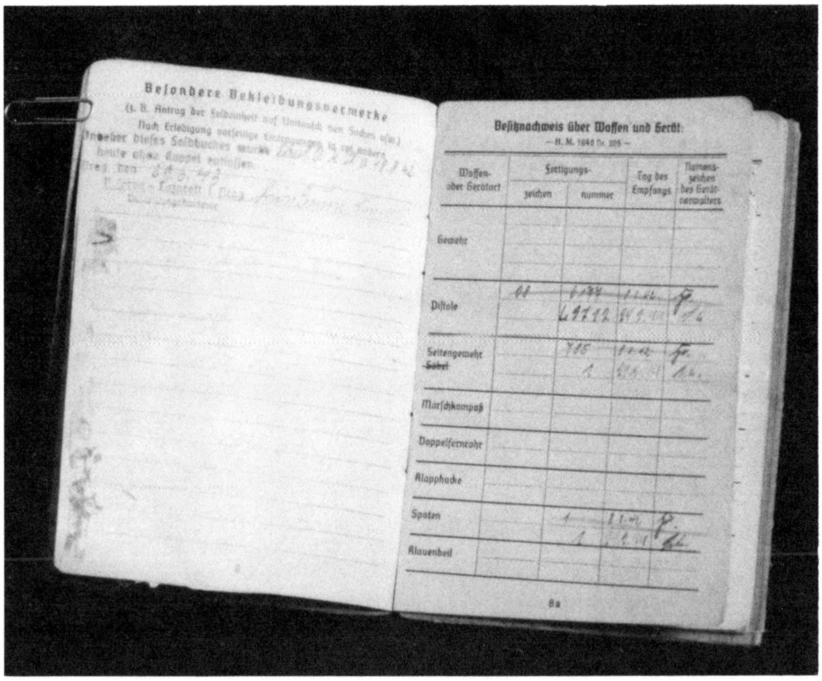
Besondere Bekleidungsvermerke
Besitznachweis über Waffen und Gerät:
— H. M. 1940 Nr. 205 —
Waffen- oder Gerätart
Fertigungs-
zeichen
nummer
Tag des Empfangs
Namenszeichen des Gerätverwalters
Gewehr
Pistole
Seitengewehr
Säbel
Marschkompaß
Doppelfernrohr
Klapphacke
Spaten
Klauenbeil
8a

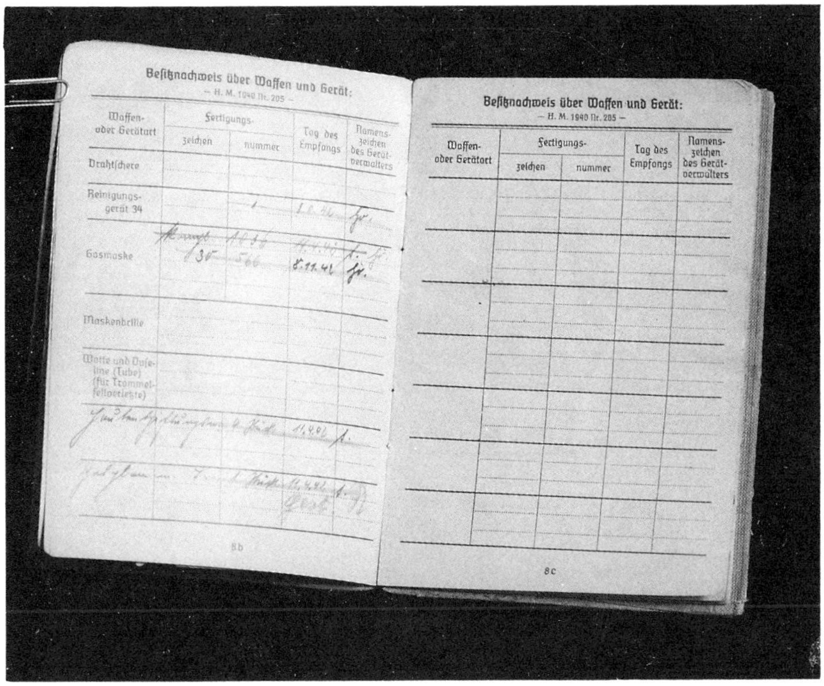
Besitznachweis über Waffen und Gerät:
— H. M. 1940 Nr. 205 —
Waffen- oder Gerätart
Fertigungs-
zeichen
nummer
Tag des Empfangs
Namenszeichen des Gerätverwalters
Drahtschere
Reinigungsgerät 34
Gasmaske
Maskenbrille
Watte und Vaseline (Tube) (für Trommelfellverletzte)
8b
Besitznachweis über Waffen und Gerät:
— H. M. 1940 Nr. 205 —
Waffen- oder Gerätart
Fertigungs-
zeichen
nummer
Tag des Empfangs
Namenszeichen des Gerätverwalters
8c

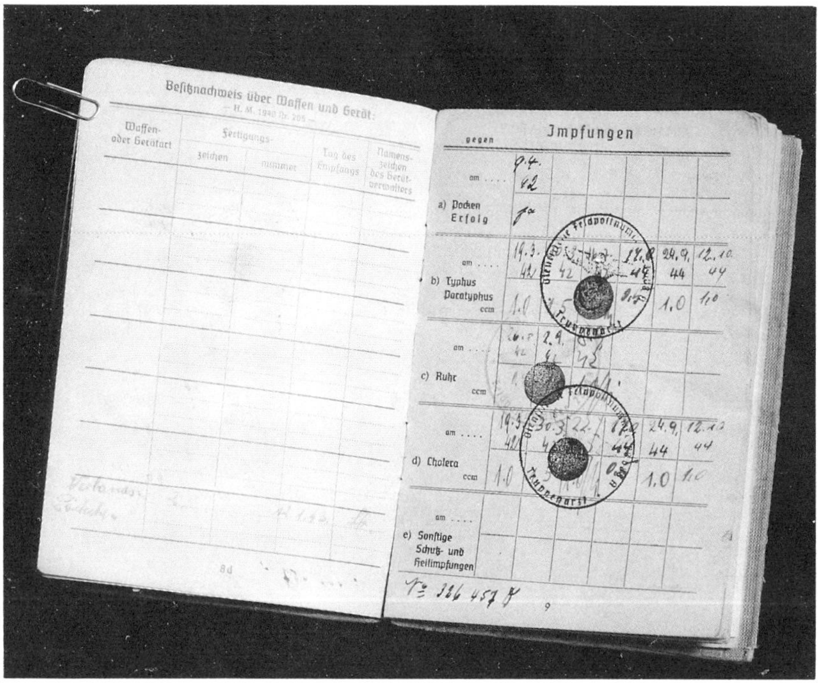
Besitznachweis über Waffen und Gerät
Impfungen
a) Pocken Erfolg
b) Typhus Paratyphus
c) Ruhr
d) Cholera
e) Sonstige Schutz- und Heilimpfungen

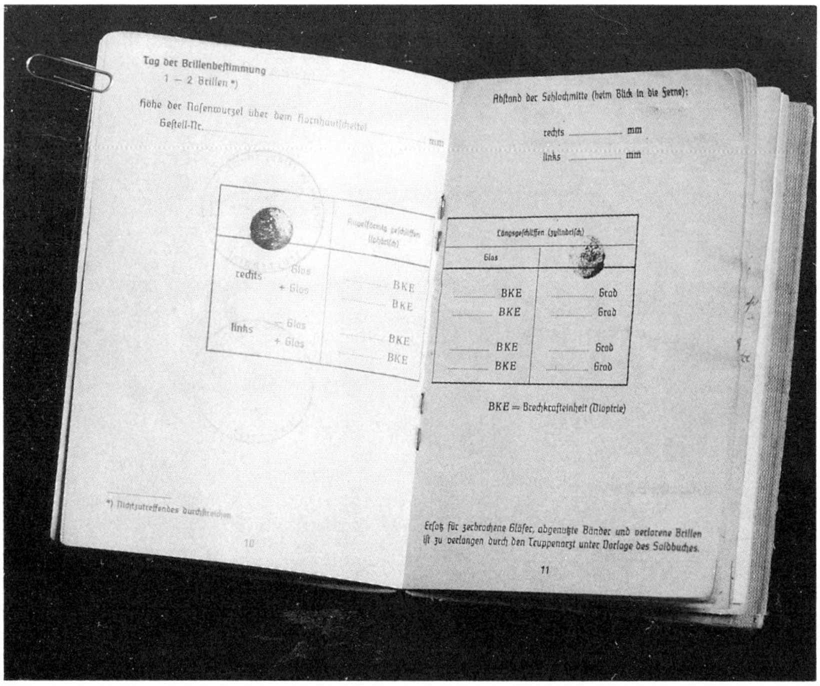
Tag der Brillenbestimmung
Abstand der Sehlochmitte (beim Blick in die Ferne):
BKE = Brechkrafteinheit (Dioptrie)
Ersatz für zerbrochene Gläser, abgenutzte Bänder und verlorene Brillen ist zu verlangen durch den Truppenarzt unter Vorlage des Soldbuches.

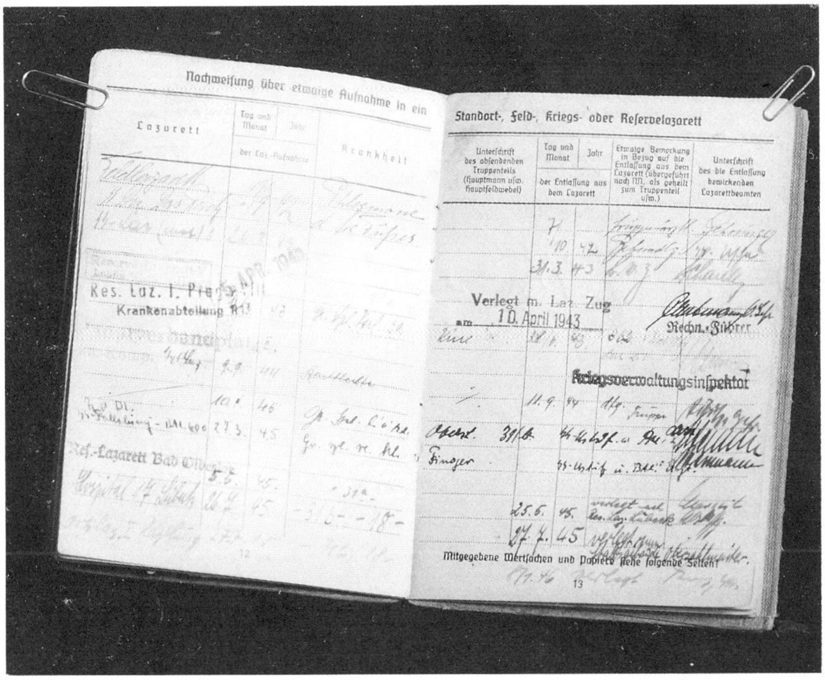

Nachweisung über etwaige Aufnahme in ein Standort-, Feld-, Kriegs- oder Reservelazarett

Lazarett	Tag und Monat der Laz.-Aufnahme	Jahr	Krankheit	Unterschrift des absendenden Truppenteils (Hauptmann usw. Hauptfeldwebel)	Tag und Monat der Entlassung aus dem Lazarett	Jahr	Etwaige Bemerkung in Bezug auf die Entlassung aus dem Lazarett (übergeführt nach ??, als geheilt zum Truppenteil usw.)	Unterschrift des die Entlassung bewirkenden Lazarettbeamten

Res. Laz. I. Prag
Krankenabteilung
Verlegt m. Laz. Zug
am 10. April 1943
Rechn.-Führer
Kriegsverwaltungsinspektor
Res.-Lazarett Bad
Mitgegebene Wertsachen und Papiere siehe folgende Seiten!
12
13

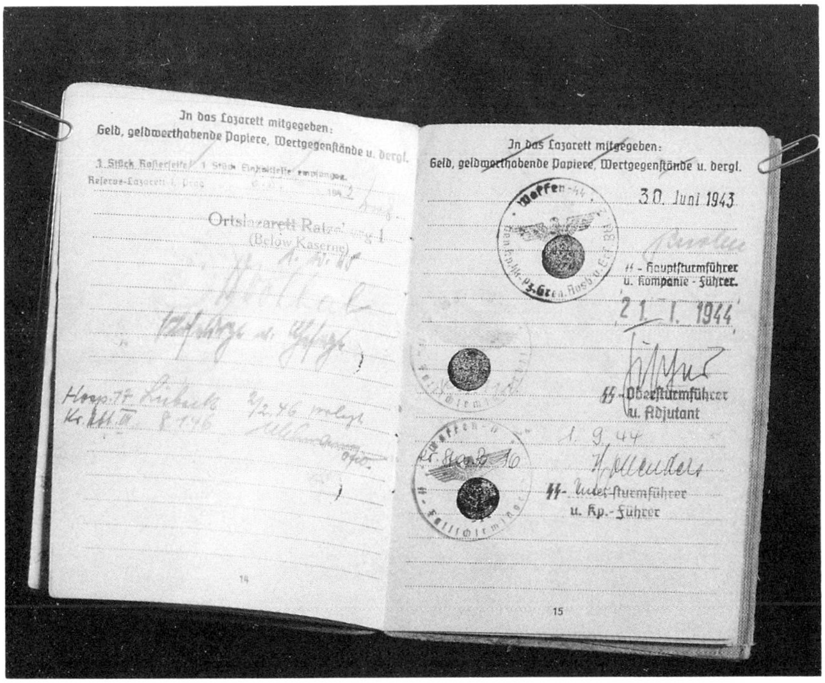

In das Lazarett mitgegeben:
Geld, geldwerthabende Papiere, Wertgegenstände u. dergl.
Ortslazarett
(Below Kaserne)
14

In das Lazarett mitgegeben:
Geld, geldwerthabende Papiere, Wertgegenstände u. dergl.
30. Juni 1943
SS-Hauptsturmführer u. Kompanie-Führer.
21. I. 1944
SS-Obersturmführer u. Adjutant
SS-Untersturmführer u. Kp.-Führer
15

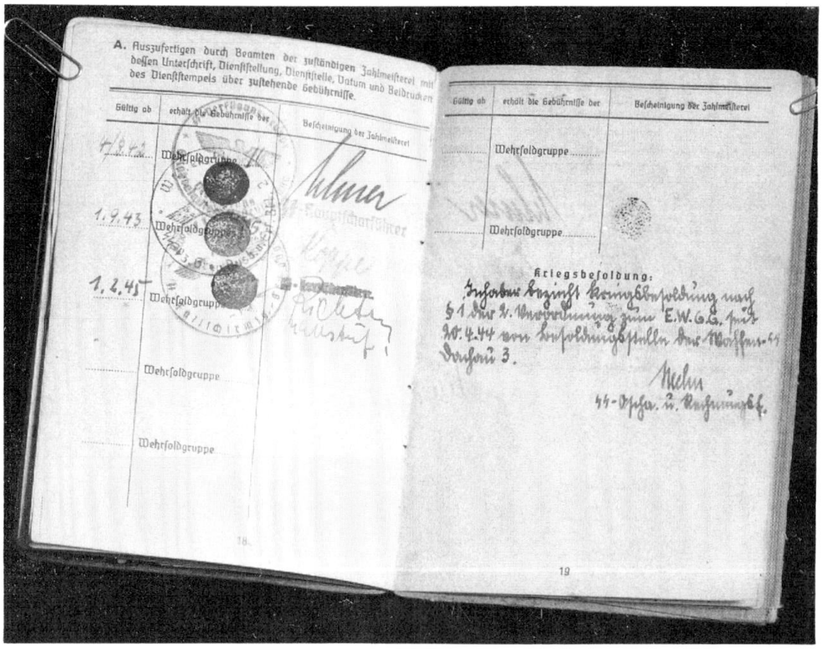

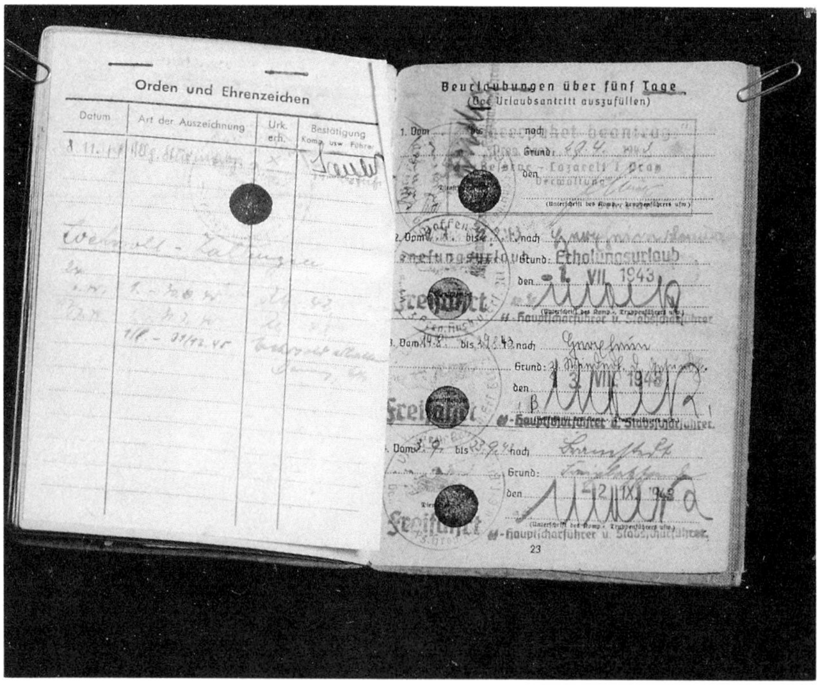

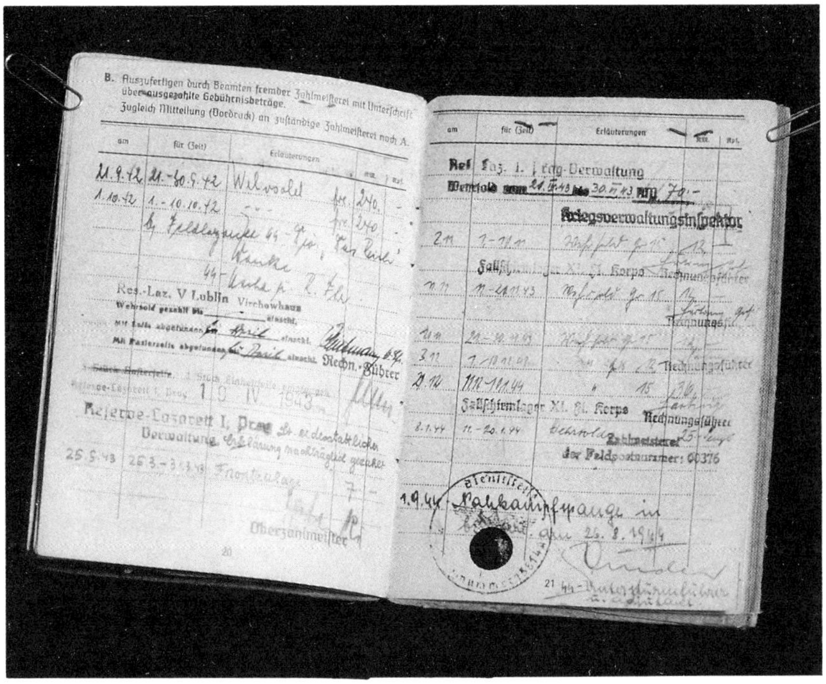

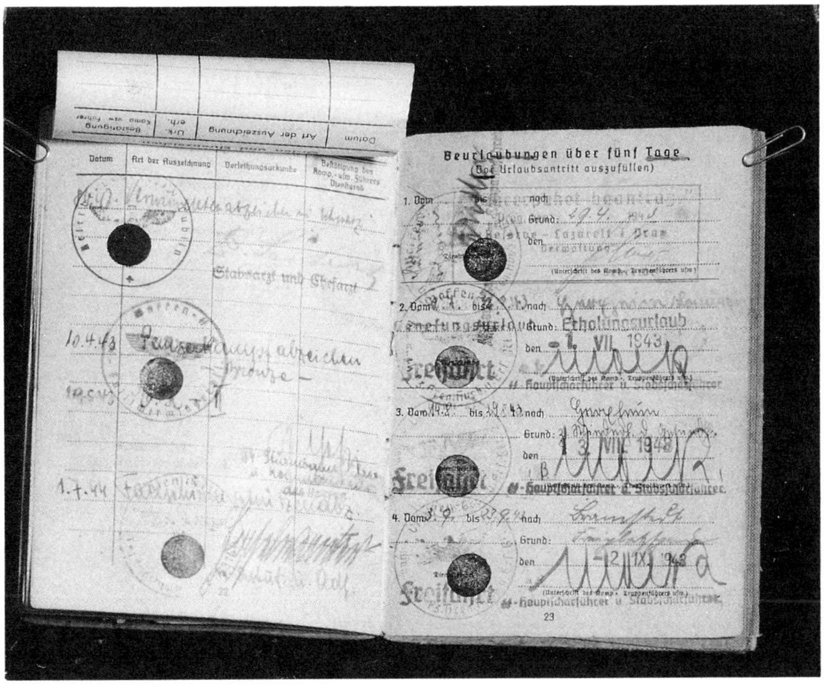

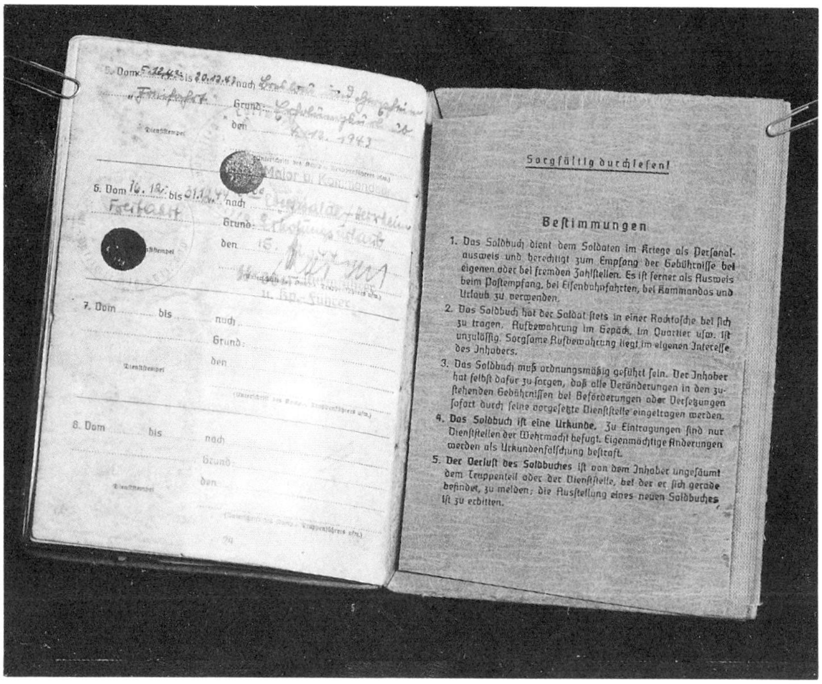
6. Vom
bis
nach
Grund:
den
7. Vom
bis
nach
Grund:
den
Dienststempel
8. Vom
bis
nach
Grund:
den
Dienststempel
Sorgfältig durchlesen!
Bestimmungen
1. Das Soldbuch dient dem Soldaten im Kriege als Personalausweis und berechtigt zum Empfang der Gebührnisse bei eigenen oder bei fremden Zahlstellen. Es ist ferner als Ausweis beim Postempfang, bei Eisenbahnfahrten, bei Kommandos und Urlaub zu verwenden.
2. Das Soldbuch hat der Soldat stets in einer Rocktasche bei sich zu tragen. Aufbewahrung im Gepäck, im Quartier usw. ist unzulässig. Sorgsame Aufbewahrung liegt im eigenen Interesse des Inhabers.
3. Das Soldbuch muß ordnungsmäßig geführt sein. Der Inhaber hat selbst dafür zu sorgen, daß alle Veränderungen in den zustehenden Gebührnissen bei Beförderungen oder Versetzungen sofort durch seine vorgesetzte Dienststelle eingetragen werden.
4. Das Soldbuch ist eine Urkunde. Zu Eintragungen sind nur Dienststellen der Wehrmacht befugt. Eigenmächtige Änderungen werden als Urkundenfälschung bestraft.
5. Der Verlust des Soldbuches ist von dem Inhaber ungesäumt dem Truppenteil oder der Dienststelle, bei der er sich gerade befindet, zu melden; die Ausstellung eines neuen Soldbuches ist zu erbitten.

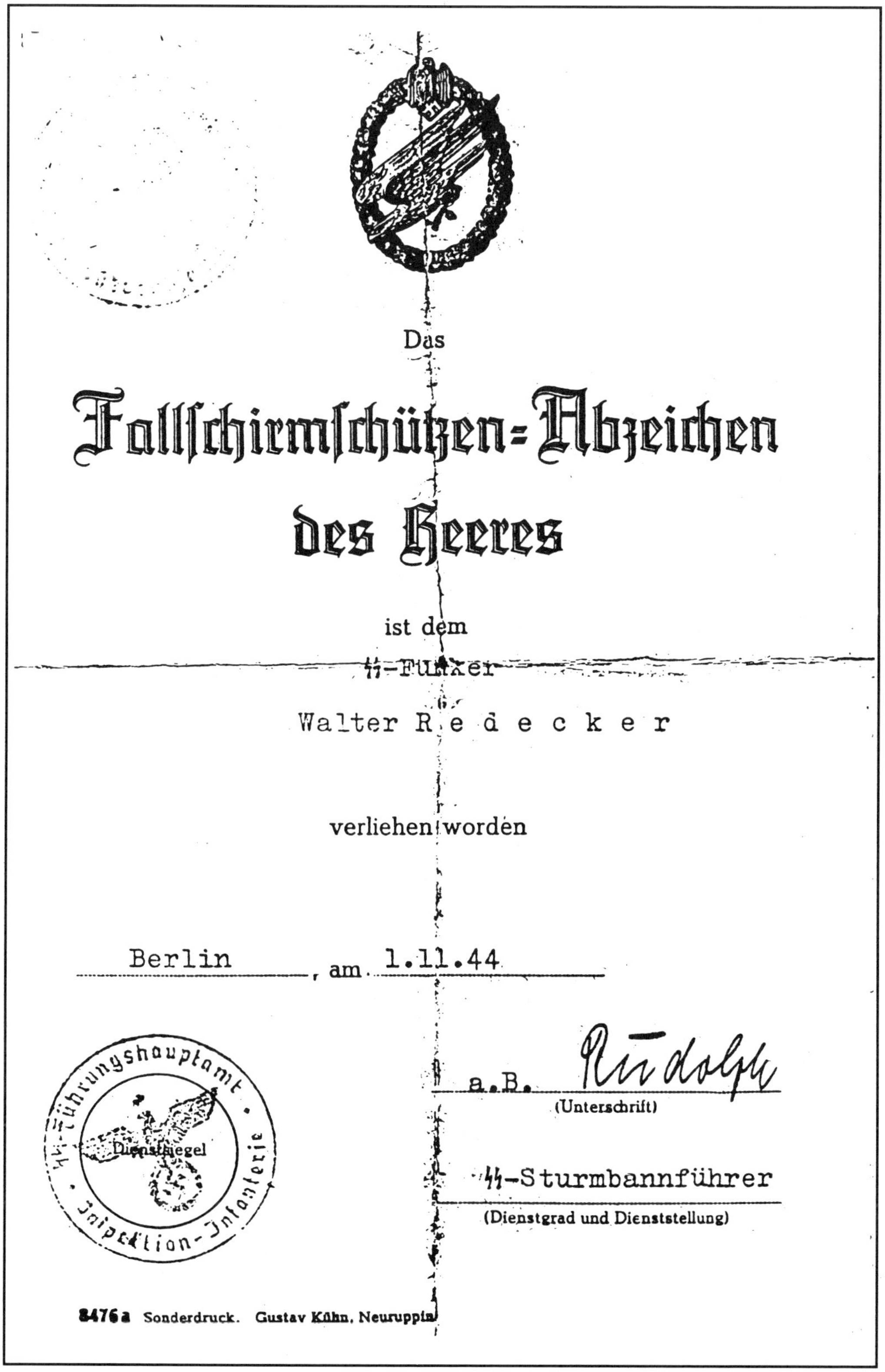

Das

Fallschirmschützen-Abzeichen

des Heeres

ist dem

SS-Funker

Walter Redecker

verliehen worden

Berlin, am 1.11.44

SS-Führungshauptamt Inspektion-Infanterie

Dienstsiegel

a.B. Rudolph

(Unterschrift)

SS-Sturmbannführer

(Dienstgrad und Dienststellung)

8476a Sonderdruck. Gustav Kühn, Neuruppin

This is a photograph of the original certificate given to SS-Radioman Walter Redecker upon completion of jump school. The certificate is dated 1 November 1944. Photo: Walter Redecker.

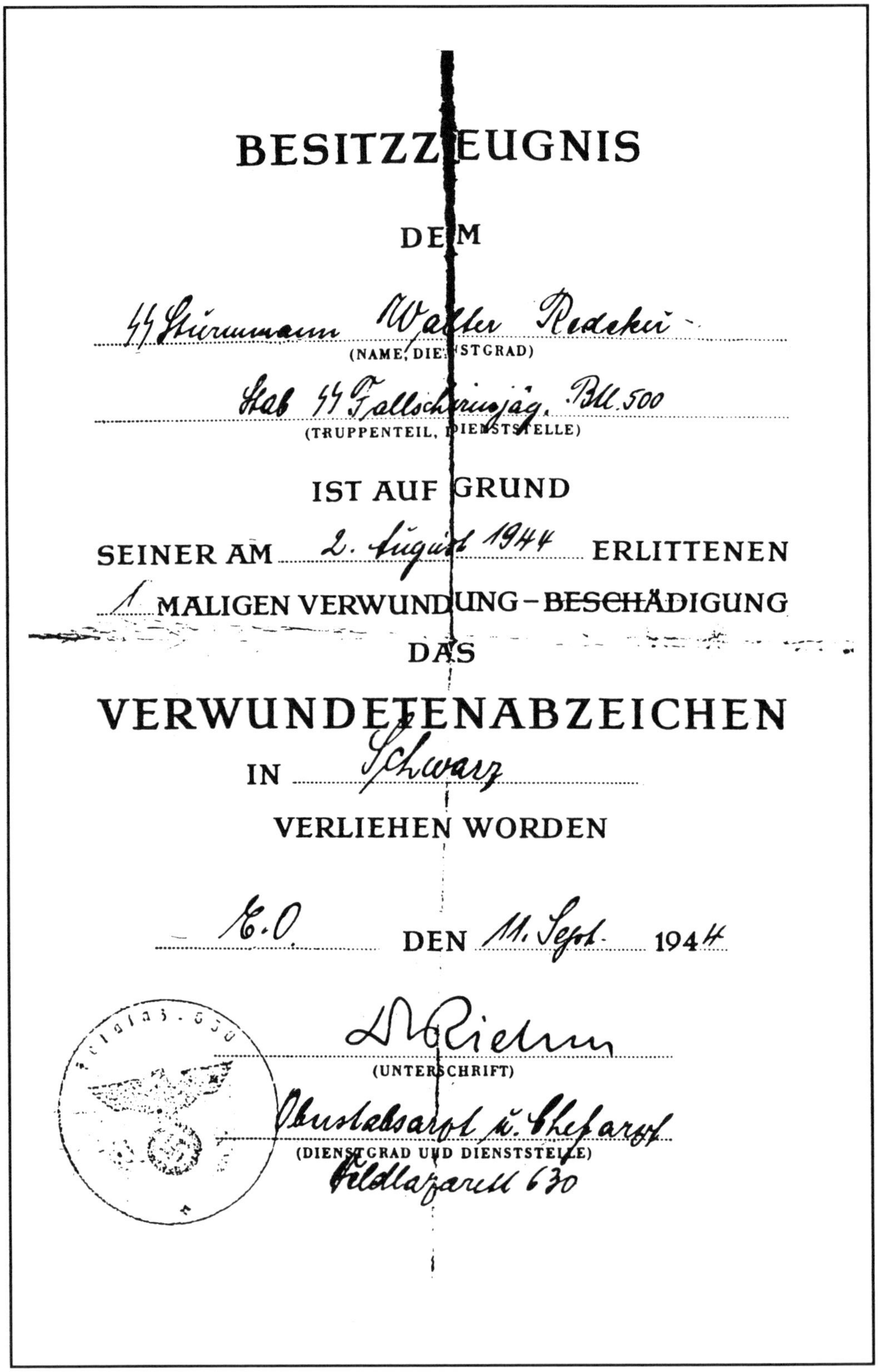

BESITZZEUGNIS

DEM

SS-Sturmmann Walter Redecker

(NAME, DIENSTGRAD)

Stab SS Fallschirmjäg. Btl. 500

(TRUPPENTEIL, DIENSTSTELLE)

IST AUF GRUND

SEINER AM 2. August 1944 ERLITTENEN

1 MALIGEN VERWUNDUNG – ~~BESCHÄDIGUNG~~

DAS

VERWUNDETENABZEICHEN

IN Schwarz

VERLIEHEN WORDEN

E.O. DEN 11. Sept. 1944

(UNTERSCHRIFT)

Oberstabsarzt u. Chefarzt

(DIENSTGRAD UND DIENSTSTELLE)

Feldlazarett 630

This is a photograph of the original certificate issued to SS-Sturmann Walter Redecker on 2 August 1944. He received the Wound Badge in Black for wounds sustained during the paratroop assault on Tito's headquarters on 25 May 1944, Drvar, Yugoslavia. Photo: Walter Redecker.

Certificate: Andreas Meyer.

Panzer-Brigade 150 | Gef.Stand, den 2.1.1945
Kampfgruppe Foelkersam

Für Tapferkeit vor dem Feinde während des Einsatzes der Panzer-Brigade 150 im Raum von Malmedy befordere ich mit Wirkung vom 21. Dezember 1944 den

SS-Röttenführer Andreas Meyer
Stab Kamfgruppe III

zum

SS-Unterscharführer

v. Foelkersam

SS-Hauptsturmführer
und Kommandeur

(This is the announcement of the promotion of SS-Rottenführer Andreas Meyer to SS-Unterscharführer on 21 December 1944.)

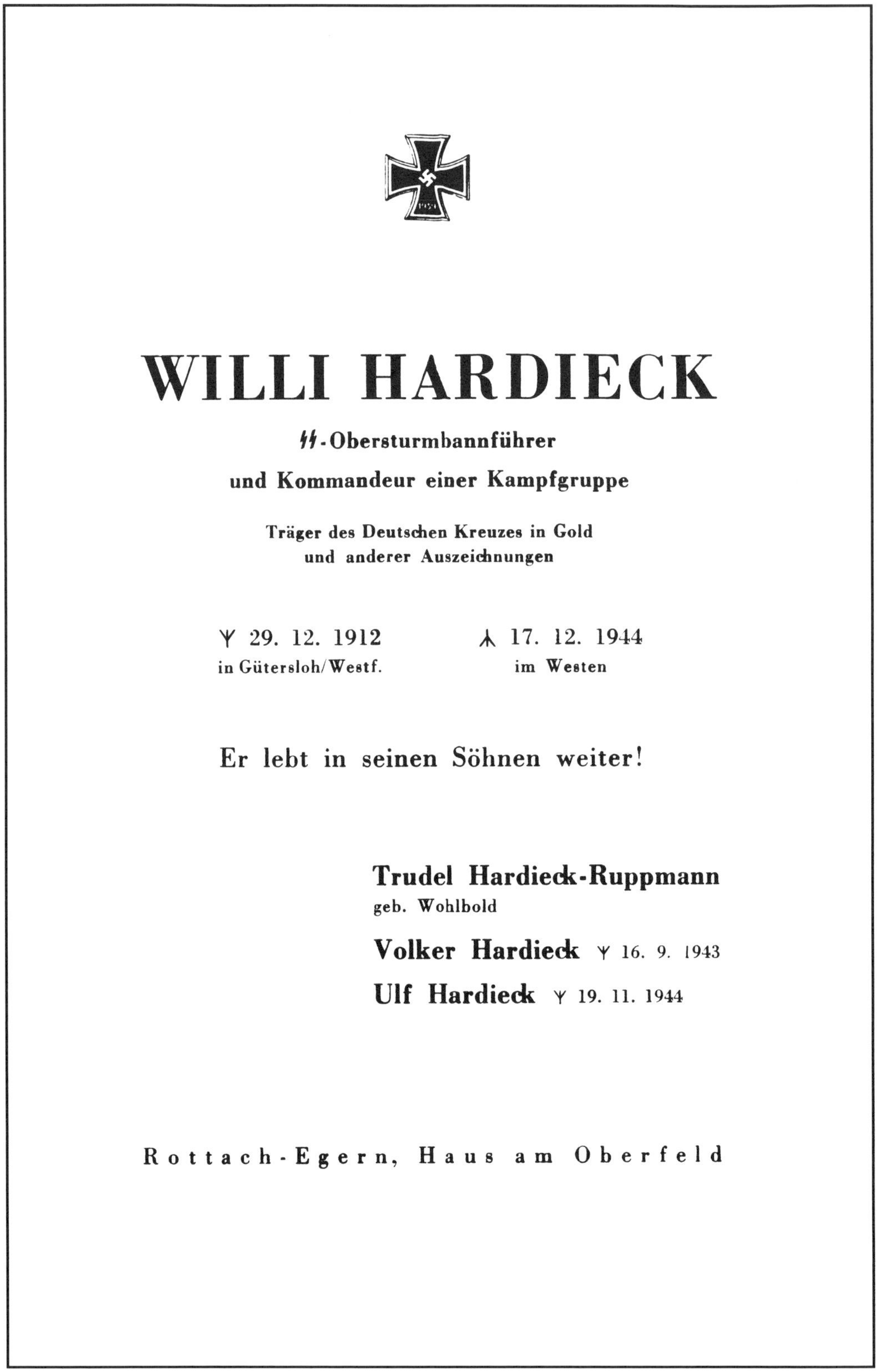
WILLI HARDIECK

SS-Obersturmbannführer

und Kommandeur einer Kampfgruppe

Träger des Deutschen Kreuzes in Gold
und anderer Auszeichnungen

ᛉ 29. 12. 1912
in Gütersloh/Westf.

ᛦ 17. 12. 1944
im Westen

Er lebt in seinen Söhnen weiter!

Trudel Hardieck-Ruppmann
geb. Wohlbold

Volker Hardieck ᛉ 16. 9. 1943

Ulf Hardieck ᛉ 19. 11. 1944

Rottach-Egern, Haus am Oberfeld

The funeral announcement for SS-Obersturmbannführer Willi Hardieck carried this epitaph: "You will live with us forever!" Hardieck left behind his wife, Trudel, and his two infant sons, Volker and Ulf. Photo: Andreas Meyer.

Besitzzeugnis

Dem SS-Schützen

(Dienstgrad)

Franz B e j o l

(Vor- und Familienname)

2./ SS-Fallschirmjäger-Btl. 600

(Truppenteil)

verleihe ich das

Allgemeine ~~Infanterie~~-Sturmabzeichen

O.U. , den 15. November 1944

(Dienstsiegel)

SS-Hauptsturmführer
und Kommandeur
(Dienstgrad und Dienststellung)

1/991/1 W. Druckerei VII, München 6. 44.

Certificate awarding medal to Franz Bejol of SS-Fallschirmjäger Bataillon 600, 15 November 1944. Certificate: Michael J. Belles Collection.

WAFFEN-SS MANPOWER STRENGTHS

The following table lists the strengths of numerous Waffen-SS units during four dates.

APPENDIX N

Unit	31 Dec. 1942	31 Dec. 1943	30 June 1944	20 Sept. 1944
SS-BeobAbt500	–	–	194	
SS-NchAbt104	–	–	663	IV.SSPzKps
SS-NchAbt501	–	–	1,196	I.SSPzKps
SS-PiPenBtl.1	–	–	207	163
SS-FlakAbt505	–	–	1,177	V.SS PzKps
SS-FlakAbt509	–	–	427	IX.SSGebKps
SS-BeobBtr105	–	–	153	
IX.SS-GebKps	–	–	416	1,217
Franz.SS-Bdge	–	800	1,688	1,100
SS-WfBtr500	–	–	417	
SS-Nch.Abt107	–	–	529	Disbanded
SS-BeobBtr101	–	–	163	I.SSPzKps
FormingTroops		105,093	127,643	
Other Duties	90,279	74,030	65,959	
SS-Schulen			10,822	
XI.SSArmeeKps	–	–	–	1,094
XII.SSArmeKps	–	–	–	979
XIIISSArmeKps	–	–	–	982
KG Kurmark	–	–	–	2,797
SS-PzAbt101	–	–	I.SS PzKps	I.SSPzKps
SS-PzAbt501	From SS-Pz	Abt101	Renamed in the fall	
SS-PzAbt102	–	–	II.SSPzKps	II.SSPzKps
SS-PzAbt502	From SS-Pz	Abt102	Renamed in Dec. 1944	
SS-PzAbt103	Never saw combat under this title. Still training.			
SS-PzAbt503	From SS-PzAbt103. It was renamed in Dec. 1944			
1 SS LSSAH	20,844	19,867	19,691	
2 SS DR	17,112	14,095	20,184	12,357
3 SS "T"	21,156	15,415	21,115	17,863
4 SS POLIZEI	13,399	16,081	16,139	
5 SS WIKING	15,928	14,647	17,368	18,016
6 SS NORD	21,247	20,129	19,355	17,041
7 SS PE	19,835	21,102	18,835	
8 SS FG	10,879	9,326	12,895	14,040
9 SS "H"	–	19,611	15,898	
10 SS "F"	–	19,313	13,552	
11 SS "N"	–	11,393	11,749	10,889
12 SS HJ	–	21,482	17,858	
13 SS HANDSCH	–	21,065	19,136	18,520
14 SS GAL	–	12,634	15,299	12,901
15 SS LETT#1	–	20,291	18,413	19,248
16 SS RF-SS	–	12,720	14,218	14,683
17 SS GB	–	11,147	16,976	14,816
18 SS HW	–	–	8,530	10,063
19 SS LETT#2	–	8,033	10,592	12,507
20 SS EST	–	–	13,423	15,382
21 SS SKNDBG	–	–	6,156	4,944
22 SS MT	–	–	4,914	12,453
23 SS KAMA	–	–	2,199	3,793

Unit –Time	31 Dec. 1942	31 Dec. 1943	30 June 1944	20 Sept. 1944
23 SS NEDLND	–	–	–	–
24 SS (KARST)	–	–	1,831	1,989
25 SS HUNYADI	–	–	–	–
26 SS HUNGARIA	–	–	–	–
27 SS LNGMRK.	–	–	–	–
28 SS WALNE.	–	–	–	–
29 SS	–	–	–	3,500
29 SS ITAL.	–	From: Ital SS Bdge		10,000
30 SS "W"	–	–	–	12,000
31 SS	–	–	–	–
32 SS 30JAN.	–	–	–	–
33 SS CHARL.	–	–	–	–
34 SS LN	–	–	–	–
II.SS Pz.Kps.	1,109	1,872	2,639	1,538
KmndoStb.RFSS	1,428	220	1,553	1,474
Bglt.Btl.RFSS	823	1,009	559	996
BgltFlak RFSS	721	744	under KmndoStb.RFSS	
BgltFlak RFSS	–	655	under KmndoStb.RFSS	
1.SS Inf.Bdge.	6,271	4,125	became part of 18 SS Div	
2.SS Inf.Bdge.	1,070	became part of Latvian SS unit		
Flandern Legn.	685	became the Langemarck SS Brigade		
Norwegen Legn.	698	became part of 11.SS Pz.Gre.Div.		
Nederlnd. Legn	1,755	became the SS Nederland PzGre.Bdge		
Lngmrk.SS Bdge	–	1,470	1,731	1,941
Ndrlnd.SS Bdge	–	6,424	6,713	6,530
LN SS Bdge.	–	1,938	3,167	3,105
Freikps.Dnmrk	642	became part of 11.SS Pz.Gre.Div.		
Btl.d.SS z.b.V	390	disbanded on 17 Sept. 1943		
SS Wrgelg.Btl.	112	992	1,447	–
SS Kriegs.Abt.	334	became the SS-Standarte KE		
SS-Stndrt."KE"	–	141	1,180	
I.SS.Pz.Kps.	–	3,311	3,729	
SS KpsWrfr.102	–	311	–	–
V.SS Geb.Kps.	–	5,068	3,530	
VI.SS ArmeeKps	–	934	1,529	2,465
Walone, SS Bdge	–	1,972	1,188	1,565
Dirlewanger SS	(Feb. 1943) 700	259	971	(Oct. 1944) 1,100
Karstwehr-SS	132	946	became the 24 SS Div.	
IV.SS Pz.Kps.	–	–	754	2,198
Ital.SS Bdge.	–			
49.SS PzGreBdg	–	–	3,886	to 17.SS Div
51.SS PzGreBdg	–	–	2,923	to 17.SS Div
SS-Jgr.Btl 500	?	?	1,271	?
SS-Jgr.Btl 501	?	?	942	?
SS-Friedenthal	–	450	became SS JägerBtl.502	
SS-Jgr.Btl 502	–	–	650	800
SS-Jgdvbng "M"	from SS-Jgr.Btl.502			(Nov. 1944) 1,170
SS-FallsBtl.	–	1,100	292	1,000
SS-Pz.Abt.103	–	–	1,037	SSPzAbt 503

These tables list the strengths of numerous Waffen-SS formations during December 1942, December 1943, June 1944, and September 1944. Each unit is abbreviated on the table. Following the tables is a list of the abbreviations, with the full name of the units. Some units had their designation/titles changed during the war, and an account of this has been made whenever possible.

MANPOWER			
Date:	Field Troops[1]	Other Troops[2]	NOTES
4 May 1940	124,199 men		[1] Actual combat troops
30 June 1941	160,405 men		[2] This number included: forming/replacement, higher-SS Headquarters, special duty, and school troops.
31 Dec. 1942	156,438 men	90,279 men	[3] "Bayonet" strength
31 Dec. 1943	257,472 men	243,577 men[3]	[4] All ranks included
30 June 1944	368,654 men	225,789 men	
15 Dec. 1944	910,000 men	(other troops included)	

Division Strengths, 23 August 1944:

2 SS-Panzer-Division Das Reich	450[3]
1 SS-Panzer-Division LSSAH	500[3]
9 SS-Panzer-Division Hohenstaufen	460[3]
10 SS-Panzer-Division Frundsberg	700[3]
12 SS-Panzer-Division Hitler Jugend	300[3]
SS-Kampfgruppe Deak	1,000[4]

Division Strengths, 16 December 1944:

1 SS LSSAH = 22,000[4]
2 SS DR = 18,000[4]
3 SS "T" = 15,400[4]
4 SS "P" = 9,000[4]
5 SS "W" = 14,800[4]
6 SS "N" = 15,000[4]
7 SS "PE" = 20,000[4]
8 SS "FG" = 13,000[4]
9 SS "H" = 19,000[4]
10 SS "F" = 15,542[4]
11 SS "N" = 9,000[4]
12 SS HJ = 19,657[4]
13 SS "H" = 12,793[4]
14 SS "G" = 22,000[4]
15 SS "L" = 16,870[4]
16 SS RFSS = 14,223[4]
17 SS GvB = 3,500[4] (1,000[3])
18 SS HW = 11,000[4]
19 SS "L" = 9,396[4]
20 SS EST = 15,382[4]
22 SS MT = 8,000[4]
Ndrlnd SS Bgde. = 6,000[4]
24 SS = 3,000[4]
25 SS Hunyadi = 15,000[4]
26 SS Hungaria = 13,000[4]
27 SS Lungmrk = 7,000[4, 5]
28 SS Walne = 4,000[4]
29 SS ITAL = 15,000[4]
Weiss SS Bgde. = 4,400[4]
31 SS = 11,000[4]
33 SS Charl. = 7,340[4]
LN SS Bgde. = 7,000[4]
Dirlewanger SS = 6,000[4]

Indian SS = 2,300[4]
Tatar SS Bgde. = 3,518[4]
Osttwvd SS = 8,000[4]
Rum SS No. 1 = 3,000[4]
Rum SS No. 2 = 2,500[4]
Bulg SS = 581[4]
XV SS Csck[4, 6]

[4] All ranks included

[5] The Langemarck Division had 7,000 trained men and an additional 8,000 undergoing training. However, there were not enough weapons to arm all of them at once.

[6] The XV. SS-Cossack Cavalry Corps had two cavalry divisions plus a Plastum Brigade. Counting those units plus the headquarters staff and support troops, the entire Cossack Corps had a strength of 50,000-plus men. At the beginning of 1945, the Plastum Brigade was redesignated as a Cossack infantry division, though its strength did not change.

LATVIAN VOLUNTEERS ASSISTING THE GERMAN WAR EFFORT
1 JULY 1944

11,537	15. Waffen-Grenadier-Division der-SS (Lett. No. 1)
19,909	19. Waffen-Grenadier-Division der-SS (Lett. Nr. 2)
31,446	Total
12,118	1st–6th Latvian Frontier Guard regiments
+ 14,884	Latvian Schuma and Police battalions
+ 5,240	Latvian local self-defense (Selbschütz) units
+ 628	Latvian air force legion (auxiliaries)
+ 927	Various other units
+ 12,159	Auxiliary assistants for the German army
+ 10,585	Various units outside of the legions

	Volunteered in Waffen-SS	Killed or Missing	Permanently Disabled
Belgian			
Walloons	15,000	3,500	?
Flemings	23,000	8,000	2,500
Danish			
Danes	11,300	3,500	?
Total	49,300	15,000	N/A
Percentage	(100)	(30+)	N/A

The following numbers are as accurate as documentation allows, but they are not exact. The smaller numbers are more exact than the higher ones, which are approximations.

Ethnic Germans:	185,000
East Europeans:	200,000
Western Europeans:	137,300
Reich Germans:	400,000

Western Europeans:
15,000 Walloons
23,000 Flemings
11,300 Danes
6,000 Norwegians
50,000 Dutch
8,000 French
20,000 Italians
1,000 Spaniards
3,000 Swiss, Swedes, Luxembourgers

Eastern Europeans:
50,000 Cossacks
12,000 Belorussians
26,000 Ukrainians
39,000 Latvians
20,000 Estonians
20,000 Croatians
15,000 Serbians
3,000 Albanians
5,000 Romanians
1,000 Finns
1,000 Bulgarians
8,000 Eastern Turks

Ethnic Germans:

United States: 5	Palestine: 2
Japan: 2	Mexico: 1
Sumatra: 2	Australia: 1
Soviet Union: 100	India: 1
Southwest Africa: 3	Croatia: 25,000[1]
Southeast Africa: 2	Hungary: 80,000[2]
France: 84	Poland: 5,000[3]
New Guinea: 1	Romania: 8,000[1]
China: 3	
Brazil: 4	Western Europe: 16,000[1]
Spain: 2	Serbia: 5,000[1]
England: 10	Czechoslovakia: 45,000
South America: 2	Scandanavia: 775[1]

[1] These figures are for ethnic Germans *only* and are separate from the purely foreign volunteers from those countries, which are listed as either Western or Eastern Europeans.

[2] Although the majority of the Hungarians serving in the Waffen-SS were ethnic Germans, thousands were "pure" Hungarians. For convenience, they have been placed together under the ethnic German section.

[3] The same applies to Poland as for Hungary.

Legion Norwegen Strength Table

700	August 1941
1,218	December 1941
1,150	16 March 1942
600	May 1942
1,000	1 December 1942
678	31 December 1942
600	May 1943

Freikorps Danmark Strength Table

900	31 March 1943
1,386	31 May 1943
1,800	15 November 1943

Approximately 500 Danes lost their lives while serving in the Freikorps Danmark, and 175 Swedish volunteers served in the Waffen-SS, out of which 45 were killed and an unspecified number wounded.

Total Waffen-SS Losses in the Field

1 September 1939 to 31 March 1942 (killed, missing)	14,213
1 April 1942 to 13 May 1945 (killed, missing)	238,787
1 September 1939 to 13 March 1945 (estimated wounded)	400,000
Total (killed, missing, and wounded)	653,000

Losses Suffered by the Totenkopf and Wiking SS Divisions

1 September 1939 to 28 February 1945	Wiking:	19,084 (killed, missing, wounded)
	Totenkopf:	59,995 (killed, missing, wounded)

A close comparison of the number of Waffen-SS troops killed, missing, and wounded between the SS Wiking and Totenkopf divisions gives a startlingly different loss ratio. Wiking Division was used to show the average normal loss incurred by each Waffen-SS division. Compare that 19,000-plus average with the high losses for the Totenkopf Division: almost three times as many losses! Since both divisions served, for the most part, alongside each other, the only reason (besides heavy fighting) for such horrendous losses is the mishandling—or rather, the rough handling—of Totenkopf troops by its commanders. It is known that Totenkopf had more requests for transfers than any other Waffen-SS combat division. Whether this had to do with the division's high attrition rate is not known. I have established one fact during my research: a large number of the men who eventually served in the SS parachute force were, at one time or another, members of Totenkopf. To put it in numbers we can all understand, while the average Waffen-SS division suffered 1.8 percent of the total losses, Totenkopf suffered 5.9 percent. Even more bluntly, chances of getting killed or wounded were three times as great if serving in the Totenkopf Division.

32. SS-FREIWILLIGEN-PANZERGRENADIER-DIVISION 30. JANUAR						
Date	**Strength**	**Dates**	**Losses**	**Killed**	**Missing**	**Wounded**
1 January 1945	6,632	1 January 1945 to 28 February 1945	1,786	314	1,385	87
15 February 1945	5,412					
17 March 1945	2,846	1 March 1945 to 31 March 1945	2,190	407	1,447	336
		1 January 1945 to 1 April 1945	**Total** 3,976	721	2,832	423

32nd SS Pz.Gre.Div.Armored Strength[1]	Operational	In Repair	Total
Sturmgeschütz-III (75 mm AT gun):	20	22	22
Sturmgeschütz-IV-H (150mm HE gun):	8	1	9
Total:	28	3	31

[1] These figures apply to 7 April 1945.

Hungarian-SS

XVII. Waffen-Armeekorps der-SS
260-member corps staff, including Ia, IIa, 01, translator, motorcycle, signal, and guard units.

Defending Waffen-SS Troop-Training Camp Neuhammer
Ungarisches SS-Alarm-Regiment No. 1: 2,600 men drawn from the 25th and 26th SS (Hungarian) divisions. 8 – 15 February 1945
Alarm Kompanie, Ungarisches SS-Schi-Bataillon No. 1: 220 men, 8 – 15 February 1945
Supply Staff and Transport Staff, 25. Waffen-Grenadier-Division der-SS Hunyadi: 13 men, 1 Waffen-Sturmbannführer der-SS, 2 chaplains, and 10 truck drivers, all killed at Neuhammer

NOTE: Of the 2,833 Hungarian SS troops who defended the training grounds Neuhammer and Strans, 707 were killed in action (including 28 officers and 2 battalion commanders), while 955 were listed as either wounded or missing in action.

	18 November 1944	November – December 1944	30 December 1944	January 1945	May 1945
25 SS Hunyadi	16,700 men		22,670 men[1]		
26 SS Hungaria			16,761 men[2]		23,500
SS KG Ney				5,000 men	
1. SS-Ski-Batln.		3,000 men			
1. SS-Sturm-Rgt.				1,500 men	

[1] Of the 22,670 men actually in the division, approximately 10,000 were in training and reserve units, and only 4,400 were fully ready.

[2] Of the 16,761 men, 351 were officers, 1,270 NCOs, and 15,140 men.

SOME 1945 WAFFEN-SS DIVISIONAL STRENGTHS
(NUMBERS IN PARENTHESIS ARE "BAYONET" STRENGTHS OF EACH DIVISION)

	1 March 1945	13 March 1945	17 March 1945	4 April 1945	12 May 1945
1. SS LSSAH	18,871	12,295	12,370	7 April	
		(6,000)	(4,594)	(4,215)	(1,582)
I. SS Pz.Kps.	3,883				
12. SS H.J.	17,423	12,075	12,128	7 April	
	(5,500)	(3,492)	(4,780)	(455)	
II. SS Pz.Kps.	3,036				
2. SS DR	19,542	9,143	11,633	7 April	
	(6,000)	(3,344)	(4,575)	(1,498)	
9. SS "H"	17,229	8,029	10,820		
	(4,500)	(4,320)	(4,614)		
6. PzAOK	1,416	?	?		
IV. SS Pz.Kps.					
3. SS "T"			43,751	(1,004)	
5. SS Wiking					
SS Rgt. Ney					
15. SS Lett No. 1	13,000			8,250	
16. SS RF-SS	10,500		9,966		
			(3,432)		
13. SS Handsch			9,228		
			(3,601)		
34. SS LN			11,000		8,956
37. SS "L"	2,500			3,500	
	(Feb. 1945)				
38. SS Nieb				3-4,000	
4. SS Polizei				7,867	
				(5,306)	

Effective Strength of IX. Waffen-Gebirgs-Armeekorps der-SS
11 February 1945

Germans: 23,900 (9,600 wounded)
Hungarians: 20,000 (2,000 wounded), plus 80,000 to 100,000 Hungarian civilians trapped in the city.[1]

[1] This SS corps was trapped in Budapest. These numbers include army *and* Waffen-SS troops.

Some Losses Between 1 – 7 January 1945				
IV. SS Pz.Kps.	**Officers**	**NCOs**	**Men**	**Total**
Dead	26	99	404	529
Wounded	76	343	2,206	2,625
Missing	1	29	328	358
5. SS Wiking	**Officers**	**NCOs**	**Men**	**Total**
Dead	10	23	133	166
Wounded	22	97	672	791
Missing	0	11	102	113
3. SS "T"	**Officers**	**NCOs**	**Men**	**Total**
Dead	10	49	183	242
Wounded	37	157	870	1,064
Missing	1	6	47	54
Total	183	814	4,945	5,942

COMPARISON OF STRENGTHS VS. LOSSES OF 3RD (GERMANIC) SS PANZER CORPS

Strengths	III. SS Pz.Kps.	II. SS "N"	Ndrlnd. SS Bgde	SS-Pz.Abt.503
December 1943	2,500	11,400	6,000	–
January 1944	2,500	11,393	6,000	500
1 July 1944	2,666	11,749	6,713	–
20 September 1944	2,544	10,889	6,530	–
February 1945	N/A	5,430[1]	3,674[2]	–

[1] Bayonet strength was 3,011 [2] Bayonet strength was 1,826

Losses	K/W/M	K/W/M	K/W/M
1 September 1943 to 28 February 1945	55/134/25	2,573/8,859/891	1,789/5,889/1,651
1 March 1945 to 31 March 1945	28/46/66	366/1,411/387	262/970/782

K = killed W = wounded M = missing

LOSSES BETWEEN 1 SEPTEMBER 1939 AND MAY 31ST 1942

Unit		Dead		Wounded		Missing		Accidental	
1. SS LSSAH		Off.	Enl.	Off.	Enl.	Off.	Enl.	Off.	Enl.
	Losses up to then :	61	1,198	91	2,598	2	63	1	–
	New Losses :	2	5	2	30	–	–	–	–
	Total :	63	1,203	93	2,628	2	63	1	–
2. SS DR		Off.	Enl.	Off.	Enl.	Off.	Enl.	Off.	Enl.
	Losses up to then :	146	3,167	334	10,126	4	274	–	–
	New Losses :	–	17	1	52	–	–	–	–
	Total :	146	3,184	335	10,178	4	274	–	–
3. SS "T"		Off.	Enl.	Off.	Enl.	Off.	Enl.	Off.	Enl.
	Losses up to then :	170	2,349	253	9,151	6	270	–	–
	New Losses :	11	232	19	608	1	20	–	–
	Total :	181	2,581	272	9,759	7	290	–	–
4. SS Polizei		Off.	Enl.	Off.	Enl.	Off.	Enl.	Off.	Enl.
	Losses :	58	1,629	150	5,215	2	166	–	–
5. SS Wiking		Off.	Enl.	Off.	Enl.	Off.	Enl.	Off.	Enl.
	Losses up to then :	74	1,613	112	4,600	2	206	1	163
	New Losses :	–	26	–	69	–	1	–	–
	Total :	74	1,639	112	4,669	2	207	1	163
6. SS Nord		Off.	Enl.	Off.	Enl.	Off.	Enl.	Off.	Enl.
	Losses up to then :	45	860	96	3,226	4	154	5	30
	New Losses :	–	131	7	559	–	21	–	1
	Total :	45	991	103	3,785	4	175	5	31
8. SS FG (A brigade at this time)		Off.	Enl.	Off.	Enl.	Off.	Enl.	Off.	Enl.
	Losses up to then :	20	334	30	889	–	91	1	1
	New Losses :	–	1	–	1	–	–	–	–
	Total :	20	335	30	890	–	91	1	1
1. SS Inf.Bdge.		Off.	Enl.	Off.	Enl.	Off.	Enl.	Off.	Enl.
	Losses up to then :	18	306	38	804	4	119	1	13
	New Losses :	2	–	–	1	–	–	–	–
	Total :	20	306	38	805	4	119	1	13
Nederlnd.Legn.		Off.	Enl.	Off.	Enl.	Off.	Enl.	Off.	Enl.
	Losses up to then :	4	191	2	398	1	36	–	–
	New Losses :	–	–	1	–	–	–	–	–
	Total :	4	191	3	398	1	36	–	–
Norwegen.Legn.		Off.	Enl.	Off.	Enl.	Off.	Enl.	Off.	Enl.
	Losses up to then :	2	36	2	91	–	–	–	–
	New Losses :	–	6	2	23	–	–	–	–
	Total :	2	42	4	114	–	–	–	–

		Off.	Enl.	Off.	Enl.	Off.	Enl.	Off.	Enl.
KG Jeckeln		Off.	Enl.	Off.	Enl.	Off.	Enl.	Off.	Enl.
	Losses up to then :	1	15	–	57	–	–	–	–
	New Losses :	–	–	–	2	–	–	–	–
	Total :	1	15	–	59	–	–	–	–
2. SS Inf. Bdge.		Off.	Enl.	Off.	Enl.	Off.	Enl.	Off.	Enl.
	Losses :	7	144	54	514	1	33	–	–
Flandern Legn.		Off.	Enl.	Off.	Enl.	Off.	Enl.	Off.	Enl.
	Losses :	3	74	2	250	–	–	–	–
SS Inf. Rgt. 4		Off.	Enl.	Off.	Enl.	Off.	Enl.	Off.	Enl.
	Losses :	22	434	26	1,124	–	65	–	–
SS Inf. Rgt. 9		Off.	Enl.	Off.	Enl.	Off.	Enl.	Off.	Enl.
	Losses :	20	339	38	1,195	1	105	–	–

	Dead		**Wounded**		**Missing**		**Accidental**	
	Off.	Enl.	Off.	Enl.	Off.	Enl.	Off.	Enl.
Total Losses :	666	13,107	1,260	41,583	28	1,624	9	208

Off. = Officers
Enl. = Enlisted

COMBAT STRENGTH OF THE 9TH ARMY
17 MARCH 1945

V. SS-Freiwilligen-Gebirgs-Armeekorps: 18,769 men
32. SS-Freiwilligen-Panzergrenadier-Division: 2,846
Infanterie-Division Raegener, later the 286. Infanterie-Division: 3,266
391. Infanterie-Division: 3,618
Festung-Frankfurt-am-Oder: 9,039 men

XI. SS-Armeekorps: 21,398 men
Panzergrenadier-Division Kurmark: 2,375
25. Panzergrenadier-Division: 5,196 men[1]
Panzer-Division Müncheberg: 2,867 men
712. Infanterie-Division: 3,699 men
303. Infanterie-Division Doberitz: 3,474 men[2]
Festung-Küstrin: 2,787 men

CI. Armeekorps: 10,349 men
309. Infanterie-Division Berlin: 5,889 men
Infanterie-Division z.b.V. 606: 4,460 men

Total 9th Army Troop Strength: 50,516 men

[1] 25th Panzer Grenadier Division lost a total of 2,151 men killed, wounded, or missing during the month of March 1945. The breakdown was: 456 killed, 1,447 wounded, and 248 men missing.

[2] Division Doberitz suffered the following losses in March 1945: 8 officers and 250 other ranks killed; 25 officers and 832 other ranks wounded; plus 305 other men missing—for a total of 1,420 men during the whole month.

SS-BeobAbt500	SS-Beobachtungs-Abteilung 500 (SS Observation Battalion 500)
SS-Nch.Abt104	SS-Nachrichten-Abteilung 104 (SS Signal Battalion 104)
SS-Nch.Abt501	SS-Nachrichten-Abteilung 501 (SS Signal Battalion 501)
SS-PiPenBtl.1	SS-Pioneer-Penal-Bataillon 1 (SS Engineer Penal Battalion 1)
SSFlakAbt505	SS-Flak-Abteilung 505 (SS Antiaircraft Battalion 505)
SSFlakAbt509	SS-Flak-Abteilung 509 (SS Antiaircraft Battalion 509)
SSBeobBtr105	SS-Beobachtungs-Batterie 105 (SS Observation Battery 105)
IX.SS-Geb.Kps	IX.Waffen-Gebirgs-Armeekorps der-SS (9th SS Mountain Corps)
Franz.SS-Bgde	8. Französich-SS Freiwilligen Sturmbrigade (8th SS French Volunteer Assault Brigade).
SS-WrfrBtr105	SS-Werfer-Batterie 105 (SS Rocket Launcher Battery 105)
SS-Nch.Abt107	Nachrichten-Abteilung 107 (SS Signal Battalion 107)
SS-BeobBtr101	SS-Beobachtungs-Batterie 101 (SS Observation Battery 101)
SS-Schulen	SS Schools
XI.SSArmeKps	XI. SS-Armeekorps (11th SS Army Corps)
XII.SSArmeKps	XII. SS-Armeekorps (12th SS ArmyCorps)
XIII.ArmKps	XIII. SS-Armekorps (13th SS Army Corps)
KG Kurmark	Kampfmarschverband Kurmark (Combat Group Kurmark)
SS-Pz.Abt.101	Schwer Panzer Abteilung 101 (Heavy Tank Battalion 101)
SS-Pz.Abt.102	Schwer Panzer Abteilung 102 (Heavy Tank Battalion 102)
SS-Pz.Abt.502	Schwer Panzer Abteilung 502 (Heavy Tank Battalion 502)
SS-pz.Abt.103	Schwer Panzer Abteilung 103 (Heavy Tank Battalion 103)
SS Pz.Abt.503	Schwer Panzer Abteilung 503 (Heavy Tank Battalion 503)
II.SSPz.Kps.	II. S-Panzerkorps (2nd SS Tank Corps)
KmndoStb RFSS	Kommando Stab Reichsführer-SS (Command Headquarters of the National SS Leader, i.e., Himmler's headquarters.)
Weiss-SS Bgde	Waffen-Grenadier-Brigade der-SS Weissruthenien (Russische Nr. 2) Armed Grenadier Brigade of the SS White Russia
Indian-SS	Indische-SS Legion (Indian SS Legion)
6.Pz.AOK	6.Panzerarmee
SS Rgt. Ney	SS Kampfgruppe Ney (SS Battle Group [Brigade] Ney)
SS Inf.Rgt.4	SS-Infanterie-Regiment Nr. 4 (4th SS Infantry Regiment)
SS Inf.Rgt.9	SS-Infanterie-Regiment Nr. 9 (9th SS Infantry Regiment)
1 SS LSSAH	1. SS-Panzer-Division Leibstandarte SS Adolf Hitler
2 SS D.R.	2. SS-Panzer-Division Das Reich
3 SS "T"	3. SS-Panzer-Division Totenkopf
4 SS Polizei	4. SS-Polizei-Panzergrenadier-Division
5 SS Wiking	5. SS-Panzer-Division Wiking
6 SS NORD	6. SS-Gebirgs-Division Nord
7 SS P.E.	7. SS-Freiwilligen-Gebirgs-Division Prinz Eugen
8 SS F.G.	8. SS-Kavallerie-Division Florian Geyer
9 SS "H"	9. SS-Panzer-Division Hohenstaufen
10 SS "F"	l0. SS-Panzer-Division Frundsberg
11 SS "N"	11. SS-Freiwil1igen-Panzergrenadier-Division Nordland
12 SS H.J.	12. SS-Panzer-Division Hitler Jugend
13 SS Handsch	13. Waffen-Gebirgs-Jäger-Division der-SS Handschar
14 SS Gal.	14. Waffen-Grenadier-Division der-SS (galizische Nr.l)
15 SS Lett#l	15. Waffen-Grenadier-Division der-SS (lettische Nr. l)
16 SS RF-SS	16. SS-Panzergrenadier-Division Reichsführer-SS
17 SS G.B.	17. SS-Panzergrenadier-Division Götz von Berlichingen
18 SS H.W.	18. SS-Freiwilligen-Panzergrenadier-Division Horst Wessel
19 SS Lett#2	19. Waffen-Grenadier-Division der-SS (lettische Nr. 2)

20 SS Est.	20. Waffen-Grenadier-Division der-SS (estnische Nr. 1)
21 SS Skndbg	21. Waffen-gebirgsjäger-Division der-SS Skanderbeg (alb.)
22 SS M.T.	22. SS-Freiwilligen-Kavallerie-Division Maria Theresa
23 SS Kama	23. Waffen-Gebirgsjäger-Division der-SS Kama (kroatische)
23 SS Nedlnd	23. SS-Freiwilligen-Panzergrenadier-Division Nederland
Bglt.Btl.RFSS	Begleit-Bataillon Reichsführer-SS (Escort Battalion RFSS)
BgltFlak RFSS	Begleit-Flak-Abteilung RFSS (Escort Flak Battalion RFSS)
1. SS Inf.Bdge.	SS-Infanterie-Brigade 1 (lst SS Infantry Brigade)
2. SS Inf.Bdge.	SS-Infanterie-Brigade 2 (2nd SS Infantry Brigade)
Flandern Legn	SS-Freiwilligen Legion Flandern (SS Volunteer Legion Flanders)
Norwegen Legn	SS-Freiwilligen Legion Norwwgen (SS Volunteer Legion Norway)
Nederlnd.Legn	SS-Freiwilligen Legion Nederland (SS Volunteer Legion Netherlands)
Lngmrk.SS Bdge	6. Langemarck-SS Freiwilligen-Sturmbrigade (6th Langemarck SS Volunteer Assault Brigade) Belgian-Flemish formation
Ndrlnd.SSBdge	4. SS-Freiwilligen-Panzergrenadier-Brigade Nederland (4.SS Volunteer Armored Infantry Brigade Netherlands)
LN SS Bdge	SS Brigade Landstorm Nederland (SS Brigade Landstorm Netherlands)
FreikpsDnmrk.	SS-Freikorps Danmark (SS Volunteer Corps Denmark)
Btl.d.SSz.b.V	Bataillon der-SS zur besonder Verwendung (SS Battalion for Special Employment)
SSWrgelg.Btl.	SS-Wehrgeolegie-Bataillon (SS Geology Battalion)
SS Kriegs.Abt.	SS-Kriegsberichter-Abteilung Kurt Eggers (SS War Reporters Battallon Kurt Eggers)
SS-Stndrt.KE	SS-Standarte Kurt Eggers (SS Regiment Kurt Eggers)
I.SS Pz.Kps.	I.SS-Panzerkorps (I.SS Tank Corps)
SS KpsWrfr.102	SS-Korps-Werfergruppe 102 (SS Corps Rocket Launcher Battery 102)
V.SSGeb.Kps.	V.SS-Freiwilligen-Gebirgs-Armeekorps (5th SS Volunteer Mountain Army Corps)
VI.SSArmeekps	VI.Waffen-Armeekorps der-SS (lettisches) (6th Armed Army Corps of the SS, Latvian)
Walone.SS Bdge	5. SS-Freiwilligen-Sturmbrigade Wallonien (5. SS Volunteer Assault Brigade Wallonien. Belgian-Wallonien formation
Dirlewanger SS	SS-Sonderkommando Dirlewanger (SS Special Commando Dirlewanger)
Karstwehr-SS -	SS-Karstjäger-Bataillon (SS Rock-Climbing Battalion) IV.SS Pz.Kps. - IV.SS-Panzerkorps (4th SS Tank Corps).
Ital. SS Bdge	9. Waffen-Grenadier-Brigade der-SS (italienische) (9th Armed Grenadier Brigade of the SS, Italian)
49.SSPzGreBdge	SS-Panzergrenadier-Brigade 49 (SS-Armored Infantry Brigade 49)
51.SSPzGreBdge	SS-Panzergrenadier-Brigade 51 (SS-Armored Infantry Brigade 51)
SS-Jgr.Btl 500	SS-Jäger-Bataillon 500 (SS Hunter Battalion 500)
SS-Jgr.Btl 501	SS-Jäger-Bataillon 500 (SS Hunter Battalion 501)
SS-Jgr.Btl 502	SS-Jäger-Bataillon 502 (SS Hunter Battalion 502)
SS-Friedenthal	SS-Sonderverband z.b.V. Friedenthal (SS Formation for Special Employment Friedenthal)
SS-Jgdvbnd "M"	SS-Jagdverband Mitte (SS Hunter Formation Center)
SS-FallsBtl.	SS-Fallschirmjäger-Bataillon 500/600 (SS Parachute Battalion 500/600)
Tartar SS BGDE	Waffen-Gebirgs-Brigade der-SS [tartarische Nr. 1] (Armed Mountain Brigade of the SS, Tartar Nr. 1)
Osttwvd-SS	Osttürkischen-Waffenverband der-SS (Eastern Turkic Armed Formation of the SS)
Rum-SS No. 1	Waffen-Grenadier-Regiment der-SS [rumänische Nr. 1] (Armed Grenadier Regiment of the SS, [Rumanian Nr. 1])

Rum-SS No. 2	Waffen-Grenadier-Regiment der-SS [rumänische Nr. 2] (Armed Grenadier Regiment of the SS [Romanian Nr. 2])
Bulg-SS	Waffen-Grenadier-Regiment der-SS [bulgarisches](Armed Grenadier Regiment of the SS [Bulgarian])
Serbian-SS	Serbisches Freiwilligen-Korps der-SS (Serbian Volunteer Corps of the SS)
XV.SS-CSCK	XV. Kosaken-Kavallerie-Korps der-SS (15th Cossack Cavalry Corps of the SS).
24. SS (Karst)	24. Waffen-Gebirgs(karst)jäger-Division der-SS (ungarische Nr. 1)
26. SS Hunyadi	26. Waffen-Grendier-Division der SS Hungaria (ungarische Nr. 2)
27. SS Hungaria	27. SS-Freiwilligen-Grendier-Division Langemarck
28. SS Walne	28. SS-Freiwilligen-Panzergrenadierdivision Wallonien
29. SS ?	29. Waffen-Grenadier-Division der-SS (russische Nr. 1)
29. SS Ital	29. Waffen-Grenadier-Division der-SS (italienische Nr. 1)
30. SS "W"	30. Waffen-Grenadier-Division der-SS (russische Nr. 2). This division was later reduced to a brigade and entitled White Russia.
31. SS	31. SS-Freiwilligen-Grenadier-Division
32. SS 30Jan	32. SS-Freiwilligen-Panzergrenadierdivision 30 Januar
33. SSChaarl	33. Waffen-Grenadier-Division der-SS Charlemagne (französische Nr. 1)
34. SS L.N.	34. SS-Freiwilligen-Grenadier-Division Landstorm Nederland. Not listed here, but it existed.
35. SS-	35. SS-Polizei-Grenadier-Division. Not listed here, but it existed.
36. SS	36. Waffen-Grenadier-Division der SS. This formation was a much expanded Dirlewanger penal SS unit.
37. SS L	37. SS-Freiwilligen-Kavallerie-Division Lützow
38. SS Nieb	38. SS-Panzergrenadier-Division Nibelungen

SS-FALLSCHIRMJÄGER-BATAILLON 500
25 MAY 1944

Battalion Strength Table

Battalion Headquarters		**(267)**
Signals Platoon	42	
Headquarters Company	66	
HQ Motor Transport Platoon	30	
Section III (Legal)	7	
Parachute Maintenance Platoon	31	
Supply Company	91	
Heavy Weapons Company		**(200)**
Company Headquarters	33	
Signals Section	11	
Flamethrower Platoon	28	
81mm Mortar Platoon	34	
Heavy Machine Gun Platoon	38	
Light 75mm Antitank Platoon	56	
1st Parachute Rifle Company		**(164)**
Company Headquarters	36	
Signals Section	11	
Parachute Rifle Platoon	39	
Parachute Rifle Platoon	39	
Parachute Rifle Platoon	39	
2nd Parachute Rifle Company		**(164)**
Company Headquarters	36	
Signals Section	11	
Parachute Rifle Platoon	39	
Parachute Rifle Platoon	39	
Parachute Rifle Platoon	39	
3rd Parachute Rifle Company		**(164)**
Company Headquarters	36	
Signals Section	11	
Parachute Rifle Platoon	39	
Parachute Rifle Platoon	39	
Parachute Rifle Platoon	39	
Training Parachute Rifle Company		**(181)**
Company Headquarters	40	
Signals Section	12	
Parachute Rifle Platoon	43	
Parachute Rifle Platoon	43	
Parachute Rifle Platoon	43	
Total:		**(1,140)**

The basic difference between the SS-Fallschirmjäger-Bataillon and the regular Luftwaffe parachute battalion lay in the relatively self-contained structure of the SS unit. The SS para battalion had its own training company, something not organic to the Luftwaffe para battalion. The Luftwaffe unit had 138 men in the battalion headquarters; the SS battalion had 267 men. Again, because the unit was self-contained, it required more people for support services.

The Heavy Weapons Company of the SS para battalion was a hybrid. The Luftwaffe para battalion either had a parachute MG (machine gun) company of 205 men or a 120mm mortar company of 163 men. While the Luftwaffe Parachute Machine Gun Company had a table of organization strength of two light 75mm antitank guns, the SS unit had four. In addition, the SS parachute unit had two flamethrowers, something missing altogether in the Luftwaffe para battalion but available in the Luftwaffe parachute division. An interesting thing found in all of the Luftwaffe parachute companies was an organic demiplatoon of seventeen men, who were in training but could be called on in case of an emergency.

Finally, the SS parachute training company was basically a para rifle company, but somewhat overstrength. The Heavy Weapons Company of the SS para battalion was the 4th Company of the battalion. Since the training company was eventually dropped at Drvar, the SS-Fallschirmjäger-Bataillon 500 employed a total of five parachute companies (one heavy weapons and four rifle companies). By 1945, the training company had been redesignated as the new 1st Company. The 4th (Heavy Weapons) Company was used as the shock troops of the battalion, being floated around at crucial points. The Heavy Weapons Company was then divided among the rifle companies for support. This was basically what happened at Drvar, as was its purpose.

APPENDIX O

SS-TOTENKOPF-STANDARTEN

The following is a complete list of the SS Death's Head regiments, including their locations on two dates and their strengths. The first three SS standarten are not listed here, since at this time they made up Totenkopf Division.

Standarte	**Commanding Officer**	**Officers/NCOs/Men**	**Location**	
			5 May 1940	**28 July 1940**
4. SS-Totenkopfstandarte	Standartenführer Reich	61/374/2,500	Prague	The Hague
5. SS-Totenkopfstandarte	Brigadeführer Breithaupt	23/294/1,561	Stettin	Organienburg
6. SS-Totenkopfstandarte	Oberführer Voss	47/502/2,407	Sandefjord, Norway	
7. SS-Totenkopfstandarte	See note below	45/343/2,383	Drammen, Norway	
8. SS-Totenkopfstandarte	See note below	48/391/2,360	Cracow	Radom
9. SS-Totenkopfstandarte	Standartenführer Nostitz	34/354/1,862	Brno	Brno
10. SS-Totenkopfstandarte	Obersturmbannführer Demme	41/254/2,458	Danzig	Cracow
11. SS-Totenkopfstandarte	Obersturmbannführer Diebitsch	55/367/2,583	Radom	Zandvoort
12. SS-Totenkopfstandarte	Obersturmbannführer Sacks	46/244/2,360	Posen-Treskaw-Litzmanstadt	
13. SS-Totenkopfstandarte	Standartenführer Klingemann	29/315/1,709	Vienna Linz	same
14. SS-Totenkopfstandarte	Obersturmbannführer Martin	40/392/2,187	Copenhagen, Denmark	
15. SS-Totenkopfstandarte	Sturmbannführer Dorheit	30/165/1,484	Plock/E.Prussia/Allenburg	
16. SS-Totenkopfstandarte	Standartenführer Hermann	33/304/2,009	Prague	same
K. SS-Totenkopfstandarte	Obersturmbannführer Reitz	24/77/913	Kirkenses, Norway	
1. SS-Totenkopf-Reiterstandarte	Standartenführer Fegelein	54/190/1,829	Warsaw	same
2. SS-Totenkopf-Reiterstandarte	Sturmbannführer Magill	31/114/1,511	Lublin	same
Sonder-Bataillon/SS-Totenkopf-Standarte Prag	Hauptsturmführer Dr. Buchegger	6/78/867	Prague	same
Inspektorate der SS-Totenkopf-Standarten	Oberführer Schwedler	18/69/169	Oranienburg	Berlin

NOTE:
7. SS-Totenkopfstandarte was led by Brigadeführer Richard Hermann on 5 May 1940 and later by Standartenführer Schneider on 28 July 1940.
8. SS-Totenkopfstandarte was led by Oberführer von Jena on 5 May 1940, then by Standartenführer Scherner on 28 July 1940, and later by Oberführer Classen.

Illustration: Eric Weisman.

APPENDIX P

ORGANIZATION OF THE SS-FHA(SS-FÜHRUNGSHAUPTAMT) MAIN SS COMMAND HEADQUARTERS

SS-Gruppenführer und Generalleutnant der Waffen-SS Hans Juttner

Adjutantur	Adjutant
Ia (Operationsoffizier)	Operations Officer
Ib (Versorgungsoffizier)	Supply Officer
Id (Ausbildungsoffizier)	Training Officer
TO (Transportoffizier)	Transport Officer
Amt I Kommandoamt der Allgemeine-SS	Command Office of the General-SS
Amt II Kommandoamt der Waffen-SS	Command Office of the Waffen-SS
Amt III Persöonaliches Amt, Abwehrbeauftrater der Waffen-SS	Personnel Office-CounteriIntelligence
Amt IV Verwaltungsamt	Administration Office
Amt V Personal=Amt	Personnel Office
Amt VI Reiter und Fahrwesen	Horse and Horse-Drawn Transport Office
Amtsgruppe-A Waffen, Gerät u. Kraftfahrwesen	Weapons, Equipment, and Transport Matters Group "A"
Amtsgruppe-B Führer u. Unterführer Ausbilding	Officer and NCO Training Group "B"
Amtsgruppe-C Waffen-Inspektorate	Arm of Service Inspectorates Group "C"
Inspektorate 2 — Infanterie	Inspectorate 2 — Infantry
Inspektorate 3 — Reiter u. Fahrwesen	Inspectorate 3 — Cavalry and Horse-Drawn
Inspektorate 4 — Artillerie und Flak	Inspectorate 4 — Artillery and Antiaircraft
Inspektorate 5 — Pioniere	Inspectorate 5 — Engineers
Inspektorate 6 — Kraftfahrer Kraftfahrer-kampftruppen	Inspectorate 6— Motor Transport Drivers and Motorized Combat Troops
Inspektorate 7 — Nachrichten Truppen	Inspectorate 7 — Signal (Communications) Troops
Inspektorate 8 — Waffen	Inspectorate 8 — Arms and Ordnance
Inspektorate 9 — Gebirg Truppen	Inspectorate 9 — Mountain Troops
Amtsgruppe -D Sanitätswesen	Medical Matters Group "D"
Amt VII Nachschub-Amt	Supply Office Ordnance Office
Amt VIII Waffen-Amt	Procurement Office
Amt IX Beschaffungs-Amt	Motor Transport Office
Amt X Kraftfahrwesen	Officer Training Office
Amt XI Unterführer-Ausbildung	NCO Training Office

APPENDIX Q

SPANISCHE-FREIWILLIGEN-KOMPANIE DER-SS 101 AND 102

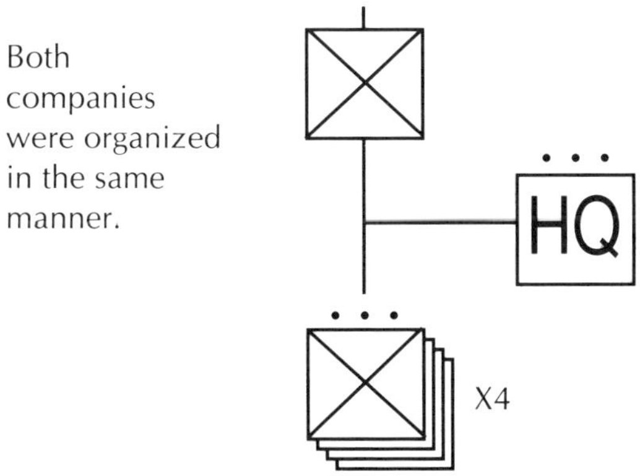

SS-JAGDVERBAND SÜDWEST WITH SUB-UNITS ITALIEN, NORDFRANKREICH, AND SÜDFRANKREICH, JANUARY 1945

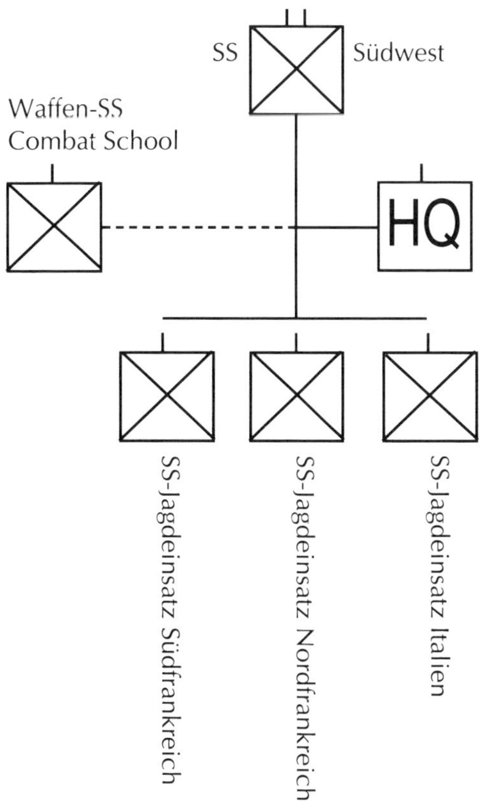

GLOSSARY

GLOSSARY OF THE MOST COMMONLY USED GERMAN MILITARY TERMS

aufklärungs	reconnaissance
abteilung	battalion or detachment
artillerie	artillery
armeekorps	army corps
armee	army
armeegruppe	army group
Abwehr	German Secret Service
albanische	Albanian
aserbaidschan	Azerbaijani
armenien	Armenian
ausbildung	forming
arbeitslager	work camp
bataillon	battalion
beobachtungs	observation
begleit	escort
bodenstandig	static (positional)
bäkerei	bakery
brigade	brigade
brücko	bridging (unit)
division	division
ersatz	reserve
freiwilligen	volunteer
festung	fortress
festungs regiment	fortress regiment
fallschirmjäger	paratrooper or parachute
feldgendarme	German Field Police
feldpost	German Military Field Post Office
flak	antiaircraft
französich	French
geschütz	gun
gebirgs	mountain
grenadier	honorary title for infantryman
Heer	German Field Army
infanterie	infantry
inf. Geschütz	infantry gun

indische	Indian
jäger	light infantryman
jagdverband	hunter group
junkerschule	officer academy
kavallerie	cavalry
km	abbreviation for kilometer
korpsabteilung	corps detachment
kompanie	company
kommando	commando or command unit
Kosaken	Cossacks
karstjäger	rock climbing
Luftwaffe	German Air Force
Luftwaffen-Feld	Luftwaffe Field formation
leichte	light
lehr	demonstration (usually an elite unit)
m	abbreviation for "miles"
Marine	naval infantry
motoriziert	motorized
nachschub	supply
nachricht(en)	signals or communications
Nord	north
Nordwest	northwest
Oberkommando	High Command
Osttürkischen	Eastern Turkic
OKH	abbreviation for German Army High Command
OKW	abbreviation for German Armed Forces High Command
Ostmuselmanisch-	Eastern Muslim
Ost	East
panzergrenadier	armored infantry
panzerjäger	antitank
panzerzug	armored train
pionier	engineer
polizei	police
russische	Russian
reiter	cavalry (see also "kavallerie")
sturmgeschütz	assault gun
sturmartillerie	assault artillery
soldbuch	German soldier's pay book or ID book
stürm	assault
scharfschützen	sharpshooter
schulen	school
Schüma	abbreviation for schützmannschaft or self-defense formation

schwer-	heavy
schi	ski
sicherung or sicherheit	security
schwadron	squadron
Süd	south
Südost	southeast
Südwest	southwest
sanitäts	sanitation or medical
stab	staff
serbisch-	Serbian
spanisch-	Spanish
SD	abbreviation for Sicherheitsdienst, the German-SS Secret Service
tatar(isch-)	Tatar
turkoman	Turkic
Truppen-Übungsplatz	Troop Training Ground
Truppen-Übugslager-	Troop Training Camp
ungarisch-	Hungarian
verband	formation (could range in size from a battalion to a brigade)
versorgungs	administration
Volksgrenadier-	People's Grenadier
veterinärs	veterinary
weissruthenien/ weissruthenisch	White Russian
Wehrmacht	German armed forces
werkstatt	workshop
Wehrkreis	military district
werfer	rocket
wach	guard
z.b.V.	abbreviation for Special Employment

BIBLIOGRAPHICAL ESSAY

History cannot be fabricated. On this assumption, I opted to escape from pages and pages of footnotes, which for academics seem the main purpose of writing. I am not an academician. Therefore, instead of offering you, the reader, a thousand pages to confirm every word in this book, I decided to denude history and present it to you without its various forms and diverse versions.

That is why you do not find hundreds of notes throughout this book. History in its barest form is not only more interesting but also more compelling. This work is not a selective or subjective version of events; neither is it a historical revision. Rather, it is a compilation of documents, remembrances, and writings that, one hopes, will shed greater light on this obscure part of history.

Nonetheless, I would like to acknowledge the main sources that aided my research. First, credit is due to several technical references, some multivolume works that, although in book form, can be considered primary sources since they were copied word for word from German military documents. Works of this nature include:

- Kurt Mehner, *Die Geheimen Tagesberichte der Deutschen Wehrmachtführung im Zweiten Weltkrieg 1939-1945*, (Osnabrück: Biblio Verlag GmbH, 1984-1990), volumes. 7-12. This is the original, unabridged secret war diary of the German army High Command.
- Percy E. Schramm, *Kriegstagebuch des Oberkommandos Der Wehrmacht 1940-1945*, (Herrsching: Manfred Pawlak Verlagsgesellschaft GmbH, 1982), volumes. 5-8. This work presents the German armed forces High Command's war diary.
- Norbert Kannapin, *Die Deutsche Feldpostübersicht 1939-1945*, (Osnabrück: Biblio Verlag GmbH, 1986), volumes 1-3. The German Field Post Numbers were the "Zip codes" used by the military postal system. They were designed to identify quickly a unit by a numbered code that was changed every six months to make it harder for enemy intelligence to obtain orders of battle and troop-transfer information by just looking at letters to frontline soldiers. For example, Field Post Number 34 445 could have belonged to U-boat 365 from January to June 1944 and then the 1st Company of the 24th SS Signal Battalion. Cross-referencing this three-volume compilation can help to formulate an order of battle. One should, however, compare the information gathered here with other sources because divisions were formed on paper and Field Post Numbers (FP No.) quickly assigned, but, in reality, the unit may not have been built up to desired strength for various reasons.

Some other books that were, for the most part, technical references but also yielded some biographical data include:

- Horst Scheibert, *Die Träger Der Ehrenblatt des Heeres, Kriegsmarine, und der Waffen SS* (Friedberg: Podzun Pallas Verlag, 1986) and *Die Träger des Deutschen Kreuzes in Gold: Kriegsmarine-Luftwaffe-Waffen-SS und die Träger des Deutschen Kreuzes in Silber: Kriegsmarine-Luftwaffe-Waffen-SS* (Friedberg: Podzun Pallas Verlag, n.d.). The former volume lists all of the army, navy and Waffen-SS recipients of the Honor Roll Clasp, while the latter lists the holders of the German Cross in gold and silver for the navy, air force, and Waffen-SS.
- Brün Meyer's *Dienstalterliste der Waffen-SS, SS-Obergruppenführer bis SS-Haupt-*

sturmführer, Stand vom 1 Juli, 1944 (Osnabrück: Biblio Verlag GmbH, 1987). This is similar to the above volume, but it lists Waffen-SS officer promotions and postings.

The following biographical histories, while subject to differing interpretations, provided invaluable insights into Waffen-SS formations and the war itself and were used throughout my book:

- Nikolaus von Preradovich, *Die Generale der Waffen-SS*, (Berg am See: Kurt Vowinkel Verlag, 1985)
- Ernst Gunther Krätschmer, *Die Ritterkreuzträger der Waffen-SS* (Preussich Oldendorf: Verlag K.W. Schütz, 1955)
- Dr. K.G. Klietmann, *Die Waffen-SS, eine Dokumentation* (Osnabrück: Verlag der Freiwillige GmbH, 1965)
- Karl Cerff, *Die Waffen-SS im Wehrmachtbericht* (Osnabrück: Munin Verlag GmbH, 1971).

Otto Skorzeny wrote numerous volumes, which were of inestimable value:

- *Skorzeny's Special Missions*, (London: Robert Hale, 1947)
- *Lebe Gefärlich and Wir Kämpften Wir Verloren*, 2 volumes.

There have also been some excellent biographies written about Skorzeny and his commandos. The ones I used most are the following:

- Charles Foley, *Commando Extraordinary* (London: Longmans Green and Co.,1954)
- Helmuth Spater, *Die Brandenburger: Eine Deutsche Kommandotruppe* (Munich: Walther Angerer Verlag, 1982).

This next masterpiece, I consider a primary reference since Tessin used official sources such as the German army and Army High Command war diaries. Additionally, Tessin took advantage of every German military unit history printed between 1945 and 1980. Although not technically a primary reference, no other work comes anywhere close to the accuracy, completeness, and detail of this 15-volume set.

- Georg Tessin, *Verbande und Truppen der deutschen Wehrmacht und Waffen-SS im Zweiten Weltkrieg, 1939-1945*, 15 volumes. (Osnabrück: Biblio Verlag GmbH, 1980).

So, as far as published works go, those mentioned above are the general ones I used throughout the book. They helped form the "skeleton," or main body, to which I later added the "tendons" (supporting material) and "muscle" (documentary evidence). I now want to discuss specific sources that helped me compile each section of the book. For the story on the SS parachute battalion, I relied heavily on numerous first-hand accounts by former members of the battalion, as well as other sources, including:

- Siegfried Milius, *Fallschirmjäger der Waffen-SS im Bild* (Osnabrück; Munin Verlag GmbH, 1986). This photo-historical journal was written by the former commander of the Fallschirmjäger battalion and is of particular note.
- Erich Busch, *Die Fallschirmjäger Chronik, 1935-1945* (Friedberg: Podzun Pallas Verlag, 1983)

• Mario D. Fenyo, *Hitler, Horthy, and Hungary: German-Hungarian Relations, 1941-1944* (New Haven: Yale University Press, 1972)

• F.H. Hinsley, editor, *British Intelligence in the Second World War*, 6 volumes. (London: Her Majesty's Stationary Office, 1984), volume 3, part 1

• James Lucas, *Kommando: German Special Forces of World War II* (London: Arms & Armour Press, 1985)

• Georg Schlaug, *Die deutschen Lastensegler-Verbande 1937-1945* (Stuttgart: Motorbuch Verlag, 1985).

For the story of the SS-Panzer-Brigade Gross, I used two volumes by the same author who thoroughly covered the story of the brigade during the Baltic battles:

• Gerd Niepold, *Mittlere Ostiront, Juni '44* (Hereford; Bonn: E.S. Mittler & Sohn, 1985) and *Panzer-Operationen: "Doppelkopf und Casar" Sommer '44* (Herford; Bonn: E.S. Mittler & Sohn, 1987).

I also made heavy use of articles that appeared in the German veterans' magazine, *Der Freiwillige*. "Recollections of a Brigade Gross Gunner" is an example of one such article. Written by a former member, it went a long way toward giving the reader a sense of what it was really like to serve in the unit.

The same system was used to write about SS-Festungs-Grenadierregiment Besslein, though in this case I was also able to get assistance from the former commanding officer of the regiment, Georg Besslein.

Some published works of great value in describing the battle for Breslau and the Silesian campaign were:

• Generalmajore Hans von Ahlfen, *Der Kampf um Schlesien 1944/1945* (Stuttgart: Motorbuch Verlag, 1980)

• Franz Kurowski, *Die Letzten Dreizic Tage* (Friedberg: Podzun Pallas Verlag, 1989).

For the history of the Osttürkischen-Waffenverband der-SS, I had the advantage of being able to look at all of the captured documents pertaining to this unit in the U.S. National Archives in Washington, D.C. (NARS Microfilm Publication T-354, roll 161, frames 3806724 through 3807091). These documents describe the Turkic unit from its inception, using the titles of Ostmuselmanische SS-Regiment, Ostmuselmanische SS-Division, and Osttürkischen-Waffenverband der-SS.

Records of the Reich Leader-SS and Chief of the German Police refer to the Turkic-SS formation and the difficulty encountered by the Germans in raising and activating the unit (NARS Microfilm Publication T-175, roll 108, frame 2631142). Roll 162 of the same publication also describes the unit in terms of conditions, operations, training, and leadership.

Some documents in this collection were so important that I translated them word for word. One such example was the testimony given by Waffen-Untersturmführer der-SS Alicham Kuliew (T-354, roll 161, beginning on frame 3807061). Kuliew testified about the events leading to the desertion of Waffen-Obersturmführer der-SS Gulam Alimov and several hundred other Turkic volunteers from the Osttürkischen-Waffenverband der-SS on 24 December 1944. Kuliew's biography was carefully detailed by the German command on the same microfilm publication (T-354, roll 161, beginning on frame 3807082).

In addition to the information on microfilm at the Archives and various articles in

Der Freiwillige and *Siegrunen* magazines, the following works provided more material on the Osttürkischen-Waffenverband der-SS:

- Hans Werner Neulen, *An deutscher Seite: Internationale Freiwilligen von Wehrmacht und Waffen-SS*
- Alexander Dallin, *German Rule in Russia, 1941-1945* (London: MacMillan & Co.,1957)

The chapter on Waffen-SS special forces was particularly difficult to piece together for various reasons. Few men really belonged to these commando formations; very few of them survived the war. Of those, fewer still lived to be interviewed by this author. I did write to several former members—among them Herr Andreas Meyer, Herr Karl-Heinz Gieseler, and Herr H. Uphoff. Along with first-hand accounts, I relied heavily on works mentioned earlier, such as those by Otto Skorzeny, Charles Foley, Ernst Gunther Krätschme, and Helmuth Spater. Other works of notable mention include:

- Georg Maier, *Drama zwischen Budapest und Wien: Der Endkampf der 6. Panzerarmee 1945* (Osnabrück: Munin Verlag GmbH, 1985)
- Gerald Reitlinger, *The House Built on Sand: The Conflicts of German Policy in Russia, 1939-1945* (London: Weidenfeld & Nicolson, 1960)
- Melton S. Davis, *Who Defends Rome? The Forty-Five Days, July 25 - September 8, 1943* (London: George Allen & Unwin Ltd., 1972).

Additionaly, both *Der Freiwillige* and *Siegrunen* provided hard information about Waffen-SS special forces. The articles in *Siegrunen* seemed to have been heavily based on an eight-part series written by Herr Horst Voigt that appeared in the German yearly *Soldaten Jährbuch*. These yearbooks include a myriad of articles dealing with German history throughout the ages. The series of articles I used were entitled, "Sondertruppen zur Frontbewährung im 2. Weltkrieg," and published between 1982 and 1989.

Finally, the Union of Soviet Socialist Republics published a large, softcover book more than thirty years ago that was very helpful. Documents form this book, *The Case against General Heusinger: Documents Illustrating the charges of the USSR against Former Lieutenant-General Adolf Heusinger, Former Operations Chief of the Wehrmacht High Command* (Chicago: Translation World Publishers, 1961), helped to piece together some of the history of the 500th and 501st SS-Jäger-Bataillonen. It's interesting to note that the Soviet Union withheld these documents until 1961 when General Heusinger became military head of the West German Bundeswehr. This book denounced Heusinger as a war criminal. Whether he was or wasn't is immaterial vis-á-vis the self-serving timing of the release of these documents.

The captured documents at the U.S. National Archives were also forthcoming about the history of the 21st, 23rd, and 30th SS divisions. For the 23. Waffen-Gebirgsjäger-Division der-SS Kama, I located some divisional papers on NARS Microfilm Publication T-354, roll 160, frames 3805952 through 3805997. For the 21. Waffen-Gebirgsjäger-Division der-SS Skanderbeg, documents are on NARS Microfilm Publication T-354, roll 160, frames 3805937 through 3805951. This source provided exceptional information about the actual condition and strength of this division in early October 1944. One particular piece of biographical data on the commander of the 21st SS Engineer Battalion was obtained from the Truppenkameradschaft—21st SS-Freiwilligen-Gebirgs-Division Prinz Eugen through an intermediary.

By far, the biggest treasure trove of documents came from papers dealing with the 30.Waffen-Grenadier-Division der-SS Weissruthenien (NARS Microfilm Publication T-354, roll 160, frames 3806172 through 3806676). I can honestly say that without these documents, I could not have done justice to the history of these three formations.

Some published works also cover these units, and I would like to mention them here:

- K.G. Klietmann, *Die Waffen-SS, eine Dokumentation* (Osnabrück: Verlag "Der Freiwillige," 1965)
- Otto Kumm, *Vorwarts Prinz Eugen! Geschichte der 7. SS-Freiwilligen-Gebirgs-Division "Prinz Eugen,"* (Osnabrück: Munin Verlag GmbH, 1984).

If my biggest research success was uncovering documents highlighting the history of the 30th SS Division, then the most difficult proved to be the history of the 24. Waffen-Gebirgs(karst)jäger-Division der-SS. I researched every conceivable work, which included:

- Richard Landwehr, *Italian Volunteers of the Waffen-SS*, (Bennington: Weapons and Warfare Press, 1988)
- Ricciotti Lazzero, *Le SS Italiane* (Italy: Rizzoli, 1982)
- Paul Hauser, *Waffen-SS im Einsatz* (Preussich Oldendorf: Verlag K.W. Schütz, 1953).

In addition, I was fortunate to be able to communicate with Herr Günther Witt, who served as an SS-Unterscharführer in the 3rd Battalion, Waffen-Gebirgs(Karst)jäger-Regiment der-SS 59. Herr Witt passed away in 1989.

Last, but definitely not least, the most helpful published work turned out to be:

- P.A. Carnier, *Lo sterminio mancato* (Mursia Publishing, 1982). Although Carnier devoted only seven pages to the history of the Karstjäger formation (pages 76 through 82), he did an excellent job of providing details I could find nowhere else.

BIBLIOGRAPHY

Unpublished Documentary Material

Miscellaneous SS Records: Einwandererzentralstelle, Waffen-SS and SS-Oberabschnitte. Washington: National Archives, Microcopy T-354. Records of the Waffen-SS, parts I and II. Washington: National Archives, Microcopy T-354.

Beförderung Urkunde: promotion certificate issued to Andreas Meyer after his promotion from SS-Röttenführer to SS-Unterscharführer.

Begräbnis Ankündigung: funeral announcement for SS-Obersturmbannführer Willi Hardieck.

Brandenburg Division letter, 14 July 1944, establishing the formation of Streikorps Südfrankreich. Bundesarchiv, Koblenz..

Fallschirmschützen Abzeichen des Heeres: Army Jump School Certificate presented to SS-Sturmmann Walter Redecker, radioman in SS-Fallschirmjäger-Bataillon 500/600.

Letter from Herr Fritz Leifheit to Herr Eric Brorup, dated "Rigelsberg, den 14.07.89." Letter from the files of Eric Brorup.

Records of the Reich Leader-SS and Chief of the German Police (Allgemeine-SS), Washington: National Archives, Microcopy T-175.

Spanish volunteers at Training Camp Stablack, documents. Bundesarchiv. Koblenz.

SS-Soldbuch: Soldier's Handbook, belonging to SS-Unterscharführer Arnim Blesinger, SS-Fallschirmjäger-Bataillon 500/600. From the Michael J. Belles Collection.

Verwundetenabzeichen in Schwarz: certificate issued to SS-Sturmann Walter Redecker acknowledging the award of the Wound Badge in black.

Vorschlag (proposal) nominating SS-Sturmbannführer Fritz Herzig to the German Cross in Gold. Courtesy of Michael J. Belles Collection.

Interviews and Correspondence

The following individuals provided personal insight, comments, photographs, and documents for this book, without which it could never have been published.

Becker, Hans	SS-Fallschirmjäger-Bataillon 500/600
Brorup, Eric	Feldausbildungskompanie/SS-Fallschirmjäger
Draeger, Hans-Joachim	SS-Fallschirmjäger-Bataillon 500/600
Gerson, Klaus	SS-Panzer-Brigade Gross
Giesele, Karl-Heinz	SS-Jagdverband Nordwest
Knöfel, Erhart	SS-Fallschirmjäger Bataillon 500/600
Meyer, Andreas	SS-Jagdverband Mitte
Redeker, Walter	SS-Fallschirmjäger-Bataillon 500/600
Schrödl, Werner	SS-Fallschirmjäger-Bataillon 500/600
Wilhelm, Helmut	SS-Fallschirmjäger-Bataillon 500/600
Witt, Günther	24.Waffen-Gebirgs(Karst)jäger-Division der-SS

Published Works

Abbot, Peter and Thomas. *Germany's Eastern Front Allies, 1941-45*. London: Osprey Publishing, 1982.

Ahlfen, Generalmajor Hans von. *Der Kampf um Schlesien 1944/1945*. Stuttgart: Motorbuch Verlag, 1980.

Alff, Wilhelm. *Der Begriff Faschismus und andere Aufsätze zur Zeitgeschichte*. Frankfurt: Suhrkanp Verlag, 1970.

Anders, Lieutenant-General Wladyslaw. *Russian Volunteers in Hitler's Army*. Metairie: Sons of Liberty, n.d.

Bender, Roger James and Taylor. *Uniforms, Organization, and History of the Waffen-SS*, 5 vols. San Jose: R. James Bender Publishing, sixth printing, May 1986. vols. 1-5.

Bennett, Ralph. *Ultra in the West: The Normandy Campaign of 1944-45*. New York: Charles Scribner's Sons, 1979.

Berthold, Will. *Brandenburg Division*. Translator Alan Neame. London: Anthony Gibbs & Phillips, 1961.

Busch, Erich. *Die Fallschirmjäger Chronik, 1935-1945*. Friedberg: PodzunPallas Verlag, 1983.

Caballero, Carlos. *El Batallón Fantasma, Españoles En La Wehrmacht Y Waffen-SS, 1943-45*. Valencia: Asociacion Cultural "Tercera Posicion" (ACTP) de Valencia, 1987.

Carnier, P.A. *Lo sterminio mancato*. Rome: Mursia Publishing, 1982.

Caroe, Olaf. *Soviet Empire: The Turks of Central Asia and Stalinism*. London: Macmillan & Co. Ltd. U.S. Edition, New York: St Martin's Press, 1954.

Carrell, Paul. *Scorched Earth: The Russo-German War, 1943-1944*. Translator Ewald Osers. Boston: Little, Brown & Co. 1970.

Cerff, Karl. *Die Waffen-SS im Wehrmachtbericht*. Osnabrück: Munin Verlag GmbH, 1971.

Dallin, Alexander. *German Rule in Russia, 1941-1945*. London: Macmillan & Co. Ltd. U.S. Edition, New York: St. Martin's Press, 1957.

Davis, Melton S. *Who Defends Rome? The Forty-Five Days, July 25 September 8, 1943*. London: George Allen & Unwin Ltd., 1972.

Dedijer, Vladimir. *Tito*. New York: Simon & Schuster, 1953.

Dieckert, Majore Kurt, and Generale Grossmann. *Der Kampf um Ostpreussen: der umfassende Dokumentarberichte uber das Kriegsgeschehen im Ostpreussen*. Stuttgart: Motorbuch Verlag, 1984.

Djilas, Milovan. *Wartime*. Translator Michael Petrovich. New York: Harcourt Brace Jovanovich, 1977.

Elstob, Peter. *Hitler's Last Offensive: The Full Story of the Battle of the Ardennes*. New York: The Macmillan Co., 1971.

Erickson, John. *The Road to Stalingrad: Stalin's War with Germany*. Vol. 1. London: Weidenfeld And Nicholson, 1979.

———. *The Road to Berlin: Stalin's War with Germany*. Vol. 2 .London: Weidenfeld And Nicholson, 1983.

Ezquerra, Miguel. *Berlin, A Vida O Muerte*. Madrid: Editorial San Martin, 1975.

Fenyo, Mario D. *Hitler, Horthy, and Hungary: German-Hungarian Relations, 1941-1944*. New Haven: Yale University Press, 1972.

Foley, Charles. *Commando Extraordinary*. London: Longmans Green & Co., 1954.

Gellerman, Günther. *Die Armee Wenck-Hitlers letzte Hoffnung*. Koblenz: Bernard & Graefe Verlag, 1984.

Haupt, Werner. *Kurland: Die Vergessene Heeresgruppe 1944/1945*. Friedberg: Podzun Pallas Verlag, 1979.

———. *Königsberg-Breslau-Wien-Berlin 1945*. Friedberg: Podzun Pallas Verlag, 1978.

———. *Leningrad-Wolchow-Kurland 1941-1945*. Friedberg: Podzun Pallas Verlag, 1976.

———. *Als die Rote Armee nach Deutschland kam*. Friedberg: Podzun Pallas Verlag, n.d.

———. *Heeresgruppe Nord*. Friedberg: Podzun Pallas Verlag, 1966.

———. *Die Schlachten der Heeresgruppe Mitte 1941-1944, Aus der Sicht der Divisionen*. Friedberg: Podzun Pallas Verlag, 1983.

———. *Die Schlachten Der Heeresgruppe Süd, Aus der Sicht der Divisionen*. Podzun Pallas Verlag, 1986.

Hausser, Paul. *Waffen-SS im Einsatz*. Preussich Oldendorf: Verlag K.W.

———. *Soldaten wie andere auch*. Osnabrück: Munin Verlag GmbH, 1966.

Hinsley, F.H. et al. *British Intelligence in the Second World War*. 5 vols. London: Her MaJesty's Stationary Office, 1984, vol. 3, part 1.

Holzmann, Walther-Karl. *Manual of the Waffen-SS*. London: Bellona Publications/Argus Books Ltd., 1976.

Housemann, Friedrich. *Die guten Glaubens waren, Geschichte der SS-Polizei-Division (4. SS-Polizei-Panzergrenadier-Division)*. 2 vols. Osnabrück: Munin Verlag GmbH, 1973, vol. 2.

Infield, Glenn B. *Skorzeny: Hitler's Commando*. New York: St. Martin's Press, 1981.

Kannapin, Norbert. *Die Feldpostubersicht 1939-1945*. 3 vols. Osnabrück:Biblio Verlag GmbH, 1986, vols 1-3.

Kaps, Dr. Johannes, editor. *The Tragedy of Silesia, 1945-1946. A Documentary Account with a Special Survey of the Archdiocese of Breslau*. Munich: Christunterwegs. 1952-53.

Keilig, Wolf. *Rangliste des Deutschen Heeres 1944/45*. Friedberg: Podzun

Kern, Erich. *Die letzte Schlacht*. Preussich Oldendorf: Verlag K.W. Schütz KG, 1972.

Kesselring, Feldmarschall Albrecht. *Soldat bis zum letzten Tag*. Bonn: Athenaum Verlag, 1953.

Klietmann, Dr. K.G. *Die Waffen-SS, eine Dokumentation*. Osnabrück: Verlag "Der Freiwillige" GmbH, 1965.

Krätschmer, Ernst Günther. *Die Ritterkreuzträger der Waffen-SS*. Preussich Oldendorf; Verlag K.W. Schütz KG, 1955.

Kumm, Otto. *Vorwarts Prinz Eugen! Geschichte der 7.SS-FreiwilligenGebirgs-Division Prinz Eugen*. Osnabrück: Munin Verlag GmbH, 1984.

Kurowski, Franz and Tornau. *Sturmartillerie*. Stuttgart: Motorbuch Verlag, 1978.

———. *Die Letzten Dreizig Tag*. Friedberg: Podzun Pallas Verlag, 1989.

Landwehr, Richard. *Italian Volunteers of the Waffen-SS*. Bennington: Weapons and Warfare Press, 1988.

———. *Fighting for Freedom: The Ukrainian Division of the Waffen-SS*. Silver Spring: Bibliophile Legion Books, Inc., 1985.

———. *Charlemagne's Legionnaires, French Volunteers of the Waffen-SS, 1943-1945*. Silver Spring: Bibliophile Legion Books, Inc., 1989.

Lasch, Generale Otto. *So Fiel Königsberg*. Stuttgart: Motorbuch Verlag, 1984.

Lazzero, Ricciotti. *Le SS Italiane*. Milano: Rizzoli Editore, 1982.

Leverkuehn, Paul. *German Military Intelligence*. London: Weidenfeld and Nicholson, 1954.

Littlejohn, David. *Foreign Legions of the Third Reich*, Vol. 2. San Jose: James R. Bender, 19981.

Lucas, James. *Alpine Elite: German Mountain Troops of World War II*. London: Jane's Publishing Co., Ltd., 1980.

———. *War on the Eastern Front 1941-1945*. London: Jane's Publishing Co., Ltd., 1979.

———. *Kommando: German Special Forces of World War II*. London: Arms & Armour Press, 1985.

———. *Last Days of the Third Reich*. New York: William Morrow and Co.,1986.

Madeja, W. Victor. *Russo-German War, Autumn 1944-25 January 1945*. Allentown: Valor Publishing, 1987.

———. *Russo-German War, 25 January to May 8 1945*. Allentown: Valor Publishing, 1987.

Maier, Georg. *Drama zwischen Budapest und Wien: Der Endkampf der 6. Panzerarmee 1945*. Osnabrück: Munin Verlag GmbH, 1985.

Mee, Charles L. *Meeting at Potsdam*. New York: M. Evans & Co., Ltd., 1975.

Mehner, Kurt. *Die Geheimen Tagesberichte der Deutschen Wehrmachtführung im Zweiten Weltkrieg 1939-1945*. 12 vols. Osnabrück: Biblio Verlag GmbH, 1984-1990, vols. 7-12.

Mellenthin, Majore-Generale F.W. von. *Panzer Battles: A Study of the Employment of Armor in the Second World War*. London: Cassell & Co., Ltd; 1955.

Meyer, Brün. *Dienstalterliste der Waffen-SS*. SS-Obergruppenführer bis SS-Hauptsturmführer, *1 Juli, 1944*. Osnabrück: Biblio Verlag GmbH, 1987.

Milius, Siegfried and Adolf Kunzmann. *Fallschirmjäger der Waffen-SS im Bild*. Osnabrück: Munin Verlag GmbH, 1986.

Milazzo, Matteo J. *The Chetnik Movement & the Yugoslav Resistance*. Baltimore: The Johns Hopkins University Press, 1975.

Mueller-Hillebrand, Burkhart. *Germany and Its Allies: A Record of Axis Collaboration Problems*. Frederick: University Publications of America, 1980.

Neulen, Hans Werner. *An Deutscher Seite: Internationale Freiwillige von Wehrmacht und Waffen-SS*. München: Universitas Verlag, 1985.

Niepold, Gerd. *Panzer-Operationen: "Doppelkopf und Casar" Sommer '44*. Hereford; Bonn: E.S. Mittler & Sohn, Herford; Bonn: E.S. Mittler & Sohn, 1987.

———. *Mittlere Ostfront*, Juni 1944, Herford.

Omrcanin, Ivo. *Military History of Croatia*. Bryn Mawr: Dorrance & Co., Inc., 1984.

Ostovic, P.D. *The Truth about Yugoslavia*. New York: Roy Publishers, 1952.

Pallud, Jean-Paul. *Battle of the Bulge—Then and Now*. London: After The Battle, Ltd., 1985.

Plato, Anton Detlev von. *Die Geschichte der 5. Panzerdivision 1938 bis 1945*. Regensburg: Walhalla u. Praetoria Verlag KG Georg Zwickenpflug, 1978.

Pohlman, Hartwig. *Wolchow 900 Tage Kampf um Leningrad 1941-1944*. Friedberg: Podzun Pallas Verlag, n.d.

Preradovich, Nikolaus von. *Die Generale der Waffen-SS*. Berg am See: Kurt Vowinkel Verlag, 1985.

Reinicke, Adolf. *Die 5. Jäger-Division 1939-1945*. Friedberg: Podzun Pallas Verlag, n.d.

Reitlinger, Gerald. *The SS: Alibi of a Nation 1922-1945*. London: 1956, rpnt. Englewood Cliffs, N.J.: Prentice-Hall, Inc., 1981.

———. *The House Built on Sand: The Conflicts of German Policy in Russia, 1939-*

1945. London: Weidenfeld & Nicolson, 1960.
Richardson, William and Freidin, editors. *The Fatal Decisions: Six Decisive Battles of the Second World War from the Viewpoint of the Vanquished.* Translator. Constantine Fitzgibbon. London: Michael Joseph, Ltd., 1956.
Rings, Werner. *Life with the Enemy: Collaboration and Resistance in Hitler's Europe 1939-1945*. Translator. J. Maxwell Brownjohn. New York: Doubleday & Co., 1982.
Rutkis, J., Chief Editor. *Latvia: Country and People*. Stockholm: Latvian National Foundation, 1967.
Ryan, Cornelius. *The Last Battle*. New York: Simon & Schuster, 1966.
Sanz, Miguel Angel. *Luchando en tierras de Francia. La participación de los españoles en la resistencia*. Madrid: Ediciones de la Torre, 1981.
Schäufler, Hans. *1945 Panzer An der Weichsel, Soldaten der letzten Stunde*. Stuttgart: Motorbuch Verlag, 1986.
Scheibert, Horst. *Die gespenster Division: Eine deutsche Panzer-Division (7.) im Zweiten Weltkrieg*. Friedberg: Podzun Pallas Verlag, n.d.
———. *Panzer-Grenadier-Division Grossdeutschland*. Dorheim: Podzun Pallas Verlag, 1973.
———. *Die Träger Der Ehrenblattspange Des Heeres, Kriegsmarine und der Waffen-SS*. Friedberg: Podzun Pallas Verlag, 1986.
———. *Die Träger des deutschen Kreuzes in Gold: Kriegsmarine-Luftwaffe-Waffen-SS und die Träger des Deutschen Kreuzes in Silber: Kriegsmarine-Luftwaffe-Waffen-SS*. Friedberg: Podzun Pallas Verlag, n.d.
Schellenberg, Walter. *The Labyrinth: The Memoirs of Hitler's Secret Service Chief*. Translator Louis Hagen. New York: Harper & Row Brothers Publishers, 1956.
Schlaug, Georg. *Die deutschen Lastensegler-Verbande 1937-1945*. Stuttgart: Motorbuch Verlag, 1985.
Schraml. Franz. *Kriegsschauplatz Kroatien. Die deutsch-kroatischen Legions Divisionen — 369., 373., 392., Inf.-Div. (kroat.) ihre Ausbildung und Ersatzformationen*. Germany: Kurt Vowinckel Verlag, 1962.
Schramm (Hrsg), Percy E. *Kriegstagebuch Des Oberkommandos der Wehrmacht 1940-1945*, vols. 1-8. Herrsching: Manfred Pawlak Verlagsgesellschaft GmbH, 1982.
Seaton, Colonel Albert. *The Russo-German War, 1941-1945*. New York: Praeger Publishers, Inc., 1971.
Senger und Etterlin, General Frido von. *Neither Fear Nor Hope: The Wartime Memoirs of the Defender of Cassino*. Novato: Presidio Press, 1989.
Shulman, Colonel Milton. *Defeat in the West*. New York: E.P. Dutton & Co., Inc., 1948.
Silgailis, Arthur. *Latvian Legion*. San Jose: James R. Bender, 1986.
Skorzeny, Otto. *Skorzeny's Special Missions*. London: Robert Hale, 1947.
———. *Lebe gefärlich and Wir kämpften Wir Verloren*, 2 vols. Preussich Oldendorf: Helmut Cramer Verlag, 1975.
Sparacino, Fausto. *Distintivi e medaglie della R.S.I. (R.S.I. Badges and Medals—1943/45)*. Milano: Editrici Militari Italiana, 1988.
Spater, Helmuth. *Die Brandenburger: Eine deutsche Kommandotruppe*. München: Walther Angerer Verlag, 1982.
Stato Maggiore Dell Esercito Italiane. *Le Operazioni delle unitea italiane in Iugoslavia (1941-1943)*. Roma: Tipografia Regionale, 1978.
Stein, George H. *The Waffen-SS: Hitler's Elite Guard at War 1939-1945*. Ithaca: Cornell University Press, 1972.

Stöber, Hans. *Die Sturmflut und das Ende: Geschichte der 17.SS-Panzergrenadier-Division Götz von Berlichingen*, 2 vols. Osnabrück: Munin Verlag GmbH, 1976 and 1987.

———, and Spiwoks. *Endkampf zwischen Mosel und Inn, XIII. SS-Armeekorps.* Osnabrück: Munin Verlag GmbH, 1976.

———. *Die Flugabwehrverbände der Waffen-SS, Aufstellung, Gliederung, Luftverteidigung und Einsätze aus den Fronten.* Preussich Oldendorf: Verlag K.W. Schütz, 1984.

Stoves, Rolf. *Die gepanzerten und motorisierten deutschen Grossverbände, 1935-1945.* Friedberg: Podzun Pallas Verlag, 1985.

Strassner, Peter. *European Volunteers. 5. SS-Panzer-Division Wiking.* Translator David Johnston. Winnipeg: J.J. Fedorowicz Publishing, 1988.

Tessin, Georg. *Verbande und Truppen der deutschen Wehrmacht und Waffen-SS im Zweiten Weltkrieg 1939-1945.* 15 vols. Osnabrück: Biblio Verlag GmbH, 1980.

Thorwald, Juergen. *Flight in the Winter: Russia Conquers—January to May 1945.* Translator Fred Wieck. New York: Pantheon Books, Inc., 1951.

Tieke, Wilhelm. Das Ende zwischen Oder und Elbe: *Der Kampf um Berlin 1945.* Stuttgart: Motorbuch Verlag, 1981.

———. *Im Feuersturm letzter Kriegsjahre: II.SS-Panzerkorps mit 9. und 10. SS-Panzer-Division Hohenstaufen und Frundsberg.* Osnabrück: Munin Verlag GmbH, 1975.

———. *Tragödie um die Treue: Kampf und Untergang des III.(germ.) SS-Panzerkorps.* Osnabrück: Munin Verlag GmbH, 1978.

Trevor-Roper, Hugh, editor. *Final Entries 1945: The Diaries of Joseph Goebbels.* New York: G.P. Putnam's & Sons, 1978.

Tully, Andrew. *Berlin: Story of a Battle.* New York: Simon & Schuster, 1963.

Union of Soviet Socialist Republics. *The Case Against General Heusinger: Documents Illustrating the Charges of the U.S.S.R. against Former Lieutenant-General Adolf Heusinger, Former Operations Chief of the Wehrmacht High Command.* Chicago: Translation World Publishers, 1961.

U.S. Army, Office of the Chief of Military History. *Special Studies: Chronology, 1941-1945.* Washington D.C.: Department of the Army, 1960.

Vincx, Jan and Schotanius. *Nederlandse vrijwillicers in europese krijgsdienst 1940-1945, "De Landstorm."* Vol. 1 . Herentals: Uitgeverij Etnika Verlag, 1987.

Warlimont, Walter. *Inside Hitler's Headquarters 1939-1945.* Translator. R. H. Barry. New York: Frederick A. Praeger, Inc., 1964.

Weidinger, Otto. *Division Das Reich. Der Weg der 2. SS-Panzer-Division Das Reich.* 5 vols. Osnabrück: Munin Verlag GmbH, 1982, vol. 5

West, Nigel. *A Threat of Deceit: Espionage Myths of World War II.* New York: Random House, 1985.

Whiting, Charles. *Hunters From the Sky: The German Parachute Corps, 1940-1945.* New York: Stein & Day, 1974.

Wilmot, Chester. *The Struggle for Europe.* New York: Harper & Row Publishers, 1952.

Winterbotham, F.W. *The Ultra Secret.* New York: Harper & Row Publishers, 1974.

Wood, Tony and Gunston. *Hitler's Luftwaffe.* Seacaucus: Chartwell Books, Inc. 1984.

Ziemke, Earl F. *The Battle for Berlin: End of the Third Reich.* New York: Ballantine Books, Inc. 1977.

———. *Moscow to Stalingrad: The German Defeat in the East.* Washington, D.C.: U.S. Government Printing Office, 1988.

———. *Stalingrad to Berlin: The German Defeat in the East.* Washington, D.C.: U.S. Government Printing Office, 1968.

Pamphlets

Department of the Army. *Historical Study: German Anti-Guerrilla Operations in the Balkans (1941-1944).* FN Nos. 20-243. Washington, D.C.: U.S. Government Printing Office, 1954.

Department of the Army. *Historical Study:The Soviet Partisan Movement (1941-1944).* FN Nos. Da Pam 20-244. Washington, D.C.: U.S. Government Printing Office, August 1956.

Department of the Army. *Historical Study: German Campaigns in the Balkans (Spring 1941).* FN Nos. Da Pam 20-260. Washington, D.C.: U.S. Government Printing Office, November 1953.

Articles and Periodicals

Siegrunen. Published monthly by Richard Landwehr. Glendale: 1976-present.

39/45. Published monthly by Editions Heimdal. Bayeux Cedex: 1985 (no.8).

Wiking-Ruf: Zeitschrift der Soldaten der ehemaligen Waffen-SS, 1955-1958. Absorbed by *Der Freiwillige/Wiking Ruf: Kameradschaftsblatt der HIAG* (Hilfsgemeinschaft aus Gegenseitigkeit). Stemmen Ober Hannover: Wiking Verlag; in 1958 and thereafter published monthly under the title *Der Freiwillige: Kameradschaftsblatt der HIAG.* Osnabrück: Verlag "Der Freiwillige."